Statistics for Industry and Technology

Advances in Distribution Theory, Order Statistics, and Inference

N. Balakrishnan
Enrique Castillo
José María Sarabia
Editors

Birkhäuser
Boston • Basel • Berlin

N. Balakrishnan
McMaster University
Department of Mathematics and Statistics
1280 Main Street West
Hamilton, Ontario L8S 4K1
Canada

Enrique Castillo
University of Cantabria
Department of Applied Mathematics
s/n Avenida de los Castros
Santander 39005
Spain

José María Sarabia
University of Cantabria
Department of Economics
s/n Avenida de los Castros
Santander 39005
Spain

Mathematics Subject Classification: 62E10, 62E15, 62E20, 62F03, 62F10, 62G30, 62J05, 62N01, 62N02, 62N05, 62P12, 62P20, 62P30

Library of Congress Control Number: 2006900999

ISBN-10: 0-8176-4361-3 e-ISBN: 0-8176-4487-3
ISBN-13: 978-0-8176-4361-4

Printed on acid-free paper.

Printed in the United States of America. (HP)

9 8 7 6 5 4 3 2 1

www.birkhauser.com

In Honor of

Barry C. Arnold

BARRY C. ARNOLD

Contents

PART V: INFERENCE

Preface

Barry Arnold has made fundamental contributions to many different areas of statistics including order statistics, distribution theory, Bayesian inference, multivariate analysis, bounds and orderings, and characterization problems. He has written numerous research articles (see the list of his Publications) in all these topics, and these have received many citations over the years.

During his illustrious career, he has contributed significantly to the statistical profession in many different ways—as a teacher (at Iowa State University, University of California at Riverside, and other places), supervisor (to many graduate students), researcher, administrator (as the Head of the Department of Statistics at University of California at Riverside), organizer (of numerous invited sessions in conferences), and editor (of *Journal of Multivariate Analysis*, managing editor of *The Annals of Statistics*, as well as being on the editorial board of many other journals).

All three of us have had a long association with Barry and have enjoyed our collaboration with him for the past two decades. Those who know him as well as we do certainly have an appreciation for his wit and humor, lively lectures, keen interest in statistics, and great enthusiasm for research. We consider him to be our friend, guide, and philosopher, and we feel that our lives have been greatly enriched by our association with him.

When Barry turned 65 last year, we therefore took the opportunity to organize an **International Conference on Distribution Theory, Order Statistics, and Inference** in his honor. This conference was held during June 16-18, 2004, at the University of Cantabria, Santander, Spain, where he has been a frequent visitor for a number of years. A number of his friends, colleagues, coauthors, and researchers participated in this event. The conference, with participation from around 140 delegates, was a great success.

Some selected papers that were presented at this conference have been included in this volume. We thank all the authors for their contributions for this volume and also the referees for helping us in the evaluation of these manuscripts. We also express our sincere gratitude to Tom Grasso (editor, **Birkhäuser**, Boston) for his support and encouragement for this project, and to Ms. Debbie Iscoe for assisting us with the preparation of this volume.

It is with great pleasure that we dedicate this volume to our friend, Barry C. Arnold!

N. Balakrishnan
McMaster University
Hamilton, Canada

Enrique Castillo
University of Cantabria
Santander, Spain

Jose Maria Sarabia
University of Cantabria
Santander, Spain

Barry C. Arnold: Career and Accomplishments

Barry C. Arnold was born on December 6, 1939, in the London borough of Lewisham to Charles and Irene Arnold. He was the second child born to his parents with his sister, Nina Arnold, born earlier on January 24, 1938.

After their house was bombed by the Germans, they were evacuated from London and then lived in Herne Bay, Barrie, and Blackpool before settling in Caterham, Surrey, a few miles south of London. In April 1952, the family emigrated to Canada. After attending St. Laurent High School, Barry joined the Engineering Program at McGill University in 1956. When the family moved to Hamilton in 1958, he transferred to McMaster University and graduated in 1961 with a Bachelor's degree in mathematics (statistics).

Barry subsequently entered the graduate program in statistics at Stanford University, the school that he selected because, not only was it highly recommended, but it also had some palm trees on campus. This was a good choice as Stanford had an all-star faculty that included Ted Anderson, Herman Chernoff, Kai Lai Chung, Shanti Gupta, M. V. Johns, Sam Karlin, Ingram Olkin, Rupert Miller, Lincoln Moses, Emmanuel Parzen, Charles Stein, Herbert Solomon, and Pat Suppes. His classmates here were a pretty impressive group, too, which included Norm Breslow, Morris Eaton, Brad Efron, Leon Gleser, Burt Holland, Myles Hollander, Jay Kadane, Carl Morris, Jim Press, Richard Royall, Steve Samuels, Galen Shorack, Muni Srivastava, David Sylwester, Grace Wahba, and Jim Zidek. Barry graduated from Stanford in 1965 after writing a doctoral dissertation under the guidance of Pat Suppes. Another event of importance that occurred while Barry was at Stanford was that he got married to Carole Revelle in September 1964. From that day on, he has had his own personal psychologist, of course!

From Stanford, Barry went to Iowa State University and joined the faculty with a joint appointment in the Departments of Mathematics and Statistics. There, he had good friends and plenty of intellectual stimulation from many statisticians of repute that included Ted Bancroft, H. A. David, H. T. David, Wayne Fuller, Dick Groeneveld, Chien-Pai Han, Dean Isaacson, B. K. Kale,

Oscar Kempthorne, Bill Kennedy, Glen Meeden, Ed Pollak, Joe Sedransk, and Vince Sposito. During 1968–1969, Barry was a visiting professor at the Colegio de Postgraduados in Chapingo, Mexico, lecturing in pretty bad Spanish. During 1974–1975, he went on a AID assignment, working with the Ministry of Agriculture in Lima, Peru. Though he was not successful in selling sampling methods there, he did improve his Spanish!

In 1979, Barry hung up his snow shovel, donated his winter coat to the Salvation Army, and moved to Riverside, California, to join Jim Press (whom he knew from Stanford) and his department there. He has been there since then. He spent two years (1982–1984) back in Mexico as the Director of the University of California Education Abroad Program.

Barry Arnold has served the Department of Statistics at the University of California, Riverside, as Chair for a number of years. In addition, he has provided distinguished service to the statistical community at large by his activities in various capacities for professional societies such as the American Statistical Association and the Institute of Mathematical Statistics. He has participated in numerous national and international conferences and delivered many invited and plenary lectures. He has provided valuable service to several research journals in various capacities including associate editor of *Journal of Multivariate Analysis*, *Journal of the American Statistical Association* and *Communications in Statistics*, editor-in-chief of *Journal of Multivariate Analysis*, and managing editor of *The Annals of Statistics*.

Barry Arnold has been elected a Fellow of the American Statistical Association and the Institute of Mathematical Statistics, and a member of the International Statistical Institute.

He has had a long list of stimulating coworkers and coauthors. Particularly noteworthy are Enrique Castillo and Jose Maria Sarabia (both at the University of Cantabria, Santander, Spain), H. N. Nagaraja (at The Ohio State University, Columbus, Ohio, USA) and N. Balakrishnan (McMaster University, Hamilton, Ontario, Canada). Numerous visits to Santander, Hamilton, and Texcoco, Mexico (where Barry has worked with Jose Villaseñor) have provided him with many pleasant productive interludes. He has never been to a foreign country he did not like, and so he rarely turns down any invitation!

Over the past 40 years, Barry Arnold, through his tremendous research in many different areas of statistics, and especially in distribution theory and ordered data, has greatly influenced the trend of research in these areas and has provided inspiration and encouragement to many young researchers. It is our wish and sincere hope that he will continue his contributions to the field with added vigor, interest, and energy!

Publications

Books

1. *Pareto Distributions*, International Cooperative Publishing House, Burtonsville, MD, 326 pages, 1983.
2. *Majorization and the Lorenz Order: A Brief Introduction*, Lecture Notes in Statistics, Vol. 43, Springer, New York, 122 pages, 1988.
3. *Relations, Bounds and Approximations for Order Statistics* (co-edited with N. Balakrishnan), Lecture Notes in Statistics, Vol. 53, Springer, New York, 173 pages, 1989.
4. *A First Course in Order Statistics* (co-edited with N. Balakrishnan and H. N. Nagaraja), John Wiley & Sons, New York, 279 pages, 1992.
5. *Conditionally Specified Distributions* (co-edited with E. Castillo and J. M. Sarabia), Lecture Notes in Statistics, Vol. 73, Springer, New York, 151 pages, 1992.
6. *Records* (co-edited with N. Balakrishnan and H.N. Nagaraja), John Wiley & Sons, New York, 312 pages, 1998.
7. *Conditional Specification of Statistical Models* (co-edited with E. Castillo and J. M. Sarabia), Springer-Verlag, New York, 411 pages, 1999.

Articles

1967

1. A generalized urn-scheme for simple learning with a continuum of responses, *Journal of Mathematical Psychology*, **4**, 301–315.
2. Response distribution for the continuous time N-element pattern model, *Journal of Mathematical Psychology*, **4**, 489–500.
3. A note on multivariate distributions with specified marginals, *Journal of the American Statistical Association*, **62**, 1460–1461.

1968

4. A modification of a result due to Moran, *Journal of Applied Probability*, **5**, 220–223.

5. Parameter estimation for a multivariate exponential distribution, *Journal of the American Statistical Association*, **63**, 648–652.

1970

6. Inadmissibility of the usual scale estimate for a shifted exponential distribution, *Journal of the American Statistical Association*, **65**, 1260–1264.

7. An alternative derivation of a result due to Srivastava and Bancroft, *Journal of the Royal Statistical Society, Series B*, **32**, 265–267.

8. Hypothesis testing incorporating a preliminary test of significance, *Journal of the American Statistical Association*, **65**, 1590–1596.

1971

9. Letter: Zero correlation and independence, *The American Statistician*, **26**, 34–36.

1972

10. Some examples of minimum variance unbiased estimates, *The American Statistician*, **26**, 34–36.

11. The waiting time until first duplication, *Journal of Applied Probability*, **9**, 841–846.

1973

12. Some characterizations of the exponential distribution by geometric compounding, *SIAM Journal of Applied Mathematics*, **24**, 242–244.

13. Response distributions for a generalized urn scheme under non-contingent reinforcement, *Journal of Mathematical Psychology*, **10**, 232–239.

14. Independence of squared order statistics, *Communications in Statistics—Theory and Methods*, **2**, 357–362.

1974

15. On estimates of the smaller of two ordered normal means which incorporate a preliminary test of significance, *Utilitas Mathematica*, **5**, 65–74.

16. Schwarz, regression and extreme deviance, *The American Statistician*, **28**, 22–23.

17. Bounds for deviations between sample and population statistics (with R. A. Groeneveld), *Biometrika*, **61**, 387–389.

1975

18. A characterization of the exponential distribution by multivariate geometric compounding, *Sankhyā: The Indian Journal of Statistics, Series A*, **37**, 164–173.

19. Characterization of distributions by sets of moments of order statistics (with G. Meeden), *Annals of Statistics*, **3**, 754–758.

20. *Estadistica Experimental, Curso Basico* (with R. Arroyo), 240 pages, Multilithed for Biometry Office, Peruvian Ministry of Food.

21. Multivariate exponential distributions based on hierarchical successive damage, *Journal of Applied Probability*, **12**, 142–147.

22. On sojourn times at particular gene frequencies (with E. Pollak), *Genetical Research*, **25**, 89–94.

23. Significant category clustering in free recall, *Psychometrika*, **40**, 579–581.

1976

24. On fitting assessment package scores with a binomial error model, National Assessment of Educational Progress, Denver, CO.

25. Background variables as predictors of package scores, National Assessment of Education Progress, Denver, CO.

26. On solutions to $\min(X,Y) \sim aX$ and $\min(X,Y) \sim aX \sim bY$, (with D. Isaacson), *Z. Wahr theorie und Verw. Gebiete*, **35**, 115–119.

27. A characterization of geometric distributions by distributional properties of order statistics (with M. Ghosh), *Scandinavian Actuarial Journal*, 232–234.

28. A characterization of the uniform distribution based on summation modulo 1, with applications to fractional backlogs (with G. Meeden), *Australian Journal of Statistics*, **18**, 173–175.

29. A stochastic mechanism leading to asymptotically Paretian distributions (with L. Leonor), *Proceedings of the Business and Economic Statistics Section of the ASA*, 208–210.

1977

30. On generalized Pareto distributions with application to income data (with L. Leonor), *International Studies in Economics, Monograph No. 10*, 48 pages, Department of Economics, Iowa State University, Ames, IA.

31. *Disenos Experimentales: Conceptos y Aplicaciones*, 240 pages, Multilithed for Biometry Office, Peruvian Ministry of Food.

32. Recurrence relations between expectations of functions of order statistics, *Scandinavian Actuarial Journal*, 169–174.

33. Distributions of times spent in various states in some absorbing processes arising arising in genetics (with E. Pollak), *Proceedings of the Washington State University Conference on Biomathematics and Biostatistics*: 145–169, May 1974, Pullman, WA.

1978

34. On characterization and decomposition of Cauchy random variables, In *Proceedings of the 3rd. National Symposium on Probability and Statistics*, pp. 139–141, Sao Paulo, Brazil.

35. Two modifications of Goodmans technique for improving estimates, *Trabajos de Estidistica y de Investigacion Operativa*, **29**, 61–70.

36. Some elementary variations of the Lyapunov inequality, *SIAM Journal of Applied Mathematics*, **35**, 117–118.

37. On normal characterizations by the distribution of linear forms, assuming finite variance (with D. L. Isaacson), *Stochastic Processes and Applications*, **7**, 227–230.

38. Bounds on deviations of estimates arising in finite population regression models (with R. A. Groeneveld), *Communications in Statistics—Theory and Methods*, **7**, 1173–1179.

39. Strong ergodicity for continuous-time Markov chains (with D. L. Isaacson), *Journal of Applied Probability*, **15**, 699–706.

1979

40. Some characterizations of the Cauchy distribution, *Australian Journal of Statistics*, **21**, 166–169.

41. Nonuniform decompositions of uniform random variables under summation modulo m, *Bolletino Unione Matematica Italiana*, **16**, 100–102.

42. The admissibility of a preliminary test estimator when the loss incorporates a complexity cost (with G. Meeden), *Journal of the American Statistical Association*, **74**, 872–874.

43. On characterizations of the uniform distribution based on identically distributed spacings (with J. S. Huang and M. Ghosh), *Sankhyā: The Indian Journal of Statistics*, **41**, 109–115.

1980

44. Bounds on expectations of linear systematic statistics based on dependent samples (with R. A. Groeneveld), *Annals of Statistics*, **8**, 1401.

45. Distribution-free bounds on the mean of the maximum of a dependent sample, *SIAM Journal of Applied Mathematics*, **38**, 163–167.

46. Two characterizations of the geometric distribution, *Journal of Applied Probability*, **17**, 570–573.

47. Some properties of the Arcsine distribution (with R. A. Groeneveld), *Journal of the American Statistical Association*, **75**, 173–175.

1981

48. Maximal deviation between sample and population means in finitc pop ulations (with R. A. Groeneveld), *Journal of the American Statistical Association*, **76**, 443–445.

49. On excess life in certain renewal processes (with R. A. Groeneveld), *Journal of Applied Probability*, **18**, 379–389.

1982

50. Strong ergodicity for continuous-time nonhomogeneous Markov chains (with M. Scott and D. L. Isaacson), *Journal of Applied Probability*, **19**, 692–694.

1983

51. When does the βth percentile residual life function determine the distribution? (with P.L. Brockett), *Operations Research*, **31**, 391–396.

52. Identifiability for dependent multiple decrement/competing risk models (with P.L. Brockett), *Scandinavian Actuarial Journal*, 117–127.

53. Bayesian inference for Pareto populations (with S. J. Press), *Journal of Econometrics*, **21**, 287–306.

54. Some limiting distributions associated with sequences of multinomial trials (with J. E. Angus), *Naval Research Logistics Quarterly*, **30**, 1–11.

1984

55. Limit laws in the best of $2n - 1$ Bernoulli trials (with R. A. Groeneveld), *Naval Logistics Research Quarterly*, **31**, 275–281.

56. On the Markov property of order statistics (with A. Becker, U. Gather and H. Zahedi), *Journal of Statistical Planning and Inference*, **9**, 147–154.

57. On the inconsistency of Bayesian nonparametric estimators in competing risks/multiple decrement models (with P. L. Brockett, W. Torrez, and A. L. Wright), *Insurance Mathematics and Economics*, **3**, 49–55.

1985

58. Pareto distributions, In *Encyclopedia of Statistical Sciences* (Eds., S. Kotz and N. L. Johnson), Volume 6, pp. 568–574, John Wiley & Sons, New York.

59. p-norm bounds on the expectation of the maximum of a possibly dependent sample, *Journal of Multivariate Analysis*, **17**, 316–332.

1986

60. Some waiting time problems, *Pakistan Journal of Statistics*, **2**, 39–45.

61. A class of hyperbolic Lorenz curves, *Sankhyā, Series B*, **48**, 427–436.

62. Bayesian analysis of censored or grouped data from Pareto populations (with S. J. Press), In *Bayesian Inference and Decision Techniques with Applications* (Eds., P. K. Goel and A. Zellner), pp. 157–173, Elsevier, Amsterdam.

63. Lorenz ordering of means and medians (with J. A. Villasenor), *Statistics & Probability Letters*, **4**, 47–49.

64. Some properties of a Pareto-type distribution (with C. A. Robertson and H. C. Yeh), *Sankhyā, Series A*, **48**, 404–408.

1987

65. Bivariate distributions with Pareto conditionals, *Statistics & Probability Letters*, **5**, 263–266.

66. Generating ordered families of Lorenz curves by strongly unimodal distributions (with C. A. Robertson, P. L. Brockett, and B. Shu), *Journal of Business and Economic Statistics*, **5**, 305–308.

67. The rating of players in racquetball tournaments (with D. Strauss), *Journal of the Royal Statistical Society, Series C*, **36**, 163–173.

1988

68. On multivariate mean remaining life functions (with H. Zahedi), *Journal of Multivariate Analysis*, **25**, 1–9.

69. Variance bounds using a theorem of Polya (with P. Brockett), *Statistics & Probability Letters*, **6**, 321–326.

72. Conditional characterizations of multivariate distributions (with M. Pouramahdi), *Metrika*, **35**, 99–108.

71. Pareto processes (with H. C. Yeh and C. A. Robertson), *Journal of Applied Probability*, **25**, 291–301.

72. Bivariate distributions with exponential conditionals (with D. Strauss), *Journal of the American Statistical Association*, **83**, 522–527.

73. Bounds on the expected maximum, *Communications in Statistics—Theory and Methods*, **17**, 2135–2150.

74. Estimation of the number of classes in a population (with R. J. Beaver), *Biometrical Journal*, **4**, 413–424.

75. Characterizations based on conditional distributions given the minimum value in the sample (with R. Shanmugam), *Sankhyā, Series A*, **50**, 452–459.

1989

76. A logistic process constructed using geometric minimization, *Statistics & Probability Letters*, **7**, 253–257.

77. Elliptical Lorenz curves (with J. Villasenor), *Journal of Econometrics*, **40**, 327–338.

78. Compatible conditional distributions (with S. J. Press), *Journal of the American Statistical Association*, **84**, 152–156.

79. Autoregressive logistic processes (with C. A. Robertson), *Journal of Applied Probability*, **26**, 524–531.

80. A characterization of the Pareto process among stationary stochastic processes of the form $X_n = c\min(X_n - 1, Y_n)$ (with T. Hallet), *Statistics & Probability Letters*, **8**, 377–380.

81. Bayesian estimation and prediction for Pareto data (with S. J. Press), *Journal of the American Statistical Association*, **84**, 1079–1084.

82. New moment identities based on the integrated survival function (with D. M. Reneau and F. J. Samaniego), *IEEE Transactions on Reliability*, **38**, 358–361.

1990

83. Sequential sampling estimation for finite populations of $N = nr$ objects (with R. A. Groeneveld), *Biometrical Journal*, **32**, 143–153.

84. The Lorenz order and the effects of taxation policies, *Bulletin of Economic Research*, **42**, 249–264.

85. On Cauchy-like distributions (with R. M. Norton), *Sankhyā, Series A*, **52**, 371–375.

86. A flexible family of multivariate Pareto distributions, *Journal of Statistical Planning and Inference*, **24**, 249–258.

1991

87. Dependence in conditionally specified distributions *Topics in Statistical Dependence, IMS Lecture Notes/Monograph Series*, **16**, 13–18.

88. Bivariate distributions with conditionals in prescribed exponential families (with D. Strauss), *Journal of the Royal Statistical Association, Series B*, **53**, 365–375.

89. Lorenz ordering of exponential order statistics (with H. N. Nagaraja), *Statistics & Probability Letters*, **11**, 485–490.

90. On some properties of bivariate weighted distributions (with H. N. Nagaraja), *Communications in Statistics—Theory and Methods*, **20**, 1853–1860.

91. Pseudo-likelihood estimation: Some examples (with D. J. Strauss), *Sankhyā, Series B*, **53**, 233–243.

92. Preservation and attenuation of inequality as measured by the Lorenzorder, *Stochastics Orders and Decision Under Risk, IMS Lecture Notes/ Monograph Series*, **19**, 25–37.

93. Lorenz ordering of order statistic (with J.A. Villasenor), *Stochastic Orders and Decision Under Risk, Lecture Notes/Monograph Series*, **19**, 38–47.

94. Centered distributions with Cauchy conditionals (with D. N. Anderson), *Communications in Statistics—Theory and Methods*, **20**, 2881–2889.

1992

95. On distributions whose component ratios are Cauchy (with P. L. Brockett), *American Statistician*, **46**, 25–26.

96. Multivariate logistic distributions, In *Handbook of the Logistic Distribution* (Ed., N. Balakrishnan), pp. 237–261, Marcel Dekker, New York.

97. Logistic and semi-logistic processes, *Journal of Computational and Applied Mathematics*, **40**, 139–149.

98. Discussion of order statistics from discrete distributions, *Statistics*, **23**, 209–211.

99. Skewness and kurtosis orderings: an introduction (with R. A. Groeneveld), *Stochastic Inequalities, IMS Lecture Notes/Monograph Series*, **22**, 17–24.

100. Classical and Bayesian analysis of fatigue strength data (with E. Castillo and J. M. Sarabia), *IABSE Reports*, **66**, 89–97.

1993

101. Logistic processes involving Markovian minimization, *Communications in Statistics—Theory and Methods*, **22**, 1699–1707.

102. Conditionally specified models: Structure and inference (with E. Castillo and J. M. Sarabia), In *Multivariate Analysis: Future Directions*, Vol. 2 (Eds., C. M. Cuadras and C. R. Rao), pp. 441–454, Elsevier, Amsterdam.

103. Linnik distributions and processes (with D. N. Anderson), *Journal of Applied Probability*, **30**, 330–340.

104. Multivariate distributions with generalized Pareto conditionals (with E. Castillo and J. M. Sarabia), *Statistics & Probability Letters*, **17**, 361–368.

105. The nontruncated marginal of a truncated bivariate normal distribution (with R. J. Beaver, R. A. Groeneveld, and W. Q. Meeker), *Psychometrika*, **58**, 471–488.

106. Nonparametric estimation of Lorenz curves (with D. N. Anderson), *International Journal of Mathematical and Statistical Sciences*, **2**, 57–72.

107. Conjugate exponential family priors for exponential family likelihoods (with E. Castillo and J. M. Sarabia), *Statistics*, **25**, 71–77.

1994

108. A conditional characterization of the multivariate normal distribution (with E. Castillo and J. M. Sarabia), *Statistics & Probability Letters*, **19**, 313–315.

109. Multivariate normality via conditional specification (with E. Castillo and J. M. Sarabia), *Statistics & Probability Letters*, **20**, 353–354.

110. On uniform marginal representation of contingency tables (with D. V. Gokhale), *Statistics & Probability Letters*, **21**, 311–316.

111. Extremal sojourn times for Markov chains, In *Extreme Value Theory and Applications, Proceedings of the Conference on Extreme Value Theory and Applications*, **3**, 19–21.

112. Conditional characterizations of the Mardia multivariate Pareto distribution (with E. Castillo and J. M. Sarabia), *Pakistan Journal of Statistics*, **10**, 143–145.

113. Where's the lack of memory? *Computational Statistics and Data Analysis*, **18**, 243–246.

114. Bayesian inference for conditionally specified models, In *Proceedings of the Section on Bayesian Statistical Science*, pp. 165–168, Annual Meeting of the American Statistical Association, Toronto, Ontario, Canada.

115. Bhattacharyya's normal conditionals distribution, In *Essays on Probability and Statistics, Festschrift in Honour of Professor Anil Kumar Bhattacharyya* (Eds., S. P. Mukherjee, A. Chaudhure, and S. K. Basu), pp. 1–13.

116. El hombre mas alto del mundo (with J. A. Villasenor), *Memoria del IX Foro Nactional de Estadistica*, pp. 113–115, Universidad Autonoma de Coahuila, Saltillo Coahuila, Mexico.

117. What price convenience? *Parisankhyan Samikkha*, **1**, 35–41.

1995

118. General conditional specification models (with E. Castillo and J. M. Sarabia), *Communications in Statistics—Theory and Methods*, **24**, 1–11.

119. The tallest man in the world (with J. Villasenor), In *Statistical Theory and Applications: Papers in Honor of Herbert A. David* (Eds., H. N. Nagaraja, P. K. Sen, and D. F. Morrison), pp. 81–88, Springer-Verlag, New York.

120. Measuring skewness with respect to the mode (with R. A. Groeneveld), *The American Statistician*, **49**, 34–38.

121. Distribution with conditionals in the Pickands-Dehaan generalized Pareto family (with J. M. Sarabia and E. Castillo), *Indian Journal for Productivity, Quality, and Reliability Transactions*, **20**, 28–35.

122. Conditional survival models, In *Recent Advances in Life-Testing and Reliability: A Volume in Honor of Alonzo Clifford Cohen, Jr.* (Ed., N. Balakrishnan), pp. 589–601, CRC Press, Boca Raton.

123. Characterizations (with J. S. Huang), In *The Exponential Distribution: Theory, Methods, and Applications* (Eds., N. Balakrishnan and A. P. Basu), pp. 185–203, Gordon and Breach, Amsterdam.

1996

124. Modeling the fatigue life of longitudinal elements (with E. Castillo and J. M. Sarabia), *Naval Research Logistics*, **43**, 885–895.

125. Conditional proportional hazards models (with Y. H. Kim), In *Lifetime Data: Models in Reliability and Survival Analysis* (Eds., N. P. Jewell, A. C. Kimber, M. L. T. Lee, and G. A. Whitmore), pp. 21–28, Kluwer Academic, Dordrecht, Netherlands.

126. Marginally and conditionally specific multivariate survival models: A survey, In *Statistics of Quality* (Eds., S. Ghosh, W. R. Schucany, and W. B. Smith), Vol. 153, pp. 233–252.

127. Specification of distributions by combinations of marginal and conditional distributions (with E. Castillo and J. M. Sarabia), *Statistics & Probability Letters*, **26**, 153–157.

128. Priors with convenient posteriors (with E. Castillo and J. M. Sarabia), *Statistics*, **28**, 347–354.

129. Modelling gas release event behavior in hazardous waste tanks (with D. N. Anderson), *Environmental and Ecological Statistics*, **3**, 281–290.

130. Multivariate distributions with Gaussian conditional structure (with J. Wesolowski), In *Stochastic Processes and Functional Analysis* (Eds., J. A. Goldstein, N. E. Gresky, and J. J. Uhl, Jr.), pp. 45–59.

131. Comparisons of means using conditionally conjugate priors (with E. Castillo and J. M. Sarabia), *Journal of the Indian Society of Agricultural Statistics*, **49**, 319–334.

132. Normal attraction of records: Variations on the theme. (with J. A. Villasenor), *Journal of Statistical Research*, **30**, 1–10.

1997

133. Variance estimation with diffuse prior information (with J. A. Villasenor), *Statistics & Probability Letters*, **33**, 35–39.

134. Specification of bivariate distributions, In *Encyclopedia of Statistical Sciences* (Eds., S. Kotz, C. B. Read, and D. L. Banks), pp. 61–63, John Wiley & Sons, New York.

135. Gumbel records and related characterizations. (with J. A. Villasenor), In *Advances in the Theory and Practice of Statistics: A Volume in Honor of Samuel Kotz* (Eds., N. L. Johnson and N. Balakrishnan), pp. 441–453, John Wiley & Sons, New York.

136. Characterizations involving conditional specification, *Journal of Statistical Planning and Inference*, **63**, 117–131.

1998

137. The use of conditionally conjugate priors in the study of ratios of gamma scale parameters (with E. Castillo and J. M. Sarabia), *Computational Statistics and Data Analysis*, **27**, 125–139.

138. Joint confidence sets for the mean and variance of a normal distribution (with R. M. Shavelle), *The American Statistician*, **52**, 133–140.

139. Lorenz ordering of order statistics and record values (with J. A. Villasenor), *Handbook of Statistics*, **16**, 75–87.

140. Distributions most nearly compatible with given families of conditional distributions (with D. V. Gokhale), *Sociedad de Estadstica e Investigación Operativa Test*, **7**, 377–390.

141. The asymptotic distributions of sums of records (with J. A. Villasenor), *Extremes*, **1**, 351–363.

142. Some alternative bivariate Gumbel models (with E. Castillo and J. M. Sarabia), *Environmetrics*, **9**, 599–616.

143. A distributional study of lifetime data from longitudinal specimens (with M. C. Sifuentes and J. A. Villasenor), *Perfiles*, pp. 32–48, Universidad Autonoma de Cahuila, Saltillo Coahuila, Mexico.

144. Bayesian analysis for classical distributions using conditionally specified priors (with E. Castillo and J. M. Sarabia), *Sankhyā, Series B*, **60**, 228–245.

145. Castillo-Galambos functional equation, In *Encyclopedia of Statistical Sciences* (Eds., S. Kotz *et al.*) Third edition, p. 70, John Wiley & Sons, New York.

1999

146. Parameter estimation under generalized ranked set sampling (with Y. Kim), *Statistics & Probability Letters*, **42**, 353–360.

147. Estimation of a distribution function under generalized ranked set sampling (with Y. Kim), *Communications in Statistics—Simulation and Computation*, **28**, 657–666.

148. Variations on a Bhattacharya theme *Calcutta Statistical Association Bulletin*, **49**, 193–194.

149. Remarks on incompatible conditional distributions (with D. V. Gokhale), *Journal of the Indian Statistical Association*, **37**, 121–140.

2000

150. Near compatibility of conditional densities (with D. V. Gokhale), *Journal of the Indian Statistical Association*, **38**, 155–170.

151. Variability ordering of functions (with H. Joe), *Internat. J. Math. Stat. Sci.*, **9**, 179–189.

152. Some skewed multivariate distributions (with R. J. Beaver), *American Journal of Mathematical and Management Sciences*, **20**, 27–38.

153. The skew-Cauchy distribution (with R. J. Beaver), *Statistics & Probability Letters*, **49**, 285–290.

154. A parametric regression model for possibly censored lifetime data (with M. Cantú-Sifuentes and J. A. Villasenor-Alva), *Agrociencia*, **34**, 453–465.

155. Multiple modes in densities with normal conditonals (with E. Castillo, J. M. Sarabia, and L. González-Vega), *Statistics & Probability Letters*, **49**, 355–363.

156. Hidden Truncation Models (with R. J. Beaver), *Sankhyā, Series A*, **62**, 23–35.

2001

157 Quantification of incompatability of conditional and marginal information (with E. Castillo and J. M. Sarabia), *Communications in Statistics—Theory and Methods*, **30**, 381–395.

158. Modeling the lifetime of longitudinal elements (with M. Cantú-Sifuentes and J. A. Villasenor-Alva), *Communications in Statistics—Simulation and Computation*, **30**, 717–741.

159. Beads, bags, Bayes, and the fundamental problem of sampling theory, *Communications in Statistics—Theory and Methods*, **30**, 1963–1967.

160. A multivariate version of Steins identity with applications to moment calculations and estimation of conditionally specified distributions (with E. Castillo and J. M. Sarabia), *Communications in Statistics—Theory and Methods*, **28**, 2517–2542.

161. Conditionally specified distributions: An introduction with discussion (with E. Castillo and J. M. Sarabia), *Statistical Science*, **16**, 249–274.

162. Pareto processes, In *Handbook of Statistics* (Eds., D. N. Shanbhag and C. R. Rao), Vol. 19, pp. 1–33, Elsevier Science.

163. Bivariate distributions compatible or nearly compatible with given conditional information (with E. Castillo and J. M. Sarabia), In *Probability and Statistical Models with Applications* (Eds., A. Charalambides, M.V. Koutras, N. Balarishnan), pp. 225-237, Chapman & Hall/CRC, Boca Raton.

164. Nonparametric Statistics: Records, *International Encyclopedia of the Social & Behavioral Sciences*, 10681–10685.

165. Conditional specification, In *A Estatistica em Movimento, Actas do VIII Congresso Annual da Sociedade Portuguesa de Estatistica* (Eds., M. M. Neves, J. Cadima, M. J. Martins, and F. Rosado), pp. 3–13.

166. Multivariate models involving conditional specification (with E. Castillo and J. M. Sarabia), *J. Ind. Soc. Prob. Statist.*, **6**, 97–121.

2002

167. Compatibility and near compatibility in multiple assessment of Bayesian networks (with E. Castillo and J. M. Sarabia), *Journal of Propagations in Probability and Statistics*, **2**, 161–176.

168. Skewed multivariate models related to hidden truncation and/or selective reporting with discussion (with R. J. Beaver), *Sociedad de Estadstica e Investigación Operativa Test*, **11**, 7–54.

169. Bayesian inference using conditionally specified priors (with E. Castillo and J. M. Sarabia), In *Handbook of Applied Econometrics and Statistical Inference* (Eds., A. Ullah, A. T. K. Wan, and A. Chaturvedi), pp. 1–26, Marcel Dekker, New York.

170. The conditional distribution of X given $X = Y$ can be almost anything! (with C. A. Robertson), In *Advances on Theoretical and Methodological Aspects of Probability and Statistics* (Ed., N. Balakrishnan), pp. 75–81, Taylor and Francis, London.

171. Parametric inference with generalized rank set data (with R. J. Beaver), In *Selected Topics in Statistics and Operations Research*, Vol. 3, Statistics and Operations Research, a mathematical sciences series of four edited volumes in honor of the Golden Jubilee of the Institute of Technology, Kharagpur, India.

172. Multivariate survival models incorporating hidden truncation (with R. J. Beaver), In *Distributions with Given Marginal and Statistical Modeling*, pp. 9–19, Kluwer Academic Publishers, Dordrecht.

173. Exact and near compatibility of discrete conditional distributions (with E. Castillo and J. M. Sarabia), *Computational Statistics & Data Analysis*, **40**, 231–252.

174. Goodness-of-fit tests based on record data and generalized ranked set data (with R. J. Beaver, E. Castillo and J. M. Sarabia), In *Goodness-of-Fit Tests and Model Validity* (Eds., C. Huber-Carol, N. Balakrishnan, M. S. Nikulin, M. Mesbah), pp. 143–157, Birkhaüser-Boston, Basel, Berlin.

175. Conditionallly specified multivariate skewed distributions (with E. Castillo and J. M. Sarabia), Selected Articles from San Antonio Conference in Honour of C. R. Rao (San Antonio, TX 2000), *Sankhyā, Series A*, **64**, 2006–2026.

2003

176. Back to Bayesics, *Journal of Statistical Planning and Inference*, **109**, 179–187.

2004

177. Some additive component to skewness models (with R. J. Beaver), *Journal of Probability and Statistical Science*, **2**, 139–147.

178. Alternative constructions of skewed multivariate distributions (with R. J. Beaver), *Acta Commentationes Universitatis Tartuensis de Mathematica*, **8**, 73–81.

179. Distributions with beta conditionals (with E. Castillo and J. M. Sarabia), In *Handbook of Beta Distributions and Its Applications* (Eds., A. K. Gupta and S. Nadarajah) pp. 255–282, Marcel Dekker, New York.

180. Elliptical models subject to hidden truncation or selective sampling (with R. J. Beaver), In *Skew Elliptical Distributions and Their Applications* (Ed., M. G. Genton), pp. 101–112, Chapman and Hall, Boca Raton.

181. Compatibility of partial or complete conditional probability specifications (with E. Castillo and J. M. Sarabia), *Journal Statistical Planning and Inference*, **123**, pp. 133–159.

182. Characterizations of the skew-normal and generalized chi distributions (with G. D. Lin), *Sankhyā*, **66**, 593–606.

183. Procrustean strategies for incompatible conditional and marginal information, *Calcutta Statistical Association Bulletin* (to appear).

184. Characterizations of multivariate distributions involving conditional specification and/or hidden truncation, In *Advances in Models, Characterizations and Applications* (Eds., N. Balakrishnan, O. Gebizlioglu and I. Bayramoglu), pp. 161–176, Taylor and Francis, Philadelphia, Pennsylvania.

2005

185. On the value of imprecise prior information (with J. A. Villasenor), *Communications in Statistics—Theory and Methods*, **34**, 807–822.

186. Distributions with conditionals in truncated weighted families (with E. Castillo and J. M. Sarabia), *Statistics*, **39**, 133–147.

Contributors

Angulo, J. M. Departamento de Estadistica e I.O., Universidad de Granada, Campus Fuente Nueva s/n, 18071 Granada, Spain

Asadi, Majid Department of Statistics, University of Isfahan, Isfahan 81744, Iran
m.asadi@sci.ui.ac.ir

Balakrishnan, N. Department of Mathematics and Statistics, McMaster University, Hamilton, ON, Canada L8K 4K1
bala@univmail.mcmaster.ca

Block, Henry Department of Statistics, University of Pittsburgh, Pittsburgh, PA 15260, USA
hwb@stat.pitt.edu

Boland, Philip J. Department of Statistics and Actuarial Science, National University of Ireland, Belfield, Dublin 4, Ireland
philip.j.boland@ucd.ie

Castillo, Carmen Department of Civil Engineering, University of Granada, Santander, Spain
mcastill@ugr.es

Castillo, Enrique Department of Applied Mathematics and Computing Sciences, University of Cantabria, Santander, Spain
castie@unican.es

Chen, Zhenmin Department of Statistics Florida International University, Miami, FL 33199, USA
chenzh@fiu.edu

David, H. A. Department of Statistics, Iowa State University, Ames, IA 50011, USA
hadavid@iastate.edu

Divinsky, Michael 94 Derech HaTayasim Rd., Apt. 37, Tel Aviv 67539, Israel
mdivinsky@hotmail.com

Dugas, Michael R. Department of Statistics, University of California, Davis, CA 95616, USA
mrdugas@ucdavis.edu

Ebrahimi, Nader Division of Statistics, Northern Illinois University, DeKalb, IL 60155, USA
nader@math.niu.edu

El-Baz, Abeer Department of Statistics, George Washington University, Washington, DC 20052, USA

García-Pérez, Alfonso Departamento de Estadística, Investigación Operativa y Cálculo Numérico, Universidad Nacional de Educación a Distancia, Paseo Senda del Rey 9, 28040 Madrid, Spain
agar-per@ccia.uned.es

Gatto, Riccardo University of Bern, Institute of Mathematical Statistics and Actuarial Science, Sidlerstrasse 5, 3012 Bern, Switzerland
gatto@stat.unibe.ch

Girón, F. Javier Departamento de Estadística e I. O., Facultad de Ciencias, Universidad de Málaga, Campus de Teatinos s/n, 29071 Malaga, Spain
fj_giron@uma.es

Guilbaud, Olivier AstraZeneca R&D, S-15185 Södertälje, Sweden
olivier.guilbaud@astrazeneca.com

Guo, Wenge Division of Epidemiology and Biostatistics, Department of Environmental Health, University of Cincinnati, Cincinnati, OH 45267, USA
wenge.guo@gmail.com

Hadi, Ali S. Department of Mathematics, The American University in Cairo, Cairo, Egypt
ahadi@aucegypt.edu

Hamedani, G.G. Department of Mathematics, Statistics, and Computer Science, Marquette University, Milwaukee, WI 53201, USA
hosseinh@mscs.mu.edu

Jammalamadaka, S. Rao Department of Statistics and Applied Probability, University of California, Santa Barbara, CA 93106, USA
rao@pstat.ucsb.edu

Joe, Harry Department of Statistics, University of British Columbia, Vancouver, BC, Canada V6T 1Z2
harry@stat.ubc.ca

Kasala, Subramanyam Mathematical Sciences Department, University of North Carolina, Wilmington, NC 28403, USA
kasalas@uncw.edu

Lai, C. D. Institute of Information Sciences and Technology, Massey University, Palmerston North, New Zealand
c.lai@massey.ac.nz

Li, T. Department of Mathematics, Statistics and Computer Science, St. Francis Xavier University, Antigonish, NS, Canada B2G 2W5
tli@stfx.ca

Lefèvre, Claude Université Libre de Bruxelles, CP210, boulevard du Triomphe, 1050 Bruxelles, Belgium
clefevre@ulb.ac.be

Martínez, M. Lina Departamento de Estadística e I. O., Facultad de Ciencias, Universidad de Málaga, Campus de Teatinos s/n, 29071 Malaga, Spain
emoreno@ugr.es

Menéndez, M. L. Department of Applied Mathematics, E.T.S.A.M., Politechnical University of Madrid, Madrid, Spain
mmenende@aq.upm.es

Mi, Jie Department of Statistics, Florida International University, Miami, FL 33199, USA
mi@fiu.edu

Moreno, Elías Departamento de Estadística e I. O., Facultad de Ciencias, Universidad de Málaga, Campus de Teatinos s/n, 29071 Malaga, Spain
mlmartinez@uma.es

Nagaraja, H. N. Department of Statistics, The Ohio State University, Columbus, OH 43210-1247, USA
hnn@stat.ohio state.cdu

Naik, Dayanand N. Department of Mathematics and Statistics, Old Dominion University, Norfolk, VA 23529, USA
dnaik@odu.edu

Navarro, Jorge Departamento de Estadistica e I.O., Universidad de Murcia, 30100 Espinardo, Murcia, Spain
jorgenav@um.es

Nayak, Tapan K. Department of Statistics, George Washington University, Washington, DC 20052, USA
tapan@gwu.edu

Pardo, L. Department of Statistics and Operations Research, Complutense University of Madrid, Madrid, Spain
lpardo@math.ucm.es

Peña, Daniel Department of Statistics, Universidad Carlos III de Madrid, Madrid, Spain
dpena@est-econ.uc3m.es

Pewsey, Arthur Departamento de Matemáticas, Universidad de Extremadura, Cáceres, Spain
apewsey@unex.es

Plungpongpun, Kusaya Silpakorn University, Bangkok, Thailand

Poncela, Pilar Departamento de Análisis Económico, Economía Cuantitativa Universidad Autónoma de Madrid, Madrid, Spain
pilar.poncela@uam.es

Rao, M. Bhaskara Center for Genome Information, Department of Environmental Health, University of Cincinnati, Cincinnati, OH 45267, USA
raomb@ucmail.uc.edu

Reed, William J. Department of Mathematics and Statistics, University of Victoria, P.O. Box 3045, Victoria, BC, Canada V8W 3P4
reed@math.uvic.ca

Ruiz, Jose M. Departamento de Estadistica e I.O., Universidad de Murcia, 30100 Espinardo, Murcia, Spain
jmruizgo@um.es

Ruiz-Medina, M. D. Departamento de Estadistica e I.O., Universidad de Granada, Campus Fuente Nueva s/n, 18071 Granada, Spain
mruiz@ugr.es

Rychlik, Tomasz Institute of Mathematics, Polish Academy of Sciences, Chopina 12, 87100 Toruń, Poland
t.rychlik@impan.gov.pl

Samaniego, Francisco J. Department of Statistics, University of California, Davis, CA 95616, USA
fjsamaniego@ucdavis.edu

Sandoval, Carlos J. Universidad Católica San Antonio de Murcia, 30701 Murcia, Spain
cjsandoval@pdi.ucam.edu

Sarabia, José María Department of Economics, University of Cantabria, Santander, Spain
sarabiaj@unican.es

SenGupta, Ashis Applied Statistics Unit, Indian Statistical Institute, 203 B.T. Road, Kolkata 700 108, India
ashis@isical.ac.in

Singh, Harshinder Department of Statistics, West Virginia University, Morgantown, WV 26506-6330, USA
hsingh@stat.wvu.edu

Soofi, Ehsan S. School of Business Administration,University of Wisconsin-Milwaukee, Milwaukee, WI 53201, USA
essofi@uwm.edu

Tucker, Brian Pracs Institute, University Drive, Fargo, ND 58105, USA

Zhou, YanYan Department of Statistics, Florida International University, Miami, FL 33199, USA
zhouy@fiu.edu

List of Tables

List of Figures

Advances in Distribution Theory,
Order Statistics, and Inference

PART I
Discrete Distributions and Applications

1

Stochastic Comparisons of Bernoulli Sums and Binomial Random Variables

Philip J. Boland and Harshinder Singh

National University of Ireland, Dublin, Ireland
West Virginia University and NIOSH, Morgantown, WV, USA

Abstract: There are many practical situations in sampling and testing, when the probability of success varies in a sequence of n independent Bernoulli trials. In many of these cases and for various reasons, we may find it useful to compare the distribution of the number of successes $X = \sum Bin(1, p_i)$ in n such trials with a binomial random variable $Y = Bin(n, p)$ for some p. For example, such a comparison might be useful in deciding whether or not stratified sampling is superior (or inferior) to simple random sampling in survey sampling, or whether or not partition (or subdomain) testing is to be preferred to simple random testing in attempting to find faults in software. We will discuss the rationale behind several methods and orders for stochastically comparing the random variables X and Y. These include comparing their means, but also comparing them with respect to the usual stochastic order, the precedence order, the ≥ 1 order and even the likelihood ratio order. It will be seen that many interesting comparisons between X and Y depend on the relationship between p and various means (harmonic, geometric, arithmetic, complimentary geometric, and complimentary harmonic) of the components in the vector $p = (p_1, p_2, \ldots, p_n)$.

Keywords and phrases: Bernoulli and binomial random variables, stochastic order, stochastic precedence, arithmetic, geometric, harmonic, complimentary geometric, complimentary harmonic means

1.1 Introduction

The binomial distribution $Y \sim Bin(n, p)$, where the variable of interest Y is the number of successes in n independent Bernoulli trials, is one of the most basic and classic probability distributions. However, in many situations of interest, the probability of success in the subsequent Bernoulli trials might vary from trial to trial, in which case the random variable of interest would be actually $X \sim \sum_1^n Bin(1, p_i)$. It is often of interest to compare the distributions of X and Y for given values of $(p_1, \ldots, p_n)$ and p. For example, such a comparison might be useful in deciding whether or not stratified random sampling (where X is the number of successes) is superior (or inferior) to simple random sampling (where the number of successes is Y) in survey sampling with replacement, or whether or not partition (or subdomain) testing is to be preferred to simple random testing in attempting to find faults in software; see Boland *et al.* (2002, 2003a).

There are of course many different ways in which one might compare the two random variables X and Y, and often an appropriate comparison is determined by the context of the application that one has in mind. For example, if we are interested in the average number of successes, we would probably prefer X to Y if the expected number of successes $E(X)$ is greater than $E(Y)$. In some cases, the probabilities of success $\{p, p_1, \ldots, p_n\}$ are all small, and hence a success is a rare event. For example, imagine a situation where one is testing for the occurrence of a rare disease in a country and where the prevalence in the ith geographical area is given by p_i, while the overall prevalence in the country is given by p. In such a situation, we might be interested in observing just one (or at least one) success (individual with the disease in question), and hence might compare X (the number found with stratified testing with one selection from each of the geographical areas) and Y (the number found from a simple random sample of the whole area) by considering the probabilities of *at least one success* with each method.

If it is desirable to observe as many successes as possible, then surely we would prefer X to Y (or conversely Y to X) if for every t, $P(X > t) \geq P(Y > t)$ (respectively, $P(X > t) \leq P(Y > t)$). In this case, we are comparing the random variables X and Y by what is commonly known as the *usual stochastic order*, which is a rather strong partial ordering on the set of all random variables. A closely related (but weaker method of comparing distributions) is the (relatively new) stochastic order known as the precedence order, whereby we prefer X to Y (and say that Y precedes X) if $P(X > Y) - P(X < Y) \geq 0$. This essentially says that the chances of X exceeding Y are at least as great as those of Y exceeding X.

We have discussed the rationale behind several methods (and implicitly stochastic orders) for comparing the random variables X and Y. In Section 1.2, we will formally define stochastic orderings corresponding to these (and some other) methods for comparing X and Y. In Section 1.3, we will see that many interesting comparisons between X and Y depend on the relationship between p and various mathematical means (harmonic, geometric, arithmetic, complimentary geometric, and complimentary harmonic) of the components in the vector $\mathbf{p} = (p_1, p_2, \ldots, p_n)$. Graphical insight into these comparisons is provided in Section 1.4 for the case when $n = 2$.

1.2 Stochastic Orders for Random Variables

The concept of **stochastic order** is often useful in comparing random variables. There are, of course, a wide variety of possible partial orders which one may consider on the set of random variables, and some of them (for example, the mean order, the $\geq_{\bar{F}(1)}$ order, and the precedence order) are basically total orders in that for these any two random variables may be compared (for the mean order this is the case if one restricts attention to those random variables with a finite mean). Many other stronger stochastic orders of interest (like the usual stochastic order or the likelihood ratio order) are partial orders. In this section we will briefly define and review some of the stochastic orders that are particularly useful in comparing sums of Bernoulli random variables. The article of Shaked and Shanthikumar (1994) provides an excellent resource on stochastic orders in general.

If U is a random variable, we use $F_U(t) = P(U \leq t)$, $\bar{F}_U(t) = 1 - F_U(t)$, and $f_U(t)$ to represent, respectively, the **distribution** function, the **survival** function, and the **density or mass** function of U. In reliability theory and survival analysis the hazard rate (or failure rate) function $r_U(t) = f_U(t)/\bar{F}_U(t)$ provides a useful characterization of the random variable U (when it exists), and represents the instantaneous rate of failure at time t given survival up to time t. We begin our list of some basic stochastic orders with the well-known and classical *usual* stochastic order.

Definition 1.2.1 U is greater than V in the **usual stochastic order** ($U \geq_{st} V$) if $\bar{F}_U(t) \geq \bar{F}_V(t)$ for all t.

Definition 1.2.2 If the hazard rate function of U is less than that of V at all points t ($r_U(t) \leq r_V(t)$), then we say that U exceeds V in the **hazard rate order** and write $U \geq_{hr} V$. U is greater than V in the **likelihood ratio order** (and we write $U \geq_{lr} V$) if $f_U(t)/f_V(t) \uparrow t$.

Generally speaking, the usual stochastic order, the hazard rate order, and the likelihood ratio order are probably the most frequently used stochastic orders, although the last two are perhaps not of much practical use when comparing sums of Bernoulli random variables. The **mean** ordering is a very weak but total stochastic order on the set of random variables with finite expectation.

Definition 1.2.3 We say that U is greater than V in the **mean order** (and write $U \geq_{mn} V$) if $E(U) \geq E(V)$.

The next (total) stochastic order we consider is called the $\bar{F}(1)$ order, and may be useful when the interest is in at least one success. This may occur when, for example, one is testing for the occurrence of a rare disease in a country or a fault in a piece of software, or even when one is sampling to find a faulty tax return in a revenue office!

Definition 1.2.4 We say that U is greater than V in the $\bar{\mathbf{F}}(\mathbf{1})$ **order** (and write $U \geq_{\bar{F}(1)} V$) if $P(U \geq 1) \geq P(V \geq 1)$.

The precedence order is a relatively new stochastic order which essentially was used by Singh and Misra (1996) to study the reliability of redundancy allocations in certain engineering systems. Arcones *et al.* (2002) provide nonparametric estimates of distribution functions that are constrained by a stochastic precedence order similar to that defined below.

Definition 1.2.5 We define U to be larger than V in the **stochastic precedence** order (or V precedes U) whenever $P(U > V) \geq P(U < V)$, and in this case we write $U \geq_{sp} V$.

One may readily establish the following implications between the above stochastic orders:

$$U \geq_{lr} V \Longrightarrow U \geq_{hr} V \Longrightarrow U \geq_{st} V \Longrightarrow \text{ both } U \geq_{mn} V \text{ and } U \geq_{\bar{F}(1)} V$$

Note that the usual stochastic ordering is stronger than both the mean and $\bar{F}(1)$ orderings, although neither of these last two orderings implies the other in general. In the case where U and V are independent random variables, then the usual stochastic order is stronger than the precedence order [see Boland *et al.* (2004)], although this is not generally true when U and V are dependent as the precedence order takes into account the joint distribution of the random variables.

1.3 Stochastic Order Comparisons for Sums of Bernoulli Random Variables

Many stochastic order comparisons between $X \sim \sum_{i=1}^{n} Bin(1, p_i)$ and $Y \sim Bin(n, p)$ can be characterized in terms of p and functions of the vector of probabilities $\mathbf{p} = (p_1, p_2, \ldots, p_n)$. We will find it useful to consider the following *means* for a vector $\mathbf{p}$:

Definition 1.3.1 (Means of p)

$$\begin{aligned} \bar{p}_a &= \sum p_i/n, \\ \bar{p}_g &= \left\{\prod p_i\right\}^{1/n}, \end{aligned}$$

$$\begin{aligned}\bar{p}_h &= n/\left(\sum 1/p_i\right),\\ \bar{p}_{ca} &= 1-\sum(1-p_i)/n,\\ \bar{p}_{cg} &= 1-\left\{\prod(1-p_i)\right\}^{1/n},\\ \bar{p}_{ch} &= 1-n/\left\{\sum 1/(1-p_i)\right\}\end{aligned}$$

which are respectively the arithmetic, geometric, harmonic, complimentary arithmetic, complimentary geometric, and complimentary harmonic means of $\mathbf{p}$.

From basic but classical results in analysis, we know that in general

$$\bar{p}_h \le \bar{p}_g \le \bar{p}_a = \bar{p}_{ca} \le \bar{p}_{cg} \le \bar{p}_{ch}.$$

In Boland *et al.* (2004), it was shown that for any vector of probabilities $\mathbf{p} = (p_1, p_2, \ldots, p_n)$ where $0 < p_i < 1$ for at least one $i \in \{1, \ldots, n\}$, there exists a unique root p of the equation

$$g(p_1, \ldots, p_n, p) = P(X > Y) - P(Y > X) = 0.$$

We will denote this unique root by $\bar{p}_{sp}$, (using *sp* for stochastic precedence). Note therefore that $\bar{p}_{sp}$ is the unique value of p, which for a given vector $\mathbf{p} = (p_1, p_2, \ldots, p_k)$ yields $X \sim \sum Bin(1, p_i) =_{sp} Bin(n, p) \sim Y$. Using the fact that the usual stochastic order implies the precedence order for independent X and Y, one may show [see Boland *et al.* (2004)] that $\bar{p}_g \le \bar{p}_{sp} \le \bar{p}_{cg}$. It is conjectured that for small values of p_i $(i = 1, \ldots, n)$, one has $\bar{p}_a \le \bar{p}_{sp}$.

Example 1.3.1 (Means for p) For the vector $\mathbf{p} = (0.15, 0.20, 0.25, 0.30, 0.35)$, one can easily establish that $\bar{p}_{sp} = 0.2507$ and

$$(\bar{p}_h,\ \bar{p}_g,\ \bar{p}_a,\ \bar{p}_{cg},\ \bar{p}_{ch}) = (0.2288, 0.2395, 0.2500, 0.2534, 0.2567).$$

The following theorem [proofs of which can be found in Boland *et al.* (2002, 2003b, 2004)] summarizes many of the known stochastic order comparisons between the random variables X and Y.

Theorem 1.3.1 *Let* $X \sim \sum_{i=1}^{n} Bin(1, p_i)$ *and* $Y \sim Bin(n, p)$. *Then*

1. $X \ge_{st} \ (\le_{st}) \ Y \Leftrightarrow p \le \bar{p}_g \ (p \ge \bar{p}_{cg})$
2. $X \ge_{hr} \ (\le_{hr}) \ Y \Leftrightarrow X \ge_{lr} \ (\le_{lr}) \ Y \Leftrightarrow p \le \bar{p}_h \ (p \ge \bar{p}_{ch})$
3. $X \ge_{\bar{F}(1)} \ (\le_{\bar{F}(1)}) \ Y \Leftrightarrow p \le \bar{p}_{cg} \ (p \ge \bar{p}_{cg})$
4. $X \ge_{mn} \ (\le_{mn}) \ Y \Leftrightarrow p \le \bar{p}_a \ (p \ge \bar{p}_{ca} = \bar{p}_a)$
5. $X \ge_{sp} \ (\le_{sp}) \ Y \Leftrightarrow p \le \bar{p}_{sp} \ (p \ge \bar{p}_{sp})$.

1.4 Graphical Insight for Two-Dimensional Stochastic Comparisons

Some interesting perspectives on stochastic comparisons for X and Y can be made in two dimensions by visualizing various contour plots.

Example 1.4.1 $X \sim Bin(1, 0.1) + Bin(1, 0.4)$ and $Y \sim Bin(2, p)$.

Let us for the moment concentrate on the vector of probabilities $\mathbf{p} = (0.1, 0.4)$. One may naturally ask for what values of p is X greater (or less) than $Y \sim Bin(2, p)$ in some stochastic order? One may readily establish that for the vector of probabilities $(p_1, p_2) = (0.1, 0.4)$, one has

$$(\bar{p}_h, \bar{p}_g, \bar{p}_a, \bar{p}_{sp}, \bar{p}_{cg}, \bar{p}_{ch}) = (0.160, 0.200, 0.250, 0.257, 0.265, 0.280).$$

The contour plots of these various (harmonic, geometric, arithmetic, precedence, complimentary geometric, and complimentary harmonic) means for $(p_1, p_2) = (0.1, 0.4)$ are given in Figure 1.1, and are respectively denoted by the letters (h, g, a, sp, cg, ch). For example, all points in the graph on the curve denoted by "g" have the same geometric mean as the coordinates of the point $(0.1, 0.4)$ (in particular, of course, the point $(0.4, 0.1)$), and all points on the line denoted by "a" have the same arithmetic mean as the coordinates of $(0.1, 0.4)$. On inspecting Figure 1.1 (paying particular attention to the curves "g" and "cg") and applying Theorem 1.3.1, it is clear for example that X is stochastically larger than $Y \sim Bin(2, p)$ for any $p \leq 0.20$, and in turn stochastically less than $Y \sim Bin(2, q)$ for any $q \geq 0.265$ in the usual stochastic order. Other similar comparisons lead us to conclude that

$$\begin{array}{llcll} Bin(2,p) \leq_{st} X & (\text{for } p \leq 0.20) & \text{and} & X \leq_{st} Bin(2,q) & (\text{for } q \geq 0.265), \\ Bin(2,p) \leq_{lr} X & (\text{for } p \leq 0.16) & \text{and} & X \leq_{lr} Bin(2,q) & (\text{for } q \geq 0.280), \\ & Bin(2, 0.265) & =_{\bar{F}(1)} & X & \\ & Bin(2, 0.250) & =_{mn} & X, & \\ & Bin(2, 0.257) & =_{sp} & X. & \end{array}$$

Example 1.4.2 $X \sim Bin(1, p_1) + Bin(1, p_2)$ and $Y \sim Bin(2, 0.25)$. Now we consider the binomial random variable $Y \sim Bin(2, 0.25)$ and see how it compares stochastically with $X = Bin(1, p_1) + Bin(1, p_2)$ for various (p_1, p_2). In Figure 1.2, the curve "g" (respectively "cg") represents those points (p_1, p_2) which have the same geometric (complimentary geometric) mean as the components in the vector $(0.25, 0.25)$. From Theorem 1.3.1, we note that any point (p_1, p_2) lying on or above the "g" contour corresponds to an $X \sim Bin(1, p_1) + Bin(1, p_2)$

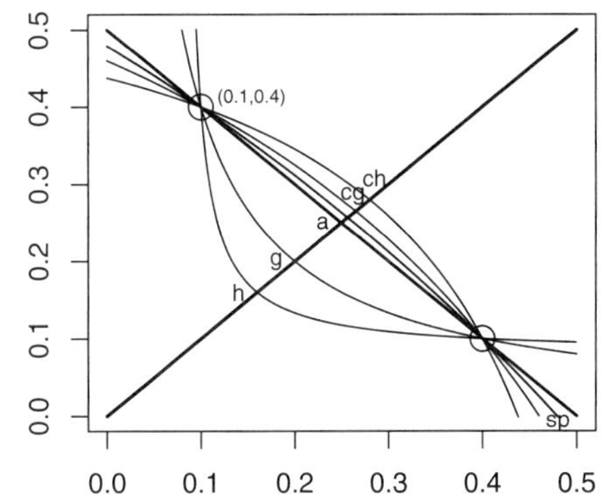

Figure 1.1: Contour means for probabilities (0.1,0.4)

which is greater in the usual stochastic order than $Y \sim Bin(2, 0.25)$, and similarly that any point (p_1, p_2) lying below the "cg" contour corresponds to an $X \sim Bin(1, p_1) + Bin(1, p_2)$ which is less than $Y \sim Bin(2, 0.25)$ in the usual stochastic order.

Figure 1.3 is an extension of Figure 1.2, in which one can clearly see the "ch," "sp," "a," and "h" contour curves for (0.25,0.25) in addition to the "g" and "cg" contours. It allows one to see clearly which X are stochastically greater (smaller) than $Bin(2, 0.25)$ for the other stochastic orders considered here. For example, X is greater (less) than $Bin(2, 0.25)$ in the stochastic precedence order if it corresponds to a point (p_1, p_2) on or above (below) the "sp" contour. Also if $X = Bin(1, 0.18) + Bin(1, 0.40)$, then $X \geq_{st} Bin(2, 0.25)$, but X and $Bin(2, 0.25)$ are not comparable in the hazard rate or likelihood ratio orders.

In conclusion, in this article we have given a reasonably thorough account of how one may stochastically compare $X = \sum Bin(1, p_i)$ in n trials with a binomial random variable $Y = Bin(n, p)$ for some p, in terms of $(p_1, p_2, \ldots, p_n)$ and p. These results suggest interesting future research should be done in extending consideration to stochastic comparisons of two Bernoulli sums of the form $X = \sum Bin(1, p_i)$ and $X^* = \sum Bin(1, p_i^*)$, and characterizing such comparisons in terms of functions of the vectors $(p_1, p_2, \ldots, p_n)$ and $(p_1^*, p_2^*, \ldots, p_n^*)$.

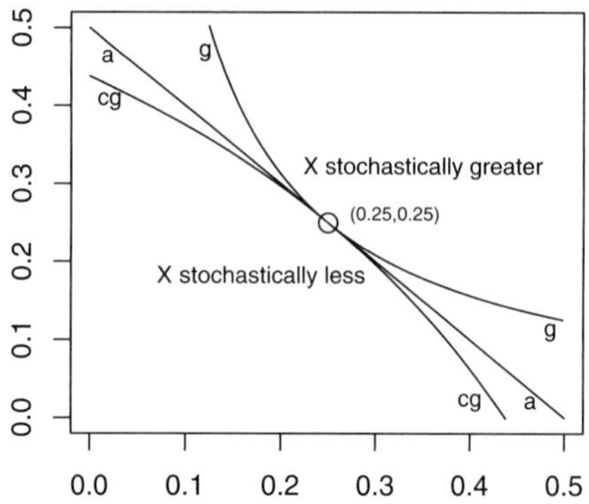

Figure 1.2: (Usual) stochastic order comparisons for X and $Y = \mathrm{Bin}(2, 0.25)$

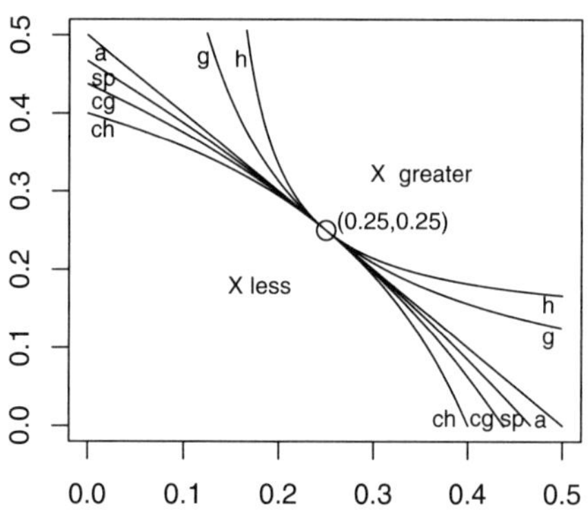

Figure 1.3: Various stochastic order comparisons for X and $Y = \mathrm{Bin}(2, 0.25)$

References

1. Arcones, M. A., Kvam, P. H., and Samaniego, F. J. (2002). Nonparametric estimation of a distribution subject to a stochastic order precedence constraint, *Journal of the American Statistical Association*, **97**, 170–182.

2. Boland, P. J., Faundez Sekirkin, S., and Singh, H. (2003a). Theoretical and practical challenges in software reliability and testing, In *Mathematical and Statistical Methods in Reliability* (Eds., B. H. Lindqvist and K. Doksum), pp. 505–520, World Scientific Publishing, Singapore.

3. Boland, P. J., and Singh, H. (2004). Comparing Bernoulli sums and binomial random variables, In *Proceedings of the International Conference on Distribution Theory, Order Statistics, and Inference in Honor of Barry C. Arnold* (Eds., N. Balakrishnan, E. Castillo, and J. M. Sarabia), University of Cantabria, Santander, Spain.

4. Boland, P. J., Singh, H., and Cukic, B. (2002). Stochastic orders in partition and random testing of software, *Journal of Applied Probability*, **39**, 555–565.

5. Boland, P. J., Singh, H., and Cukic, B. (2003b). Comparing partition and random testing via majorization and Schur functions, *IEEE Transactions in Software Engineering*, **29**, 88–94.

6. Boland, P. J., Singh, H., and Cukic, B. (2004). The stochastic precedence ordering with applications in sampling and testing, *Journal of Applied Probability*, **41**, 73–82.

7. Shaked, M., and Shanthikumar, J. G. (1994). *Stochastic Orders and Their Applications*, Academic Press, San Diego.

8. Singh, H., and Misra, N. (1996). On redundancy allocations in systems, *Journal of Applied Probability*, **31**, 1004-1014.

2

Stopped Compound Poisson Process and Related Distributions

Claude Lefèvre

Université Libre de Bruxelles, Bruxelles, Belgium

Abstract: This chapter considers the first-crossing problem of a *compound Poisson process* with positive integer-valued jumps in a nondecreasing *lower boundary*. The cases where the boundary is a given linear function, a standard renewal process, or an arbitrary deterministic function are successively examined. Our interest is focused on the *exact distribution* of the *first-crossing level* (or time) of the compound Poisson process. It is shown that, in all cases, this law has a simple remarkable form which relies on an underlying polynomial structure. The impact of a raise of a lower deterministic boundary is also discussed.

Keywords and phrases: Compound Poisson process, first-crossing, lower boundary, ballot theorem, generalized Abel–Gontcharoff polynomials, generalized Poisson distribution, quasi-binomial distribution, damage model

2.1 Introduction

Many questions in probability and statistics can be formulated as first-crossing problems between the trajectory of a random process and a nondecreasing boundary, fixed or random, that starts either below or above the trajectory. Applications arise, for instance, in the modelling of queues, dams, and storage, in the theory of risk and ruin in insurance and finance, in the planning of sequential statistical procedures, and in the study of order statistics and empirical processes. The mathematical analysis of first-crossing problems is often focused on asymptotic approximations when explicit formulae are not available.

The present paper deals with the problem of the first-crossing of a *compound Poisson process* with positive integer-valued jumps in a nondecreasing *lower boundary*. We will successively examine the cases where the boundary is a given linear function, a standard renewal process, or an arbitrary deterministic function (Sections 2.2–2.4). Our purpose is to determine, for each of these situations, the *exact* (i.e., nonasymptotic) *distribution* of the corresponding *first-crossing level* (or time). We will also discuss the impact of a raise of the boundary in the deterministic case (Section 2.5).

To tackle these questions, it would be possible to have recourse to a standard technique based on Laplace transforms. Here, however, we will follow a different method which has the advantage that it leads to a simple and efficient evaluation of the distribution by recursion. This approach relies on the property that the first-crossing level probabilities have an underlying polynomial component with a remarkable structure. For the general case (i.e., with a deterministic boundary), the polynomials involved correspond to the so-called *generalized Abel–Gontcharoff polynomials.* Recently, in a few joint papers with Picard, we have developed a general theory on polynomials (and even functions) that enjoy such a structure, and we have used it to study several first-crossing problems in epidemic and risk theories; see, for example, Lefèvre and Picard (1990, 1999) and Picard and Lefèvre (1994, 1996, 2003).

Moreover, this work will allow us to point out several nonstandard discrete distributions of own interest. In particular, we will present extensions of the Poisson law and the binomial law which are based on the generalized Abel–Gontcharoff polynomials (Section 2.5). We mention that various special cases of these distributions have been previously derived in a context of urn models; see, for example, Kotz and Balakrishnan (1997).

It is worth noticing that the first-crossing problem of a compound Poisson in a nondecreasing *upper boundary* is of different nature. Indeed, with a lower boundary, crossing occurs necessarily on a continuous part of the trajectory (and so corresponds to a meeting), while with an upper boundary, crossing arises always at a jump time of the trajectory.

First-crossing problems for a Poisson or compound Poisson process in a given lower or upper boundary, linear or arbitrary, is the object of many papers in the literature. We refer to, for example, Pyke (1959), Daniels (1963), Durbin (1971), Zacks (1991), Stadje (1993), Gallot (1993), Zacks (1997), Böhm and Mohanty (1997), Picard and Lefèvre (1997), De Vylder (1999), Perry *et al.* (1999), Zacks *et al.* (1999), Perry (2000), Ignatov *et al.* (2001), Perry *et al.* (2002), Stadje and Zacks (2003) and Ignatov and Kaishev (2004).

Throughout the chapter, the compound Poisson process is generated from a Poisson process with parameter $\lambda > 0$, and the successive jump sizes W_i, $i \geq 1$, are i.i.d. r.v.'s with positive integer values. Initially, the process starts at a *positive* integer level k; the case $k = 0$, however, will be also considered in some special places (clearly marked).

2.2 The Boundary Is Linear

Let us assume that the boundary is a linear increasing straight line with slope $1/b > 0$. To begin with, we examine the case where the line passes at the origin. Obviously, crossing can only occur at positive integer levels j. Thus, the effective boundary reduces to the discrete set of points $\{(bj, j), j \geq 1\}$, denoted by $\mathcal{B}_b$.

From the structure of the Poisson process, one can discretize the possible crossing times bj and focus on these instants, denoted by $t = 1, 2, \ldots$. So, the boundary $\mathcal{B}_b$ becomes the bisectrix $\{(t, t), t \geq 1\}$. Moreover, the compound Poisson process is then described by the sequence $\{k + S_t, t \in \mathbb{N}\}$ where

$$S_0 = 0, \quad \text{and} \quad S_t = Y_1 + \cdots + Y_t, \quad t \geq 1,$$

the r.v.'s Y_i, $i \geq 1$, being i.i.d. with compound Poisson laws of parameter λb and jump sizes W_i; thus, for $t \geq 1$,

$$P(S_t = s) = e^{-\lambda bt} \sum_{l=1}^{s} \frac{(\lambda bt)^l}{l!} P(W_1 + \cdots + W_l = s), \quad s \in \mathbb{N}.$$

For our purpose, we prefer to rewrite the law of S_t, $t \geq 1$, as

$$P(S_t = s) = e^{-\lambda bt} e_s(\lambda bt), \quad s \in \mathbb{N}, \tag{2.1}$$

where, using an argument x say,

$$e_s(x) = \sum_{l=0}^{s} \frac{x^l}{l!} q_s^{*l}, \quad s \in \mathbb{N}, \tag{2.2}$$

the set $\{q_s^{*l}, s \geq 1\}$ denoting the l-th convolution of the distribution of W_1, for any $l \geq 1$, and $q_s^{*0} \equiv \delta_{s,0}$. Note that the generating function of these e_s's is

$$\sum_{s=0}^{\infty} e_s(x) z^s = e^{xg(z)}, \quad z \in [0, 1], \tag{2.3}$$

where $g(z)$ is the p.g.f. of the law of W_1.

From (2.2), we see that for each $s \in \mathbb{N}$, $e_s(x)$ is a polynomial of degree s in x, for $x \in \mathbb{N}$ and, by extension, for $x \in \mathbb{R}$. The family of polynomials $\{e_s(x), s \in \mathbb{N}\}$ is linearly independent when $P(W_1 = 1) > 0$ (since then, $q_s^{*s} > 0$). For clarity, this condition is assumed to hold true; otherwise, passing to the limit is allowed.

A basic property satisfied by the e_s's is the following convolution property, which is a direct consequence of (2.3):

$$e_s(x_1 + x_2) = \sum_{i=0}^{s} e_i(x_1)e_{s-i}(x_2), \qquad \text{for any } \; x_1, x_2 \in \mathbb{R}. \tag{2.4}$$

Now, we want to determine the distribution of the first-crossing level, denoted by N, of the bisectrix by the compound Poisson (starting at k). Thus,

$$N = S_T \qquad \text{such that} \qquad k + S_T = T. \tag{2.5}$$

Theorem 2.2.1 *For a linear boundary $\mathcal{B}_b$,*

$$P(N = n|k) = \frac{k}{k+n} e^{-\lambda b(k+n)} e_n[\lambda b(k+n)], \quad n \in \mathbb{N}. \tag{2.6}$$

PROOF. By the ballot theorem [see, e.g., Takács (1967)], if $\{Y_i, i \geq 1\}$ is a sequence of i.i.d. $\mathbb{N}$-valued r.v.'s, then their partial sums S_t, $t \geq 1$, satisfy the relations below:

$$P(k + S_t > t, \; k \leq t \leq n-1, \;\; \text{and} \;\; k + S_n = n) = \frac{k}{n} P(k + S_n = n), \quad n \geq k.$$

By (2.1) and (2.5), this is equivalent to

$$P(N = n-k|k) = \frac{k}{n} P(S_n = n-k) = \frac{k}{n} e^{-\lambda bn} e_{n-k}(\lambda bn), \quad n \geq k,$$

hence (2.6). ■

Let us suppose that the compound Poisson process starts at level $k = 0$, and the linear boundary does not pass at the origin but is of the form $\{(a+bj, j), j \in \mathbb{N}\}$ with $a > 0$ and $b \geq 0$; this boundary is denoted by $\mathcal{B}_{a,b}$.

Corollary 2.2.2 *For a linear boundary $\mathcal{B}_{a,b}$,*

$$P(N = n|0) = \frac{a}{a+bn} e^{-\lambda(a+bn)} e_n[\lambda(a+bn)], \quad n \in \mathbb{N}. \tag{2.7}$$

PROOF. Counting the number of events arising during the time interval $[0, a]$, we get from (2.1) (with $b = 1$, $t = a$) and (2.6) that

$$\begin{aligned} P(N = n|0) &= \sum_{k=0}^{n} e^{-\lambda a} e_k(\lambda a) P(N = n-k|k) \\ &= e^{-\lambda(a+bn)} \frac{\lambda a}{n} \sum_{k=0}^{n} k \frac{e_k(\lambda a)}{\lambda a} e_{n-k}(\lambda bn). \end{aligned}$$

But it can be shown from (2.3) that

$$n\frac{e_n(x_1+x_2)}{x_1+x_2} = \sum_{k=0}^{n} k\frac{e_k(x_1)}{x_1}e_{n-k}(x_2), \qquad \text{for any } x_1 \neq 0 \neq x_1+x_2,$$

which yields (2.7). ■

Special case. In the particular case of a simple Poisson process, (2.2) yields $e_s(x) = x^s/s!$, $s \in \mathbb{N}$. So, for $\mathcal{B}_b$, (2.6) becomes

$$P(N=n|k) = k\,\frac{(k+n)^{n-1}(\lambda b)^n}{n!}\,e^{-\lambda b(k+n)}, \quad n \in \mathbb{N}, \tag{2.8}$$

i.e., $k+N$ has a Borel–Tanner distribution [see, e.g., Johnson *et al.* (1992)]. For $\mathcal{B}_{a,b}$, (2.7) becomes

$$P(N=n|0) = \lambda a\,\frac{[\lambda(a+bn)]^{n-1}}{n!}\,e^{-\lambda(a+bn)}, \quad n \in \mathbb{N}, \tag{2.9}$$

that is, N has a generalized Poisson distribution [see, e.g., Consul (1989)]. Thus, the laws derived in (2.6) and (2.7) provide compound extensions of the more classical laws given in (2.8) and (2.9).

Going back to the original situation, we know by the SLLN that $S_t/t \rightarrow_{a.s.} E(Y_1) = \lambda b m_1$ where $m_1 = E(W_1)$. Thus, if $\lambda b m_1 < 1$, we see by (2.5) that $N < \infty$ a.s. The moments of N can then be obtained by standard methods (in terms of the moments $m_j = E(W_1^j)$, $j \geq 1$).

Property 2.2.3 *If $\lambda b m_1 < 1$, then the first two moments of N, for example, are given by*

$$E(N) = k\lambda b m_1/(1-\lambda b m_1), \tag{2.10}$$
$$Var(N) = k\lambda b m_2/(1-\lambda b m_1)^3. \tag{2.11}$$

PROOF. From the conditional p.g.f. of S_{t+1} given S_t (with argument $z \in [0,1]$), we see that the process $\{z^{S_t}e^{\lambda bt[1-g(z)]}, t \in \mathbb{N}\}$ forms a martingale. Applying the optional stopping theorem with respect to time T (which is allowed because $\lambda b m_1 < 1$), we obtain a Wald identity:

$$E\left(\{ze^{\lambda b[1-g(z)]}\}^N\right) = e^{-k\lambda b[1-g(z)]}. \tag{2.12}$$

Put $z = e^v$ with $v \in \mathbb{R}_-$, define the function $\phi(v) = v + \lambda b[1-g(e^v)]$ and let ψ be the inverse function ϕ^{-1}, that is, $\psi(u) = v$ when $u = \phi(v)$ $(u \in \mathbb{R}_-)$ (it exists because $\lambda b m_1 < 1$). Then, (2.12) becomes

$$\sum_{n=0}^{\infty} P(N=n)e^{nu} = e^{k[\psi(u)-u]}. \tag{2.13}$$

By successive differentiations and putting $u = 0$, we obtain, for example,

$$E(N) = k[\psi^{(1)}(0) - 1], \quad \text{and} \quad Var(N) = k\psi^{(2)}(0), \tag{2.14}$$

and since by definition of ϕ and ψ,

$$\psi^{(1)}(u) = 1/\phi^{(1)}(v), \quad \text{with} \quad \phi^{(1)}(0) = 1 - \lambda b m_1,$$
$$\psi^{(2)}(u) = -\phi^{(2)}(v)/[\phi^{(1)}(v)]^3, \quad \text{with} \quad \phi^{(2)}(0) = -\lambda b m_2,$$

(2.14) yields (2.10) and (2.11). ■

2.3 The Boundary Is of Renewal Type

Let $\{X_j, j \geq 1\}$ be a sequence of i.i.d. r.v.'s, and denote their partial sums by $D_t = X_1 + \cdots + X_t$, $t \geq 1$. The boundary under consideration here is the renewal process $\{(D_j, j), j \geq 1\}$, starting at level 0; it is denoted by $\mathcal{B}_r$.

Adopting the above time change, one may still come back to a bisectrix boundary, but this time, for the actualized process $\{k+S_t, t \in \mathbb{N}\}$ where $S_0 = 0$, $S_t = Y_1 + \cdots + Y_t$, $t \geq 1$, and the r.v.'s Y_i, $i \geq 1$, are i.i.d. with compound Poisson laws of parameters, this time random, λX_i and jump sizes W_i. Thus, for $t \geq 1$,

$$P(S_t = s) = E[e^{-\lambda D_t} e_s(\lambda D_t)], \quad s \in \mathbb{N}. \tag{2.15}$$

It is clear that the ballot theorem is again applicable; this leads to the formula (2.16) below. Inserting (2.2) in (2.16), we then deduce the more explicit formula (2.17).

Theorem 2.3.1 *For a renewal type boundary $\mathcal{B}_r$,*

$$P(N = n|k) = \frac{k}{k+n} E[e^{-\lambda D_{k+n}} e_n(\lambda D_{k+n})], \quad n \in \mathbb{N}. \tag{2.16}$$

Denoting by $h(\lambda)$ the Laplace transform $E(e^{-\lambda X})$,

$$P(N = n|k) = \frac{k}{k+n} \sum_{l=0}^{n} \frac{(-\lambda)^l}{l!} q_n^{*l} \left([h(\lambda)]^{k+n}\right)^{(l)}, \quad n \in \mathbb{N}, \tag{2.17}$$

where $(.)^{(l)}$ is the l-th derivative of $(.)$.

Now, let us suppose that the compound Poisson process starts at level $k = 0$, and the renewal process is shifted in time by a quantity a; the boundary is denoted by $\mathcal{B}_{a,r}$. Following the proof of Corollary 2.2.2, we deduce from (2.16) the formula (2.18) below. Note, however, that (2.18) is less tractable than (2.16).

Corollary 2.3.2 *For a renewal type boundary* $\mathcal{B}_{a,r}$,

$$P(N = n|0) = a\, E\left\{ \frac{e^{-\lambda(a+D_n)} e_n[\lambda(a + D_n)]}{a + D_n} \right\}, \quad n \in \mathbb{N}. \tag{2.18}$$

Special case. For a simple Poisson process with the boundary $\mathcal{B}_r$, (2.17) gives

$$P(N = n|k) = \frac{k}{k+n} \frac{(-\lambda)^n}{n!} \left([h(\lambda)]^{k+n}\right)^{(n)}, \quad n \in \mathbb{N}. \tag{2.19}$$

For instance, if X has a gamma distribution with $E(e^{-\lambda X}) = [\mu/(\mu+\lambda)]^{-c}$, for given parameters $\mu, c > 0$, then (2.19) becomes

$$P(N = n|k) = \frac{ck}{ck + (c+1)n} \binom{ck + (c+1)n}{n} \left(\frac{\lambda}{\lambda+\mu}\right)^n \left(\frac{\mu}{\lambda+\mu}\right)^{ck+cn}, \; n \in \mathbb{N}, \tag{2.20}$$

that is, N has a generalized negative binomial distribution [see, e.g., Johnson *et al.* (1992)].

Finally, by the same argument as for Property 2.2.3, we can find the moments of N under the condition $\lambda d_1 m_1 < 1$ where $d_1 = E(X_1)$.

Property 2.3.3 *If* $\lambda d_1 m_1 < 1$, *then the first two moments of* N, *for example, are given by*

$$E(N) = k\lambda d_1 m_1/(1 - \lambda d_1 m_1), \tag{2.21}$$

$$Var(N) = k\lambda[d_1 m_2 + \lambda m_1^2 Var(X_1)]/(1 - \lambda d_1 m_1)^3. \tag{2.22}$$

2.4 The Boundary Is Any Deterministic Function

Let us examine the case of an arbitrary nondecreasing deterministic boundary, denoted by $\mathcal{B}_d$, and which is represented as a set of points $\{(u_j, j), j \geq 1\}$ where the u_j's, $j \geq 1$, form a given sequence of nondecreasing non-negative reals.

As before, we may apply a time change which allows us to consider a bisectrix boundary, for the compound Poisson process $\{k + S_t, t \in \mathbb{N}\}$ where $S_0 = 0$, $S_t = Y_1 + \cdots + Y_t$, $t \geq 1$ and the r.v.'s Y_i, $i \geq 1$, are i.i.d. with compound Poisson laws of parameters, this time nonhomogeneous, $\lambda(u_i - u_{i-1})$ (with $u_0 \equiv 0$) and jump sizes W_i. Thus, for $t \geq 1$,

$$P(S_t = s) = e^{-\lambda u_t} e_s(\lambda u_t), \quad s \in \mathbb{N}. \tag{2.23}$$

For convenience, we denote $U = \{u_1, u_2, \ldots\}$ and $E^l U = \{u_{l+1}, u_{l+2}, \ldots\}$ the l-shifted family, for any $l \in \mathbb{N}$.

To determine the law of N, we will follow a nonstandard approach of algebraic and computational nature. The mathematical tool used is a remarkable family of polynomials, which is called of generalized Abel–Gontcharoff (in short, AG) type. We refer the reader to Lefèvre and Picard (1990) and Picard and Lefèvre (1996) for a general presentation of these polynomials.

Let us briefly recall their construction. For that, the two basic elements are an arbitrary family of reals $U = \{u_1, u_2, \ldots\}$, and any family of linearly independent polynomials $\{e_n(x), n \in \mathbb{N}\}$ of degree n in x ($\in \mathbb{R}$), with $e_0(x) = 1$. Then, an associated family of generalized AG polynomials, $\{G_n(x|U), n \in \mathbb{N}\}$, of degree n in x, is defined univocally by the following recursion:

$$G_n(x|U) = e_n(x) - \sum_{s=0}^{n-1} e_{n-s}(u_{s+1}) G_s(x|U), \quad n \in \mathbb{N}; \tag{2.24}$$

in particular, $G_0(x|U) = 1$ and $G_n(u_1|U) = \delta_{n,0}$. We underline that this recursion being quite direct, the G_n's can be numerically computed in a direct and efficient way. Notice that (2.24) can also be viewed as an Abelian-type expansion of e_n with respect to the family $\{G_s, s \in \mathbb{N}\}$. The generalized AG polynomials enjoy various other nice properties. So, the identity $G_n(x|U+a) = G_n(x-a|U)$ holds for any real a. Moreover, a Taylor-type expansion of G_n about y ($\in \mathbb{R}$) and with respect to the family $\{e_s, s \in \mathbb{N}\}$ yields

$$G_n(x|U) = \sum_{s=0}^{n} e_s(x-y) G_{n-s}(y|E^s U), \quad n \in \mathbb{N}. \tag{2.25}$$

Clearly, for $y = u_1$, (2.25) provides another possible recursion for the G_n's, using the previous border conditions $G_n(u_1|U) = \delta_{n,0}$.

Theorem 2.4.1 *For a deterministic boundary* $\mathcal{B}_d$,

$$P(N = n|k) = e^{-\lambda u_{k+n}}\, G_n(0|\{-\lambda u_j,\ j \geq k\}), \quad n \in \mathbb{N}. \tag{2.26}$$

PROOF. It will be easier hereafter to take $u_0 \in [0, u_1]$, instead of $u_0 = 0$. In other words, at time u_0 the compound Poisson process is at level k and the successive parameters of S_t, $t \geq 1$, are given by $\lambda(u_t - u_0), t \geq 1$. So, it is natural to denote $P(N = n|k) = p_n[k, \lambda(U - u_0)]$, $n \in \mathbb{N}$. Firstly, using a renewal argument, we obtain

$$p_n[k, \lambda(U - u_0)] = \sum_{s=0}^{n} P(Y_1 = s) p_{n-s}[k + s - 1, \lambda(EU - u_1)], \quad n \in \mathbb{N}, \tag{2.27}$$

where we put, for $k = 0$, $p_n[0, \lambda(U - u_0)] = \delta_{n,0}$. Let us try to find an expression of the form

$$p_n[k, \lambda(U - u_0)] = e^{-\lambda(u_{k+n} - u_0)} R_n(-\lambda u_0| - \lambda E^{k-1} U), \quad n \in \mathbb{N}, \tag{2.28}$$

with $E^{-1}U \equiv \{u_j, j \in \mathbb{N}\}$, for some (so far) mysterious function R_n. Then, (2.27) becomes, after simplification,

$$R_n(-\lambda u_0| - \lambda E^{k-1}U) = \sum_{s=0}^{n} e_s(\lambda u_1 - \lambda u_0) R_{n-s}(-\lambda u_1| - \lambda E^{k+s-1}U), \quad n \in \mathbb{N}, \tag{2.29}$$

where for $k = 0$, $R_n(-\lambda u_0| - \lambda E^{-1}U) = \delta_{n,0}$. Now, we observe that the recursion (2.29) is equivalent to the recursion (2.25) written for the polynomial $G_n(-\lambda u_0| - \lambda E^{k-1}U)$. Both recursions being also based on the same border conditions, we deduce that $R_n(-\lambda u_0| - \lambda E^{k-1}U) = G_n(-\lambda u_0| - \lambda E^{k-1}U)$ for all $n \in \mathbb{N}$. Therefore, (2.28) with $u_0 = 0$ yields (2.26). ■

With a compound Poisson process starting at $k = 0$, we suppose that the boundary begins with a delay of $a \geq 0$, that is, is the set $\{(a + u_j, j), j \in \mathbb{N}\}$ where, as before, $u_0 = 0$ and $U = \{u_j, j \geq 1\}$; it is denoted by $\mathcal{B}_{a,d}$. From the proof of Theorem 2.4.1, we see that (2.26) is valid too when $k = 0$, hence the following result.

Corollary 2.4.2 *For a deterministic boundary* $\mathcal{B}_{a,d}$,

$$P(N = n|0) = e^{-\lambda(a+u_n)}\, G_n[0|\{-\lambda(a + u_j),\ j \in \mathbb{N}\}], \quad n \in \mathbb{N}. \tag{2.30}$$

One can show that (2.26) and (2.30) reduce to (2.6) and (2.7) for a linear boundary, because when the u_j's depend linearly on j,

$$G_n(x|U) = \frac{x - u_1}{x - u_{n+1}}\, e_n(x - u_{n+1}), \quad n \in \mathbb{N}. \tag{2.31}$$

Also, as expected, a randomization of (2.26), (2.30) yields (2.16), (2.18) for a renewal type boundary. These verifications are omitted.

Special case. For a simple Poisson process, the G_n's correspond to the standard (nongeneralized) AG polynomials. Then, the identity $G_n(ax|aU) = a^n G_n(x|U)$ holds for any real a, which yields a minor simplification in (2.26) and (2.30).

Illustration. Motivated by applications in queueing and graph theory, Takács (1989) derived in his Example 2 an expression for the law of the first-crossing level of the bisectrix line by a particular process $\{k + S_t, t \in \mathbb{N}\}$ where the r.v.'s Y_i, $i \geq 1$, are independent and exponentially distributed with parameters of geometric form $\lambda p q^{i-1}$ $(0 < p = 1 - q < 1)$.

Let us consider the more general situation where the Y_i's are compound Poisson distributed with the same geometric parameters and jump sizes W_i. Thus, this corresponds to the first-crossing problem of a compound Poisson process with a boundary $\mathcal{B}_d$ as above where $u_j - u_{j-1} = pq^{j-1}$, $j \geq 1$, that is, with $u_j = 1 - q^j$, $j \geq 0$. By (2.26), we then have

$$P(N = n|k) = e^{-\lambda(1-q^{k+n})}\, G_n(\lambda|\{\lambda q^j,\ j \geq k\}), \quad n \in \mathbb{N}. \tag{2.32}$$

To compare with the approach of Takàcs, we observe from (2.24) that G_n in (2.32) may also be viewed as a polynomial in q of degree $n(n+2k-1)/2$. In the Poisson case, one knows that $G_n(\lambda|\{\lambda q^j, j \geq k\}) = (\lambda p)^n G_n(1/p\,|\{q^j/p, j \geq k\})$. Thus, the factor denoted by $\Phi^{(k)}_{k+n}(q)$ in Takàcs corresponds in our notation to $(n!)G_n(1/p\,|\{q^j/p, j \geq k\})$. Takács exploited the property that $\Phi^{(k)}_{k+n}(q)$ is a polynomial in q of degree $n(n+2k-3)/2$, to derive a (different) recurrence formula for its determination. The method proposed, however, is rather intricate; moreover, it relies on the particular geometric form of the parameters.

Now, as proposed by Takács, let us examine the asymptotic distribution of N when $\lambda \to \infty$ and $p \to 0$ in such a way that $\lambda p \to a$ $(0 < a < \infty)$. Thus, $q \sim 1 - a/\lambda$, yielding $q^j \sim 1 - aj/\lambda$, $j \geq 0$, so that by (2.32) with (2.31),

$$P(N = n|k) \to \frac{k}{k+n} e^{-a(k+n)} e_n[a(k+n)], \quad n \in \mathbb{N}. \tag{2.33}$$

In the Poisson case, (2.33) becomes the formula (38) given in Takács. Other limiting behaviours could be of some interest. For example, suppose that $k \to \infty$ and $q \sim 1 - a/k$ $(0 < a < \infty)$. Then, $q^{k+j} \sim e^{-a}$, $j \geq 0$, and we get

$$P(N = n|k) \to e^{-\lambda(1-e^{-a})}\, e_n[\lambda(1 - e^{-a})], \quad n \in \mathbb{N}.$$

2.5 A Higher Deterministic Boundary

Pursuing the analysis made in Section 2.4, let us introduce a second lower deterministic boundary that is situated above the first one. More precisely, the new boundary, $\mathcal{B}_{d,h}$ say, corresponds to a set of points $\{(v_j, j), j \geq 1\}$ where, as for $\mathcal{B}_d$, $0 \leq v_1 \leq v_2 \leq \cdots$, but in addition, $v_j \leq u_j$ for all $j \geq 1$. The first-crossing level of the compound Poisson process with $\mathcal{B}_{d,h}$ is denoted by N_h.

Our goal is to point out the distributional impact of raising the boundary $\mathcal{B}_d$ to $\mathcal{B}_{d,h}$. For that, we will determine the conditional law of N_h given N.

Theorem 2.5.1 *For a deterministic boundary $\mathcal{B}_{d,h}$ above $\mathcal{B}_d$,*

$$\begin{aligned} P(N_h = m|k, N = n) = {} & \frac{G_m(0|\{-\lambda v_j,\ j \geq k\})}{G_n(0|\{-\lambda u_j,\ j \geq k\})} \\ & \times G_{n-m}(0|\{-\lambda(u_{m+j} - v_{m+k}),\ j \geq k\}),\ 0 \leq m \leq n. \end{aligned} \tag{2.34}$$

PROOF. By definition,

$$P(N_h = m|k, N = n) = \frac{P(N_h = m|k)\, P(N = n|k, N_h = m)}{P(N = n|k)}, \quad 0 \leq m \leq n. \tag{2.35}$$

The probabilities $P(N = n|k)$ and $P(N_h = m|k)$ are provided by (2.26). Moreover, the event $(N = n|k, N_h = m)$ is equivalent to the event that starting at level 0, the compound process crosses the following boundary, $\{(u_{m+k+j} - v_{m+k}, j), j \in \mathbb{N}\}$ (which thus begins with a delay), for the first time at level $n - m$. So, by (2.30), we get

$$P(N = n|k, N_h = m) = e^{-\lambda(u_{k+n} - v_{m+k})}\, G_{n-m}(0|\{-\lambda(u_{m+j} - v_{m+k}),\ j \geq k\}). \tag{2.36}$$

Substituting (2.26) and (2.36) in (2.35) then yields (2.34). ■

Linear boundaries. (1) To begin with, let us consider two vertical lines for the boundaries after level k, that is, $\mathcal{B}_d = \{(u, j), j \geq k\}$ and $\mathcal{B}_{d,r} = \{(v, j), j \geq k\}$ with $v \leq u$. By (2.34) and using (2.31), we find that

$$P(N_h = m|k, N = n) = \frac{e_m(\lambda v)\, e_{n-m}[\lambda(u - v)]}{e_n(\lambda u)}, \quad 0 \leq m \leq n. \tag{2.37}$$

In particular, for a simple Poisson process, N_h has a binomial law:

$$P(N_h = m|k, N = n) = \binom{n}{m} \left(\frac{v}{u}\right)^m \left(1 - \frac{v}{u}\right)^{n-m}. \tag{2.38}$$

Note that k has an indirect role in these formulae (and the next ones) through the definition of the boundaries.

(2) Suppose now that these boundaries are two parallel lines after level k, i.e. $\mathcal{B}_d = \{[u + b(j - k), j], j \geq k\}$ and $\mathcal{B}_{d,r} = \{[v + b(j - k), j], j \geq k\}$ with $b \geq 0$ and $v \leq u$. Again by (2.34) and (2.31),

$$P(N_h = m|k, N = n) = \frac{e_m[\lambda(v + bm)]\, e_{n-m}\{\lambda[u - v + b(n - m)]\}}{e_n[\lambda(u + bn)]} \times \frac{v(u - v)(u + bn)}{u(v + bm)[u - v + b(n - m)]}, \quad 0 \leq m \leq n. \tag{2.39}$$

For a Poisson process, N_h has a quasi-binomial law of kind II [in the sense of Consul and Mittal (1975)]:

$$P(N_h = m|k, N = n) = \frac{v(u - v)}{u} \binom{n}{m} \frac{(v + bm)^{m-1}\, [u - v + b(n - m)]^{n-m-1}}{(u + bn)^{n-1}}. \tag{2.40}$$

(3) Finally, suppose that the boundaries are, after level k, two nonintersecting lines until level $k + n$, that is, $\mathcal{B}_d = \{[u + b(j - k), j], k \leq j \leq k + n\}$ and

$\mathcal{B}_{d,r} = \{[v+d(j-k), j], k \leq j \leq k+n\}$ with $b, d \geq 0$ and $v+d(j-k) \leq u+b(j-k)$ for $k \leq j \leq k+n$. Then,

$$P(N_h = m|k, N = n) = \frac{e_m[\lambda(v+dm)]\, e_{n-m}[\lambda(u-v+bn-dm)]}{e_n[\lambda(u+bn)]} \times \frac{v[u-v+(b-d)m](u+bn)}{u(v+dm)(u-v+bn-dm)}, \quad 0 \leq m \leq n. \tag{2.41}$$

For a Poisson process,

$$P(N_h = m|k, N = n) = \frac{v}{u}\binom{n}{m}\frac{[u-v+(b-d)m]\,(v+dm)^{m-1}}{(u+bn)^{n-1}} \times (u+bn-v-dm)^{n-m-1}; \tag{2.42}$$

in particular, if $\mathcal{B}_d$ is vertical, that is, when $b = 0$, (2.42) corresponds to a quasi-binomial law of kind I [following Consul (1974)]:

$$P(N_h = m|k, N = n) = v\binom{n}{m}\frac{(v+dm)^{m-1}\,(u-v-dm)^{n-m}}{u^n},$$

whilst if $\mathcal{B}_{d,h}$ is vertical, that is, when $d = 0$, (2.42) becomes

$$P(N_h = m|k, N = n) = \frac{1}{u}\binom{n}{m}\frac{v^m\,(u-v+bm)(u-v+bn)^{n-m-1}}{(u+bn)^{n-1}}.$$

Appellations. It might be worth giving a name to the remarkable distributions (2.26) and (2.34). In view of the special cases (2.9) and (2.40), and the central role of the generalized AG polynomials, we suggest calling (2.26) a *generalized AG Poisson law* and (2.34) a *generalized AG binomial law*—the word "generalized" being omitted for a simple Poisson process.

As the usual Poisson and binomial distributions, these two generalized AG laws are linked by various properties. This can be examined in a context of damage models [see, Bhaskara Rao and Shanbhag (1982)], and is the object of a work in preparation. We only present here the following simple result.

Property 2.5.2 *Let N be a random variable with generalized AG Poisson law. Suppose that N is a.s. finite and can be decomposed into the sum of two random variables, N_h and N_g say, valued in $\mathbb{N}$. Then, N_h has a generalized Poisson law if, and only if, N_h given $N = n$ has a generalized binomial law, for all $n \in \mathbb{N}$.*

PROOF. Let us go back to the previous first-crossing type representation. The necessity part follows directly from Theorem 2.5.1. For the sufficiency part, we write

$$P(N_h = m|k) = \sum_{n=m}^{\infty} P(N = n|k)\; P(N_h = m|k, N = n), \quad m \in \mathbb{N}. \tag{2.43}$$

Inserting in (2.43) the laws of N and N_h given $N = n$ such as specified by hypothesis yields

$$\begin{aligned} P(N_h = m|k) &= \sum_{n=m}^{\infty} e^{-\lambda u_{k+n}} G_m(0|\{-\lambda v_j,\ j \geq k\}) \\ &\qquad \times G_{n-m}(0|\{-\lambda(u_{m+j} - v_{m+k}),\ j \geq k\}) \\ &= c\, e^{-\lambda v_{k+m}} G_m[0|\{-\lambda v_j,\ j \geq k\}], \quad m \in \mathbb{N}, \end{aligned}$$

where

$$c = \sum_{n=0}^{\infty} e^{-\lambda(u_{k+m+n} - v_{k+m})} G_n(0|\{-\lambda(u_{m+j} - v_{m+k}),\ j \geq k\}.$$

It then remains to note that $c = 1$ because the distribution of N is assumed to be nondefective. ■

Acknowledgement. I thank the referee for a careful reading of the chapter. This research was supported in part by the Banque Nationale de Belgique.

References

1. Bhaskara Rao, M., and Shanbhag, D. N. (1982). Damage models, In *Encyclopedia of Statistical Sciences* (Eds., S. Kotz and N. L. Johnson), **2**, pp. 262–265, John Wiley & Sons, New York.

2. Böhm, W., and Mohanty, S. G. (1997). On the Karlin–McGregor theorem and applications, *The Annals of Applied Probability*, **7**, 314–325.

3. Consul, P. C. (1974). A simple urn model dependent upon predetermined strategy, *Sankhyã, Series B*, **36**, 391–399.

4. Consul, P. C. (1989). *Generalized Poisson Distributions: Properties and Applications*, Marcel Dekker, New York.

5. Consul, P. C., and Mittal, S. P. (1975). A new urn model with predetermined strategy, *Biometrische Zeitung*, **17**, 67–75.

6. Daniels, H. E. (1963). The Poisson process with a curved absorbing boundary, *Bulletin of the International Statistical Institute*, **40**, 994–1008.

7. De Vylder, F. E. (1999). Numerical finite-time ruin probabilities by the Picard–Lefèvre formula, *Scandinavian Actuarial Journal*, **2**, 97–105.

8. Durbin, J. (1971). Boundary-crossing probabilities for the Brownian motion and Poisson processes and techniques for computing the power of the Kolmogorov–Smirnov, *Journal of Applied Probability*, **8**, 431–453.

9. Gallot, S. F. L. (1993). Absorption and first-passage times for a compound Poisson process in a general upper boundary, *Journal of Applied Probability*, **30**, 835–850.

10. Ignatov, Z. G., Kaishev, V. K., and Krachunov, R. S. (2001). An improved finite-time ruin probability formula and its *Mathematica* implementation, *Insurance: Mathematics and Economics*, **29**, 375–386.

11. Ignatov, Z. G., and Kaishev, V. K. (2004). A finite-time ruin probability formula for continuous claim severities, *Journal of Applied Probability*, **41**, 570–578.

12. Johnson, N. L., Kotz, S., and Kemp, A. W. (1992). *Univariate Discrete Distributions*, 2nd ed., John Wiley & Sons, New York.

13. Kotz, S., and Balakrishnan, N. (1997). Advances in urn models during the past two decades, In *Advances in Combinatorial Methods and Applications to Probability and Statistics* (Ed., N. Balakrishnan), pp. 203–257, Birkhäuser, Boston.

14. Lefèvre, Cl., and Picard, Ph. (1990). A nonstandard family of polynomials and the final size distribution of Reed–Frost epidemic processes, *Advances in Applied Probability*, **22**, 25–48.

15. Lefèvre, Cl., and Picard, Ph. (1999). Abel–Gontchareoff pseudopolynomials and the exact final outcome of SIR epidemic models (III), *Advances in Applied Probability*, **31**, 532–549.

16. Perry, D. (2000). Stopping problems for compound processes with applications to queues, *Journal of Statistical Planning and Inference*, **91**, 65–75.

17. Perry, D., Stadje, W., and Zacks, S. (1999). Contributions to the theory of first-exit times of some compound processes in queueing theory, *Queueing Systems*, **33**, 369–379.

18. Perry, D., Stadje, W., and Zacks, S. (2002). Hitting and ruin probabilities for compound Poisson processes and the cycle maximum of the M/G/1 queue, *Communications in Statistics: Stochastic Models*, **18**, 553–564.

19. Picard, Ph., and Lefèvre, Cl. (1994). On the first crossing of the surplus process with a given upper boundary, *Insurance: Mathematics and Economics*, **14**, 163–179.

20. Picard, Ph., and Lefèvre, Cl. (1996). First crossing of basic counting processes with lower non-linear boundaries: a unified approach through pseudopolynomials (I), *Advances in Applied Probability*, **28**, 853–876.

21. Picard, Ph., and Lefèvre, Cl. (1997). The probability of ruin in finite time with discrete claim size distribution, *Scandinavian Actuarial Journal*, **1**, 58–69.

22. Picard, Ph., and Lefèvre, Cl. (2003). On the first meeting or crossing of two independent trajectories for some counting processes, *Stochastic Processes and their Applications*, **104**, 217–242.

23. Pyke, R. (1959). The supremum and infimum of the Poisson process, *Annals of Mathematical Statistics*, **30**, 568–576.

24. Stadje, W. (1993). Distributions of first exit times for empirical counting and Poisson processes with moving boundaries, *Communications in Statistics: Stochastic Models*, **9**, 91–103.

25. Stadje, W., and Zacks, S. (2003). Upper first-exit times of compound Poisson processes revisited, *Probability in the Engineering and Informational Sciences*, **17**, 459–465.

26. Takács, L. (1967). *Combinatorial Methods in the Theory of Stochastic Processes*, John Wiley & Sons, New York.

27. Takács, L. (1989). Ballots, queues and random graphs, *Journal of Applied Probability*, **26**, 103–112.

28. Zacks, S. (1991). Distributions of stopping times for Poisson processes with linear boundaries, *Communications in Statistics: Stochastic Models*, **7**, 233–242.

29. Zacks, S. (1997). Distributions of first exit times for Poisson processes with lower and upper linear boundaries, In *Advances in the Theory and Practice of Statistics: A Volume in Honor of Samuel Kotz* (Eds. N. L. Johnson and N. Balakrishnan), pp. 339–350, John Wiley & Sons, New York.

30. Zacks, S., Perry, D., Bshouty, D., and Bar-Lev, S. (1999). Distributions of stopping times for compound Poisson processes with positive jumps and linear boundaries, *Communications in Statistics: Stochastic Models*, **15**, 89–101.

3

Constructions of Discrete Bivariate Distributions

C. D. Lai

Massey University, Palmerston North, New Zealand

Abstract: Various techniques for constructing discrete bivariate distributions are scattered in the literature. We review these methods of construction and group them into some loosely defined clusters.

Keywords and phrases: Bernoulli, bivariate distributions, conditioning; canonical correlation, clustering, constructions, compound, discrete, extreme points, Fréchet bounds, marginal transformation, mixing, sampling, trivariate, truncations, urn models, weighting functions

3.1 Introduction

Over the last two or three decades, a vast amount of literature on discrete bivariate and multivariate distributions has been accumulated. For an extensive account of these distributions, we refer our readers to the books by Kocherlakota and Kocherlakota (1992) and Johnson *et al.* (1997), and the review articles by Papageorgiou (1997), Kocherlakota and Kocherlakota (1998), and Balakrishnan (2004, 2005).

In this chapter, we restrict ourselves to reviewing methods of constructing discrete bivariate distributions. A review on constructions of continuous bivariate distributions is given by Lai (2004). Unlike their continuous analogues, discrete bivariate distributions appear to be harder to construct. One of the problems is highlighted in Kemp and Papageorgiou (1982) in which they said, "Various authors have discussed the problem of constructing meaningful and useful bivariate versions of a given univariate distribution, the main difficulty being the impossibility of producing a standard set of criteria that can always be applied to produce a unique distribution which could unequivocally be called the bivariate version." Many bivariate distributions arise without having prespecified the marginals. There is no satisfactory unified mathematical scheme of classifying these methods. What we hope to achieve is to group them into semicoherent clusters. The clusters may be listed as

- Mixing and compounding
- Trivariate reduction
- One conditional and one marginal given
- Conditionally specified method
- Construction of discrete bivariate distributions with given marginals and correlation
- Sums and limits of Bernoulli trials models
- Sampling from urn models
- Clustering (bivariate distributions of order k)
- Construction of finite bivariate distributions via extreme points of convex sets
- Generalized distributions method
- Canonical correlation coefficients and semi-groups
- Distributions arising from accident theory
- Bivariate distributions generated from weight functions
- Marginal transformations method
- Truncation method
- Constructions of positively dependent discrete bivariate distributions.

Several of these are also common methods for constructing continuous bivariate distributions. We refer the reader to Lai (2004) for a review of these and other methods of constructing continuous bivariate distributions. We note that for discrete bivariate distributions, the probability generating function is often used as a tool for construction as well as for studying their properties.

We have not discussed computer generation of discrete bivariate random variables. We refer interested readers to the works by Professors A. W. Kemp and C. D. Kemp on this subject. Kocherlakota and Kocherlakota (1992) present several such references by the Kemps.

3.2 Mixing and Compounding

3.2.1 Mixing

As for continuous bivariate distributions, an easy way to construct a discrete bivariate distribution is to use the method of mixing two or more distributions. Suppose H_1 and H_2 are two discrete bivariate distributions; then

$$H(x, y) = \alpha H_1(x, y) + (1 - \alpha) H_2(x, y) \tag{3.1}$$

($0 \leq \alpha \leq 1$) is a new bivariate distribution.

Example: Consider the problem of describing the sex distribution of twins. Twin pairs fall into three classes: MM, MF, and FF where M denotes male and F female. This leads to the trinomial distribution. As twins may be dizygotic or monozygotic, a mixture of trinomials results. For more details, see Blischke (1978), Goodman and Kruskal (1959), and Strandskov and Edelen (1946).

Papageorgiou and David (1994) studied several countable mixtures of binomial distributions.

3.2.2 Compounding

Compounding is perhaps the most common method of constructing discrete bivariate distributions. Let X and Y be two random variables with parameters θ_1 and θ_2, respectively. For a given value of (θ_1, θ_2), X and Y may be either independent or correlated.

(i) X and Y are conditionally independent.

If θ_1 and θ_2 are independent, then the resulting pair $X$and Y are also independent. For example, for given (θ_1, θ_2), X and Y are independent Poissons. If θ_1 and θ_2 are independent gammas, then the resulting X and Y are independent negative binomials.

- θ_1 and θ_2 may have a bivariate distribution such as the case of Consael's bivariate Poisson distribution [Consael (1952)].
- David and Papageorgiou (1994) presented several compounded bivariate Poisson distributions that can be derived in this manner.

(ii) X and Y are dependent for given values of the compounding parameters.

- The compounded bivariate Poisson distributions given by Kocherlakota (1988) are obvious examples.
- Another example is the generalized Consael distribution obtained by

$$(X, Y) \sim \operatorname{Biv} \mathrm{P}(\lambda_1, \lambda_2, \lambda_3) \underset{(\lambda_1,\lambda_2,\lambda_3)}{\wedge} F(\lambda_1, \lambda_2, \lambda_3)$$

where the symbol $\wedge$ denotes compounding. Here $\operatorname{Biv} \mathrm{P}(\lambda_1, \lambda_2, \lambda_3)$ has a bivariate Poisson distribution with a probability-generating function given by

$$g(s, t) = \exp\{\lambda_1(s-1) + \lambda_2(t-1) + \lambda_3(st-1)\}, \qquad (3.2)$$

and $(\lambda_1, \lambda_2, \lambda_3)$ has a trivariate distribution function F.

For example, H_8 distribution [Kemp and Papageorgiou (1982)] is obtained when $(\lambda_1,\lambda_2, \lambda_3)$ has a trivariate normal distribution.

There are other variants of compounding; see, for example, Chapter 8 of Kocherlakota and Kocherlakota (1992).

3.3 Trivariate Reduction

This is also known as "the variables in common method." The idea here is to create a pair of dependent random variables from three or more random variables. In many cases, these initial random variables are independent, but occasionally they may be dependent. An important aspect of this method is that the functions connecting these random variables to the two dependent random variables are generally elementary ones; random realizations of the latter can therefore be generated easily from random realizations of the former. A broad definition of the variables-in-common technique is as follows. Set

$$\left.\begin{aligned} X &= \tau_1(X_1, X_2, X_3), \\ Y &= \tau_2(X_1, X_2, X_3), \end{aligned}\right\} \tag{3.3}$$

where X_1, X_2, X_3 are not necessarily independent or identically distributed. A narrow definition is

$$\left.\begin{aligned} X &= X_1 + X_3, \\ Y &= X_2 + X_3, \end{aligned}\right\} \tag{3.4}$$

with X_1, X_2, X_3 being i.i.d. Another possible definition is

$$\left.\begin{aligned} X &= \tau(X_1, X_3), \\ Y &= \tau(X_2, X_3), \end{aligned}\right\} \tag{3.5}$$

with (i) the X_i being independently distributed and having c.d.f. $F_0(x_i; \lambda_i)$, and (ii) X and Y having distributions $F_0(x; \lambda_1{+}\lambda_2)$ and $F_0(y; \lambda_1{+}\lambda_3)$, respectively.

Example: Suppose $X_i \sim \text{Poisson}(\lambda_i)$, $i = 1, 2, 3$. Define $X = X_1 + X_3$, $Y = X_2 + X_3$ so that the joint pgf of (X, Y) is given by

$$g(s,t) = \exp\{\lambda_1(s-1) + \lambda_2(t-1) + \lambda_3(st-1)\} \tag{3.6}$$

which is called the bivariate Poisson distribution. This distribution is often used as a basis for obtaining a compound bivariate Poisson distribution. More specifically, if each independent $\lambda_i \sim \text{Gamma}(\alpha_i, \beta)$, then the resulting distribution is a bivariate negative binomial [see, e.g., Stein and Juritz (1987)]. If each independent $\lambda_i \sim \text{GIG}(\alpha_i, \zeta_i, \frac{1}{2})$ (GIG = generalized inverse Gaussian), then (X, Y) has a bivariate inverse Gaussian–Poisson distribution.

(**Note**: $\lambda_1 + \lambda_2 \sim \text{GIG}(\alpha_1 + \alpha_2, \zeta_1 + \zeta_2, \frac{1}{2})$. The inverse Gaussian–Poisson distribution is a special case of Sichel distribution.)

An obvious disadvantage of this method is that the correlation is restricted to be strictly positive.

Zheng and Matis (1993) generalized the trivariate reduction method by considering a random rewarding system so that

$$X = \begin{cases} X_1 + X_2 & \text{with prob } \pi_1 \\ X_1 & \text{with prob } 1 - \pi_1 \end{cases}$$

and

$$Y = \begin{cases} X_1 + X_3 & \text{with prob } \pi_2 \\ X_3 & \text{with prob } 1 - \pi_2. \end{cases}$$

Several discrete bivariate distributions were constructed, whose marginal distributions are mixtures of negative binomial distributions.

Lai (1995) proposed an extension to the model of Zheng and Matis (1993) by setting

$$\left.\begin{aligned} X &= X_1 + I_1 X_2, \\ Y &= X_3 + I_2 X_2, \end{aligned}\right\} \tag{3.7}$$

where $I_i \, (i = 1, 2)$ are indicator random variables which are independent of X_i, but (I_1, I_2) has a joint probability function.

3.4 One Conditional and One Marginal Given

A discrete bivariate distribution can be expressed as the product of a marginal distribution and a conditional distribution as

$$\Pr\{X = x, Y = y\} = \Pr\{Y = y | X = x\} \Pr\{X = x\}. \tag{3.8}$$

This is an intuitively appealing approach, especially when Y can be thought of caused by, or predictable from, X.

Moreover, given positive $\Pr\{X = x | Y = y\}$ for all x, y, and $\Pr\{Y = y | X = x_0\}$, for all y and a fixed x_0, the joint distribution can be determined uniquely [Patil (1965)]:

$$\Pr\{X = x, Y = y\} \propto \frac{\Pr\{X = x | Y = y\} \Pr\{Y = y | X = x_0\}}{\Pr\{X = x_0 | Y = y\}}. \tag{3.9}$$

Normalization determines the proportional constant; see, for example, Gelman and Speed (1993).

Furthermore, discrete bivariate distributions can be generated from given conditional distributions and regression functions. We will discuss this in the next section dealing with conditionally specified distributions below.

Examples: Korwar (1975), Dahiya and Korwar (1977), Cacoullos and Papageorgiou (1983), Papageorgiou (1983, 1984, 1985a), Kyriakoussis (1988), and Kyriakoussis and Papageorgiou (1989).

3.5 Conditionally Specified Method

Suppose in the preceding section, both $\Pr(Y = y|X = x)$ and $\Pr(X = x|Y = y)$ are given for all x and y. We may have then overspecified the conditions as the two conditional distributions may not be compatible. In cases in which compatibility is confirmed, the question of possible uniqueness of the compatible distribution need to be addressed. The book of Arnold *et al.* (1999) has revolutionized this subject area as it provides a rich mechanism for generating bivariate distributions. This book focuses on those conditional distributions that are members of some well-defined parametric families such as the exponential families. Three discrete distributions are from exponential families, that is, binomial, geometric, and Poisson. Section 4.12 of the above mentioned monograph devotes a discussion to constructions of bivariate binomial, geometric, and Poisson distributions.

Section 7.7 of Arnold *et al.* (1999) discusses generation of bivariate discrete distributions (as well as continuous bivariate distributions) for a given conditional distribution of X given Y and the regression function of Y on X. In particular, Wesolowski (1995) has shown that if $X|Y = y$ has a power series distribution, that is,

$$\Pr(X = x|Y = y) = c(x)y^x/c^*(y),$$

then the joint distribution of (X, Y) will be uniquely determined by the regression function of Y on X provided $c(\cdot)$ is reasonably well behaved.

3.6 Construction of Discrete Bivariate Distributions with Given Marginals and Correlation

3.6.1 Discrete Fréchet bounds

For given marginals F and G, Hoeffding (1940) and Fréchet (1951) have proved that there exist bivariate distribution functions, H_L and H_U, called the lower and upper Fréchet bounds, respectively, having minimum and maximum correlation. Specifically, we have

$$H_L(x, y) = \max[F(x) + G(y) - 1, 0] \tag{3.10}$$

$$H_U(x, y) = \min[F(x), G(y)] \tag{3.11}$$

satisfying

$$H_L(x, y) \leq H(x, y) \leq H_U(x, y) \tag{3.12}$$

and that

$$\rho_L \leq \rho \leq \rho_U \tag{3.13}$$

where ρ_L, ρ and ρ_U denote the Pearson product-moment correlation coefficients for H_L, H and H_U, respectively.

3.6.2 Probability functions of Fréchet bounds

We now assume that X and Y are discrete with ranges that are subsets of $N = \{0, 1, 2, \ldots\}$. Let h, f, and g be the probability functions that correspond to H, F, and G, respectively. Our aim now is to construct the probability functions h_L and h_U that correspond to H_L and H_U, respectively. In the following, we adopt the notations given in Nelsen (1987).

Let D denote the portion of N^2 where $H_L(x, y) > 0$, D' denote the complement of D in N^2, and ∂D denote the border of D; that is,

$$D = \{(x, y) \in N^2 \,|F(x) + G(y) - 1 > 0\}$$
$$D' = \{(x, y) \in N^2 \,|F(x) + G(y) - 1 = 0\},$$

and

$$\partial D = \{(x, y) \in D|(x-1, y), (x, y-1) \text{ or } (x-1, y-1) \notin D\}.$$

Nelsen (1987) has shown that

$$h_L(x, y) = \begin{cases} f(x) & (x, y) \in \partial D, (x, y-1) \notin D, (x-1, y) \in \partial D \\ g(y) & (x, y) \in \partial D, (x-1, y) \notin D, (x, y-1) \in \partial D \\ F(x) + G(y) - 1 & (x, y) \in \partial D, (x, y-1) \notin D, (x-1, y) \notin D \\ 1 - F(x-1) - G(y-1) & (x, y) \in \partial D, (x, y-1) \in \partial D, (x-1, y) \in \partial D \\ 0 & \text{otherwise.} \end{cases}$$

In order to obtain h_U, we set

$$S = \{(x, y) \in N^2 | F(x) = G(y)\},$$
$$T = \{(x, y) \in N^2 | F(x) > G(y)\},$$

and

$$\partial S = \{(x, y) \in S | (x, y - 1) \notin S\},$$
$$\partial T = \{(x, y) \in T | (x - 1, y) \notin T\}.$$

Nelsen (1987) has shown that

$$h_U(x, y) = \begin{cases} f(x) & (x, y) \in \partial S, (x-1, y-1) \in T, \text{ or } y = 0 \\ g(y) & (x, y) \in \partial S, (x-1, y-1) \in S, \text{ or } x = 0, \ y \neq 0 \\ F(x) + G(y) - 1 & (x, y) \in \partial T, (x-1, y-1) \in S, \text{ or } x = 0 \\ 1 - F(x-1) - G(y-1) & (x, y) \in \partial T, (x-1, y-1) \in T, \text{ or } y = 0, \ x \neq 0 \\ 0 & \text{otherwise.} \end{cases}$$

The author has also presented two examples of finding h_L and another two of finding h_U.

3.6.3 Construction of bivariate distributions

Having obtained h_L and h_U, we are now in a position to generate one-parameter or two-parameter families of bivariate distributions with given marginals:

$$h_{\theta,\phi} = \theta h_L(x, y) + (1 - \theta - \phi) f(x) g(y) + \phi h_U(x, y), \theta, \phi \geq 0, \theta + \phi \leq 1. \quad (3.14)$$

Upon setting $\theta = 0, \phi > 0$, we obtain a one-parameter family with positive correlation; and upon setting $\phi = 0, \theta > 0$, a one-parameter family with negative correlation; and correlation coefficients for members of these families are functions of θ, ϕ, ρ_{L} and ρ_{U}.

Mardia (1970, p. 33) has noted that if we let $\theta^2 = \frac{\gamma^2}{2}(1-\gamma)$ and $\phi = \frac{\gamma^2}{2}(1+\gamma)$, then (3.14) becomes

$$h_\gamma = \frac{\gamma^2}{2}(1 - \gamma) h_L(x, y) + (1 - \gamma^2) f(x) g(y) + \frac{1}{2}\gamma^2 (1 + \gamma) h_U(x, y) \quad (3.15)$$

It is worth noting that for $\phi = 0$,

$$h_\theta = \theta h_L(x, y) + (1 - \theta) f(x) g(y) \quad (3.16)$$

and that the correlation coefficient ρ is given by

$$\rho = \theta \rho_L, \ 0 \leq \theta \leq 1 \quad (3.17)$$

which has values between ρ_L and 0. Thus for any desired correlation ρ between ρ_L and 0, we can find the required value of θ in [0, 1] to satisfy (3.17).

Similarly, for $\theta = 0, \phi > 0$, we have

$$h_\phi = (1-\phi)f(x)g(y) + \phi h_U(x,y) \tag{3.18}$$

and that the correlation coefficient ρ is given by

$$\rho = \phi\rho_U. \tag{3.19}$$

For any desired correlation between 0 and ρ_U, we can find the required value of ϕ in [0, 1].

Nelsen (1987) presented two examples:

1. both marginals are Poisson but with different parameters, $\rho = -0.5$ and
2. one marginal is binomial $(n = 4, p = 0.8)$ and the other discrete uniform on $\{1, 2, 3, 4, 5\}$; and ρ positive.

If we wish to use Mardia's one-parameter family (3.15), then the correlation coefficient ρ for h_γ is given by

$$\rho = \frac{\gamma^2}{2}(1-\gamma)\rho_L + \frac{\gamma^2}{2}(1+\gamma)\rho_U.$$

To find the required value ρ between ρ_L and ρ_U, we need to solve for γ in the following cubic equation

$$(\rho_U - \rho_L)\gamma^3 + (\rho_U + \rho_L)\gamma^2 - 2\rho = 0.$$

Then, we can construct the probability function by substituting the value γ into (3.14).

3.6.4 Construction of bivariate Poisson distributions

Griffiths *et al.* (1979) gave procedures for constructing bivariate Poisson distributions having negative correlations when the two marginals are specified. For given Poisson marginals F and G having parameters λ_1 and λ_2, respectively, they calculated and tabulated the minimum and maximum correlation coefficients (i.e., the correlation coefficients of H_L and H_U defined, respectively, by (3.10) and (3.11)).

3.7 Sums and Limits of Bernoulli Trials

3.7.1 The bivariate Bernoulli distribution

Suppose (X, Y) has Bernoulli marginals; then it has only four possible values: $(1, 1), (1, 0), (0, 1), (0, 0)$ with probabilities $p_{11}, p_{10}, p_{01}, p_{00}$, respectively. The marginal probabilities are given by

$$\left.\begin{array}{l} p_{1+} = p_{11} + p_{10} = 1 - p_{0+}, \\ p_{+1} = p_{11} + p_{01} = 1 - p_{+0} \end{array}\right\}. \tag{3.20}$$

It is easy to show that the correlation coefficient is given by

$$\rho = \frac{p_{11}p_{1+}p_{+1}}{\sqrt{p_{1+}p_{0+}p_{+1}p_{+0}}}. \tag{3.21}$$

It takes on values -1 and +1 when $\rho_{11} = \rho_{00} = 0$ and $\rho_{10} = \rho_{01} = 0$, respectively. Here, $\rho = 0$ implies X and Y are independent.

3.7.2 Construction of bivariate Bernoulli distributions

It is well known that in the univariate case, the binomial, negative binomial (including geometric), hypergeometric and Poisson distributions are obtainable from the univariate Bernoulli distribution. Marshall and Olkin (1985) showed that these methods of derivation (using sums and limits) can be extended to twodimensions to obtain many bivariate distributions with binomial, negative binomial, geometric, hypergeometric, or Poisson marginals.

3.8 Sampling from Urn Models

Many discrete bivariate distributions are constructed by sampling from urn models. There are two types of sampling: (i) direct sampling and (ii) inverse sampling. By inverse sampling, we mean the sampling is continued until k individuals of a certain type are observed. For both types, sampling may be with or without replacement.

Suppose a population has three distinct characters and let the population size be N. Let N_i, $i = 0, 1, 2$, be the number of individuals having character i, for $i = 0, 1, 2$ such that $N_0 + N_1 + N_2 = N$ (alternatively, an urn contains N balls of three different colours, N_i being of ith colour $(i = 0, 1, 2)$ such that $N_0 + N_1 + N_2 = N$). Suppose that n individuals (balls) are drawn from the population (urn) with various forms of sampling schemes, and let X and Y

Table 3.1: Bivariate distributions from direct and inverse samplings

No	Name	Type of Sampling	Rcplace (Yes/No)	Special Features
(i)	Bivariate Binomial	Direct	Yes	N_i finite
(ii)	Bivariate Negative Binomial	Inverse	Yes	N_i infinite
(iii)	Bivariate Hypergeometric	Direct	No	—
(iv)	Bivariate Inverse Hypergeometric	Inverse	No	—
(v)	Bivariate Negative Hypergeometric	Direct	—	Trinomial compounded by bivariate beta
(vi)	Bivariate Inverse Negative Hypergeometric	Inverse	—	Negative trinomial compounded by bivariate beta
(vii)	Bivariate Polya	Direct		Add c additional individuals
(viii)	Bivariate Inverse Polya	Inverse		Add c additional individuals

denote the number of type 1 character and type 2 character, respectively, in the sample. We can then construct various kinds of bivariate distributions which are summarized below:

- Distribution (i) is also known as type 1 bivariate binomial distribution; see, for example, Section 3.3 of Kocherlakota and Kocherlakota (1992).
- For distribution (ii), see, for example, Section 5.2 of Kocherlakota and Kocherlakota (1992).
- For distributions (iii)–(vi), see Janardan (1972, 1973, 1975, 1976), Janardan and Patil (1970, 1971, 1972). See also Chapter 6 of Kocherlakota and Kocherlakota (1992).
- For distributions (vii) and (viii), see Janardan and Patil (1970, 1971) and Patil *et al.* (1986).

For other references and other distributions generated from urn models, see Johnson and Kotz (1977), Korwar (1988), and Marshall and Olkin (1990).

3.9 Clustering (Bivariate Distributions of Order k)

In recent years, several bivariate generalizations of the binomial, negative binomial, hypergeometric, Poisson, logarithmic, and other distributions were obtained. These are often called bivariate distributions of order k or bivariate cluster distributions; see Balakrishnan and Koutras (2002). As they bear the names binomial, negative binomial, hypergeometric, and negative hypergeometric, it is not surprising that they also have the origin of sampling from an urn with and without replacements.

3.9.1 Preliminary

Consider an urn that contains balls of $k+1$ types such that α balls bear the number 0 and $\beta_{\rm i}$ balls bear the number $i, i = 1, 2, \ldots, k$.

(i) Suppose a sample of n balls is drawn with replacement. Let X denote the sum of the numbers shown on the balls drawn and p_i, $i = 1, 2, \ldots, k$ be the probability that a ball bearing the number i will be drawn: $\sum\limits_{i=1}^{k} p_i = p$ and $q = 1 - p$ is the probability that a ball bearing a zero will be drawn. Then, X has a cluster binomial distribution.

(ii) If the sampling scheme above is without replacement, then a cluster hypergeometric distribution results.

(iii) If as in (i) above, but with n not fixed and letting X be the sum of numbers sampled before the $r^{\rm th}$ zero, then X has a cluster negative binomial distribution.

(iv) If as in (ii) above but the compositions of balls is to be altered at each stage by adding a ball of the same type as the sampled one before the next draw is made, then X has a cluster Polya distribution.

3.9.2 Bivariate Distributions of order k

Now we may generalize this idea to the bivariate case.

Suppose an urn contains balls of two different colours (say colour 1 and colour 2). The balls of colour i are numbered from 0 to k_i, $i = 1, 2$. n balls are drawn with replacement. Let p_{ij} denote the probability that a ball of colour i will bear number j, $j = 0, 1, 2, \ldots, k_i$. Let X and Y denote the sum of the numbers of the first and second colour, respectively; then (X, Y) has a cluster bivariate binomial distribution [Panaretos and Xekalaki (1986)].

Suppose now in the above example, $k_1 = k_2$ and another ball is added and labelled by (0, 0) with proportion p such that $p + \sum_{i=1}^{2} \sum_{j=1}^{k} p_{ij} = 1$. Balls are drawn with replacement until the r balls ($r \geq 1$) bearing the number (0, 0) appear. Let X and Y denote the sum of the numbers on colour 1 and colour 2, respectively. Then (X, Y) has the bivariate negative binomial distribution of order k [Philippou *et al.* (1989) and Antzoulakos and Philippou (1991)].

Philippou *et al.* (1989) obtained a bivariate Poisson distribution of order k by taking limits from the above model such that

$$p_{ij} \to 0 \quad \text{and} \quad rp_{ij} \to \lambda_{ij} \quad (0 < \lambda_{ij} < \infty, \ \text{for } 1 \le i \le 2, 1 \le j \le k).$$

For construction of bivariate logarithmic series distribution of order k, also a limiting case of bivariate negative binomial of order k, see Philippou *et al.* (1989, 1990). For constructions of bivariate Polya and inverse Polya distributions of order k, see Philippou and Tripsiannis (1991).

Aki and Hirano (1994, 1995) have constructed multivariate geometric distributions of order k. For a review on this subject, see Chapter 42 of Johnson *et al.* (1997) and Balakrishnan and Koutras (2002).

Philippou and Antzoulakos (1990) have obtained several bivariate distributions of order k through a "generalised sequence of order k" which was first introduced by Aki (1985). For other types of bivariate binomial distributions of order k, see Ling and Tai (1990).

3.10 Construction of Finite Bivariate Distributions via Extreme Points of Convex Sets

In this section, we consider the construction of bivariate distributions with finite support. The key reference for the following discussion is that of Rao and Subramanyam (1990).

Let $M(F, G)$ be the collection of all bivariate distributions with finite support and marginals F and G. Then M is a compact convex set. In order to give an insight of the problem, we begin by considering joint probabilities of X and Y: $p_{ij} = \Pr(X = i, Y = j)$, $p_i = \Pr(X = i)$, $q_j = \Pr(Y = j)$, $i, = 1, 2$; $j = 1, 2, 3$.

It is easy to see that the following set of equations hold (assuming for the time being that p_{11} and p_{12} are known):

$$\left.\begin{array}{lcl} p_{13} & = & p_1 - p_{11} - p_{12} \\ p_{21} & = & q_1 - p_{11} \\ p_{22} & = & q_2 - p_{12} \\ p_{13} + p_{23} & = & q_3 \end{array}\right\}. \tag{3.22}$$

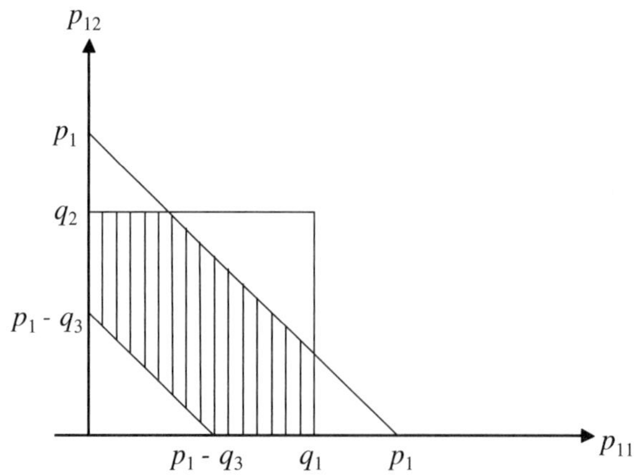

Figure 3.1: Feasible region

There are five equations and four unknowns. As $p_{ij} \geq 0$, it follows that

$$\left.\begin{array}{r} p_{13} = p_1 - p_{11} - p_{12} \geq 0 \\ p_{21} = q_1 - p_{11} \geq 0 \\ p_{22} = q_2 - p_{12} \geq 0 \\ p_{23} = q_3 - p_1 + p_{11} + p_{12} \geq 0 \end{array}\right\}. \tag{3.23}$$

The above may be expressed as four inequalities for p_{11} and p_{12}. These are

$$\left.\begin{array}{r} p_{11} + p_{12} \leq p_1 \\ p_{11} \leq q_1 \\ p_{12} \leq q_2 \\ p_{11} + p_{12} \geq p_1 - q_3 \end{array}\right\}. \tag{3.24}$$

In addition, we have two obvious inequalities which are

$$p_{11} \geq 0 \qquad \text{and} \qquad p_{12} \geq 0.$$

These six inequalities may be illustrated by the diagram above. The feasible region of bivariate distributions is a hexagon. However, if either q_1 or q_2 exceeds p_1, the region is then reduced to a pentagon. If both q_1 and q_2 exceed p_1, then the region is a quadrilateral. If both q_1 and q_2 are smaller or equal to $p_1 - q_3$, then the region is a triangle. If one of q_1 and q_2 is less than $p_1 - q_3$ whereas the other one exceeds $p_1 - q_3$, then the resulting region is a quadrilateral.

Note that the intersections of the boundary lines are the extreme points. In this example, there are three to six extremal points.

Well-established mathematical fact: Let $A_i (i = 1, 2, \ldots, n)$ be the extreme points of a compact convex set M. Then any element B of M can be written as $B = \sum_{i=1}^{n} \alpha_i A_i$ where $\sum_{i=1}^{n} \alpha_i = 1$.

It follows that we can generate a discrete bivariate distribution after the extreme points are identified.

3.10.1 Finding extreme points

From the above discussion, it is clear that it is easy to generate a bivariate distribution with specified marginals if we can identify the extremal points of M. For example, suppose we have $p_1 = \frac{1}{3}, p_2 = \frac{2}{3}; q_1 = \frac{1}{4}, q_2 = \frac{1}{2}, q_3 = \frac{1}{4}$. As $q_2 = \frac{1}{2} > p_1 = \frac{1}{3}$, the region is a pentagon. It follows from the above diagram that one of the intersections is $p_{11} = p_1 - q_3 = \frac{1}{12}, p_{12} = 0$. It follows from (3.22) and (3.23) that one of the extreme points of M is

$$\begin{bmatrix} \frac{1}{12} & 0 & \frac{1}{4} \\ \\ \frac{1}{6} & \frac{1}{2} & 0 \end{bmatrix}.$$

The other four extreme points can be found similarly.

Let m be the number of $p_i > 0$ and n be the number of $q_i > 0$. This contingency table has $(m-1)(n-1)$ degrees of freedom. In general, we have $(m+n)$ equations and $(m+n-1)$ unknowns (i.e., one equation is always redundant). These $(m+n-1)$ equations are expressed in terms of p_i, q_i and the $(m-1)(n-1)$ free parameters (with dependent parameters on the left of the equations, and p_i, q_i and free parameters on the right). Also as the free parameters $p_{ij} \geq 0$, we therefore have $(m+n-1)+(m-1)(n-1) = mn$ inequalities. Hence they form a polygon with a maximum of mn sides.

Oluyede (1994) obtained a family of bivariate binomial distributions generated by extreme bivariate Bernoulli distributions.

3.11 Generalized Distributions

The adjective "generalized" has often been used for discrete distributions, however, its meaning is not uniquely defined. In the literature, there is no clear-cut discrimination between the terms "compound" and "generalized." Moreover, the word "generalized" in this discussion is also used with other meanings such as extension. For example, we used the term "generalized inverse Gaussian" in Section 3.3 to denote a distribution which includes the inverse Gaussian as its special case.

We now define "generalized" in a restricted sense.

Suppose the pgf (probability-generating function) of a distribution F_1 is $g_1(s)$. If the argument s is replaced by the pgf $g_2(s)$ of another distribution F_2, then the resulting generating function $g_1(g_2(s))$ is also a probability-generating function. This distribution is called a generalized F_1 distribution. More precisely, it is called an F_1 distribution generalized by the generalizer (or generalizing distribution) F_2. It may be written in the symbolic form

$$F_1 \vee F_2; \tag{3.25}$$

see Johnson and Kotz (1969, p. 202).

In the univariate case, the generalized distribution is simply a compound distribution.

3.11.1 Generalized bivariate distributions

The above idea may be extended to the bivariate case. In a general setting, there are at least two ways of "generalizing."

(i) Let $G(s)$ be the pgf of the original distribution F_1 and $\pi(s,t)$ be the joint pgf of the bivariate distribution of F_2. Then a generalized bivariate distribution can be obtained by replacing s of G by $\pi(s,t)$ to give

$$g(s,t) = G(\pi(s,t)). \tag{3.26}$$

(ii) Let $G(s,t)$ be the original pgf of a bivariate distribution F_1. Replace the arguments s and t of G by the univariate pgf's $\pi_1(s)$ and $\pi_2(t)$, respectively, so that the resulting generalized distribution has pgf

$$g(s,t) = G(\pi_1(s), \pi_2(t)). \tag{3.27}$$

(iii) The third way may be obtained by combining the trivariate reduction technique together with the "generalized" method. Let π_i be the pgf of the generalizer X_i and G_i be the pgf of the distribution that generalizes $X_i, i = 1,2,3$. Let $(X,Y) = (X_1+X_3, X_2+X_3)$. Then the resulting generalized bivariate distribution of (X,Y) has pgf given by

$$g(s,t) = G_1(\pi_1(s))G_2(\pi_2(t))G_3(\pi_3(st)). \tag{3.28}$$

3.11.2 Generalized bivariate Poisson distributions

(i) Bivariate Neyman type A distributions

Holgate (1966) constructed three types of bivariate Neyman A distributions.

Type I: This corresponds to (3.26) with G being the pgf of a Poisson and $\pi(s,t)$, the pgf of the bivariate Poisson given by

$$\pi(s,t) = \exp\{\lambda_1(s-1) + \lambda_2(t-1) + \lambda_3(st-1)\}. \tag{3.29}$$

Type II: This corresponds to (3.27) where G is the pgf of the bivariate Poisson given by (3.29) and $\pi_1(s) = \exp\{\phi_1(s-1)\}$ and $\pi_2(t) = \exp\{\phi_2(t-1)\}$.

Type III: This is obtained via the trivariate reduction method such that $X = X_1+X_3$ and $Y = X_2+X_3$ where X_i $(i = 1, 2, 3)$ are independent Neyman A distributions. Alternatively, let $G_i(s) = \exp\{\lambda_i(s-1)\}$ and $\pi_i(s) = \exp\{\phi_i(s-1)\}$. By applying (3.28), we obtain this distribution.

(ii) Bivariate Poisson binomial distributions

Charalambides and Papageorgiou (1981a) also derived three types of bivariate Poisson binomial distributions based on the "generalized" method.

3.11.3 Generalized bivariate general binomial distributions

Three types of bivariate generalized general binomials were derived by Charalambides and Papageorgiou (1981b).

For other examples, see Papageorgiou and Kemp (1983).

3.12 Canonical Correlation Coefficients and Semigroups

3.12.1 Diagonal expansion

The diagonal expansion of a bivariate distribution involves representing it as

$$dH(x, y) = dF(x)dG(y)\sum_{i=1}^{\infty} \rho_i \xi_i(x)\eta_j(y), \tag{3.30}$$

ξ_i and η_i being known as the canonical variables and the ρ_i as the canonical correlations.

When X and Y have finite moments of all orders, sets of orthonormal polynomials $\{P_n\}$ and $\{Q_n\}$ can be constructed with respect to F and G; for example, the Krawtchouk polynomials for binomial marginals, the Meixner polynomials for negative binomial marginals, and the Poisson–Charlier polynomials for Poisson marginals.

If

$$\left.\begin{aligned} E[X^n|Y = y] &= \text{ a polynomial of degree } n \\ E[Y^n|X = x] &= \text{ a polynomial of degree } n \end{aligned}\right\}, \tag{3.31}$$

then H has a diagonal expression in terms of F and G and their respective orthonormal polynomials.

3.12.2 Canonical correlation coefficients and positive definite sequence

Suppose now X and Y are two exchangeable variables so that the two sets of orthonormal polynomials $\{P_n\}$ and $\{Q_n\}$ are identical. A sequence $\{t_n\}$ is said to be positive definite with respect to $\{Q_n\}$ if for all M (integer), all $x = 0, 1, 2, \ldots$ and all sequences $\{a_n\}$ of real numbers, $\sum_{n=0}^{M} a_n Q_n(x)$ implies that $\sum_{n=0}^{M} a_n t_n Q_n(x)$. (We assume here $t_0 = 1$.)

For finite discrete bivariate distributions, Eagleson (1969) showed that every canonical sequence $\{\rho_n : \sum_{i=0}^{\infty} \rho_i^2 < \infty\}$ is a positive definite sequence. Griffiths (1970) generalized the result to the case when the support of X is unbounded.

3.12.3 Moment sequence and canonical correlation coefficient

A sequence $\{b_n\}$ is said to be a moment sequence if it can be expressed as $b_n = \int t^n dG(t)$ for some distribution function G. Assume again that the support of X is unbounded and X and Y are exchangeable. Tyan and Thomas (1975) showed that every sequence of canonical correlation coefficients is a moment sequence on $[0, 1]$ or $[-1, 1]$. If X is non-negative, then the moment sequence is defined on $[0, 1]$. Conversely, if $\{\rho_n = \rho^n\}$ is a sequence of canonical correlation coefficients, it is easy to show that every moment sequence is a sequence of canonical correlation coefficients. For the binomial and Poisson, the sequence $\{\rho^n\}$ is indeed a sequence of canonical correlation coefficients; see, for example, Lancaster (1983).

3.12.4 Constructions of bivariate distributions via canonical sequences

Let C denote the set of all sequences of canonical correlation coefficients.

- It is easy to see that C is convex. Hence, if $\{a_n\}$ and $\{b_n\}$ are two sequences of canonical correlation coefficients, then $\{\rho_n = \lambda a_n + (1-\lambda) b_n\}$ is also a sequence of canonical correlation coefficients for a new bivariate distribution having the same set of marginals.

- As positive definite sequences are closed under termwise multiplication, C forms a semigroup with respect to termwise multiplication. For finite discrete distribution, this result was proved by Vere-Jones (1971). Vere-Jones's result can be easily generalized to the case with unbounded support. In other words, $\{\rho_n = a_n b_n\}$ is a sequence of canonical correlation coefficients if $\{a_n\}$ and $\{b_n\}$ are. In this way, numerous bivariate distributions can be constructed.

3.13 Bivariate Distributions from Accident Models

In Section 20.3 and Section 21, Hutchinson and Lai (1990) considered the joint distribution of the severities of injury to two people in the same road accident. It was found that a bivariate normal distribution, generated by the method of variables in common, may be used to model such injury. Here, we are concerned with the number of injury accidents rather than the amount of injury.

Let X denote the number of injury accidents on a given stretch of highway and Z_i denote the number of fatalities in the ith accident, $i = 1, 2, \ldots, X$. Also, let Y denote the total number of fatalities recorded among the X accidents. In other words, we may represent them in the following manner:

$$Y = Z_1 + Z_2 + \cdots + Z_X \tag{3.32}$$

The question of interest is to find the joint distribution of X and Y. Unlike the bivariate distributions we have discussed so far, the two marginals are, in general, of different types of univariate distributions.

Following the pioneering work of Edwards and Gurland (1961) in using a discrete bivariate distribution (i.e., a bivariate negative binomial) to model accident data, Leiter and Hamdan (1973), Cacoullos and Papageorgiou (1980, 1982) and others developed several models to represent the joint distribution of (X, Y) as specified in (3.32).

3.13.1 The Poisson-Poisson, Poisson-binomial, and Poisson-Bernoulli methods

Suppose X has a Poisson distribution. By letting Z_i (assuming they are i.i.d), we obtain

- Poisson-Bernoulli model when Z_i has a Bernoulli distribution [Leiter and Hamdan (1973)].

- Poisson-Binomial model when Z_i has a binomial distribution [Cacoullos and Papageorgiou (1980)].

- Poisson-Poisson model when Z_i has a Poisson distribution [Leiter and Hamdan (1973)].

- Poisson-geometric model when Z_i has a geometric distribution [Papageorgiou (1985b)].

3.13.2 Negative binomial-Poisson and negative binomial-Bernoulli models

It has been pointed out by many authors [see Kemp (1970)] that the number of accidents is more adequately described by a negative binomial (i.e., the Poisson distribution whose parameter λ has a gamma distribution). For this reason, Cacoullos and Papageorgiou (1982) constructed the following bivariate distribution assuming X to have a negative binomial distribution.

- Negative binomial-Poisson model where Z_i has a Poisson distribution.
- Negative binomial-Bernoulli models where Z has a Bernoulli distribution. The joint distribution of (X, Y) is a special case of the bivariate negative binomial of Edwards and Gurland (1961).

3.14 Bivariate Distributions Generated from Weight Functions

Let $f(x, y)$ be the probability function of (X, Y). Kocherlakota (1995), and Gupta and Tripathi (1996) defined the probability function of the weighted distribution with the weight function $W(x, y)$ as

$$h_W(x, y) = \frac{f(x, y)W(x, y)}{E[W(X, Y)]}.$$

In particular, they considered the multiplicative weight function of the form

$$W(x, y) = x^{(\alpha)}y^{(\beta)},$$

where $x^{(\alpha)} = x(x-1)\cdots(x-\alpha+1)$. The weighted bivariate Poisson, weighted bivariate binomial, weighted bivariate negative binomial, and weighted bivariate logarithmic series distributions were obtained by this method; see also Section 43.5 of Johnson *et al.* (1997) for other details.

3.15 Marginal Transformations Method

The marginal transformation method to generate a continuous bivariate distribution from another continuous bivariate distribution can be implemented easily. Suppose (X, Y) has a joint cumulative distribution function $H(x, y)$

with marginal $F(x)$ and $G(y)$. If we transform $X \to X^*$ and $Y \to Y^*$, then the joint distribution function of (X^*, Y^*) is given by

$$H^*(x^*, y^*) = H(F^{-1}[F^*(x^*)], G^{-1}[G^*(y^*)]), \tag{3.33}$$

where F^* and G^* are the distribution functions of X^* and Y^*, respectively. The key to this method lies on the fact that $U = F(X), V = G(Y)$ as well as $U' = F^*(X^*), V' = G^*(Y^*)$ are all uniformly distributed for continuous marginals. Thus, the method cannot be readily applied to construct discrete bivariate distributions as discrete random variables cannot be transformed into uniform random variables.

It appears that the method can be transportable if $H(x, y)$ is continuous, whereas X^* and Y^* are two discrete random variables with finite or countable values. Then, the $H^*(x^*, y^*)$ can be expressed as

$$H^*(x^*, y^*) = \int_{-\infty}^{F^{-1}(x^*)} \int_{-\infty}^{G^{-1}(y^*)} h(x, y)\, dxdy, \tag{3.34}$$

where $h(x, y)$ is the joint density function of (X, Y).

Van Ophem (1999) has constructed a discrete bivariate distribution in this manner assuming $h(x, y)$ to be the standard bivariate normal density function with correlation coefficient ρ. Lee (2001) derived the range of correlation coefficients of a discrete bivariate distribution and showed that the discrete bivariate distribution of Van Ophem (1999) has a flexible correlation coefficient.

3.16 Truncation Methods

Similar to its continuous counterpart, discrete bivariate distributions may be obtained through truncations. Truncations may be necessary where certain values are missing or may not be recorded in the data sets. Piperigou and Papageorgiou (2003) gave a unified treatment of three types of zero class truncation:

- The zero cell $(0, 0)$ is not recorded.
- The zero class for the variable X, $\{(0, y),\ y = 0, 1, \ldots\}$, is not recorded.
- The zero class for both X and Y, $\{(0, y),\ y = 0, 1, \ldots; (x, 0),\ x = 0, 1, \ldots\}$, is not recorded.

Using the probability-generating function approach, various properties of the truncated discrete bivariate distributions are then examined.

3.17 Construction of Positively Dependent Discrete Bivariate Distributions

There are various concepts of positive dependence for a bivariate distribution. We consider only two of these here.

A pair of random variables, X and Y, are said to be positively quadrant dependent (PQD) if the following inequality holds, that is, if

$$\Pr(X \le x, Y \le Y) \ge \Pr(X \le x)\Pr(Y \le y). \tag{3.35}$$

The variable Y is said to be positive regression dependent (PRD) on X if $\Pr(Y > x|X = x)$ is increasing in x for every y.

For other concepts of stochastic dependence, one may see, for example, Chapter 12 of Hutchinson and Lai (1990).

3.17.1 Positive quadrant dependent distributions

We shall begin with construction of a pair of PQD binary variables. A binary random variable may be used to indicate the state of a component (or a system) which is either functioning or not functioning. More specifically, we let the binary variable X_i denote the state of the ith component such that

$$X_i = \begin{cases} 1 & \text{if it is functioning} \\ 0 & \text{otherwise.} \end{cases} \tag{3.36}$$

Then, $\Pr(X_i = 1)$ is the static reliability of the component at a given time instant.

Suppose X and Y are two identically distributed binary random variables having the joint probability function given as follows:

$$\Pr(X = 0) = a + b, \quad \Pr(X = 1) = 1 - a - b$$

and

$$\Pr(Y = 0) = a + b, \quad \Pr(Y = 1) = 1 - a - b.$$

Table 3.2: Joint probabilities

$\Pr(X = 0, Y = 0) = a$	$\Pr(X = 0, Y = 1) = b$
$\Pr(X = 1, Y = 0) = b$	$\Pr(X = 1, Y = 1) = 1 - a - 2b$

We now proceed to construct a pair of PQD binary variables as follows:

Clearly, for $(x, y) = (0, 1), (1, 0)$, or $(1, 1)$, inequality (3.35) readily holds without requiring any condition. Thus, the binary pair X and Y are positively quadrant dependent if and only if

$$\Pr(X = 0, Y = 0) \geq \Pr(X = 0) \Pr(Y = 0) \tag{3.37}$$

which is equivalent to the condition

$$(a + b)^2 \leq a. \tag{3.38}$$

It is clear that for a given b, $0 < b < 1$, we can solve for a so that (3.38) holds. It is easy to show that

$$0 \leq \frac{(1 - 2b) - \sqrt{1 - 4b}}{2} < a < \frac{(1 - 2b) + \sqrt{1 - 4b}}{2}. \tag{3.39}$$

Now, let X and Y be two discrete non-negative integer valued random variables with $\Pr(X = i, Y = j) = p_{ij}$, $i = 1, 2, \ldots, r$ and $j = 1, 2, \ldots, c$.

Holzsager (1996) has proved that if

$$p_{i+1\,j+1} \Pr(X \leq i, Y \leq j) \geq \Pr(X \leq i, Y = j+1) \Pr(X = i+1, Y \leq j), \tag{3.40}$$

then X and Y are PQD. Thus, (3.40) provides a mechanism to construct a pair of discrete PQD random variables.

Rao and Subramanyam (1990) provided a mechanism to identify the extreme points of the set of all discrete PQD bivariate distributions when the marginal distributions have finite support. It is easy to see that we can utilize this idea to generate PQD discrete distributions with finite marginals.

3.17.2 Positive regression dependent distributions

Subramanyam and Rao and (1996) also provided an algorithm to identify the extreme points of the set of all discrete PRD bivariate distributions when the marginal distributions have finite support. After identifying these points, positive regression dependent discrete bivariate distributions can be constructed.

References

1. Aki, S. (1985). Discrete distributions of order k on a binary sequence, *Annals of the Institute of Statistical Mathematics*, **37**, 205–224.

2. Aki, S., and Hirano, K. (1994). Distributions of number of failures and successes until the first k consecutive successes, *Annals of the Institute of Statistical Mathematics*, **46**, 193–202.

3. Aki, S., and Hirano, K. (1995). Joint distributions of numbers of success-runs and failures until the first k consecutive successes, *Annals of the Institute of Statistical Mathematics*, **47**, 225–235.

4. Antzoulakos, D. L., and Philippou, A. N. (1991). A note on multivariate negative binomial distribution of order k, *Communications in Statistics—Theory and Methods*, **20**, 1389–1399.

5. Arnold, B. C., Castillo, E., and Sarabia, J. M. (1999). *Conditional Specification of Statistical Methods*, Springer-Verlag, New York.

6. Balakrishnan, N. (2004). Discrete multivariate distributions, In *Encyclopedia of Actuarial Sciences* (Eds., J. L. Tuegels and B. Sundt), pp. 549–571, John Wiley & Sons, New York.

7. Balakrishnan, N. (2005). Discrete multivariate distributions, In *Encyclopedia of Statistical Sciences*, 2nd ed., (Eds., N. Balakrishnan, C. Read, and B. Vidakovic), John Wiley & Sons, Hoboken, NJ (to appear).

8. Balakrishnan, N., and Koutras, M. V. (2002). *Runs and Scans with Applications*, John Wiley & Sons, New York.

9. Bates, G. E., and Neyman, J. (1952). Contribution to the theory of accident proneness I, *University of California Publications in Statistics*, **1**, 215–254.

10. Blischke, W. R. (1978). Mixtures of distributions, *International Encyclopedia of Statistics*, Vol 1, 174-179.

11. Cacoullos, T., and Papageorgiou, H. (1980). On some bivariate probability models applicable to traffic accidents and fatalities, *International Statistical Review*, **48**, 345–356.

12. Cacoullos, T., and Papageorgiou, H. (1982). Bivariate negative binomial-Poisson and negative binomial-Bernoulli models with an application to accident data. In *Statistics and Probability: Essays in Honor of C. R. Rao* (Eds., G. Kallianpur, P. R. Krishnaiah, and J. K. Ghosh), pp. 155–168, North-Holland, Amsterdam.

13. Cacoullos, T., and Papagergiou, H. (1983). Characterisations of discrete distributions by a conditional distribution and a regression function, *Annals of the Institute of Statistical Mathematics*, **35**, 95–104.

14. Charalambides, Ch. A., and Papageroiou, H. (1981a). Bivariate Poisson binomial distributions, *Biometrical Journal*, **23**, 437–450.

15. Charalambides, Ch. A., and Papageorgiou, H. (1981b). On bivariate generalised binomial and negative binomial distributions, *Metrika*, **28**, 83–92.

16. Consael, R. (1952). Sur les processus composes de Poisson a deux variables aleatories, *Academie Royale de Belgigue, Classe des Sciences, Memoires*, **27**, 4–43.

17. Dahiya, R. C., and Korwar, R. M. (1977). On characterising some bivariate discrete distributions by linear regression, *Sankhyā, Series A*, **39**, 124–129.

18. David, K. M., and Papageorgiou, H. (1994). On compounded bivariate Poisson distributions, *Naval Research Logistics*, **41**, 203–214.

19. Eagleson, G. K. (1969). A characterization theorem for positive definite sequences on the Krawtchouk polynomials, *Australian Journal of Statistics*, **11**, 29–38.

20. Edwards, C. B., and Gurland, J. (1961). A class of distributions applicable to accidents, *Journal of the American Statistical Association*, **56**, 503–517.

21. Fréchet, M. (1951). Sur le tableaux de correlation dont les marges sont donness, *Annales de l'Universite de Lyon, Serie 3*, **14**, 53–77.

22. Gelman, A., and Speed, T. P. (1993). Characterizing a joint probability distributions by conditionals, *Journal of the Royal Statistical Society, Series B*, **55**, 185–188.

23. Goodman, L. A., and Kruskal, W. H. (1959). Measures of association for cross classifications: II, Further discussion and references, *Journal of the American Statistical Association*, **54**, 123–163.

24. Griffiths, R. C. (1970). Positive definite sequences and canonical correlation coefficients, *Australian Journal of Statistics*, **12**, 162–165.

25. Griffiths, R. C., Milne, R. K., and Wood, R. (1979). Aspects of correlation in bivariate Poisson distributions and processes, *Australian Journal of Statistics*, **21**, 238–255.

26. Gupta, R. C., and Tripathi, R. C. (1996). Weighted bivariate logarithmic series distributions, *Communications in Statistics—Theory and Methods*, **25**, 2517–2539.

27. Hoeffding, W. (1940). Masstabinvariate Korrelations-theorie, *Schriffen des Mathematischen Instituts und des Instituts fur Angewandte Mathematik der Universitat Berlin*, **5**, 179–233.

28. Holgate, P. (1966). Bivariate generalisations of Neyman's Type A distribution, *Biometrika*, **53**, 241–244.

29. Holzsager, R. (1996). Positive quadrant dependent random variables, *American Mathematical Monthly*, **103**, 350–351.

30. Hutchinson, T. P., and Lai, C. D. (1990). *Continuous Bivariate Distributions, Emphasising Applications*, Rumsby Scientific Publishing, Adelaide, Australia.

31. Janardan, K. G. (1972). A unified approach for a class of multivariate hypergeometric models, *Sankhyā, Series A*, **35**, 363–376.

32. Janardan, K. G. (1973). Chance mechanisms for multivariate hypergeometric models, *Sankhyā, Series A*, **35**, 465–478.

33. Janardan, K. G. (1975). Certain inference problems for multivariate hypergeometric models, *Communications in Statistics*, **4**, 375–388.

34. Janardan, K. G. (1976). Certain estimation problems for multivariate hypergeometric models, *Annals of the Institute of Statistical Mathematics*, **28**, 429–444.

35. Janardan, K. G., and Patil, G. P. (1970). On the multivariate Polya distribution: a model of contagion for data with multiple counts, In *Random Counts in Scientific Work* (Ed., G. P. Patil), Vol. 3, pp. 143–162, The Pennsylvania State University Press, University Park, PA.

36. Janardan, K. G., and Patil, G. P. (1971). The multivariate inverse Polya distribution: a model of contagion for data with multiple counts on inverse sampling, *Studi di Probabilita Statistica e Ricera Operative in Onore de G. Pompilj, Toreno*, 327–341.

37. Janardan, K. G., and Patil, G. P. (1972). A unified approach for a class of multivariate hypergeometric models, *Sankhyā, Series A*, **34**, 363–376.

38. Johnson, N. L. and Kotz, S. (1969). *Distributions in Statistics: Discrete Distributions*, Houghton Mifflin, Boston.

39. Johnson, N. L., and Kotz, S. (1977). *Urn Models and Their Applications*, John Wiley & Sons, New York.

40. Johnson, N. L., Kotz, S., and Balakrishnan, N. (1997). *Discrete Multivariate Distributions*, John Wiley & Sons, New York.

41. Kemp, C. D. (1970). Accident proneness and discrete distribution theory, In *Random Counts in Scientific Work* (Ed., G. P. Patil), pp. 41–64, The Pennsylvania University Press, Philadelphia, PA.

42. Kemp, C. D., and Papageorgiou, H. (1982). Bivariate Hermite distributions, *Sankhyā, Series A*, **44**, 269–280.

43. Kocherlakota, S. (1988). On the compounded bivariate Poisson distribution: a unified approach, *Annals of the Institute of Statistical Mathematics*, **40**, 61–76.

44. Kocherlakota, S. (1995). Discrete bivariate weighted distributions under multiplicative weight function, *Communications in Statistics—Theory and Methods*, **25**, 533–551.

45. Kocherlakota, S., and Kocherlakota, K (1992). *Bivariate Discrete Distributions*, Marcel Dekker, New York.

46. Kocherlakota, S., and Kocherlakota, K (1998). *Bivariate Discrete Distributions.* In *Encyclopedia of Statistical Sciences* (Eds., S. Kotz, C. B. Read and D. L. Banks), Update Vol. 2, pp. 68–83, John Wiley & Sons, New York.

47. Korwar, R. M. (1975). On characterizing some discrete distributions by linear regression, *Communication in Statistics*, **4**, 1133-1147.

48. Korwar, R. M. (1988). On the observed number of classes from multivariate distributions, *Sankhyā, Series B*, **50**, 39–59.

49. Kyriakoussis, A. (1988). Characterizations of bivariate discrete distributions, *Sankhyā, Series A*, **50**, 286–287.

50. Kyriakoussis, A., and Papageorgiou, H. (1989). On characterization of power series distribution by a marginal distribution and a regression function, *Annals of the Institute of Statistical Mathematics*, **41**, 671–676.

51. Lai, C. D. (1995). Construction of bivariate distributions by a generalised trivariate reduction technique, *Statistics & Probability Letters*, **25**, 265–270.

52. Lai, C. D. (2004). Constructions of continuous bivariate distributions, *Journal of the Indian Society for Probability and Statistics*, **8**, 21–43.

53. Lancaster, H. O. (1983). Special joint distributions of Meixner variables, *Australian Journal of Statistics*, **25**, 298–309.

54. Lee, L. F. (2001). On the range of correlation coefficients of bivariate ordered discrete random variables, *Econometric Theory*, **17**, 247–256.

55. Leiter, R. E., and Hamdan, M. A. (1973). Some bivariate probability models applicable to traffic accidents and fatalities, *International Statistical Review*, **41**, 87-100.

56. Ling, K. D., and Tai, T. H. (1990). On bivariate binomial distributions of order k, *Soochow Journal of Mathematics*, **16**, 211–220.

57. Marida, K. V. (1970). *Families of Bivariate Distributions*, Charles Griffin, London.

58. Marshall, A. W., and Olkin, I. (1985). A family of bivariate distributions generated by the bivariate Bernoulli distribution, *Journal of the American Statistical Association*, **80**, 332–338.

59. Marshall, A. W., and Olkin, I. (1990). Bivariate distributions generated from Polya-Eggenberger urn models, *Journal of Multivariate Analysis*, **35**, 48–65.

60. Nelsen, R. B. (1987). Discrete bivariate distributions with given marginals and correlation. *Communications in Statistics—Simulation and Computation*, **16**, 199–208.

61. Oluyede, B. O. (1994). A family of bivariate binomial distributions generated by extreme Bernoulli distributions, *Communications in Statistics—Theory and Methods*, **23**, 1531–1547.

62. Panaretos, J., and Xekalaki, E. (1986). On generalised binomial and multinomial distributions and their applications to generalised Poisson distributions, *Annals of the Institute of Statistical Mathematics*, **38**, 223–231.

63. Papageorgiou, H. (1983). On characterizing some bivariate discrete distributions, *Australian Journal of Statistics*, **25**, 136–144.

64. Papageorgiou, H. (1984). Characterizations of multinomials and negative multinomial mixtures by regression, *Australian Journal of Statistics*, **26**, 25-29.

65. Papageorgiou, H. (1985a). On characterizing some discrete distributions by a conditional distribution and a regression function, *Biometrical Journal*, **27**, 473–479.

66. Papageorgiou, H. (1985b). On a bivariate Poisson-geometric distribution. *Zastowania Mathematyki*, **18**, 541–547.

67. Papageorgiou, H. (1997). Multivariate discrete distributions, In *Encyclopedia of Statistical Sciences* (Eds., S. Kotz, C. B. Read and D. L. Banks), Update Vol. 1, pp. 408–419, John Wiley & Sons, New York.

68. Papageorgiou, H., and David, K. M. (1994). On the countable mixture of bivariate binomial distributions, *Biometrical Journal*, **36**, 581–601.

69. Papageorgiou, H., and Kemp, C. D. (1983). Conditionality in bivariate generalized distributions, *Biometrical Journal*, **25**, 757–763.

70. Patil, G. P. (1965). On a characterization of multivariate distribution by a set of its conditional distributions, In *Handbook of 35th International Statistical Institute Conference in Belgrade*, International Statistical Institute.

71. Patil, G. P. (1986). Polya distribution, multivariate, In *Encyclopedia of Statistical Sciences* (Eds., S. Kotz and N. L. Johnson), Vol. 7, pp. 59–63, John Wiley & Sons, New York.

72. Patil, G. P., Rao, C. R., and Ratnaparkhi, M. V. (1986). On discrete weighted distributions and their use in model choice for observed data, *Communications in Statistics—Theory and Methods*, **15**, 907–918.

73. Philippou, A. N., and Antzoulakos, D. L. (1990). Multivariate distributions of order k on a generalised sequence, *Statistics & Probability Letters*, **9**, 453–463.

74. Philippou, A. N., Antzoulakos, D. L., and Tripsiannis, G. A. (1989). Multivariate distributions of order k, *Statistics & Probability Letters*, **7**, 207–216.

75. Philippou, A. N., Antzoulakos, D. L., and Tripsiannis, G. A. (1990). Multivariate distributions of order k, Part II, *Statistics & Probability Letters*, **10**, 29–35.

76. Philippou, A. N., and Tripsiannis, G. K. (1991). Multivariate Polya and inverse Polya distributions of order k, *Biometrical Journal*, **33**, 225–236.

77. Piperigou, V. E., and Papageorgiou, H. (2003). On truncated bivariate discrete distributions: A unified treatment, *Metrika*, **58**, 221–233.

78. Rao, M. B., and Subramanyam, K. (1990). The structure of some classes of bivariate distributions and some applications, *Computational Statistics & Data Analysis*, **10**, 175–187.

79. Stein, G. Z., and Juritz, J. M. (1987). Bivariate compound distributions, *Communications in Statistics—Theory and Methods*, **16**, 3591–3607.

80. Strandskov, H. H., and Edelen, E. W. (1946). Monozygotic and dizygotic twin birth frequencies in the total of the "white" and the "coloured" US population, *Genetics*, **31**, 438–446.

81. Subramanyam, K., and Rao, M. B. (1996). Analysis of regression dependence in $2 \times n$ bivariate distributions and some applications in contingency tables, *Mathematics and Applications*, Vol. 359, pp. 385–400, Kluwer Academic Publishers, Dordrecht.

82. Tyan, S., and Thomas, J. B. (1975). Characterization of a class of bivariate distribution functions, *Journal of Multivariate Analysis*, **5**, 227–235.

83. Van Ophem, H. (1999). A general method to estimate correlated discrete random variables, *Econometric Theory*, **15**, 228–237.

84. Vere-Jones, D. (1971). Finite bivariate distributions and semigroups of non-negative matrices, *The Quarterly Journal of Mathematics*, **22**, 247–270.

85. Wesolowski, J. (1995). Bivariate discrete measures via power series conditional distribution and a regression function, *Journal of Multivariate Analysis*, **55**, 219–229.

86. Zheng, Q., and Matis, J. H. (1993). Approximating discrete multivariate distributions from known moments, *Communications in Statistics—Theory and Methods*, **22**, 3553–3567.

PART II
Continuous Distributions and Applications

4

The Normal-Laplace Distribution and Its Relatives

William J. Reed

University of Victoria, Victoria, BC, Canada

Abstract: The *normal-Laplace* (NL) distribution results from convolving independent normally distributed and Laplace distributed components. It is the distribution of the stopped state of a Brownian motion with a normally distributed starting value if the stopping hazard rate is constant. Properties of the NL distribution discussed in the article include its shape and tail behaviour (fatter than the normal), its moments, and its infinite divisibility. The *double Pareto-lognormal* distribution is that of an exponentiated normal-Laplace random variable and provides a useful parametric form for modelling size distributions. The *generalized normal-Laplace* (GNL) distribution is both infinitely divisible and closed under summation. It is possible to construct a Lévy process whose increments follow the GNL distribution. Such a Lévy motion can be used to model the movement of the logarithmic price of a financial asset. An option pricing formula is derived for such an asset.

Keywords and phrases: Fat tails, generalized normal-Laplace distribution, double Pareto-lognormal distribution, Brownian-Laplace motion, Lévy process, financial returns, option value

4.1 Introduction

Although the normal (Gaussian) distribution plays a central role in basic statistics, it has long been recognized that the empirical distributions of many phenomena modelled by the normal distribution sometimes do not closely follow the Gaussian shape. For example, Wilson (1923) in a paper in the *Journal of the American Statistical Association* stated that "the frequency we actually meet in everyday work in economics, biometrics, or vital statistics often fails to conform closely to the so-called normal distribution." In recent years, the huge burst of research interest in financial modelling along with the availability of high-frequency price data and the concomitant realisation that logarithmic price returns do not follow exactly a normal distribution [see, for example, Rydberg (2000)], as previously assumed, has led to a search for more realistic alternative parametric models.

Distributions can of course differ from one another in myriad ways, but those for which empirical distributions modelled by the normal tend to differ from the normal can be broadly classified into two kinds, viz., the presence of skewness, and having fatter tails than the normal (leptokurtosis).

A number of alternative parametric forms have been used to deal with the presence of leptokurtosis, ranging from the *Student-t* (including the $t_{(1)}$ or *Cauchy*) distribution to the *logistic* and *Laplace* distributions. The Laplace distribution can be extended to an asymmetric form (*skew-Laplace*) as well as to the *generalized Laplace* distribution [Kotz *et al.* (2001)]. Other distributions of this type, which are parameter rich and can incorporate both skewness and kurtosis, are the *generalized hyperbolic* distributions [Barndorff Nielsen (1977) and Eberlein and Keller (1995)] and its subclass the *normal inverse Gaussian* distribution [Barndorff Nielsen (1997)]. These latter distributions have all been used recently in finance to model logarithmic price returns.

It is the purpose of this chapter to present a new distribution which (in its symmetric form) behaves somewhat like the normal distribution in the middle of its range, and like the Laplace distribution in its tails. This distribution, named herein as the *normal-Laplace* distribution, results from convolving independent normal and Laplace components. Skewness can be introduced into the distribution by using a skew-Laplace component in the convolution.

In Section 4.2 the distribution is defined and its genesis and properties are discussed. In Section 4.3 the *double Pareto-lognormal* distribution (which is that of an exponentiated normal-Laplace random variable) is briefly discussed along with its use in modelling the size distribution of various phenomena. Also in this section the *generalized normal-Laplace* distribution is introduced and some of its properties are discussed. In Section 4.4 the construction of a Lévy process (termed *Brownian-Laplace motion*), whose increments follow the generalized normal-Laplace distribution, is described along with its potential use in financial modelling. This includes the determination of the option value of a European call option for an asset whose logarithmic price follows Brownian-Laplace motion. In Section 4.5 parameter estimation for the normal-Laplace and generalized normal-Laplace distributions is discussed.

4.2 The Normal-Laplace Distribution

Definition

The basic normal-Laplace distribution can be defined in terms of its cumulative distribution function (cdf) which for all real y is

$$F(y) = \Phi\left(\frac{y-\mu}{\sigma}\right) - \phi\left(\frac{y-\mu}{\sigma}\right)\frac{\beta R(\alpha\sigma - (y-\mu)/\sigma) - \alpha R(\beta\sigma + (y-\mu)/\sigma)}{\alpha+\beta}, \tag{4.1}$$

where Φ and ϕ are the cdf and probability density function (pdf) of a standard normal random variable and R is *Mills' ratio*:

$$R(z) = \frac{\Phi^c(z)}{\phi(z)} = \frac{1-\Phi(z)}{\phi(z)}.$$

The location parameter μ can assume any real value while the scale parameter σ and the other two parameters α and β, which determine tail behaviour, are assumed to be positive.

The corresponding density (pdf) is

$$f(y) = \frac{\alpha\beta}{\alpha+\beta}\,\phi\left(\frac{y-\mu}{\sigma}\right)\,\left[R\left(\alpha\sigma - (y-\mu)/\sigma\right) + R\left(\beta\sigma + (y-\mu)/\sigma\right)\right]. \tag{4.2}$$

We shall write

$$Y \sim \mathrm{NL}(\mu, \sigma^2, \alpha, \beta) \tag{4.3}$$

to indicate that a random variable Y has such a distribution.

Genesis

The distribution arises as the convolution of a normal distribution and an asymmetric Laplace, that is, $Y \sim \mathrm{NL}(\mu, \sigma^2, \alpha, \beta)$ can be represented as

$$Y \stackrel{d}{=} Z + W, \tag{4.4}$$

where Z and W are independent random variables with $Z \sim \mathrm{N}(\mu, \sigma^2)$ and W following an asymmetric Laplace distribution with pdf

$$f_W(w) = \begin{cases} \frac{\alpha\beta}{\alpha+\beta}\, e^{\beta w}, & \text{for } w \le 0 \\ \frac{\alpha\beta}{\alpha+\beta}\, e^{-\alpha w}, & \text{for } w > 0. \end{cases} \tag{4.5}$$

Such a convolution might naturally occur if a Brownian motion

$$dX = \nu dt + \tau dw \tag{4.6}$$

with initial state $X_0 \sim \mathrm{N}(\mu, \sigma^2)$ were to be observed at an exponentially distributed time T; or, put another way, if such a Brownian motion were stopped (or "killed," or observed) with a constant hazard rate λ, and the stopped state $X(T)$ observed. This follows from the fact that the state of the Brownian motion (4.6) with fixed (nonrandom) initial state after an exponentially distributed time follows an asymmetric Laplace distribution [see Kotz *et al.* (2001, p. 145)].

Thus, for example, if the logarithmic price of a stock or other financial asset $\{\log P_t\}_{t \geq 0}$ followed Brownian motion, as has been widely assumed, the log(price) *at the time of the first trade* on a fixed day n, say, could be expected to follow a distribution close to a normal-Laplace. This is because the log(price) at the start of day n would be normally distributed, while under the assumption that trades on day n occur in a Poisson process, the time until the first trade would be exponentially distributed.

Some properties

Because a Laplace random variable can be represented as the difference between two exponentially distributed variates [Kotz *et al.* (2001)] it follows from (4.4) that an $\mathrm{NL}(\mu, \sigma^2, \alpha, \beta)$ random variable can be expressed as

$$Y \stackrel{d}{=} \mu + \sigma Z + E_1/\alpha - E_2/\beta, \tag{4.7}$$

where E_1, E_2 are independent standard exponential deviates and Z is a standard normal deviate independent of E_1 and E_2. This provides a convenient way to simulate pseudo-random numbers from the NL distribution.

Kotz *et al.* (2001, p. 149) provide several other representations of asymmetric Laplace random variables. With suitable adjustment (addition of a $\mathrm{N}(\mu, \sigma^2)$ component), these all carry over for normal-Laplace random variables. Some other properties are:

• *Shape and tail behaviour.* The normal-Laplace pdf is smooth (differentiable) and has a single mode. It decays to zero as $y \to \pm\infty$. In the case $\alpha = \beta$ it is symmetric and bell-shaped, occupying a intermediate position between a normal and a Laplace distribution. Figure 4.1 shows the NL(0, 1/3, $1/\sqrt{3}$, $1/\sqrt{3}$) distribution (solid curve), which has mean zero and variance 1 along with the normal (dot-dash) and Laplace (dashed) distributions with the same mean and variance. The parameters α and β determine the behaviour in the right and left tails, respectively. Small values of either of these parameters correspond to heaviness in the corresponding tail. Figure 4.2 shows the NL(0,1,1,β) pdf for values of $\beta = 1, 1/2, 1/3, 1/4$ and $1/5$, while Figure 4.3 shows the symmetric NL(0,1,α, α) pdf for values of $\alpha = 2, 1, 3/4$ and $1/2$.

In comparison with the $\mathrm{N}(\mu, \sigma^2)$ distribution, the $\mathrm{NL}(\mu, \sigma^2, \alpha, \beta)$ distribution will always have more weight in the tails, in the sense that for y suitably

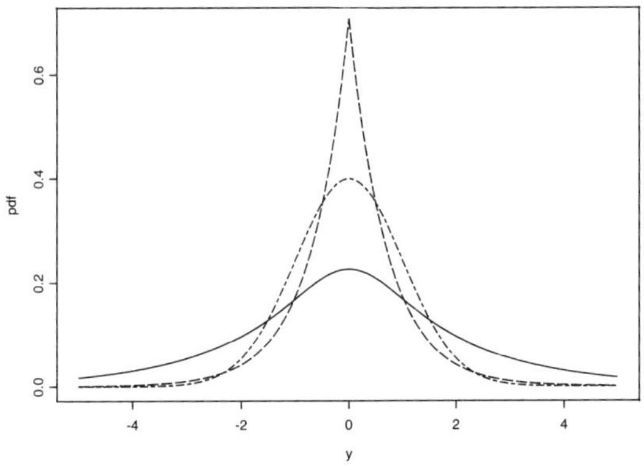

Figure 4.1: Solid curve—the normal-Laplace density with $\mu = 0, \sigma^2 = 1/3, \alpha = 1/\sqrt{3}, \beta = 1/\sqrt{3}$, which has mean 0 and variance 1; dot-dash curve—standard normal density; and dashed curve—the Laplace density with mean zero and variance 1

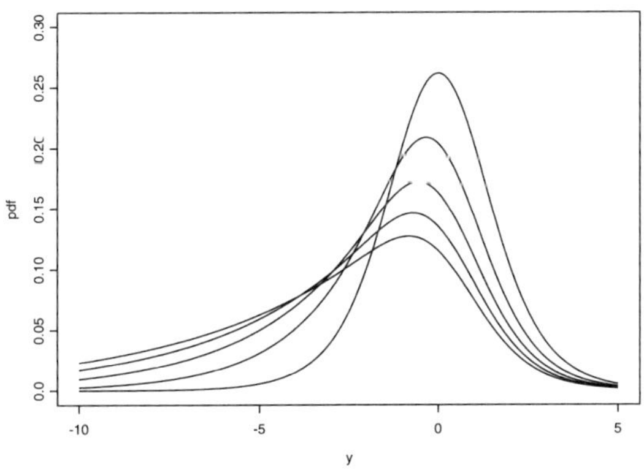

Figure 4.2: The density of the NL(0,1,1,β) for (moving down the peaks) $\beta = 1, 1/2, 1/3, 1/4$, and $1/5$

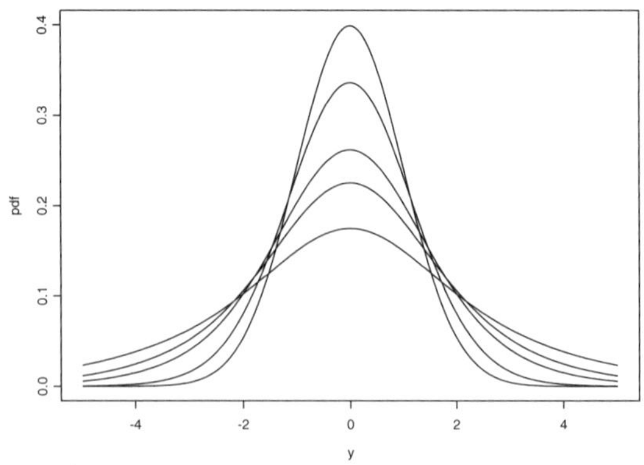

Figure 4.3: Densities of the standard normal and symmetric normal-Laplace distribution. The curve with the highest peak is the density of $N(0,1)$ and (moving down the peaks) the densities of $N(0,1,\alpha,\alpha)$ with $\alpha = 2, 1, 3/4$ and $1/2$

small $F(y) > \Phi((y-\mu)/\sigma)$, while for y suitably large $1 - F(y) > 1 - \Phi((y-\mu)/\sigma)$. This follows from the expression (4.1) for the cdf, because the term $\beta R(\alpha\sigma - (y-\mu)/\sigma) - \alpha R(\beta\sigma + (y-\mu)/\sigma)$ is decreasing in y from ∞ to $-\infty$ over the interval $(-\infty, \infty)$.

If the NL distribution is thought of as a convolution of normal and Laplace components, it is the Laplace component that dominates in the tails in the sense that the tails decay exponentially, that is,

$$f(y) \sim k_1\, e^{-\alpha y} \quad (y \to \infty), \qquad f(y) \sim k_2\, e^{\beta y} \quad (y \to -\infty),$$

where $k_1 = \alpha \exp[\alpha\sigma + \alpha^2\sigma^2/2]$ and $k_2 = \beta \exp[-\beta\sigma + \beta^2\sigma^2/2]$.

•*Moment generating function (mgf).* From the representation (4.4), it follows that the mgf of $NL(\alpha, \beta, \mu, \sigma^2)$ is the product of the mgfs of its normal and Laplace components. Specifically, it is given by

$$M_Y(s) = \frac{\alpha\beta \exp(\mu s + \sigma^2 s^2/2)}{(\alpha - s)(\beta + s)}. \tag{4.8}$$

•*Mean, variance, and cumulants.* Expanding the cumulant generating function, $K_Y(s) = \log M_Y(s)$, we obtain

$$E(Y) = \mu + 1/\alpha - 1/\beta \quad \text{and} \quad \mathrm{Var}(Y) = \sigma^2 + 1/\alpha^2 + 1/\beta^2. \tag{4.9}$$

Higher-order cumulants are

$$\kappa_r = (r-1)!\left(\alpha^{-r} + (-\beta)^{-r}\right), \qquad \text{for integer } r > 2. \tag{4.10}$$

In particular,

$$\kappa_3 = 2/\alpha^3 - 2/\beta^3; \qquad \kappa_4 = 6/\alpha^4 + 6/\beta^4. \tag{4.11}$$

• *Closure under linear transformation.* The NL distribution is closed under linear transformation. Specifically, if $Y \sim NL(\alpha, \beta, \mu, \sigma^2)$ and a and b are any constants, then $aY + b \sim NL(\alpha/a, \beta/a, a\mu + b, a^2\sigma^2)$.

• *Infinite divisibility.* The NL distribution is infinitely divisible. This follows from writing its mgf as

$$M_Y(s) = \left[\exp\left(\frac{\mu}{n}s + \frac{\sigma^2}{2n}s^2\right)\left(\frac{\alpha}{\alpha - s}\right)^{1/n}\left(\frac{\beta}{\beta + s}\right)^{1/n}\right]^n$$

for any integer $n > 0$ and noting that the term in square brackets is the mgf of a random variable formed as $Z + G_1 - G_2$, where Z, G_1 and G_2 are independent and $Z \sim N(\frac{\mu}{n}, \frac{\sigma^2}{n})$ and G_1 and G_2 have gamma distributions with parameters $1/n$ and α and $1/n$ and β, respectively.

Some special cases

From the representation (4.4) of the NL as a convolution of normal and Laplace components, it is clear that as $\sigma \to 0$, the distribution tends to an asymmetric Laplace distribution; and as $\alpha, \beta \to \infty$, it tends to a normal distribution. If only $\beta = \infty$, the distribution is that of the sum of independent normal and exponential components and has a fatter tail than the normal only in the upper tail. In this case, the pdf is

$$f_1(y) = \alpha\phi\left(\frac{y - \mu}{\sigma}\right) R(\alpha\sigma - (y - \mu)/\sigma). \tag{4.12}$$

Similarly if only $\alpha = \infty$, the distribution exhibits extra-normal variation only in the lower tail and the pdf is

$$f_2(y) = \beta\phi\left(\frac{y - \mu}{\sigma}\right) R(\beta\sigma + (y - \mu)/\sigma). \tag{4.13}$$

Clearly the general NL$(\mu, \sigma^2, \alpha, \beta)$ pdf (4.2) can be represented as a mixture of the above pdfs as

$$f(y) = \frac{\beta}{\alpha + \beta} f_1(y) + \frac{\alpha}{\alpha + \beta} f_2(y). \tag{4.14}$$

A special case of some importance already mentioned (Fig. 4.3) is the *symmetric normal-Laplace* distribution arising when $\alpha = \beta$, with pdf

$$f(y) = \frac{\alpha}{2}\,\phi\left(\frac{y - \mu}{\sigma}\right)\ \left[R\left(\alpha\sigma - (y - \mu)/\sigma\right) + R\left(\alpha\sigma + (y - \mu)/\sigma\right)\right]. \tag{4.15}$$

4.3 Related Distributions

4.3.1 The double Pareto-lognormal distribution

The double Pareto-lognormal distribution is related to the normal-Laplace distribution in the same way as the lognormal is related to the normal, that is, a random variable X for which $\log X \sim \mathrm{NL}(\mu, \sigma^2, \alpha, \beta)$ is defined as following the double Pareto-lognormal distribution. As such it can be termed the "log normal-Laplace." However, the name "double Pareto-lognormal" (which was coined because the distribution results from the product of double Pareto and lognormal components) has already been used [Reed and Jorgensen (2004)]. The double Pareto-lognormal (or *dPlN*) distribution shares many characteristics with the log-hyperbolic distribution (Barndorff-Nielsen, 1977). For example, it exhibits power-law behaviour in both tails and has an approximately hyperbolic shape when the pdf is plotted on logarithmic axes. Like the log-hyperbolic distribution, the dPlN distribution has proved useful in modelling size distributions. It has been shown to provide a very good fit to a variety of empirical size distribution data [such as incomes and wealth, city sizes, particle sizes, oil field sizes, etc.; see Reed and Jorgensen (2004)].

4.3.2 The generalized normal-Laplace distribution

While the NL distribution is infinitely divisible, it is not closed under the convolution operation, that is, sums of independent NL random variables do not themselves follow NL distributions. The generalized normal-Laplace is an extension of the NL distribution for which a closure property of this type holds. The advantage of this is that for such a class of distributions one can construct a Lévy motion for which the increments follow the given distribution. This is useful in financial applications for obtaining an alternative stochastic process model to Brownian motion for logarithmic prices, in which the increments (logarithmic returns) exhibit fatter tails than the normal distribution (something that has been widely observed in high-frequency finance data).

The *generalized-normal Laplace* (GNL) distribution is defined as that of a random variable X with characteristic function

$$\phi_{GNL}(s) = \left[\frac{\alpha\beta\exp(i\mu s - \sigma^2 s^2/2)}{(\alpha - is)(\beta + is)}\right]^{\rho} \tag{4.16}$$

and hence moment generating function

$$M_{GNL}(s) = \left[\frac{\alpha\beta\exp(\mu s + \sigma^2 s^2/2)}{(\alpha - s)(\beta + s)}\right]^{\rho}, \tag{4.17}$$

where α, β, ρ and σ are positive parameters, $-\infty < \mu < \infty$. Let

$$X \sim \text{GNL}(\mu, \sigma^2, \alpha, \beta, \rho)$$

denote the random variable X following such a distribution.[1] Writing the mgf as

$$\exp(\rho\mu s + \rho\sigma^2 s^2/2)\left(\frac{\alpha}{\alpha - s}\right)^\rho \left(\frac{\beta}{\beta + s}\right)^\rho,$$

it can be seen that X can be represented as

$$X \stackrel{d}{=} \rho\mu + \sigma\sqrt{\rho}Z + \frac{1}{\alpha}G_1 - \frac{1}{\beta}G_2, \tag{4.18}$$

where Z, G_1, and G_2 are independent with $Z \sim \text{N}(0,1)$ and G_1, G_2 are gamma random variables with scale parameter 1 and shape parameter ρ, that is, with probability density function (pdf)

$$\gamma(u) = \frac{1}{\Gamma(\rho)} u^{\rho - 1} e^{-u}, \qquad u > 0.$$

From (4.16) it is easily established that the GNL is infinitely divisible. Furthermore, sums of independent and identically distributed (iid) GNL random variables, with common α and β parameters, also follow a GNL distribution.

The mean and variance of the $\text{GNL}(\mu, \sigma^2, \alpha, \beta, \rho)$ distribution are

$$\text{E}(Y) = \rho\left(\mu + \frac{1}{\alpha} - \frac{1}{\beta}\right) \quad \text{and} \quad \text{Var}(Y) = \rho\left(\sigma^2 + \frac{1}{\alpha^2} + \frac{1}{\beta^2}\right),$$

while the higher-order cumulants are (for $r > 2$)

$$\kappa_r = \rho(r-1)!\left(\frac{1}{\alpha^r} + (-1)^r \frac{1}{\beta^r}\right). \tag{4.19}$$

Note that the coefficient of kurtosis

$$\kappa_4/\kappa_2^2 = \frac{1}{\rho}\,\frac{3!(\alpha^4 + \beta^4)}{(\sigma^2\alpha^2\beta^2 + \alpha^2 + \beta^2)^2}$$

is decreasing in ρ.

The parameters μ and σ^2 influence the central location and spread of the distribution, while α, β and ρ affect the tail behaviour. *Ceteris paribus* decreasing α (or β) puts more weight into the upper (or lower) tail. When $\alpha = \beta$ the distribution is symmetric and in the limiting case $\alpha = \beta = \infty$ the GNL reduces to a normal distribution. Also increasing ρ moves the shape of the distribution towards normality. In the case $\rho = 1$, the GNL becomes an ordinary normal-Laplace (NL) distribution. For finite values of α and β the GNL distribution, like the NL distribution, has fatter tails than a normal distribution.

[1] The distribution with the above mgf with $\mu = \sigma^2 = 0$ has been called the generalized Laplace distribution by Kotz *et al.* (2001) (it has also been called the *Bessel function distribution* and the *variance-gamma* distribution by other authors). The generalized normal-Laplace distribution defined above bears the same relation to the normal-Laplace distribution as does the generalized Laplace to the Laplace.

4.4 A Lévy Motion Based on the GNL Distribution

We now consider a Lévy process $\{X_t\}_{t\geq 0}$, say for which the increments $X_{t+\tau} - X_\tau$ have characteristic function $(\phi_{GNL}(s))^t$, where ϕ_{GNL} is the characteristic function of the GNL$(\mu, \sigma^2, \alpha, \beta, \rho)$ defined in (4.16) [such a construction is always possible for an infinitely divisible distribution; see, for example, Schoutens (2003)]. It is not difficult to show that the Lévy triplet for this process is $(\rho\mu, \rho\sigma^2, \Lambda)$ where Λ is the Lévy measure of asymmetric Laplace motion [see Kotz *et al.* (2001, p. 198)]. Laplace motion has an infinite number of jumps in any finite time interval (a pure jump process). The extension considered here adds a continuous Brownian component to Laplace motion. We shall thus call the process $\{X_t\}_{t\geq 0}$ defined above *Brownian-Laplace motion.*

The increments $X_{t+\tau} - X_\tau$ of this process will follow a GNL$(\mu, \sigma^2, \alpha, \beta, \rho t)$ distribution and will have fatter tails than the normal. However, as t increases the kurtosis of the distribution drops. Exactly this sort of behaviour has been observed in various studies on high-frequency financial data [see Rydberg (2000)] — very little kurtosis in the distribution of logarithmic returns over long intervals but increasingly fat tails as the reporting interval is shortened. Thus, Brownian-Laplace motion seems to provide a good model for the movement of logarithmic prices.

4.4.1 Option pricing for assets with logarithmic prices following Brownian-Laplace motion

We consider an asset whose price S_t is given by

$$S_t = S_0 \exp(X_t),$$

where $\{X_t\}_{t\geq 0}$ is a Brownian-Laplace motion with $X_0 = 0$ and parameters $\mu, \sigma^2, \alpha, \beta, \rho$. We wish to determine the risk-neutral valuation of a European call option on the asset with strike price K at time T and a discount rate r.

It can be shown using the Escher equivalent martingale measure [see Schoutens (2003, p. 77)] that the option value can be expressed in a form similar to that of the Black-Scholes formula. Precisely,

$$OV = S_0 \int_\gamma^\infty d^{*T}_{GNL}(x; \theta + 1) dx - e^{-rT} K \int_\gamma^\infty d^{*T}_{GNL}(x; \theta) dx, \tag{4.20}$$

where $\gamma = \log(K/S_0)$ and

$$d^{*T}_{GNL}(x; \theta) = \frac{e^{\theta x} d^{*T}_{GNL}(x)}{\int_{-\infty}^{\infty} e^{\theta y} d^{*T}_{GNL}(y) dy} \tag{4.21}$$

is the pdf of X_T under the risk-neutral measure. Here, d_{GNL}^{*T} is the pdf of the T-fold convolution of the generalized normal-Laplace, GNL$(\mu, \sigma^2, \alpha, \beta, \rho)$, distribution and θ is the unique solution to the following equation involving its mgf

$$\log M_{GNL}(\theta + 1) - \log M_{GNL}(\theta) = r. \tag{4.22}$$

The T-fold convolution of GNL$(\mu, \sigma^2, \alpha, \beta, \rho)$ is GNL$(\mu, \sigma^2, \alpha, \beta, \rho T)$ and so its moment generating function is (4.17) with ρ replaced by ρT. This provides the denominator of the expression (4.21) for the risk-neutral pdf.

Now let

$$I_\theta = \int_\gamma^\infty d_{GNL}^{*T}(x;\theta)dx = \frac{1}{[M_{GNL}(\theta)]^T} \int_\gamma^\infty e^{\theta x} d_{GNL}^{*T}(x) \tag{4.23}$$

so that

$$OV = S_0 I_{\theta+1} - e^{-rT} K I_\theta.$$

Thus, to evaluate the option value, we need to evaluate only the integral in (4.23). This can be done using the representation (4.18) of a GNL random variable as the sum of normal, positive, and negative gamma components. The integral can be written as

$$\int_0^\infty g(u;\alpha) \int_0^\infty g(v;\beta) \int_\gamma^\infty e^{\theta x} \frac{1}{\sigma\sqrt{\rho T}} \phi\left(\frac{x - u + v - \mu\rho T}{\sigma\sqrt{\rho T}}\right) dx dv du, \tag{4.24}$$

where

$$g(x;a) = \frac{a^{\rho T}}{\Gamma(\rho T)} x^{\rho T - 1} e^{-ax}$$

is the pdf of a gamma random variable with scale parameter a and shape parameter ρT, and ϕ is the pdf of a standard normal deviate. After completing the square in x and evaluating the x integral in terms of Φ^c, the complementary cdf of a standard normal, the integral can be expressed as

$$\int_0^\infty g(u;\alpha - \theta) \int_0^\infty g(v;\beta + \theta) \Phi^c \left(\frac{\gamma - u + v - \mu\rho T - \theta\sigma^2\rho T}{\sigma\sqrt{\rho T}}\right) dv du. \tag{4.25}$$

For given parameter values, the double integral in (4.25) can be evaluated numerically quite quickly and thence via (4.24) and (4.23) the option value can be computed.

Figure 4.4 shows the difference (vertical axis) between the Black-Scholes option value (assuming a normal distribution for logarithmic daily returns) and the option value assuming a GNL distribution for various values of the current stock price (horizontal axis). The strike price was set at $K = 1$ and the discount rate at $r = 0.05$ per annum. The distribution of daily logarithmic returns was assumed to be GNL$(\mu = 0, \sigma^2 = 0.02, \alpha = 17.5, \beta = 17.5, \rho = 0.1)$. This has

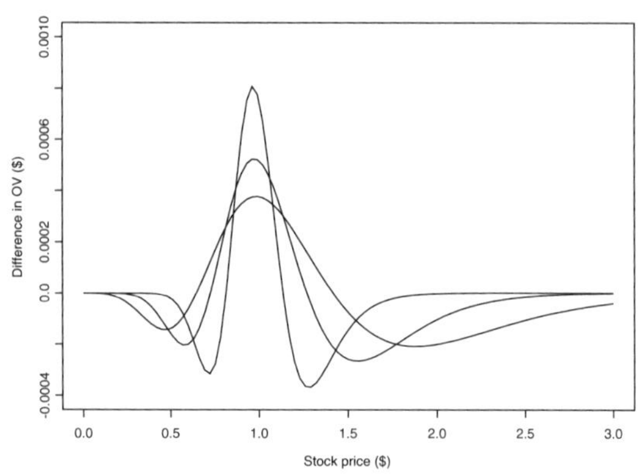

Figure 4.4: The difference between option values for a European call option using a normal distribution (Black-Scholes option value) and a generalized normal-Laplace (GNL) distribution for the log(price) increments. The horizontal axis shows the current stock price, S, and the vertical axis the difference in option values. The strike price was set at $K = 1$; the per-annum discount rate at $r = 0.05$; the GNL parameter values at $\mu = 0, \sigma^2 = 0.02, \alpha = 17.5, \beta = 17.5, \rho = 0.1$; and the normal distribution for computing the Black-Scholes option value had mean 0 and variance 0.00165, the same as those of the GNL. The three curves correspond to exercise dates (moving down the peaks) $T = 10, 30$, and 60 days ahead

mean zero and variance of 0.00165, which was used in computing the Black-Scholes option value. The coefficient of kurtosis is 4.68, which is close to the value of 4.73 observed for a sequence of 929 logarithmic returns for IBM common stock over the period Jan. 1999–Sept. 2003. The three curves correspond to exercise dates $T = 10$, 30, and 60 days in advance.

It can be seen in Figure 4.4 that "at the money" ($S = 1$) the Black-Scholes price is too high. Although the difference is less than one-tenth of one cent it amounts to about 1.5 percent (for $T = 10$) of the Black-Scholes option value. The corresponding percentages for $T = 30$ and $T = 60$ are about 0.5 percent and about 0.3 percent. The reason why the difference decreases as T increases is that the distribution of log-returns ($\text{GNL}(\mu, \sigma^2, \alpha, \beta, \rho T)$) is closer to normality for larger T (a central limit effect).

Far enough "in the money" ($S > 1$) or "out of the money" ($S < 1$), the Black-Scholes valuation is too low. This is because the normal model fails to anticipate more extreme fluctuations, which are slightly more likely to occur with the GNL distributed daily returns.

4.5 Estimation for NL and GNL Distributions

For the NL distribution, maximum likelihood estimation of parameters can be carried out numerically because there is a closed-form expression for the pdf. In fact, it is shown in Reed and Jorgensen (2003) how one can estimate μ analytically and then maximize numerically the concentrated (profile) log-likelihood over the remaining three parameters. Another approach, also discussed by Reed and Jorgensen, uses the EM-algorithm (considering an NL random variable as the sum of normal and Laplace components, with one regarded as missing data).

Things are more difficult for the GNL distribution, because there is no apparent closed-form expression for the pdf. It may be possible to use the EM-algorithm, but calculating the required conditional expectations appears to be a formidable task. Parameter estimates can be obtained by the method of moments (solving the equations produced by setting the first five sample cumulants equal to their theoretical counterparts, using (4.19)). This can be achieved by solving numerically a pair of equations (in α and β) and then obtaining the solutions for the other parameters by substitution. One drawback with the method of moments is that it is difficult to impose constraints on parameters (such as requiring estimates of α, β, ρ, and σ^2 be positive) and estimates that are unsatisfactory in this respect may sometimes occur.

References

1. Barndorff-Nielsen, O. E. (1977). Exponentially decreasing distributions for the logarithm of particle size, *Proceedings of the Royal Society of London, Series A*, **353**, 401–419.

2. Barndorff-Nielsen, O. E. (1997). Normal inverse Gaussian distributions and stochastic volatility models, *Scandinavian Journal of Statistics*, **24**, 1–13.

3. Eberlein, E., and Keller, U. (1995). Hyperbolic distributions in finance, *Bernoulli*, **1**, 281–289.

4. Kotz, S., Kozubowski, T. J., and Podgórski, K. (2001). *The Laplace Distribution and Generalizations*, Birkhäuser, Boston.

5. Reed, W. J., and Jorgensen, M. (2004). The double Pareto-lognormal distribution: A new parametric model for size distributions, *Communications in Statistics—Theory and Methods*, **33**, 1733–1753.

6. Rydberg, T. H. (2000). Realistic statistical modelling of financial data, *International Statistical Review*, **68**, 233–258.

7. Schoutens, W. (2003). *Lévy Processes in Finance*, John Wiley & Sons, Chichester.

8. Wilson, E. B. (1923). First and second laws of error, *Journal of the American Statistical Association*, **18**, 841–852.

5

Some Observations on a Simple Means of Generating Skew Distributions

Arthur Pewsey

Universidad de Extremadura, Cáceres, Spain

Abstract: During the last decade, a substantial part of Barry Arnold's research effort has been directed towards developing models capable of describing the forms of asymmetry manifested by real data. One general and seemingly elegant means of constructing skew distributions is provided by a lemma presented in Azzalini (1985). The now widely known skew-normal distribution is just one special case belonging to the family of distributions generated using the construction implicit in that lemma. In this paper, a simple alternative proof of the lemma is given, and reflections are made upon how the construction arising from it has been employed in the literature. The densities of various special cases are presented, which highlight both the flexibility and limitations of the construction. Likelihood-based inference for the parameters of the location-scale extensions of classes arising from the construction is also considered. General results are given for the solutions to the score equations and for the observed information matrix. For the special case of the skew-normal distribution, it is shown that, for one of the solutions to the score equations, the observed information matrix is always singular.

Keywords and phrases: Asymmetry, boundary estimates, location-scale family, observed information matrix, reparametrisation, score equations, skew-normal distribution

5.1 Introduction

The mainspring for this chapter is the following lemma from Azzalini (1985).

Lemma 5.1.1 *Let f be a density function that is symmetric about 0, and G an absolutely continuous distribution function such that G' is symmetric about 0.*

Then

$$2f(z)G(\lambda z) \qquad (-\infty < z < \infty)$$

is a density function for any real λ.

Rather than reproduce the proof of Azzalini (1985), we present the following alternative, which we consider to be simpler and more direct.

PROOF. Given the definitions of f and G, the proof only requires us to show that $2f(z)G(\lambda z)$ integrates to 1. Thus, making use of the assumed symmetry of f and G about 0, we have

$$\begin{aligned}
\int_{-\infty}^{\infty} 2f(z)G(\lambda z)dz &= 2\left\{\int_{-\infty}^{0} f(z)G(\lambda z)dz + \int_{0}^{\infty} f(z)G(\lambda z)dz\right\} \\
&= 2\left[\int_{-\infty}^{0} f(z)G(\lambda z)dz + \int_{-\infty}^{0} f(z)\{1 - G(\lambda z)\}dz\right] \\
&= 2\int_{-\infty}^{0} f(z)dz = 1.
\end{aligned}$$

■

In what follows we will refer to any density generated using the construction implicit in the lemma as belonging to the family $S(\lambda)$ ("s" being the first letter of "skew").

The remainder of the chapter is divided into two main sections. In the first, we consider the flexibility and limitations of the construction arising from Azzalini's lemma. In Section 5.3, we discuss issues of inference and present new results for the score equations and observed information matrix for location-scale extensions of any class in $S(\lambda)$. These results lead to interesting observations regarding likelihood-based inference for location-scale extensions of classes in $S(\lambda)$ generated using $f(z) = \phi(z)$.

5.2 Flexibility and Limitations of the Construct

Azzalini (1985, 1986) and Henze (1986) considered in detail the case where f and G are the density function and distribution function, respectively, of the standard normal distribution. The resulting class of distributions is referred to in the literature as the skew-normal class. Using an obvious notation, we will denote the skew-normal class as $S_{\phi\Phi}(\lambda)$.

Surprisingly, the inherent flexibility of the construction in Azzalini's lemma has been little exploited. Indeed, authors have generally limited themselves to cases such as the skew-normal class where f and G are the density and distribution function, respectively, of some common distribution. For instance,

Mukhopadhyay and Vidakovic (1995) refer to the $S_{\phi\Phi}(\lambda)$, $S_{t_3T_3}(\lambda)$, $S_{lL}(\lambda)$ and $S_{dD}(\lambda)$ classes obtained using the standard normal, t_3, logistic and double exponential distributions, respectively. DiCiccio *et al.* (1997), Azzalini and Capitanio (2003), and Jones and Faddy (2003) consider the general $S_{t_\nu T_\nu}(\lambda)$ class which has $S_{\phi\Phi}(\lambda)$ as a limiting class. Gupta, Chang and Huang (2002) consider all the classes to which we have referred so far, as well as the $S_{uU}(\lambda)$ class generated using a uniform distribution.

In Figure 5.1, we present the densities of $S_{\phi\Phi}(\lambda)$, $S_{t_2T_2}(\lambda)$, $S_{lL}(\lambda)$, and $S_{dD}(\lambda)$ distributions for λ-values of 0, 2, 5, 20, and 100. The $S_{t_2T_2}(\lambda)$ class results on using the t_2 distribution, proposed as being the simplest t distribution by Jones (2002), as the common distribution. The densities of the last three of these classes are:

$$\varphi_{t_2T_2}(z;\lambda) = \frac{1}{(2+z^2)^{3/2}}\left\{1+\frac{\lambda z}{(2+\lambda^2 z^2)^{1/2}}\right\},$$

$$\varphi_{lL}(z;\lambda) = \frac{2e^z}{(1+e^z)^2(1+e^{-\lambda z})}$$

and

$$\varphi_{dD}(z;\lambda) = \begin{cases} e^{z(1+\lambda)}/2, & z<0, \quad \lambda \geq 0, \\ e^{-z}(1-e^{-\lambda z}/2), & z \geq 0, \quad \lambda \geq 0, \\ e^{z}(1-e^{-\lambda z}/2), & z<0, \quad \lambda<0, \\ e^{-z(1-\lambda)}/2, & z \geq 0, \quad \lambda<0, \end{cases}$$

respectively.

The plots in Figure 5.1 provide an indication of the range of distributions that can be generated using the construction of Azzalini's lemma with well-known distributions defined on $\Re$. Clearly, the four classes are capable of modelling different ranges of skewness and kurtosis, and one major consideration in choosing between them in practice would be the weights in their tails. As is evident from these plots, even for only moderately skew members of a given $S_{fF}(\lambda)$ class (with $\lambda > 0$), the right-hand tail behaviour is essentially that of the limiting half-f distribution obtained as $\lambda \to \infty$.

As an example of combining a density and a distribution function from different distributions, Mukhopadhyay and Vidakovic (1995) refer to the $S_{t_\nu L}(\lambda)$ class obtained using a t_ν density and the distribution function of the logistic distribution. More recently, Nadarajah and Kotz (2003) presented results for the moment properties of certain $S_{\phi G}(\lambda)$ classes, while Nadarajah (2003) studied classes of the form $S_{uG}(\lambda)$. In fact, for the $S_{\phi\Phi}(\lambda)$, $S_{t_\nu T_\nu}(\lambda)$, $S_{lL}(\lambda)$, and $S_{dD}(\lambda)$ classes, little flexibility is gained by replacing F in the $S_{fF}(\lambda)$ formulation by the distribution function, G say, of any one of the other three classes, as the ranges of densities generated using the different combinations differ only very marginally. However, the flexibility of the construction arising out of Azzalini's lemma improves considerably on widening the set of possible component distributions.

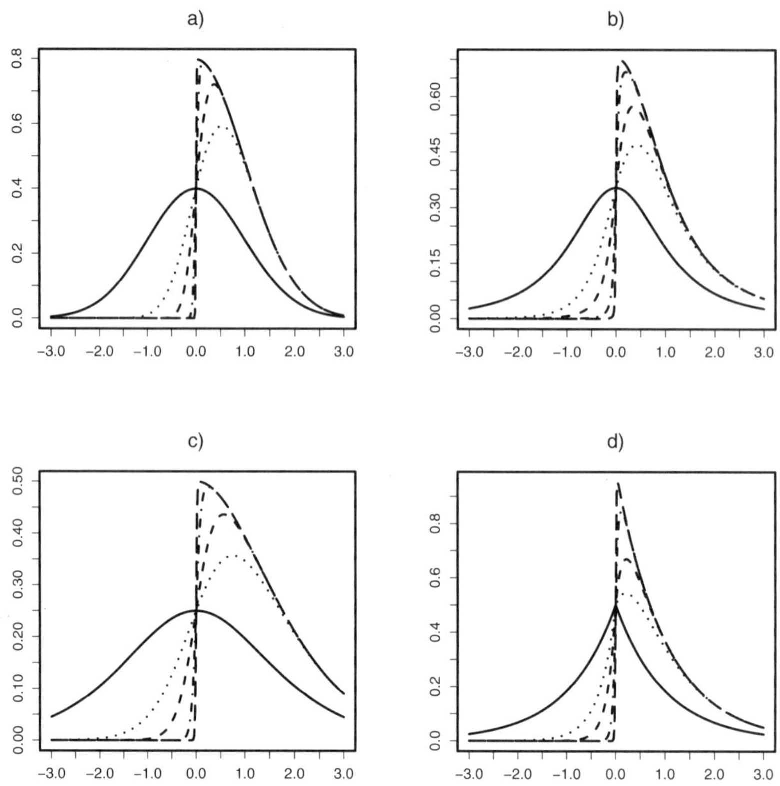

Figure 5.1: Densities of: (a) $S_{\phi\Phi}(\lambda)$, (b) $S_{t_2T_2}(\lambda)$, (c) $S_{lL}(\lambda)$ and (d) $S_{dD}(\lambda)$ distributions for λ-values of 0 (unbroken), 2 (dot), 5 (dash), 20 (dot dash), and 100 (long dash)

As two examples of classes generated using somewhat nonstandard densities with finite interval support, in Figure 5.2 we present some densities from the $S_{tT}(\lambda)$ and $S_{qL}(\lambda)$ classes. The first of these results on combining the triangular density

$$t(z) = \begin{cases} 0, & z < -1,\ z > 1, \\ 1 + z, & -1 \leq z < 0, \\ 1 - z, & 0 \leq z \leq 1, \end{cases}$$

and the corresponding distribution function

$$T(z) = \begin{cases} 0, & z \leq -1, \\ \frac{1}{2} + z(1 + \frac{z}{2}), & -1 < z \leq 0, \\ \frac{1}{2} + z(1 - \frac{z}{2}), & 0 < z < 1, \\ 1, & z \geq 1. \end{cases}$$

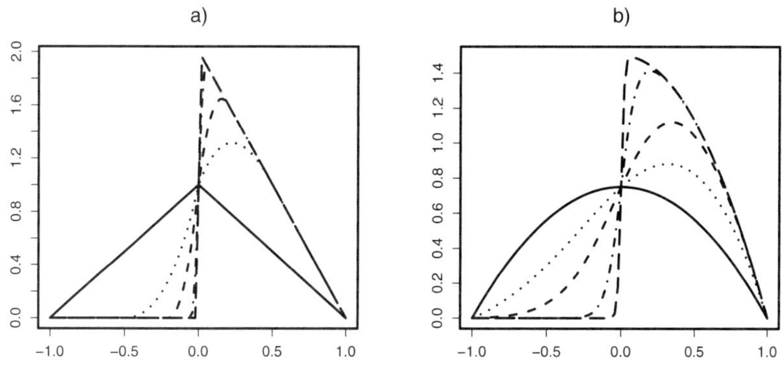

Figure 5.2: Densities of: (a) $S_{tT}(\lambda)$ and (b) $S_{qL}(\lambda)$ distributions for λ-values of 0 (unbroken), 2 (dot), 5 (dash), 20 (dot dash), and 100 (long dash)

The second combines the logistic distribution function with the quadratic density

$$q(z) = \begin{cases} \frac{3}{4}(1 - z^2), & -1 \le z \le 1, \\ 0, & z < -1,\ z > 1. \end{cases}$$

From a practical point of view, it is debatable whether these two classes provide useful models for real data. Nevertheless, the potential of the construction is manifest and it is conceivable that for a given application a suitable density and distribution function combination might be found that would provide an adequate model.

5.3 Inference

5.3.1 General considerations

Generalising what has been observed for the classes considered so far, any $S_{fG}(\lambda)$ class includes densities ranging from the symmetric density f $(\lambda = 0)$ through to the (generally highly skew) positive and negative half-f densities $(\lambda = \pm\infty)$.

Of course, in practice, we will usually be interested in fitting some member of the location-scale extension of a $S_{fG}(\lambda)$ class to data, rather than a member of the $S_{fG}(\lambda)$ class itself. Introducing some extra notation, if $Z \sim S_{fG}(\lambda)$, then $X = \xi + Z\eta \sim S_{fG}(\xi, \eta, \lambda)$, where $S_{fG}(\xi, \eta, \lambda)$ denotes the extended class.

Now, we might envisage that inference for any $S_{fG}(\xi, \eta, \lambda)$ class will potentially be fraught as precisely what ξ represents depends on the value of λ. For instance, when $\lambda = 0$, ξ pinpoints the centre of a symmetric distribution, whereas when $\lambda = \infty$, ξ is the lower bound for the support of a half-f distribution. Clearly, the maximum likelihood (ML) estimate of ξ in the second of these two scenarios will be very different from that in the first [see, e.g., Pewsey (2002, 2004)]. For (at the very least) the $S_{\phi\Phi}(\xi, \eta, \lambda)$ class, this is not merely an observation of academic concern. It is known [see Azzalini (1985), Azzalini and Capitanio (1999), and Pewsey (2000)] that, for this class, maximum likelihood estimation often results in a solution on the boundary of the parameter space corresponding to a half-normal distribution, with the probability of such a solution occurring being greatest for small-sized samples drawn from highly skew cases of the $S_{\phi\Phi}(\xi, \eta, \lambda)$ class. Moreover, the usual regularity conditions underpinning likelihood inference do not apply for solutions on the boundary of a parameter space.

Other known problems associated with ML estimation for the $S_{\phi\Phi}(\xi, \eta, \lambda)$ class are those of:

1. multiple maxima on the likelihood surface [Pewsey (2000)],
2. a solution to the score equations always exists associated with $\lambda = 0$ [Azzalini (1985), Arnold *et al.* (1993), and Chiogna (1997)],
3. the expected information matrix is singular when $\lambda = 0$ [Azzalini (1985)].

The last of these problems can be circumvented using reparametrisation [Azzalini (1985)]. As we will show, the second problem is not unique to the $S_{\phi\Phi}(\xi, \eta, \lambda)$ class. We will also demonstrate that the observed information matrix is in fact *always* singular for *any* $S_{\phi G}(\xi, \eta, \lambda)$ class for which $G''(0) = g'(0)$ is 0.

5.3.2 Score equations for any $S_{fG}(\xi, \eta, \lambda)$ class

Consider $Z \sim S_{fG}(\lambda)$ for which $\varphi_{fG}(z; \lambda) = 2f(z)G(\lambda z)$. Then $X = \xi + Z\eta \sim S_{fG}(\xi, \eta, \lambda)$ with density

$$\varphi_{fG}(x; \xi, \eta, \lambda) = \frac{2}{\eta} f\left(\frac{x-\xi}{\eta}\right) G\left\{\lambda\left(\frac{x-\xi}{\eta}\right)\right\},$$

where x, ξ and $\lambda \in \Re$ and $\eta \in \Re^+$. Thus, for a random sample, $\mathbf{x} = (x_1, \ldots, x_n)$, drawn from $S_{fG}(\xi, \eta, \lambda)$, the log-likelihood function is

$$l(\xi, \eta, \lambda; \mathbf{x}) = n \log 2 - n \log \eta + \sum_{i=1}^{n} \log f\left(\frac{x_i - \xi}{\eta}\right) + \sum_{i=1}^{n} \log G\left\{\lambda\left(\frac{x_i - \xi}{\eta}\right)\right\}. \tag{5.1}$$

Assuming f' exists, and denoting G' as g, the first-order partial derivatives of the log-likelihood are:

$$\frac{\partial l}{\partial \xi} = -\frac{1}{\eta}\left\{\sum_{i=1}^{n}\frac{f'(z_i)}{f(z_i)} + \lambda\sum_{i=1}^{n}\frac{g(\lambda z_i)}{G(\lambda z_i)}\right\},$$

$$\frac{\partial l}{\partial \eta} = -\frac{1}{\eta}\left\{n + \sum_{i=1}^{n} z_i\frac{f'(z_i)}{f(z_i)} + \lambda\sum_{i=1}^{n} z_i\frac{g(\lambda z_i)}{G(\lambda z_i)}\right\},$$

$$\frac{\partial l}{\partial \lambda} = \sum_{i=1}^{n} z_i\frac{g(\lambda z_i)}{G(\lambda z_i)},$$

where $z_i = (x_i - \xi)/\eta$. Setting $v_i = f'(z_i)/f(z_i)$ and $w_i = g(\lambda z_i)/G(\lambda z_i)$, the solutions to the score equations satisfy $-\overline{v} = \lambda\overline{w}$, $(1 + \overline{zv} + \lambda\overline{zw}) = 0$ and $\overline{zw} = 0$. So, for any solution, $\overline{zv} = -1$. Solving for ξ, η and λ, any solution to the score equations satisfies $\xi = \overline{xw}/\overline{w}$, $\eta = \xi\overline{v} - \overline{vx}$ and $\lambda = -\overline{v}/\overline{w}$. Clearly, for $\lambda = 0$ to be a solution to the score equations requires $\overline{v}$ to equal 0. However, if $\lambda = 0$, $\overline{w} = 2g(0)$, $\xi = \overline{x}$ and $\eta = -\overline{vx}$. Then, η is the solution to

$$n\eta = -\sum_{i=1}^{n}\frac{x_i f'\left(\frac{x_i-\overline{x}}{\eta}\right)}{f\left(\frac{x_i-\overline{x}}{\eta}\right)}.$$

However, we repeat, $\lambda = 0$, $\xi = \overline{x}$ and $\eta = -\overline{vx}$ will only be a solution to the score equations if, for this choice of η, $\overline{v}$ also equals 0, that is, if

$$\sum_{i=1}^{n}\frac{f'\left(\frac{x_i-\overline{x}}{\eta}\right)}{f\left(\frac{x_i-\overline{x}}{\eta}\right)} = 0.$$

These findings generalise results given by Arnold *et al.* (1993) and Chiogna (1997) for the skew-normal distribution. As $\phi'(z) = -z\phi(z)$, our results confirm that, as shown by Arnold *et al.* (1993), $\lambda = 0$, $\xi = \overline{x}$, and $\eta^2 = \sum_{i=1}^{n}(x_i - \overline{x})^2/n$ is always a solution to the score equations for the $S_{\phi\Phi}(\xi, \eta, \lambda)$ class. Moreover, it is evident that this combination will *always* be a solution to the score equations for *any* $S_{\phi G}(\xi, \eta, \lambda)$ class, whatever the choice of G.

5.3.3 Observed information matrix for any $S_{fG}(\xi, \eta, \lambda)$ class

Assuming g' and f'' exist, and letting $u_i = f''(z_i)/f(z_i)$ and $t_i = g'(\lambda z_i)/G(\lambda z_i)$, the second-order partial derivatives of the log-likelihood (5.1) can be expressed as follows:

$$\frac{\partial^2 l}{\partial \xi^2} = -\frac{n}{\eta^2}\left\{\overline{v^2} - \overline{u} + \lambda^2(\overline{w^2} - \overline{t})\right\},$$

$$\frac{\partial^2 l}{\partial \xi \partial \eta} = -\frac{n}{\eta^2}\left\{-\overline{v} - \lambda\overline{w} + \overline{zv^2} - \overline{zu} + \lambda^2(\overline{zw^2} - \overline{zt})\right\},$$

$$\frac{\partial^2 l}{\partial\xi\partial\lambda} = -\frac{n}{\eta}\left\{\overline{w} + \lambda(\overline{zt} - \overline{zw^2})\right\},$$

$$\frac{\partial^2 l}{\partial\eta^2} = -\frac{n}{\eta^2}\left\{\overline{z^2v^2} - \overline{z^2u} - 2\overline{zv} + \lambda^2\overline{z^2w^2} - \lambda^2\overline{z^2t} - 2\lambda\overline{zw} - 1\right\},$$

$$\frac{\partial^2 l}{\partial\eta\partial\lambda} = -\frac{n}{\eta}\left\{\overline{zw} + \lambda(\overline{z^2t} - \overline{z^2w^2})\right\}, \quad \frac{\partial^2 l}{\partial\lambda^2} = n(\overline{z^2t} - \overline{z^2w^2}).$$

For any solution to the score equations, $\overline{zw} = 0$ and $\overline{zv} = -1$. Also, if there is a solution to the score equations for which $\lambda = 0$, then (as $\overline{v} = 0$, $\xi = \overline{x}$, $\overline{z} = 0$, $w_i = 2g(0)$, $t_i = 2g'(0)$ and $\eta = -\overline{vx}$) for any such solution:

$$\frac{\partial^2 l}{\partial\xi^2} = -\frac{n}{\eta^2}(\overline{v^2} - \overline{u}), \quad \frac{\partial^2 l}{\partial\xi\partial\eta} = -\frac{n}{\eta^2}(\overline{zv^2} - \overline{zu}), \quad \frac{\partial^2 l}{\partial\xi\partial\lambda} = -\frac{2ng(0)}{\eta},$$

$$\frac{\partial^2 l}{\partial\eta^2} = -\frac{n}{\eta^2}(\overline{z^2v^2} - \overline{z^2u} + 1), \quad \frac{\partial^2 l}{\partial\eta\partial\lambda} = 0, \quad \frac{\partial^2 l}{\partial\lambda^2} = 2n\overline{z^2}\{g'(0) - 2g^2(0)\}.$$

We note that $g'(0) = 0$ for any differentiable density g that is symmetric about 0. This certainly holds for the standard normal, logistic and t densities, but not for the density of the double exponential distribution.

As we have stated previously, for a skew-normal distribution $\lambda = 0$, $\xi = \overline{x}$, and $\eta^2 = \sum_{i=1}^n (x_i - \overline{x})^2/n$ always provides a solution to the score equations. For this solution, $w_i = 2\phi(0) = \sqrt{2/\pi}$, $z_i = (x_i - \overline{x})/\eta$ and hence $\overline{z} = 0$ and $\overline{z^2} = \sum_{i=1}^n (x_i - \overline{x})^2/n\eta^2 = 1$. Also, $v_i = \phi'(z_i)/\phi(z_i) = -z_i$ and thus $\overline{v} = -\overline{z} = 0$, $\overline{v^2} = \overline{z^2} = 1$, $\overline{zv^2} = \overline{z^3}$ and $\overline{z^2v^2} = \overline{z^4}$. Moreover, as $u_i = \phi''(z_i)/\phi(z_i) = (z_i^2 - 1)$, $\overline{u} = \overline{z^2} - 1 = 0$, $\overline{zu} = \overline{z(z^2-1)} = \overline{z^3} - \overline{z} = \overline{z^3}$ and $\overline{z^2u} = (\overline{z^4} - 1)$. For this solution then, the second-order partial derivatives become

$$\frac{\partial^2 l}{\partial\xi^2} = -\frac{n}{\eta^2}, \quad \frac{\partial^2 l}{\partial\xi\partial\eta} = 0, \quad \frac{\partial^2 l}{\partial\xi\partial\lambda} = -\frac{n}{\eta}\sqrt{2/\pi},$$

$$\frac{\partial^2 l}{\partial\eta^2} = -\frac{2n}{\eta^2}, \quad \frac{\partial^2 l}{\partial\eta\partial\lambda} = 0, \quad \frac{\partial^2 l}{\partial\lambda^2} = -\frac{2n}{\pi},$$

and the observed information matrix is therefore

$$n\begin{pmatrix} 1/\eta^2 & 0 & \sqrt{2/\pi}/\eta \\ 0 & 2/\eta^2 & 0 \\ \sqrt{2/\pi}/\eta & 0 & 2/\pi \end{pmatrix}$$

which is obviously *always* singular.

Similarly, for any $S_{\phi G}(\xi, \eta, \lambda)$ class, the observed information matrix for the solution $\lambda = 0$, $\xi = \overline{x}$ and $\eta^2 = \sum_{i=1}^n (x_i - \overline{x})^2/n$ is,

$$n\begin{pmatrix} 1/\eta^2 & 0 & 2g(0)/\eta \\ 0 & 2/\eta^2 & 0 \\ 2g(0)/\eta & 0 & -2g'(0) + 4g^2(0) \end{pmatrix}$$

which is singular if $g'(0) = 0$, and undefined if g is not differentiable at the origin. This suggests that, for any $S_{\phi G}(\xi, \eta, \lambda)$ class of distributions, it would be advisable to reparametrise [see Azzalini (1985)].

References

1. Arnold, B. C., Beaver, R. J., Groeneveld, R. A., and Meeker, W. Q. (1993). The nontruncated marginal of a truncated bivariate normal distribution, *Psychometrika*, **58**, 471–488.

2. Azzalini, A. (1985). A class of distributions which includes the normal ones, *Scandinavian Journal of Statistics*, **12**, 171–178.

3. Azzalini, A. (1986). Further results on a class of distributions which includes the normal ones, *Statistica*, **46**, 199–208.

4. Azzalini, A., and Capitanio, A. (1999). Statistical applications of the multivariate skew normal distribution, *Journal of the Royal Statistical Society, Series B*, **61**, 579–602.

5. Azzalini, A., and Capitanio, A. (2003). Distributions generated by perturbation of symmetry with emphasis on a multivariate skew t-distribution, *Journal of the Royal Statistical Society, Series B*, **65**, 367–389.

6. Chiogna, M. (1997). Notes on estimation problems with scalar skew-normal distributions, Technical Report 1997.15, Department of Statistical Science, University of Padua, Italy.

7. DiCiccio, T. J., Kass, R. E., Raftery, A., and Wasserman, L. (1997). Computing Bayes factors by combining simulation and asymptotic approximations, *Journal of the American Statistical Association*, **92**, 903–915.

8. Gupta, A. K., Chang, F. C., and Huang, W. J (2002). Some skew-symmetric models, *Random Operators and Stochastic Equations*, **10**, 133–140.

9. Henze, N. (1986). A probabilistic representation of the "skew-normal" distribution, *Scandinavian Journal of Statistics*, **13**, 271–275.

10. Jones, M. C. (2002). Student's simplest distribution, *The Statistician*, **51**, 41–49.

11. Jones, M. C., and Faddy, M. J. (2003). A skew extension of the t-distribution, with applications, *Journal of the Royal Statistical Society, Series B*, **65**, 159–174.

12. Mukhopadhyay, S., and Vidakovic, B. (1995). Efficiency of linear Bayes rules for a normal mean: skewed priors class, *The Statistician*, **44**, 389–397.

13. Nadarajah, S. (2003). Skewed distributions generated by the uniform kernel, *Random Operators and Stochastic Equations*, **11**, 297–305.

14. Nadarajah, S., and Kotz, S. (2003). Skewed distributions generated by the normal kernel, *Statistics & Probability Letters*, **65**, 269–277.

15. Pewsey, A. (2000). Problems of inference for Azzalini's skew-normal distribution, *Journal of Applied Statistics*, **27**, 859–870.

16. Pewsey, A. (2002). Large-sample inference for the general half-normal distribution, *Communications in Statistics: Theory and Methods*, **31**, 1045–1054.

17. Pewsey, A. (2004). Improved likelihood based inference for the general half-normal distribution, *Communications in Statistics—Theory and Methods*, **33**, 197–204.

6

Bivariate Distributions Based on the Generalized Three-Parameter Beta Distribution

José María Sarabia and Enrique Castillo

University of Cantabria, Santander, Spain

Abstract: The generalized three-parameter beta distribution with pdf proportional to $x^{a-1}(1-x)^{b-1}/\{1-(1-\lambda)x\}^{a+b}$ is a flexible extension of the classical beta distribution with interesting applications in statistics. In this chapter, several bivariate extensions of this distribution are studied. We propose models with given marginals: a first model consists of a transformation with monotonic components of the Dirichlet distribution and a second model that uses the bivariate Sarmanov–Lee distribution. Next, the class of distributions whose conditionals belong to the generalized three-parameter beta distribution is considered. Two important subfamilies are studied in detail. The first one contains as a particular case the models of Libby and Novick (1982) and Olkin and Liu (2003). The second family is more general, and contains among others, the model proposed by Arnold, Castillo and Sarabia (1999). In addition, using two different conditional schemes, we study conditional survival models. Multivariate extensions are also discussed. Finally, an application to Bayesian analysis is given.

Keywords and phrases: Generalized three-parameter beta distribution, Gauss hypergeometric distribution, Dirichlet and Sarmanov-Lee distributions, conditionally specified models

6.1 Introduction

The purpose of this paper is to study several classes of bivariate distributions whose conditionals and/or marginals belong to the generalized three-parameter beta distribution, and to one of their extensions. There are several reasons that justify the study of these classes of distributions. Bivariate or multivariate versions of the generalized three-parameter beta distribution will clearly be useful tools for data analysts and modelers. For example, in the analysis of income data, we are interested in the study of the evolution of the proportion of expenses of a certain departure of goods (e.g., health, foods, etc.) in several periods of time. In this case, we seek multivariate distributions whose marginals and/or conditionals have at least three parameters, in order to model, mean, variance and skewness, and with unlimited correlations of any sign. Another important application arises in Bayesian statistics. The well-known Dirichlet distribution with probability density function:

$$f(x_1, \ldots, x_m) \propto x_1^{a_1-1} \cdots x_m^{a_m-1}(1 - x_1 - \cdots - x_m)^{a_0-1},$$

defined over $x_i \geq 0$, $i = 1, 2, \ldots, m$, and $\sum x_i \leq 1$, is a natural prior distribution for the parameters of a multinomial distribution; see Kotz, Balakrishnan and Johnson (2000). However, if we deal with an independent or correlated binomial distribution, we need a density defined over the m-dimensional unit cube $0 \leq x_i \leq 1$, $i = 1, 2, \ldots, m$. Recently, Olkin and Liu (2003) proposed a distribution of this kind. This distribution possesses marginal distributions of classical beta type and conditionals of the generalized three-parameter beta type. However, it is not conjugate for likelihoods that are the product of independent or correlated binomial distributions. This fact suggests multivariate distributions whose conditional distributions are of generalized three-parameter beta type. The use of conjugate prior distributions with conditional specification has been proposed by Arnold, Castillo and Sarabia (1998, 1999).

The paper is organized as follows. Section 6.2 presents a brief review of the generalized three-parameter beta distribution. Section 6.3 proposes models with given marginals. A first model consists of a transformation with monotonic components of the Dirichlet distribution and a second model uses the bivariate Sarmanov–Lee distribution. In Section 6.4 the class of distributions whose conditionals belong to the generalized three-parameter beta distribution is considered. Two important subfamilies are studied in detail. The first one contains as a particular case the models of Libby and Novick (1982) and Olkin and Liu (2003). The second family is more general, and contains among others, the model proposed by Arnold, Castillo and Sarabia (1999). Some extensions are discussed in Section 6.6. Applications to Bayesian analysis are given in Section 6.7. In Section 6.8, using two different conditional schemes, conditional survival models are studied. Finally, some multivariate extensions are also discussed.

6.2 The Generalized Three-Parameter Beta Distribution

The generalized three-parameter beta distribution has pdf

$$f(x;a,b,\lambda)=\begin{cases}\dfrac{\lambda^a x^{a-1}(1-x)^{b-1}}{B(a,b)\{1-(1-\lambda)x\}^{a+b}} & \text{if } 0\le x\le 1,\\ 0 & \text{otherwise,}\end{cases}\tag{6.1}$$

where $B(a,b)=\Gamma(a)\Gamma(b)/\Gamma(a+b)$ represents the beta function. We denote by $X\sim\mathcal{GB}(a,b,\lambda)$ the random variable with pdf (6.1). When $\lambda=1$, (6.1) reduces to the standard beta distribution. If $X\sim\mathcal{GB}(a,b,\lambda)$, then $1-X\sim\mathcal{GB}(b,a,\lambda^{-1})$, which is a property shared with the standard beta distribution. The cdf can be expressed in terms of the incomplete beta function. When $a=1$, then (6.1) becomes

$$F(x;b,\lambda)=\begin{cases}0 & \text{if } x<0\\ 1-\left\{\dfrac{1-x}{1-(1-\lambda)x}\right\}^b & \text{if } 0\le x<1,\\ 1 & \text{if } x\ge 1.\end{cases}\tag{6.2}$$

If $b=1$, (6.1) becomes

$$F(x;a,\lambda)=\begin{cases}0 & \text{if } x<0\\ \left\{\dfrac{\lambda x}{1-(1-\lambda)x}\right\}^a & \text{if } 0\le x<1,\\ 1 & \text{if } x\ge 1.\end{cases}\tag{6.3}$$

In the case $a=b=1/2$, the cdf corresponding to (6.1) is

$$F(x;\lambda)=\begin{cases}0 & \text{if } x<0\\ \dfrac{2}{\pi}\tan^{-1}\left(\sqrt{\dfrac{\lambda x}{1-x}}\right) & \text{if } 0\le x<1\\ 1 & \text{if } x\ge 1.\end{cases}$$

The generalized three-parameter beta distribution is the distribution of the ratio $X_1/(X_1+X_2)$, where $X_1\sim\mathcal{G}(a,\lambda_1)$ and $X_2\sim\mathcal{G}(b,\lambda_2)$ are independent gamma variables, and where $\lambda=\lambda_1/\lambda_2$. Alternatively, we can obtain (6.1) from a standard beta distribution; if $Z\sim\mathcal{B}(a,b)$, then,

$$\frac{Z}{\lambda+(1-\lambda)Z}\sim\mathcal{GB}(a,b,\lambda).\tag{6.4}$$

Libby and Novick (1982) studied these distributions in a multivariate setting and use them for fitting utility functions. Chen and Novick (1984) used them as priors for binomial sampling models. The kth moment of (6.1) is:

$$E(X^k) = \lambda^a \frac{B(a+k,b)}{B(a,b)} \, {}_2F_1(a+k, a+b; a+b+k; 1-\lambda);$$

here, ${}_2F_1$ represents the Gauss hypergeometric function defined by

$${}_2F_1(a,b;c;z) = \sum_{n=0}^{\infty} \frac{(a)_n (b)_n}{(c)_n} \frac{z^n}{n!}, \tag{6.5}$$

where $(a)_0 = 1$ and $(a)_n = a(a+1)\cdots(a+n-1) = \Gamma(a+n)/\Gamma(a)$, $n \geq 1$, is called the Pochammer coefficient. According to Pham-Gia and Duong (1989) and Johnson, Kotz and Balakrishnan (1995), the presence of the parameter λ allows $\mathcal{GB}$ to take a variety of shapes wider than the standard beta distribution. For example, a $\mathcal{GB}(a,a,\lambda)$ random variable can be positively or negatively skewed according to $\lambda > 1$ or $\lambda < 1$, respectively. In relation with the kurtosis coefficient, there exists a region of λ where the kurtosis is smaller than the kurtosis of the normal distribution, and for other values of λ the kurtosis is larger than the kurtosis of a normal distribution.

6.2.1 Relationships with other distributions and extensions

The generalized three-parameter beta random variable can be related with well-known probability distributions by means of simple transformations. These results will be applied in later sections. We consider a random variable $Z \sim \mathcal{GB}(a,b,\lambda)$. The monotone transformation $X = Z/(1-Z)$ leads to the random variable with pdf

$$f(x;a,b,\lambda) = \begin{cases} \dfrac{\lambda^a}{B(a,b)} \dfrac{x^{a-1}}{(1+\lambda x)^{a+b}} & \text{if } \; 0 < x < \infty, \\ 0 & \text{otherwise.} \end{cases} \tag{6.6}$$

This distribution corresponds to the Pearson type VI distribution, sometimes called second-kind beta distribution or beta-prime distribution, with scale parameter λ; Stuart and Ord (1987, Chapter 6) and Johnson, Kotz and Balakrishnan (1995, Chapter 27). A random variable with pdf (6.6) will be denoted by $X \sim \mathcal{B}2(a,b,\lambda)$. Now, if we consider the transformation $X = \log(Z) - \log(1-Z)$, we obtain the pdf

$$f(x;a,b,\lambda) = \frac{\lambda^a}{B(a,b)} \frac{e^{ax}}{(1+\lambda e^x)^{a+b}}, \tag{6.7}$$

which corresponds to the logarithm of a F distribution with location parameter $\log \lambda$. If $X_1 \sim \chi^2_{2a}$ and $X_2 \sim \chi^2_{2b}$, the random variable $\log \lambda + \log \left\{ \frac{X_1/2a}{X_2/2b} \right\}$ is distributed according to (6.7), that is, the log of a F distribution with a location parameter [Fisher (1924)]. This distribution is called type III generalized logistic distribution by Balakrishnan (1992). This distribution has been recently introduced by Jones (2004) in a different way.

A natural extension of (6.1) is the Gauss hypergeometric distribution. This distribution was considered by Armero and Bayarri (1994) in a queuing theory context, and its probability density function is given by

$$f(x; a, b, c, \lambda) = \begin{cases} n(a, b, c, \lambda) \dfrac{x^{a-1}(1-x)^{b-1}}{[1-(1-\lambda)x]^c} & \text{if } 0 \leq x \leq 1 \\ 0 & \text{otherwise,} \end{cases} \tag{6.8}$$

where the normalizing constant is

$$n(a, b, c, \lambda)^{-1} = B(a, b)\, {}_2F_1(a, c; a+b; 1-\lambda). \tag{6.9}$$

It will be denoted $X \sim \mathcal{GH}(a, b, c, \lambda)$. The Gauss hypergeometric distribution reduces to the classical beta distribution when $c = 0$ or $\lambda = 1$, and reduces to the generalized beta distribution when $c = a + b$ or when $\lambda \to 0$, with $b > c$. If $a = b = 1$, we have the cdf

$$F(x; c, \lambda) = \frac{1}{1 - \lambda^{-(c-1)}} \left[1 - \frac{1}{\{1-(1-\lambda)x\}^{c-1}} \right], \quad 0 \leq x \leq 1.$$

If $X \sim \mathcal{GH}(a, b, c, \lambda)$ we have

$$E(X^k) = \frac{B(a+k, b)}{B(a, b)} \cdot \frac{{}_2F_1(a+k, c; a+b+k; 1-\lambda)}{{}_2F_1(a, c; a+b; 1-\lambda)}.$$

Because it has an additional parameter, it possesses a better flexibility for data fitting, and it is possible to match the first four moments. On the other hand, it is a conjugate prior distribution for several likelihoods, including the binomial case, the geometric case, and the negative binomial.

6.3 Models with Generalized Three-Parameter Beta Marginals

In this section, we propose distributions whose marginal distributions are of the generalized three-parameter beta type. Because the generalized three-parameter beta is related to the classical beta distribution by the monotonic transformation (6.4), we will use models of distributions whose marginal distributions are of the classic beta type.

6.3.1 Model based on the Dirichlet distribution

We begin with a Dirichlet distribution for (Z_1, Z_2). The first model has the following stochastic representation

$$(X, Y) = \left(\frac{Z_1}{\lambda_1 + (1-\lambda_1)Z_1}, \frac{Z_2}{\lambda_2 + (1-\lambda_2)Z_2} \right), \tag{6.10}$$

$$(Z_1, Z_2) \sim \mathcal{D}ir(\theta_1, \theta_2, \theta_3), \tag{6.11}$$

where $\mathcal{D}ir(\theta_1, \theta_2, \theta_3)$ represents a bivariate Dirichlet distribution with pdf

$$f(z_1, z_2) = \frac{1}{B(\theta_1, \theta_2, \theta_3)} z_1^{\theta_1 - 1} z_2^{\theta_2 - 1} (1 - z_1 - z_2)^{\theta_3 - 1},$$

defined on the set $z_1 + z_2 \leq 1$, $z_1, z_2 \geq 0$, where $B(\theta_1, \theta_3, \theta_3) = \prod \Gamma(\theta_i)/\Gamma(\sum \theta_i)$. The properties of the model in (6.10)–(6.11) can be derived using the properties of the Dirichlet distribution. The joint probability density function is given by

$$f(x, y) = \frac{\lambda_1^{\theta_1} \lambda_2^{\theta_2}}{B(\theta_1, \theta_2, \theta_3)} \frac{x^{\theta_1 - 1} y^{\theta_2 - 1} [1 - x - y + (1 - \lambda_1 \lambda_2) xy]^{\theta_3 - 1}}{[1 - (1 - \lambda_1)x]^{\theta_1 + \theta_3} [1 - (1 - \lambda_2)y]^{\theta_2 + \theta_3}} \tag{6.12}$$

with support

$$0 \leq x,\ y \leq 1;\ \ x + y - (1 - \lambda_1 \lambda_2) xy \leq 1.$$

The marginal distributions of (6.10) are

$$X \sim \mathcal{GB}(\theta_1, \theta_2 + \theta_3, \lambda_1),$$
$$Y \sim \mathcal{GB}(\theta_2, \theta_1 + \theta_3, \lambda_2),$$

and the conditional distributions are not standard, and are given by ($\tilde{\lambda}_i = 1 - \lambda_i$):

$$f(x|y) = \frac{1 - \tilde{\lambda}_2 y}{1 - y} f_{\mathcal{B}(\theta_1, \theta_3)} \left(\frac{1 - \tilde{\lambda}_2 y}{1 - y} \cdot \frac{\lambda_1 x}{1 - \tilde{\lambda}_1 x} \right) \frac{\lambda_1}{(1 - \tilde{\lambda}_1 x)^2},$$

$$f(y|x) = \frac{1 - \tilde{\lambda}_1 x}{1 - x} f_{\mathcal{B}(\theta_2, \theta_3)} \left(\frac{1 - \tilde{\lambda}_1 x}{1 - x} \cdot \frac{\lambda_2 y}{1 - \tilde{\lambda}_2 y} \right) \frac{\lambda_2}{(1 - \tilde{\lambda}_2 y)^2},$$

which have been written in this way for the sake of easy comparison with the Dirichlet case. If $\lambda_i = 1$, $i = 1, 2$, $f(x|y)$ and $f(y|x)$ are scale beta distributions. Note that, because the Dirichlet distribution has negative correlation and because the marginal transformations in (6.10) are both monotone, the correlations in the new model are also negative. Figure 6.1 shows the joint pdf, the contour plot and the marginal distributions with positive skewness. The graph shows a negative correlation coefficient.

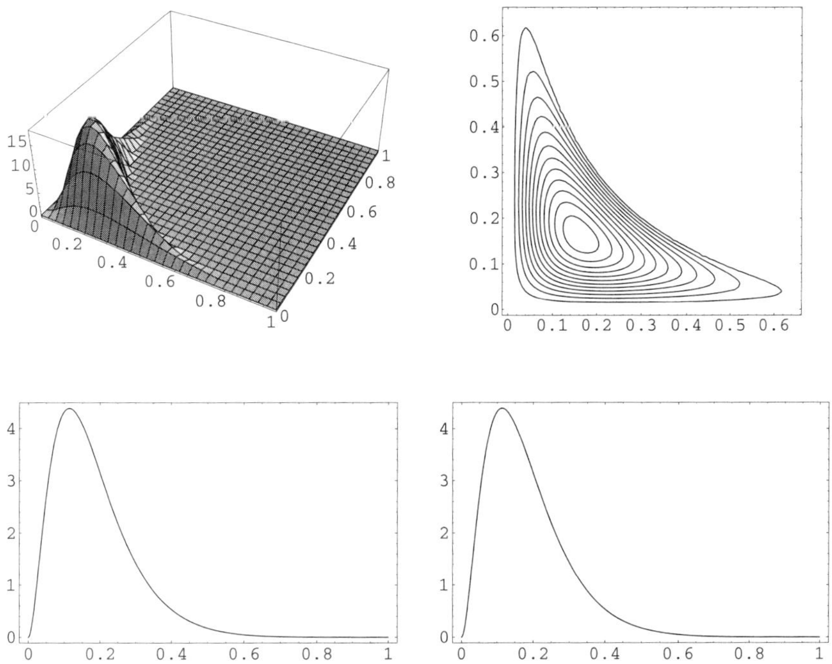

Figure 6.1: Bivariate Dirichlet-GBeta distribution with pdf (6.12) and parameters $\theta_1 = \theta_2 = 3$, $\theta_3 = 2$, $\lambda_1 = \lambda_2 = 3$, and marginals with positive skewness

6.3.2 Model based on the Sarmanov-Lee distribution

Let $f_1(x)$ and $f_2(y)$ be univariate pdf with supports A_i, and let $\phi_i(z)$ be bounded nonconstant functions such that

$$\int_{\mathbb{R}} \phi_i(z) f_i(z) dz = 0, \quad i - 1, 2.$$

Sarmanov (1966) defined the following bivariate pdf with given marginals $f_1(x)$ and $f_2(y)$

$$f(x,y) = f_1(x) f_2(y) \{1 + w\phi_1(x)\phi_2(y)\}, \tag{6.13}$$

where w is a real number such that $1 + w\phi_1(x)\phi_2(y) \geq 0$, $\forall (x, y)$. Lee (1996) studied some properties of this family and proposed a multivariate version. In our case, f_1 and f_2 are of the generalized three-parameter beta distribution type. In order to specify formula (6.13), we need to determine the mixing functions $\phi_i(x)$ for this type of marginals, and to know the constraints to be satisfied by w. In this situation, because of $A_i \subset [0, 1]$, it is possible to use

Corollary 1 of Lee (1996). Consequently, we propose the bivariate distribution:

$$f(x,y;\underline{a},\underline{b},\underline{\lambda},w)=f_1(x;a_1,b_1,\lambda_1)f_2(y;a_2,b_2,\lambda_2)\{1+w\phi_1(x)\phi_2(y)\}, \quad (6.14)$$

where

$$\begin{aligned} f_i(z;a_i,b_i,\lambda_i) &\sim \mathcal{GB}(a_i,b_i,\lambda_i),\ \ i=1,2,\\ \phi_i(z) &= z-\mu_i,\ \ i=1,2, \end{aligned}$$

$$\max\left\{\frac{-1}{\mu_1\mu_2},\frac{-1}{(1-\mu_1)(1-\mu_2)}\right\}\le w\le \min\left\{\frac{1}{\mu_1(1-\mu_2)},\frac{1}{(1-\mu_1)\mu_2}\right\},$$

and μ_1 and μ_2 represent the mathematical expectations of X and Y, respectively. Several properties of this model have been studied by Lee (1996). For example, the regression of Y on X is linear and is given by

$$E(Y|X=x)=\mu_2+w\nu_2(x-\mu_1),$$

where $\nu_2=E[Y\phi_2(Y)]$. The model presents a range of correlation wider than the Farlie-Gumbel-Morgenstern model with given marginals. A property of the proposed model is that it can be expressed as a linear combination of products of univariate $\mathcal{GB}$ and weighted $\mathcal{GB}$ as follows:

$$\begin{aligned} f(x,y;\underline{a},\underline{b},\underline{\lambda},w) &= (1+w\mu_1\mu_2)f_1(x)f_2(y)+w\mu_1\mu_2 f_1^w(x)f_2^w(y)\\ &\quad -w\mu_1\mu_2 f_1(x)f_2^w(y)-w\mu_1\mu_2 f_1^w(x)f_2(y), \end{aligned}$$

where $f_i^w(z)=zf_i(z)/\mu_i$, $i=1,2$, represent the weighted version of the $\mathcal{GB}$ distribution. With some changes, this model can be adapted to obtain a two-dimensional distribution with marginals of the type (6.8).

6.4 The Generalized Three-Parameter Beta Conditionals Distribution

Let (X,Y) be a two-dimensional random variable with support on the unit square. We want to consider all possible joint distributions for (X,Y) with the following properties:

(a) For each $y\in(0,1)$, the conditional distribution of X given $Y=y$ is a generalized three-parameter beta distribution with parameters $a_1(y)$, $b_1(y)$ and $\lambda_1(y)$, which may depend on y.

(b) For each $x\in(0,1)$, the conditional distribution of Y given $X=x$ is a generalized three-parameter beta distribution with parameters $a_2(x)$, $b_2(x)$ and $\lambda_2(x)$, which may depend on x.

Thus, we seek the most general random variable (X, Y) such that the conditional distributions satisfy

$$\begin{aligned} X|Y=y &\sim \mathcal{GB}(a_1(y), b_1(y), \lambda_1(y)), & (6.15)\\ Y|X=x &\sim \mathcal{GB}(a_2(x), b_2(x), \lambda_2(x)), & (6.16)\end{aligned}$$

where $a_i(x) : [0,1] \to \mathbb{R}^+$, $b_i(x) : [0,1] \to \mathbb{R}^+$ and $\lambda_i(x) : \mathbb{R}^+ \to \mathbb{R}$ are unknown functions. Now, writing the density as product of marginals and conditionals, we obtain the functional equation

$$\frac{u_1(y)x^{a_1(y)-1}(1-x)^{b_1(y)-1}}{\{1-\tilde{\lambda}_1(y)x\}^{a_1(y)+b_1(y)}} = \frac{u_2(x)y^{a_2(x)-1}(1-y)^{b_2(x)-1}}{\{1-\tilde{\lambda}_2(x)y\}^{a_2(x)+b_2(x)}}, \tag{6.17}$$

where

$$\begin{aligned} u_1(y) &= \frac{\lambda_1(y)f_Y(y)}{B(a_1(y), b_1(y))},\\ u_2(x) &= \frac{\lambda_2(x)f_X(x)}{B(a_2(x), b_2(x))},\\ \tilde{\lambda}_i(z) &= 1-\lambda_i(z),\ \ i=1,2,\end{aligned}$$

and $f_X(x)$, $f_Y(y)$ represent the marginal densities. The solution of the functional equation (6.17) is not trivial. In this paper, we consider two important particular cases. The first case corresponds to constants and known $\lambda_i(u) = \lambda_i$ for $i = 1, 2$. In this case, the generalized three-parameter beta distribution belongs to the two-parameter exponential family and so we can use some well known results. The second case corresponds to the choice $a_i(u) = a_i$ and $b_i(u) = b_i$, $\forall u \in (0,1)$, $i = 1, 2$. In this case, the generalized beta distribution does not belong to the exponential family, but (6.17) becomes a Stephanos-Levi-Civita-Suto functional equation type, that can be easily solved. In the following sections, we will study these two cases.

6.4.1 The Generalized Beta conditionals distribution with $\lambda_i(\cdot)$ constant

If λ is known and if we write (6.1) in the form

$$f(x; a, b) \propto x^{-1}(1-x)^{-1}\exp\left[a\log\{x/(1-\tilde{\lambda}x)\} + b\log\{(1-x)/(1-\tilde{\lambda}x)\}\right],$$

we have a two-parameter exponential family, and we can make use of a theorem due to Arnold and Strauss (1991), dealing with bivariate distributions with conditionals in prescribed exponential families. Then, we consider two different exponential families of densities $\{f_1(x;\underline{\theta}) : \underline{\theta} \in \Theta \subset \mathbb{R}^{\ell_1}\}$ and $\{f_2(y;\underline{\tau}) : \underline{\tau} \in T \subset \mathbb{R}^{\ell_2}\}$, where

$$f_1(x;\underline{\theta}) = r_1(x)\beta_2(\underline{\theta}) \exp\left\{\sum_{i=1}^{\ell_1} \theta_i q_{1i}(x)\right\} \tag{6.18}$$

and

$$f_2(y;\underline{\tau}) = r_2(y)\beta_2(\underline{\tau}) \exp\left\{\sum_{j=1}^{\ell_2} \tau_j q_{2j}(y)\right\}. \tag{6.19}$$

The class of all bivariate pdf $f(x, y)$ with conditionals in these prescribed exponential families can be obtained as follows.

Theorem 6.4.1 *Let $f(x, y)$ be a bivariate density whose conditional densities satisfy*

$$f(x|y) = f_1(x;\underline{\theta}(y))$$

and

$$f(y|x) = f_2(y;\underline{\tau}(x))$$

for every x and y for some functions $\underline{\theta}(y)$ and $\underline{\tau}(x)$, where f_1 and f_2 are as defined in (6.18) and (6.19). It follows that $f(x, y)$ is of the form

$$f(x, y) = r_1(x)r_2(y) \exp\left\{\underline{q}^{(1)}(x) M \underline{q}^{(2)}(y)^T\right\} \tag{6.20}$$

in which

$$\underline{q}^{(1)}(x) = (1, q_{11}(x), \ldots, q_{1\ell_1}(x))$$

and

$$\underline{q}^{(2)}(y) = (1, q_{21}(y), \ldots, q_{2\ell_2}(y))$$

and M is a matrix of parameters of dimension $(\ell_1 + 1) \times (\ell_2 + 1)$ subject to the requirement that

$$\int\int_{\mathbb{R}^2} f(x, y)dxdy = 1. \tag{6.21}$$

The term $e^{m_{00}}$ is the normalizing constant that is a function of the other m_{ij}'s determined by the constraint (6.21).

Note that the class of densities with conditionals in the prescribed family is itself an exponential family with $(\ell_1 + 1) \times (\ell_2 + 1) - 1$ parameters. Upon partitioning the matrix M in (6.20) in the following manner

$$M = \begin{pmatrix} m_{00} & | & m_{01} & \cdots & m_{0\ell_2} \\ -- & + & -- & -- & -- \\ m_{10} & | & & & \\ \vdots & | & & \tilde{M} & \\ m_{\ell_1 0} & | & & & \end{pmatrix}, \tag{6.22}$$

it can be verified that independent marginals will be encountered iff the matrix $\tilde{M} \equiv 0$. The elements of $\tilde{M}$ determine the dependence structure in $f(x,y)$.

Now, we may apply Theorem 6.4.1 to the case of the generalized three-parameter beta distribution, where (6.18) and (6.19) are of the form (6.1). In this case, we have $\ell_1 = \ell_2 = 2$ and the functions $r_1, r_2, q_{11}, q_{12}, q_{21}$, and q_{22} are of the form ($\tilde{\lambda}_i = 1 - \lambda_i$, $i = 1, 2$):

$$\begin{aligned}
r_1(x) &= \{x(1-x)\}^{-1} I(0 < x < 1), \\
r_2(y) &= \{y(1-y)\}^{-1} I(0 < y < 1), \\
q_{11}(x) &= \log\{x/(1-\tilde{\lambda}_1 x)\}, \\
q_{12}(x) &= \log\{(1-x)/(1-\tilde{\lambda}_1 x)\}, \\
q_{21}(y) &= \log\{y/(1-\tilde{\lambda}_2 y)\}, \\
q_{22}(y) &= \log\{(1-y)/(1-\tilde{\lambda}_2 y)\}.
\end{aligned}$$

Finally, substituting these functions in the general expression (6.20), we obtain the class of bivariate densities with generalized three-parameter beta conditionals (assuming constant λ_i), which is given by

$$f(x,y) \propto f_X(x) f_Y(y) \exp\{u(x,y)\}, \tag{6.23}$$

where

$$\begin{aligned}
X &\sim \mathcal{GB}(m_{10}, m_{20}, \lambda_1), \\
Y &\sim \mathcal{GB}(m_{10}, m_{20}, \lambda_1),
\end{aligned}$$

and

$$u(x,y) = m_{11} q_{11}(x) q_{21}(y) + m_{12} q_{11}(x) q_{22}(y) + m_{21} q_{12}(x) q_{21}(y) + m_{22} q_{12}(x) q_{22}(y).$$

The parameters m_{11}, m_{12}, m_{21}, and m_{22} are the dependence parameters. In order for it to be a proper density (integrate to 1), we need to impose the following restrictions to their parameters:

$$m_{10}, m_{20}, m_{01}, m_{02} > 0, \tag{6.24}$$

$$m_{11}, m_{12}, m_{21}, m_{22} \leq 0, \tag{6.25}$$

$$\lambda_1, \lambda_2 > 0. \tag{6.26}$$

If we denote

$$\begin{aligned}
a_1(y) &= m_{10} + m_{11} \log\{y/(1-\tilde{\lambda}_2 y)\} + m_{12} \log\{(1-y)/(1-\tilde{\lambda}_2 y)\}, \\
b_1(y) &= m_{20} + m_{21} \log\{y/(1-\tilde{\lambda}_2 y)\} + m_{22} \log\{(1-y)/(1-\tilde{\lambda}_2 y)\}, \\
a_2(x) &= m_{01} + m_{11} \log\{x/(1-\tilde{\lambda}_1 x)\} + m_{21} \log\{(1-x)/(1-\tilde{\lambda}_1 x)\}, \\
b_2(x) &= m_{02} + m_{12} \log\{x/(1-\tilde{\lambda}_1 x)\} + m_{22} \log\{(1-x)/(1-\tilde{\lambda}_1 x)\},
\end{aligned}$$

we have

$$
\begin{aligned}
X|Y=y &\sim \mathcal{GB}(a_1(y), b_1(y), \lambda_1), \\
Y|X=x &\sim \mathcal{GB}(a_2(x), b_2(x), \lambda_2),
\end{aligned}
$$

and the marginal densities of X and Y are given by

$$
\begin{aligned}
f_X(x) &= \exp(m_{00}) \times \frac{x^{m_{10}-1}(1-x)^{m_{20}-1}}{[1-(1-\lambda_1)x]^{m_{10}+m_{20}}} \times \frac{B(a_2(x), b_2(x))}{\lambda_2^{a_2(x)}}, \\
f_Y(y) &= \exp(m_{00}) \times \frac{y^{m_{01}-1}(1-y)^{m_{02}-1}}{[1-(1-\lambda_2)y]^{m_{01}+m_{02}}} \times \frac{B(a_1(y), b_1(y))}{\lambda_1^{a_1(y)}}.
\end{aligned}
$$

Note that the marginal distributions are not generalized three-parameter beta distributions, except in the independence case. If we define two auxiliary random variables

$$Z_1 \sim \mathcal{GB}(m_{10}, m_{20}, \lambda_1)$$

and

$$Z_2 \sim \mathcal{GB}(m_{01}, m_{02}, \lambda_2),$$

we can then write the normalizing constant of two alternative forms:

$$
\begin{aligned}
\exp(m_{00}) &= \{B(m_{10}, m_{20}) E[B(a_2(Z_1), b_2(Z_1))/\lambda_2^{a_2(Z_1)}\}/\lambda_1^{m_{10}}]^{-1}, \\
&= \{B(m_{01}, m_{01}) E[B(a_1(Z_2), b_1(Z_2))/\lambda_1^{a_1(Z_2)}\}/\lambda_2^{m_{02}}]^{-1}.
\end{aligned}
$$

The moments of X and Y can be written in terms of expectations of the random variables Z_1 and Z_2. Then, for $n = 1, 2, \ldots$, we have

$$
\begin{aligned}
E(X^n) &= \frac{E[Z_1^n B(a_2(Z_1), b_2(Z_1))/\lambda_2^{a_2(Z_1)}]}{E[B(a_2(Z_1), b_2(Z_1))/\lambda_2^{a_2(Z_1)}]}, \\
E(Y^n) &= \frac{E[Z_2^n B(a_1(Z_2), b_1(Z_2))/\lambda_1^{a_1(Z_2)}]}{E[B(a_1(Z_2), b_1(Z_2))/\lambda_1^{a_1(Z_2)}]}.
\end{aligned}
$$

The modes of (6.23) are given by the solution in (x, y) to the system:

$$
\begin{aligned}
& f'_X(x) + f_X(x)\{q_{21}(y)[m_{11}q'_{11}(x) + m_{21}q'_{12}(x)] \\
& \qquad + q_{22}(y)[m_{12}q'_{11}(x) + m_{22}q'_{12}(x)]\} = 0, \\
& f'_Y(y) + f_Y(y)\{q_{11}(x)[m_{11}q'_{21}(y) + m_{12}q'_{22}(y)] \\
& \qquad + q_{12}(x)[m_{21}q'_{21}(y) + m_{22}q'_{22}(y)]\} = 0.
\end{aligned}
$$

It seems that one, two and four modes are possible. Figure 6.2 shows a joint pdf with two modes. Multimodality appears in other models with conditional specification; see Arnold, Castillo and Sarabia (2000, 2001). Some simplified submodels can be obtained invoking symmetry, and/or exchangeability assumptions.

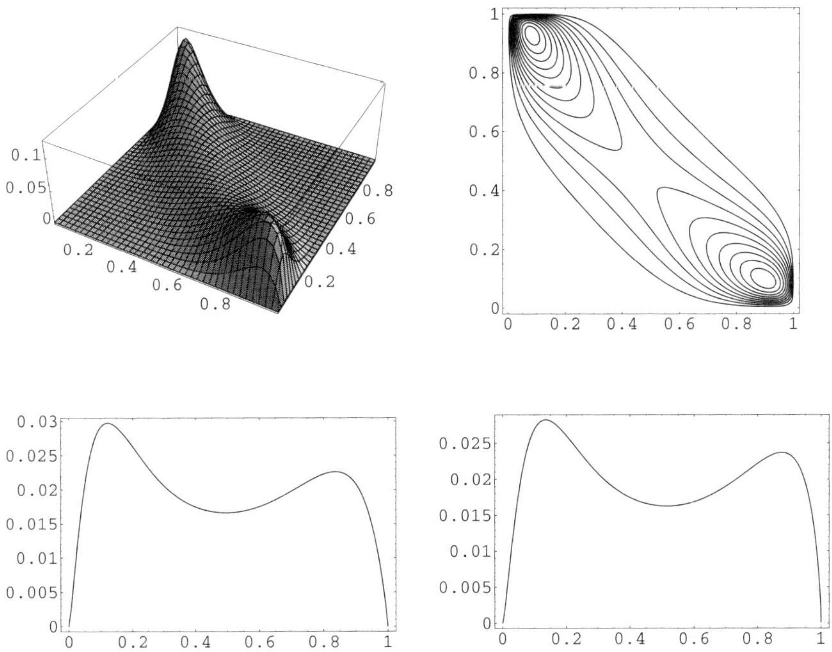

Figure 6.2: Bimodal distribution with generalized three-parameter beta conditionals

6.4.2 The Generalized Beta conditionals distribution with constant $a_i(\cdot)$ and $b_i(\cdot)$

Assume now that a and b are fixed and known parameters, and that λ is unknown. In this case, (6.1) does not belong to the exponential family. Then we seek the more general bivariate random variable, such that their conditional distributions are of the type

$$\begin{aligned} X|Y=y &\sim \mathcal{GB}(a_1, b_1, \lambda_1(y)), && (6.27)\\ Y|X=x &\sim \mathcal{GB}(a_2, b_2, \lambda_2(x)). && (6.28)\end{aligned}$$

Then, the functional equation (6.17) becomes

$$\frac{\lambda_1(y)^{a_1} x^{a_1-1}(1-x)^{b_1-1} f_Y(y)}{B(a_1,b_1)\{1-\tilde{\lambda}_1(y)x\}^{a_1+b_1}} = \frac{\lambda_2(x)^{a_2} y^{a_2-1}(1-y)^{b_2-1} f_X(x)}{B(a_2,b_2)\{1-\tilde{\lambda}_2(x)y\}^{a_2+b_2}}. \quad (6.29)$$

Denoting

$$u_1(x) = \frac{x^{a_1-1}(1-x)^{b_1-1}}{B(a_1,b_1)\lambda_2(x)^{a_2} f_X(x)}, \quad (6.30)$$

$$u_2(y) = \frac{y^{a_2-1}(1-y)^{b_2-1}}{B(a_2,b_2)\lambda_1(y)^{a_1} f_Y(y)}, \tag{6.31}$$

we obtain the functional equation

$$\frac{u_1(x)}{\{1-\tilde{\lambda}_1(y)x\}^{a_1+b_1}} = \frac{u_2(y)}{\{1-\tilde{\lambda}_2(x)y\}^{a_2+b_2}}, \tag{6.32}$$

which is solved in the following lemma.

Lemma 6.4.1 *Under constraint $a_1+b_1=a_2+b_2$, the solutions of equation (6.32) are:*

$$u_1(x) = (m_{11}-m_{12}x)^{a_1+b_1}, \tag{6.33}$$
$$\tilde{\lambda}_1(y) = \frac{m_{12}+m_{22}y}{m_{11}-m_{21}y}, \tag{6.34}$$
$$u_2(y) = (m_{11}-m_{21}y)^{a_2+b_2}, \tag{6.35}$$
$$\tilde{\lambda}_2(x) = \frac{m_{21}+m_{22}x}{m_{11}-m_{12}x}, \tag{6.36}$$

where m_{ij} are constants.

PROOF. Raising to the power $1/(a_1+b_1) = 1/(a_2+b_2)$ both sides of the equation and denoting $v_i(x) = u_i(x)^{1/(a_i+b_i)}$, we obtain the functional equation

$$v_1(x) - v_1(x)\tilde{\lambda}_2(x)y - v_2(y) + v_2(y)\tilde{\lambda}_1(y)x = 0,$$

which is a functional equation of the form

$$\sum_{i=1}^{k} f_i(x)g_i(y) = 0,$$

which is a functional equation of the type Stephanos-Levi-Civita-Suto. The solution of this equation appears in Theorem 1.3 on page 13 in Arnold, Castillo and Sarabia (1999). ■

The joint and the marginal pdfs are obtained from (6.29)–(6.31) and (6.33)–(6.36), and are given by

$$f(x,y) \propto \frac{x^{a_1-1}(1-x)^{b_1-1}y^{a_2-1}(1-y)^{a_1+b_1-a_2-1}}{(m_{11}-m_{12}x-m_{21}y-m_{22}xy)^{a_1+b_1}},$$

$$f_X(x) \propto \frac{x^{a_1-1}(1-x)^{b_1-1}}{(m_{11}-m_{12}x)^{a_1+b_1-a_2}\{m_{11}-m_{21}-(m_{12}+m_{22})x\}^{a_2}},$$

$$f_Y(y) \propto \frac{y^{a_2-1}(1-y)^{a_1+b_1-a_2-1}}{(m_{11}-m_{21}y)^{b_1}\{m_{11}-m_{12}-(m_{21}+m_{22})y\}^{a_1}}.$$

Their properties are studied in the following section.

The basic model

Without lost of generality, we assume $m_{11} = 1$. Then, we work with the joint pdf $(0 < x, y < 1)$

$$f(x,y) \propto \frac{x^{a_1-1}(1-x)^{b_1-1}y^{a_2-1}(1-y)^{a_1+b_1-a_2-1}}{(1-m_{12}x-m_{21}y-m_{22}xy)^{a_1+b_1}} \tag{6.37}$$

whose conditional distributions are (6.27) and (6.28), with

$$\lambda_1(y) = \frac{1-m_{12}-(m_{21}+m_{22})y}{1-m_{21}y},$$
$$\lambda_2(x) = \frac{1-m_{21}-(m_{12}+m_{22})x}{1-m_{12}x},$$

and the marginal pdfs are:

$$f_X(x) \propto \frac{x^{a_1-1}(1-x)^{b_1-1}}{(1-m_{12}x)^{a_1+b_1-a_2}\{1-m_{21}-(m_{12}+m_{22})x\}^{a_2}},$$
$$f_Y(y) \propto \frac{y^{a_2-1}(1-y)^{a_1+b_1-a_2-1}}{(1-m_{21}y)^{b_1}\{1-m_{12}-(m_{21}+m_{22})y\}^{a_1}}.$$

In general, the marginals are not generalized beta distributions. The parameter constraints for it to be a genuine joint density are

$$a_1, a_2, b_1 > 0, \quad a_1+b_1-a_2 > 0,$$
$$m_{12}, m_{21} \le 1, \quad m_{12}+m_{21}+m_{22} \le 1.$$

The new model (6.37) includes the following important particular cases.

- **The independence case**:

$$m_{12}m_{21}+m_{22} = 0.$$

  In this case, the marginals are generalized beta distributions.

- **The Libby and Novick (1982) model**, which corresponds to the choice

$$m_{12}+m_{21}+m_{22} = 1.$$

  This model presents generalized three-parameter beta marginals and conditionals.

- **The Olkin and Liu (2003) model**, which corresponds to the choice

$$m_{12} = m_{21} = 0, \quad m_{22} = 1.$$

  This model presents classical beta marginals and generalized three-parameter beta conditionals.

- **Gauss Hypergeometric Model**. This model corresponds to the choice

$$m_{12} = m_{21} = 0$$

and contains the Olkin and Liu (2003) model that will be studied in the next section.

A Gauss hypergeometric marginals model

If we choose $m_{12} = m_{12} = 0$, we obtain a model that depends on four parameters, and its joint pdf is

$$f(x, y; a_1, a_2, b_1, m) = n(a_1, a_2, b_1, m) \frac{x^{a_1-1}(1-x)^{b_1-1} y^{a_2-1}(1-y)^{a_1+b_1-a_2-1}}{(1-mxy)^{a_1+b_1}}, \tag{6.38}$$

where the normalizing constant is given by

$$n(a_1, a_2, b_1, m)^{-1} = B(a_1, b_1) B(a_2, a_1 + b_1, a_2) \; {}_2F_1(a_1, a_2; a_1 + b_1; m). \tag{6.39}$$

For (6.38) to be a genuine probability density function, it is necessary that

$$a_1, b_1, a_1 + b_1 - a_2 > 0, \quad m \leq 1.$$

This model contains as a particular case the Olkin and Liu (2003) proposal, for $m = 1$. This model satisfies sign$\rho(X, Y) = \text{sign}(m)$. Consequently, if $0 < m \leq 1$ we have positive correlation and if $m < 0$, negative correlation. The marginal distributions are of the Gauss hypergeometric type and are given by

$$\begin{aligned} f_X(x) &= B(a_2, a_1 + b_1 - a_2) n(a_1, a_2, b_1, m) \frac{x^{a_1-1}(1-x)^{b_1-1}}{(1-mx)^{a_2}}, \\ f_Y(y) &= B(a_1, b_1) n(a_1, a_2, b_1, m) \frac{y^{a_2-1}(1-y)^{a_1+b_1-a_2-1}}{(1-my)^{a_1}}. \end{aligned}$$

Figure 6.3 shows the bivariate pdf and contour plots corresponding to the model with Gauss hypergeometric marginals with parameters $a_1 = a_2 =$, $b_1 = 4$, and $m = -3$ and $m = 1/20$.

6.4.3 Dependence conditions

In this section, we study some dependence conditions corresponding to the conditional models. A distribution is said to be positively ratio likelihood dependent (or positive quadrant dependence) if the density $f(x, y)$ satisfies the condition

$$f(x_1, y_1) f(x_2, y_2) \geq f(x_1, y_2) f(x_2, y_1) \tag{6.40}$$

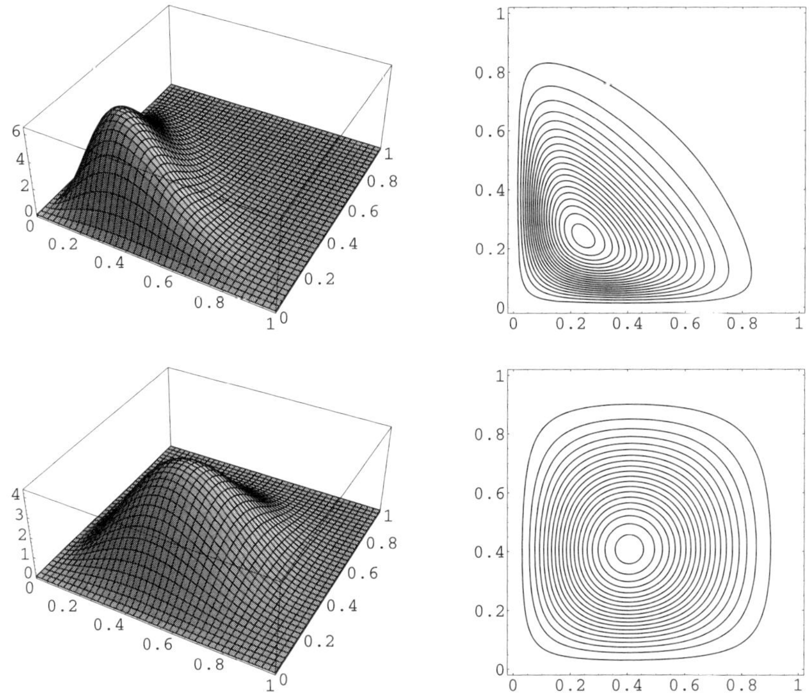

Figure 6.3: Bivariate pdf and contour plots corresponding to a model with Gauss hypergeometric marginals with parameters $a_1 = a_2 =$, $b_1 = 4$, $m = -3$ (upper) and $m = 1/20$ (lower)

for every $x_1 < x_2$, $y_1 < y_2$ in $S(X)$ and $S(Y)$, respectively; see Barlow and Proschan's (1981, Theorem 5.4.2). By substituting the general pdf (6.20) in (6.40), we obtain the condition

$$[\underline{\tilde{q}}^{(1)}(x_1) - \underline{\tilde{q}}^{(1)}(x_2)]' \tilde{M} [\underline{\tilde{q}}^{(2)}(y_1) - \underline{\tilde{q}}^{(2)}(y_2)] \geq 0. \tag{6.41}$$

In the case of model (6.23), it is not possible to obtain a general condition about the parameters m_{ij} for (6.41) to hold. In general, it is quite possible to encounter both positive and negative correlations for this model. With respect to model (6.37), we can obtain more explicit results. Considering the function $g(y) = \Pr(X > x|Y = y)$, it can be proved that the sign of its first derivative depends on the sign of $m_{12}m_{21} + m_{22}$. Thus, according to Barlow and Proschan (1981), we conclude that X is stochastically increasing or decreasing with Y. So, for values of m_{ij} such that the correlation coefficient exists, we have

$$\text{sign } \rho(X, Y) = \text{sign}(m_{12}m_{21} + m_{22}).$$

Scalar measures of dependence such as the correlation coefficient, do not always tell everything of the dependence properties of a bivariate distribution. The local dependence function [see, e.g., Holland and Wang (1987) and Jones (1996)] defined by

$$\gamma(x,y) = \frac{\partial^2 \log f(x,y)}{\partial x \partial y} \tag{6.42}$$

gives more detailed information. For the joint pdf (6.23), the local dependence function is

$$\gamma(x,y) = \frac{a_0 - a_1 x - a_2 y + a_3 xy}{x(1-x)y(1-y)\{1-(1-\lambda_1)x\}\{1-(1-\lambda_2)y\}},$$

where

$$\begin{aligned} a_0 &= m_{11}, \\ a_1 &= m_{11} + \lambda_1 m_{21}, \\ a_2 &= m_{11} + \lambda_2 m_{12}, \\ a_3 &= m_{11} + \lambda_1 m_{21} + \lambda_2 m_{12} + \lambda_1 \lambda_2 m_{22}. \end{aligned}$$

Similarly, the local dependence function of the model (6.37) is given by

$$\gamma(x,y) = \frac{(a_1+b_1)(m_{12}m_{21}+m_{22})}{(1-m_{12}x-m_{21}y-m_{22}xy)^2}.$$

Note that the local dependence function has the same sign as the correlation coefficient.

6.5 Bivariate Distributions with Gauss Hypergeometric Conditionals

In this section, we obtain some interesting classes of bivariate distributions with Gauss hypergeometric conditionals of kind (6.8). We seek the most general bivariate density of (X,Y) such that the associated conditionals satisfy

$$\begin{aligned} X|Y=y &\sim \mathcal{GH}(a_1, b_1, c, \lambda_1(y)), \\ Y|X=x &\sim \mathcal{GH}(a_2, b_2, c, \lambda_2(x)), \end{aligned}$$

where $\lambda_i(z)$, $i = 1,2$, are unknown functions and now the parameters a_i, b_i, $i=1,2$ and c are fixed and known. Defining

$$u_1(x) = \frac{x^{a_1-1}(1-x)^{b_1-1}}{n(a_2,b_2,c,\lambda_2(x))f_X(x)}, \tag{6.43}$$

$$u_2(y) = \frac{y^{a_2-1}(1-y)^{b_2-1}}{n(a_1,b_1,c,\lambda_1(y))f_Y(y)}, \tag{6.44}$$

we have the functional equation

$$\frac{u_1(x)^{1/c}}{1-\{1-\lambda_1(y)\}x}=\frac{u_2(y)^{1/c}}{1-\{1-\lambda_2(x)\}y}$$

whose solution is given in Lemma 6.4.1. Then, the joint pdf becomes

$$\begin{aligned} f(x,y) &= n(a_1,b_1,c,\lambda_1(y))\frac{x^{a_1-1}(1-x)^{b_1-1}}{\{1-\tilde{\lambda}_1(y)\}^c}f_Y(y) \\ &= \frac{x^{a_1-1}(1-x)^{b_1-1}y^{a_2-1}(1-y)^{b_2-1}}{\{1-\tilde{\lambda}_1(y)\}^c u_2(y)} \\ &= \frac{x^{a_1-1}(1-x)^{b_1-1}y^{a_2-1}(1-y)^{b_2-1}}{(m_{11}-m_{12}x-m_{21}y-m_{22}xy)^c}\,. \end{aligned} \tag{6.45}$$

Using (6.43), we get the marginal distributions as

$$\begin{aligned} f_X(x) &= \frac{x^{a_1-1}(1-x)^{b_1-1}}{n(a_2,b_2,c,\lambda_2(x))u_1(x)} \\ &\propto {}_2F_1(a_2,c;a_2+b_2;1-\lambda_1(x))\frac{x^{a_1-1}(1-x)^{b_1-1}}{(m_{11}-m_{12}x)^c} \end{aligned}$$

and

$$f_Y(y)\propto {}_2F_1(a_1,c;a_1+b_1;1-\lambda_2(y))\frac{y^{a_2-1}(1-y)^{b_2-1}}{(m_{11}-m_{21}y)^c}.$$

For this family, the local dependence function (6.42) is

$$\gamma(x,y)=\frac{c(m_{12}m_{21}+m_{22})}{(1-m_{12}x-m_{21}y-m_2xy)^2}$$

which shows that the sign of the correlation coefficient is determined by the sign of $c(m_{12}m_{21}+m_{22})$.

6.5.1 A flexible model

A simple and flexible model with six parameters is

$$f(x,y)=n(a,b,c,m)\frac{x^{a_1-1}(1-x)^{b_1-1}y^{a_2-1}(1-y)^{b_2-1}}{(1-mxy)^c}, \tag{6.46}$$

which has been obtained by letting $m_{11}=1$ and $m_{12}=m_{21}=0$ in (6.45). The normalizing constant is given by

$$n(a,b,c,m)^{-1}=B(a_1,b_1)B(a_2,b_2)\,{}_3F_2(\{a_1,a_2,c\};\{a_1+b_1,a_2+b_2,c\};m),$$

where ${}_pF_q(\underline{a};\underline{b};z)$ denotes the generalized hypergeometric function. The marginal distributions are

$$\begin{aligned} f_X(x) &= B(a_2,b_2)n(a,b,c,m)x^{a_1-1}(1-x)^{b_1-1}{}_2F_1(a_2,c;a_2+b_2;mx), \\ f_Y(y) &= B(a_1,b_1)n(a,b,c,m)y^{a_2-1}(1-y)^{b_2-1}{}_2F_1(a_1,c;a_1+b_1;my). \end{aligned}$$

This model admits positive correlations for $0 < cm \leq 1$ and negative correlations for $cm < 0$. Several moments can be obtained from

$$E[X^{r_1}(1-X)^{s_1}Y^{r_2}(1-Y)^{s_2}] = \frac{n(a_1+r_1,a_2+r_2,b_1+s_1,b_2+s_2,c,m)}{n(a_1,a_2,b_1,b_2,c,m)}, \tag{6.47}$$

and its local dependence function is

$$\gamma(x,y) = \frac{cm}{(1-mxy)^2}.$$

6.6 Other Bivariate Distributions with Specified Conditionals

By means of Jacobians, we can obtain new families of two-dimensional distributions whose conditional distributions are of certain types. Consider a bivariate distribution with joint pdf $f_{Z_1,Z_2}(z_1,z_2)$, with conditionals of the generalized three-parameter beta type. Then, the bivariate random variable (X_1,X_2) with joint pdf

$$f_{X_1,X_2}(x_1,x_2) = f_{Z_1,Z_2}\left(\frac{x_1}{1+x_1},\frac{x_2}{1+x_2}\right)\frac{1}{(1+x_1)^2(1+x_2)^2}$$

has conditional distributions of the Pearson type VI, as in (6.6). For example, if we begin with the bivariate distribution (6.38), we obtain the distribution with Pearson type VI conditionals

$$f_{X_1,X_2}(x_1,x_2) = n(a_1,a_2,b_1,m)\frac{x_1^{a_1-1}x_2^{a_2-1}}{\{1+x_1+x_2+(1-m)x_1x_2\}^{a_1+b_1}}.$$

This model was considered by Castillo and Sarabia (1990). In this way, we can obtain distributions whose conditionals are of the type $\log F$. Again, using the basic distribution (6.38), we obtain the class of distributions

$$f_{X_1,X_2}(x_1,x_2) = n(a_1,a_2,b_1,m)\frac{e^{a_1x_1+a_2x_2}}{\{1+e^{x_1}+e^{x_2}+(1-m)e^{x_1+x_2}\}^{a_1+b_1}}.$$

6.7 Application to Bayesian Inference

In Bayesian inference, when a bivariate prior distribution in needed, a family of distributions that can model both positive and negative associations and also allow one to easily obtain the posterior density is usually preferred. Correlated binary data occur in many applications. In the simplest case, assume that our model for the data is formed by two independent binomial random variables, with likelihood

$$\ell(p_1, p_2) \propto p_1^{x_1}(1-p_1)^{n_1-x_1} p_2^{x_2}(1-p_2)^{n_2-x_2}. \tag{6.48}$$

For the prior specification of (p_1, p_2), note that a natural conjugate prior for p_1, assuming that p_2 is known, is a beta prior or any beta extension. The same is of course true for p_2, assuming p_1 is known. It is then natural to look for the most general density for (p_1, p_2) whose conditionals satisfy

$$\begin{aligned} p_1|p_2 &\sim \mathcal{GB}(a_1(p_2), b_1(p_2), \lambda_1), \\ p_2|p_1 &\sim \mathcal{GB}(a_2(p_1), b_2(p_1), \lambda_2) \end{aligned} \tag{6.49}$$

or

$$\begin{aligned} p_1|p_2 &\sim \mathcal{GH}(a_1, b_1, c, \lambda_1(p_2)), \\ p_2|p_1 &\sim \mathcal{GH}(a_2, b_2, c, \lambda_2(p_1)) \end{aligned} \tag{6.50}$$

that is, a conditionally conjugate prior in the terminology of Arnold, Castillo and Sarabia (1998, 1999).

The prior distributions corresponding to the specifications (6.49) and (6.50) are given by (6.23) and (6.46), respectively. Both models can be used as conjugate prior distributions for the likelihood (6.48). Both priors allow us to accommodate dependent as well as independent prior beliefs. In the model (6.49), we need the elicitation of ten hyperparameters, and in the (6.50) we need only six. When combining (6.49) or (6.50) with the data, it is evident that only four of the parameters are affected by the data. Specifically, if

$$(p_1, p_2) \sim \mathcal{BGHC}(a_1, b_1, a_2, b_2, c, m),$$

where $\mathcal{BGHC}(a_1, b_1, a_2, b_2, c, m)$ denotes the joint pdf (6.46), then

$$(p_1, p_2)|\underline{x} \sim \mathcal{BGHC}(a_1 + x_1, b_1 + n_1 - x_1, a_2 + x_2, b_2 + n_2 - x_2, c, m).$$

If we use the prior (6.49) or (6.50), the resulting posterior density is readily implemented using the Gibbs sampler; for example, with model (6.50), we have

$$\begin{aligned} p_1|(p_2, \underline{x}) &\sim \mathcal{GH}(a_1 + x_1, b_1 + n_1 - x_1, c, 1 - mp_2), \\ p_2|(p_1, \underline{x}) &\sim \mathcal{GH}(a_2 + x_2, b_2 + n_2 - x_2, c, 1 - mp_1). \end{aligned}$$

If we are interested in the ratio of the corresponding odds ratios, that is, in the cross-product ratio

$$\Phi(p_1, p_2) = \frac{p_1/(1-p_1)}{p_2/(1-p_2)},$$

and (p_1, p_2) is distributed as in (6.50), the mean value according to formula (6.47) is

$$E[\Phi(p_1, p_2)] = \frac{n(a_1+1, a_2-1, b_1-1, b_2+1, c, m)}{n(a_1, a_2, b_1, b_2, c, m)}.$$

6.8 Conditional Survival Models

In this section, we consider new models for the generalized three-parameter beta distribution with conditional specification, by conditioning on events of the type $\{X > x\}$ and $\{Y > y\}$. This problem was considered initially by Arnold (1992). We study the case corresponding to (6.2). Then, let (X, Y) be a two-dimensional random variable with support $[0,1] \times [0,1]$ such that for each $y \in (0,1)$,

$$\Pr(X > x | Y > y) = \left\{\frac{1-x}{1-\tilde{\lambda}_1(y)x}\right\}^b, \quad 0 < x < 1, \tag{6.51}$$

and for each $x \in (0,1)$

$$\Pr(Y > y | X > x) = \left\{\frac{1-y}{1-\tilde{\lambda}_2(x)y}\right\}^b, \quad 0 < y < 1. \tag{6.52}$$

The corresponding joint survival function compatible with (6.51) and (6.52) must be of the form

$$\Pr(X > x, Y > y) = \frac{(1-x)^b(1-y)^b}{(1+\lambda_{12}x + \lambda_{21}y + \lambda_{22}xy)^b}. \tag{6.53}$$

Note that $X \sim \mathcal{GB}(1, b, 1+\lambda_{12})$ and $Y \sim \mathcal{GB}(1, b, 1+\lambda_{21})$. Consequently, this model presents marginal and conditional distributions of the same kind. The independence case corresponds to the choice $\lambda_{22} = \lambda_{12}\lambda_{21}$. This type of models can be viewed as having proportional hazard functions. From model (6.53), we can build a copula taking $b = 1$ and $\lambda_{12} = \lambda_{12} = 0$. After some computations, we obtain

$$C_\lambda(x, y) = \frac{xy(1-\lambda+\lambda x+\lambda y)}{1+\lambda xy}, \tag{6.54}$$

with $0 \leq \lambda < 1$. The Spearman correlation coefficient of (6.54) is

$$\begin{aligned} \rho_S(\lambda) &= 12 \int_0^1 \int_0^1 \{C_\lambda(x,y) - xy\} dx dy \\ &= \frac{6}{\lambda^2} \left\{ \frac{1}{2}(12-\lambda)\lambda - 4(1+\lambda)\log(1+\lambda) + 2(1-\lambda)\mathrm{Polylog}(2,-\lambda) \right\}, \end{aligned}$$

where Polylog(n, z) represents the nth polylogarithm function of z, and the Kendall's τ coefficient is

$$\begin{aligned} \tau(\lambda) &= 1 - 4 \int_0^1 \int_0^1 \frac{\partial C_\lambda(x,y)}{\partial x} \frac{\partial C_\lambda(x,y)}{\partial y} dx dy \\ &= 1 - \frac{2}{3\lambda^2} \left\{ (1+\lambda)^2 \log(1+\lambda) - \lambda \right\}. \end{aligned}$$

6.9 Multivariate Extensions

A straightforward k-dimensional extension of Theorem 6.4.1 can be obtained. It may be used to generate k-dimensional joint densities with generalized beta conditionals. We consider a k-dimensional random vector $\underline{X} = (X_1, \ldots, X_k)$ and introduce the notation $\underline{X}_{(i)}$ to denote the vector $\underline{X}$ with the ith coordinate deleted. An analogous notation is used to define $\underline{x}_{(i)}$. We are then led to consider joint densities for $\underline{X}$ for which, for each i and each $\underline{x}_{(i)} \in \mathbb{R}^{k-1}$,

$$X_i | \underline{X}_{(i)} = \underline{x}_{(i)} \sim \mathcal{GB}(a_i(\underline{x}_{(i)}), b_i(\underline{x}_{(i)}), \lambda_i), \quad i = 1, 2, \ldots, k$$

for some functions $a_i(\cdot), b_i(\cdot)$ and $\lambda_i(\cdot)$, $i = 1, 2, \ldots, k$. The resulting class of k-dimensional generalized three-parameter conditional is of the form

$$f_{\underline{X}}(\underline{x}) = \left[\prod_{i=1}^k r_i(x_i) \right] \exp \left\{ \sum_{i_1=0}^{\ell_1} \sum_{i_2=0}^{\ell_2} \cdots \sum_{i_k=0}^{\ell_k} m_{i_1, i_2, \ldots, i_k} \left[\prod_{j=1}^k q_{ij}(x_j) \right] \right\},$$

where

$$\begin{aligned} r_i(x_i) &= \{x_i(1-x_i)\}^{-1}, \quad i = 1, 2, \ldots, k \\ q_{i0}(x_i) &= 1, \quad i = 1, 2, \ldots, k \\ q_{i1}(x_i) &= \log\{x_i/(1-\tilde{\lambda}_i x_i)\}, \quad i = 1, 2, \ldots, k \\ q_{i2}(x_i) &= \log\{(1-x_i)/(1-\tilde{\lambda}_i x_i)\}, \quad i = 1, 2, \ldots, k, \end{aligned}$$

and where $m_{\underline{0}}$ is a function of the other $m_{\underline{i}}$, chosen so that the density integrates to 1. There are constraints on the $m_{\underline{i}}$, needed to ensure that the conditional densities are proper beta densities.

Acknowledgments. The authors are indebted to the Spanish Ministry of Science and Technology (Projects SEJ2004-02810, DPI2002-04172-C04-02, and DPI2003-01362) for partial support.

References

1. Armero, C., and Bayarri, M. J. (1994). Prior assessments for prediction in queues, *Journal of the Royal Statistical Society, Series D*, **43**, 139–153.

2. Arnold, B. C. (1992). Conditional survival models, In *Recent Advances in Life-Testing and Reliability: A Volume in Honor of Alonzo Clifford Cohen Jr.* (Ed., N. Balakrishnan), pp. 589–601, CRC Press, Boca Raton.

3. Arnold, B. C., Castillo, E., and Sarabia, J. M. (1992). *Conditionally Specified Distributions*, Lecture Notes in Statistics, Vol. 73, Springer-Verlag, New York.

4. Arnold, B. C., Castillo, E., and Sarabia, J. M. (1998). Bayesian analysis for classical distributions using conditionally specified prior, *Sankhyā, Series B*, **60**, 228–245.

5. Arnold, B. C., Castillo, E., and Sarabia, J. M. (1999). *Conditional Specification of Statistical Models*, Springer-Verlag, New York.

6. Arnold, B. C., Castillo, E., and Sarabia, J. M. (2001). Conditionally specified distributions: An introduction (with discussion), *Statistical Science*, **16**, 249–274.

7. Arnold, B. C., Castillo, E., Sarabia, J. M., and González-Vega, L. (2000). Multiple modes in densities with normal conditionals, *Statistics and Probability Letters*, **49**, 355–363.

8. Arnold, B. C. and Strauss, D. (1991). Bivariate distributions with conditionals in prescribed exponential families, *Journal of the Royal Statistical Society, Series B*, **53**, 365–375.

9. Balakrishnan, N. (Ed.) (1992). *Handbook of Logistic Distribution*, Marcel Dekker, New York.

10. Barlow, R. E., and Proschan, F. (1981). *Statistical Theory of Reliability and Life Testing: Probability Models*, To Begin With, Silver Springs, Maryland.

11. Castillo, E., and Sarabia, J. M. (1990). Bivariate distributions with second kind beta conditionals, *Communications in Statistics—Theory and Methods*, **19**, 3433–3445.

12. Chen, J. J., and Novick, M. R. (1984). Bayesian analysis for binomial models with generalized beta prior distributions, *Journal of Educational Statistics*, **9**, 163–175.

13. Fisher, R. A. (1924). On a distribution yielding the error functions of several well known statistics, In *Proceedings of the International Mathematical Congress, Toronto*, Vol. 2, pp. 805–813.

14. Holland, P. W., and Wang, Y. L. (1987). Dependence function for continuous bivariate densities, *Communications in Statistics—Theory and Methods*, **16**, 863–876.

15. Johnson, N. L., Kotz, S., and Balakrishnan, N. (1995). *Continuous Univariate Distributions*, Vol. 2, Second edition, John Wiley & Sons, New York.

16. Jones, M. C. (1996). The local dependence function, *Biometrika*, **83**, 899–904.

17. Jones, M. C. (2004). Families of distributions arising from distributions of order statistics, *Test*, 13, 1–43.

18. Kotz, S., Balakrishnan, N., and Johnson, N. L. (2000). *Continuous Multivariate Distributions*, Vol. 1, Second edition, John Wiley & Sons, New York.

19. Lee, M.-L. T. (1996). Properties and applications of the Sarmanov family of bivariate distributions, *Communications in Statistics—Theory and Methods*, **25**, 1207–1222.

20. Libby, D. L., and Novick, M. R. (1982). Multivariate generalized beta distributions with applications to utility assessment, *Journal of Educational Statistics*, **7**, 271–294.

21. Olkin, I. and Liu, R. (2003). A bivariate beta distribution, *Statistics and Probability Letters*, **62**, 407–412.

22. Pham-Gia, T., and Duong, Q. P. (1989). The generalized beta and F-distributions in statistical modelling, *Mathematical and Computer Modelling*, **12**, 1613–1625.

23. Sarmanov, O. V. (1966). Generalized normal correlation and two-dimensional Frechet classes, *Doklady, Soviet Mathematics*, **168**, 596–599.

24. Stuart, A., and Ord, J. K. (1987). *Kendall's Advanced Theory of Statistics–Vol. 1: Distribution Theory*, Oxford University Press, New York.

7

A Kotz-Type Distribution for Multivariate Statistical Inference

Dayanand N. Naik and Kusaya Plungpongpun

Old Dominion University, Norfolk, VA, USA
Silpakorn University, Bangkok, Thailand

Abstract: In this chapter, we consider a Kotz-type distribution (of a p-variate random vector $\boldsymbol{X}$) which has fatter tail regions than that of multivariate normal distribution, and its probability density function (*pdf*) is given by

$$f(\mathbf{x}, \boldsymbol{\mu}, \boldsymbol{\Sigma}) = c_p \mid \boldsymbol{\Sigma} \mid^{-\frac{1}{2}} \exp \{-[(\mathbf{x}-\boldsymbol{\mu})'\boldsymbol{\Sigma}^{-1}(\mathbf{x}-\boldsymbol{\mu})]^{\frac{1}{2}}\},$$

where $\boldsymbol{\mu} \in \Re^p$, $\boldsymbol{\Sigma}$ is a positive definite matrix and $c_p = \frac{\Gamma(\frac{p}{2})}{2\pi^{\frac{p}{2}}\Gamma(p)}$. We review various characteristics and provide a simulation algorithm to simulate samples from this distribution. Estimation of the parameters using the maximum likelihood method is discussed. An interesting fact is that the maximum likelihood estimators under this distribution are the generalized spatial median (GSM) estimators as defined by Rao (1988). Using the asymptotic distribution of the estimates, statistical inferences on the parameters of the distribution are illustrated with an example.

Keywords and phrases: Generalized spatial median, Kotz-type distribution, simulation algorithm, simultaneous confidence intervals

7.1 Introduction

Our focus in this chapter is the probability density function

$$f(\mathbf{x}, \boldsymbol{\mu}, \boldsymbol{\Sigma}) = c_p \mid \boldsymbol{\Sigma} \mid^{-\frac{1}{2}} e^{-[(\mathbf{x}-\boldsymbol{\mu})'\boldsymbol{\Sigma}^{-1}(\mathbf{x}-\boldsymbol{\mu})]^{\frac{1}{2}}}, \quad \boldsymbol{\mu} \in \Re^p, \ \boldsymbol{\Sigma} \ p.d., \tag{7.1}$$

where $c_p = \frac{\Gamma(\frac{p}{2})}{2\pi^{\frac{p}{2}}\Gamma(p)}$. This *pdf* of $p \times 1$ random vector $\boldsymbol{X}$ has appeared in the literature in different forms. For example, it is a special case of *pdf* proportional to

$$\exp\left\{-\frac{1}{r}[(\mathbf{x}-\boldsymbol{\xi})'\mathbf{A}(\mathbf{x}-\boldsymbol{\xi})]^{\frac{r}{2}}\right\},$$

where $\mathbf{A}$ is p. d. and $r \geq 1$ [Simoni (1968)]. Further, it is a special case of an *elliptically symmetric distribution* denoted by $EC_p(\boldsymbol{\mu}, \boldsymbol{\Sigma}, g)$, having the *pdf*

$$f(\mathbf{x}) = c_p|\boldsymbol{\Sigma}|^{-\frac{1}{2}}g[(\mathbf{x}-\boldsymbol{\mu})'\boldsymbol{\Sigma}^{-1}(\mathbf{x}-\boldsymbol{\mu})],$$

where g is a one-dimensional real-valued function independent of p and c_p is a normalizing constant. The function g is usually referred to as density generator. See Muirhead (1982) and Fang *et al.* (1990) for details on elliptically symmetric distributions. For the distribution in (7.1), $g(t) = \exp\{-\sqrt{t}\}$.

Kotz (1975) and Fang *et al.* (1990) studied a special class of elliptical distributions and named this class as *Kotz-type distributions.* If $\boldsymbol{X} \sim EC_p(\boldsymbol{\mu}, \boldsymbol{\Sigma}, g)$ and the density generator g is of the form $g(u) = c_p u^{N-1}\exp(-ru^s), \quad r, s > 0, \; 2N + p > 2,$ then we say that $\boldsymbol{X}$ possesses a symmetric Kotz distribution. The *pdf* of $\boldsymbol{X}$ is given by

$$f(\mathbf{x}, \boldsymbol{\mu}, \boldsymbol{\Sigma}) = c_p \mid \boldsymbol{\Sigma} \mid^{-\frac{1}{2}} [(\mathbf{x}-\boldsymbol{\mu})'\boldsymbol{\Sigma}^{-1}(\mathbf{x}-\boldsymbol{\mu})]^{N-1} \exp\,\{-r[(\mathbf{x}-\boldsymbol{\mu})'\boldsymbol{\Sigma}^{-1}(\mathbf{x}-\boldsymbol{\mu})]^s\},$$

where $c_p = \frac{s\Gamma(\frac{p}{2})}{\pi^{\frac{p}{2}}\Gamma(\frac{2N+p-2}{2s})} r^{\frac{2N+p-2}{2s}}$. See Nadarajah (2003) for a recent exposition and applications of Kotz-type distributions. The *pdf* (7.1) is obtained when $N = 1, s = \frac{1}{2}$ and $r = 1$.

Kano (1994) and Gómez *et al.* (1998) studied a special class of elliptical distributions called *power exponential distributions.* A random vector $\boldsymbol{X}$ is said to have a p-dimensional power exponential distribution with parameters $\boldsymbol{\mu}, \boldsymbol{\Sigma}$ and β, denoted by $PE_p(\boldsymbol{\mu}, \boldsymbol{\Sigma}, \beta)$, where $\boldsymbol{\mu} \in \Re^p$, $\boldsymbol{\Sigma}$ is a $p \times p$ positive definite symmetric matrix and $\beta \in (0, \infty)$, if its density function is

$$f(\mathbf{x}, \boldsymbol{\mu}, \boldsymbol{\Sigma}, \beta) = c_p|\boldsymbol{\Sigma}|^{-\frac{1}{2}} \exp\left\{-\frac{1}{2}[(\mathbf{x}-\boldsymbol{\mu})'\boldsymbol{\Sigma}^{-1}(\mathbf{x}-\boldsymbol{\mu})]^{\beta}\right\},$$

where $c_p = \frac{p\Gamma(\frac{p}{2})}{\pi^{\frac{p}{2}}\Gamma(1+\frac{p}{2\beta})2^{1+\frac{p}{2\beta}}}$. See Lindsey (1999) for an application of power exponential distributions to analyze repeated measures data. For $\beta = \frac{1}{2}$, the distribution (7.1) is obtained after adjusting the scaling to absorb the 1/2 in the exponent.

The distribution (7.1) has heavier tail regions than the multivariate normal distribution and hence can be useful in providing robustness against "outliers" [Lindsey (1999)]. The pdf in (7.1) can be written as a normal mixture; see Kariya and Sinha (1989) and Kano (1994). For $p = 1$, the *pdf* in (7.1) reduces to that of a double exponential distribution. Hence, we may treat this distribution as a multivariate generalization of double exponential distribution. However, this is not a multivariate double exponential distribution because its marginal distributions are not double exponential distributions. See Kotz *et al.* (2001) for several multivariate double exponential (Laplace) distributions.

In the following subsections we will provide various characteristics of a Kotz-type distribution, such as moments and the marginal and conditional distributions. A simulation algorithm to simulate data from this distribution is provided in Section 7.2. Estimation of parameters using maximum likelihood method will be discussed in Section 7.3. We show that the MLE of the location parameter under the assumption of a Kotz-type distribution is same as the generalized spatial median (GSM) defined by Rao (1988). Multivariate analysis of variance is performed in Section 7.4 and illustrated with an example.

7.1.1 Moments and other properties

In the following, we provide the expected value, variance covariance matrix, and Mardia's measures of skewness and kurtosis [Mardia (1970)] of the distribution given in (7.1) using some formulae in Baringhaus and Henze (1992).

The expected value, $E(\boldsymbol{X}) = \boldsymbol{\mu}$, the variance covariance matrix, $Var(\boldsymbol{X}) = (p+1)\boldsymbol{\Sigma}$, Mardia's multivariate skewness measure, $\beta_{1p} = 0$, and Mardia's multivariate kurtosis measure, $\beta_{2p} = \frac{p(p+2)(p+3)}{(p+1)}$. Suppose $\mathbf{x}_1, \ldots, \mathbf{x}_n$ are a set of sample multivariate data. Then, Mardia's multivariate skewness and kurtosis measures are defined, respectively, as $b_{1p} = n^{-2}\sum_{i=1}^n \sum_{j=1}^n g_{ij}^3$ and $b_{2p} = n^{-1}\sum_{i=1}^n g_{ii}^2$, where $g_{ij} = (\mathbf{x}_i - \bar{\mathbf{x}})'\mathbf{S}_n^{-1}(\mathbf{x}_j - \bar{\mathbf{x}})$, $i, j = 1, \ldots, n$, $\bar{\mathbf{x}} = \frac{1}{n}\sum_{i=1}^n \mathbf{x}_i$, and $\mathbf{S}_n = \frac{1}{n}\sum_{i=1}^n (\mathbf{x}_i - \bar{\mathbf{x}})(\mathbf{x}_i - \bar{\mathbf{x}})'$. The asymptotic distribution of Mardia's skewness measure, b_{1p}, under any elliptically symmetric distribution, is a weighted sum of two independent χ^2 random variables [Baringhaus and Henze (1992)]. That is,

$$nb_{1p} \xrightarrow{\mathcal{D}} \alpha_1 \chi_p^2 + \alpha_2 \chi^2_{\frac{p(p-1)(p+4)}{6}}, \tag{7.2}$$

where $\alpha_1 = \frac{3}{p}\left[\frac{m_6}{p+2} - 2m_4 + p(p+2)\right]$ and $\alpha_2 = \frac{6m_6}{p(p+2)(p+4)}$. For the Kotz-type distribution in (7.1), $m_4 = \frac{p(p+2)(p+3)}{p+1}$, and $m_6 = \frac{p(p+2)(p+3)(p+4)(p+5)}{(p+1)^2}$. Also, from Henze (1994), the asymptotic distribution of Mardia's kurtosis measure, b_{2p}, under any elliptically symmetric distribution, is

$$\sqrt{n}\Big(b_{2p} - p(p+2)(p+3)/(p+1)\Big) \xrightarrow{\mathcal{D}} N(0, \tau^2), \tag{7.3}$$

where $\tau^2 = r_8 - r_4^2 + \frac{4}{p} r_4 (\frac{r_4^2}{p} - r_6)$. For the distribution in (7.1),

$$r_k = (\frac{1}{\sqrt{p+1}})^k E[R^k] = \frac{p(p+1)(p+2)(p+3)\cdots(p+(k-1))}{(p+1)^{k/2}}, \quad k \geq 1.$$

7.1.2 Marginal and conditional distributions

Suppose $\boldsymbol{X}$ is partitioned as $\boldsymbol{X} = (\boldsymbol{X}'_{(1)},\ \boldsymbol{X}'_{(2)})'$, where $\boldsymbol{X}_{(1)} = (x_1, \ldots, x_k)'$, $\mathbf{x}_{(2)} = (x_{k+1}, \ldots, x_p)'$ with $k < p$ and similarly, $\boldsymbol{\mu} = (\boldsymbol{\mu}'_{(1)}, \boldsymbol{\mu}'_{(2)})'$ with $\boldsymbol{\mu}_{(1)} = (\mu_1, \ldots, \mu_k)'$ and $\boldsymbol{\mu}_{(2)} = (\mu_{k+1}, \ldots, \mu_p)'$. Further suppose $\boldsymbol{\Sigma}$ is accordingly partitioned as

$$\boldsymbol{\Sigma} = \begin{pmatrix} \boldsymbol{\Sigma}_{11} & \boldsymbol{\Sigma}_{12} \\ \boldsymbol{\Sigma}_{21} & \boldsymbol{\Sigma}_{22} \end{pmatrix},$$

where $\boldsymbol{\Sigma}_{11}$ is a $k \times k$ p.d. matrix and $\boldsymbol{\Sigma}_{22}$ is a $(p-k) \times (p-k)$ p.d. matrix and $\boldsymbol{\Sigma}_{12} = \boldsymbol{\Sigma}'_{21}$. Then, $\boldsymbol{X}_{(1)}$ has an elliptically symmetric $EC_k(\boldsymbol{\mu}_{(1)}, \boldsymbol{\Sigma}_{11}, g_1)$ distribution with

$$g_1(t) = t^{\frac{p-k}{2}} \int_0^1 \omega^{\frac{k-p}{2}\ -1} (1-\omega)^{\frac{p-k}{2}\ -1}\ e^{-\sqrt{\frac{t}{\omega}}}\ d\omega.$$

The marginal characteristics of $\boldsymbol{X}_{(1)}$ are $E(\boldsymbol{X}_{(1)}) = \boldsymbol{\mu}_{(1)}$, $\mathrm{Var}(\boldsymbol{X}_{(1)}) = (p+1)\boldsymbol{\Sigma}_{11}$, $\beta_{1p}(\boldsymbol{X}_{(1)}) = 0$, and $\beta_{2p}(\boldsymbol{X}_{(1)}) = \frac{k(k+2)(p+3)}{p+1}$.

The conditional distribution of $\boldsymbol{X}_{(2)}$ given $\boldsymbol{X}_{(1)}$ is elliptically contoured $EC_{p-k}(\boldsymbol{\mu}_{2.1},\ \boldsymbol{\Sigma}_{22.1},\ g_{2.1})$, where

$$\begin{aligned} \boldsymbol{\mu}_{2.1} &= \boldsymbol{\mu}_{(2)} + \boldsymbol{\Sigma}_{21}\boldsymbol{\Sigma}_{11}^{-1}(\mathbf{x}_{(1)} - \boldsymbol{\mu}_{(1)}), \\ \boldsymbol{\Sigma}_{22.1} &= \boldsymbol{\Sigma}_{22} - \boldsymbol{\Sigma}_{21}\boldsymbol{\Sigma}_{11}^{-1}\boldsymbol{\Sigma}_{12}, \quad \text{and} \\ g_{2.1}(t) &= \exp\ \{-[t + (\mathbf{x}_{(1)} - \boldsymbol{\mu}_{(1)})'\boldsymbol{\Sigma}_{11}^{-1}(\mathbf{x}_{(1)} - \boldsymbol{\mu}_{(1)})]^{\frac{1}{2}}\}. \end{aligned}$$

7.2 An Algorithm for Simulation

In this section, we provide an algorithm for simulating data from the distribution in (7.1). Naik and Patwardhan (1991) have used a method for simulating data from a bivariate Kotz-type distribution. We shall use a similar method to generate a random sample from a p-variate Kotz-type distribution. The proposed algorithm is given as follows.

Step 1. Simulate $\mathbf{y}' = (y_1, \ldots, y_p)$ having the density

$$f(\mathbf{y}) = c_p\ \exp\Big\{-\sqrt{\mathbf{y}'\mathbf{y}}\Big\},$$

where $-\infty < y_i < \infty$, and $c_p = \frac{\Gamma(\frac{p}{2})}{2\pi^{\frac{p}{2}}\Gamma(p)}$. Note that $f(\mathbf{y})$ is the standardized version of Kotz-type distribution given in (7.1) and also $E(\mathbf{y}) = \mathbf{0}$ and $Var(\mathbf{y}) = (p+1)\mathbf{I}_p$.

The simulation of $\mathbf{y}$ is achieved by using the polar coordinate transformation as follows:

$$
\begin{aligned}
y_1 &= R\cos\theta_1 \\
y_2 &= R\sin\theta_1\cos\theta_2 \\
&\vdots \\
y_{p-1} &= R\sin\theta_1\sin\theta_2\cdots\sin\theta_{p-2}\cos\theta_{p-1} \\
y_p &= R\sin\theta_1\sin\theta_2\cdots\sin\theta_{p-2}\sin\theta_{p-1},
\end{aligned}
$$

where $R = \sqrt{\mathbf{y}'\mathbf{y}}$, $\theta_j \in [\,0,\pi)$ for $1 \le j \le p-2$ and $\theta_{p-1} \in [\,0, 2\pi)$. The Jacobian of the transformation is $R^{p-1}\prod_{j=1}^{p-2}\sin^{p-j-1}(\theta_j)$. For the *pdf* $f(\mathbf{y})$, $R \sim G(p,1)$. See Koutras (1986) for the distribution of the quadratic form under an elliptical gamma law.

For an odd p, R and θ_j, $j = 1,\ldots,p-1$, are independently distributed with the probability density function given by

$$
\begin{aligned}
g(r) &= \frac{1}{\Gamma(p)}\, r^{p-1}e^{-r}, \text{ that is, } \mathrm{R} \sim \mathrm{G(p,1)} \text{ and} \\
g(\theta_1) &= \frac{p-2}{2}\left[\frac{(p-4)\cdots 3\cdot 1}{(p-3)\cdots 4\cdot 2}\right]\sin^{p-2}(\theta_1), \\
g(\theta_2) &= \frac{2^{p-3}}{\pi}\,\frac{\left[(\frac{p-3}{2})!\right]^2}{(p-3)!}\,\sin^{p-3}(\theta_2), \\
g(\theta_3) &= \frac{p-4}{2}\left[\frac{(p-6)\cdots 3\cdot 1}{(p-5)\cdots 4\cdot 2}\right]\sin^{p-4}(\theta_3), \\
g(\theta_4) &= \frac{2^{p-5}}{\pi}\,\frac{\left[(\frac{p-5}{2})!\right]^2}{(p-5)!}\,\sin^{p-5}(\theta_4), \\
&\vdots \\
g(\theta_{p-2}) &= \frac{1}{2}\sin(\theta_{p-2}), \\
g(\theta_{p-1}) &= \frac{1}{2\pi}.
\end{aligned}
$$

For an even p, R and θ_j, $j = 1,\ldots,p-1$, are independently distributed with the probability density function given by

$$
\begin{aligned}
g(r) &= \frac{1}{\Gamma(p)}\, r^{p-1}e^{-r} \text{ and} \\
g(\theta_1) &= \frac{2^{p-2}}{\pi}\,\frac{\left[(\frac{p-2}{2})!\right]^2}{(p-2)!}\,\sin^{p-2}(\theta_1), \\
g(\theta_2) &= \frac{p-3}{2}\left[\frac{(p-5)\cdots 3\cdot 1}{(p-4)\cdots 4\cdot 2}\right]\sin^{p-3}(\theta_2),
\end{aligned}
$$

$$\begin{aligned}
g(\theta_3) &= \frac{2^{p-4}}{\pi} \frac{\left[(\frac{p-4}{2})!\right]^2}{(p-4)!} \sin^{p-4}(\theta_3), \\
g(\theta_4) &= \frac{p-5}{2} \left[\frac{(p-7)\cdots 3\cdot 1}{(p-6)\cdots 4\cdot 2}\right] \sin^{p-5}(\theta_4), \\
&\vdots \\
g(\theta_{p-2}) &= \frac{1}{2}\sin(\theta_{p-2}), \\
g(\theta_{p-1}) &= \frac{1}{2\pi}.
\end{aligned}$$

Of course, any uniform random number generating algorithm can be successfully used with the inverse cumulative distribution function to generate pseudo-random numbers from a nonuniform distribution.

To simulate $\theta \sim g(\theta)$, we use the bisection method, which is one of the popular numerical inversion algorithms. See Devroye (1986) for details.

Algorithm: Find an initial interval $[a, b]$ to which the solution belongs.

REPEAT
$\theta \leftarrow \frac{(a+b)}{2}$
IF $G(\theta) \le U$ THEN $a \leftarrow \theta$
ELSE $b \leftarrow \theta$
UNTIL $b - a \le 2\delta$
RETURN θ

Here, $\delta > 0$ is a small number.

Step 2. Obtain $\mathbf{x}' = (x_1, \ldots, x_p)$ having the distribution in (7.1) by making the transformation $\mathbf{x} = \mathbf{\Gamma y} + \boldsymbol{\mu}$, where $\boldsymbol{\mu}' = (\mu_1, \ldots, \mu_p)$ and $\mathbf{\Gamma}'\mathbf{\Gamma} = \mathbf{\Sigma}$. Note that $E(\boldsymbol{X}) = \boldsymbol{\mu}$ and $V(\boldsymbol{X}) = (p+1)\mathbf{\Sigma}$.

For example, to generate a 5-variate ($p = 5$) random vector $\mathbf{x}' = (x_1, \ldots, x_5)$ having the distribution in (7.1), first we simulate $\mathbf{y}' = (y_1, \ldots, y_5)$ which has the density

$$f(\mathbf{y}) = \frac{1}{64\pi^2} \exp\left\{-\sqrt{\mathbf{y}'\mathbf{y}}\right\}, \quad -\infty < y_i < \infty,$$

where

$$\begin{aligned}
y_1 &= R\cos\theta_1, \\
y_2 &= R\sin\theta_1\cos\theta_2, \\
y_3 &= R\sin\theta_1\sin\theta_2\cos\theta_3, \\
y_4 &= R\sin\theta_1\sin\theta_2\sin\theta_3\cos\theta_4, \\
y_5 &= R\sin\theta_1\sin\theta_2\sin\theta_3\sin\theta_4,
\end{aligned}$$

and R and θ_j are independently distributed with

$$g(r) = \frac{1}{24} r^4 e^{-r}, \text{ that is, } \mathrm{R} \sim \mathrm{G}(5, 1) \text{ and}$$

$$\begin{aligned} g(\theta_1) &= \frac{3}{4}\sin^3(\theta_1), \\ g(\theta_2) &= \frac{2}{\pi}\sin^2(\theta_2), \\ g(\theta_3) &= \frac{1}{2}\sin(\theta_3), \\ g(\theta_4) &= \frac{1}{2\pi}, \end{aligned}$$

where $\theta_j \in [\,0,\pi)$ for $j = 1,2,3$ and $\theta_4 \in [\,0,2\pi)$. Then, we obtain $\mathbf{x}$ by making the transformation $\mathbf{x} = \boldsymbol{\Gamma}\mathbf{y} + \boldsymbol{\mu}$ for fixed $\boldsymbol{\mu}$ and $\boldsymbol{\Gamma}$.

7.3 Estimation of Parameters

Many researchers have discussed statistical inference for elliptical distributions. For example, see Fang and Anderson (1990) and the references therein. However, the maximum likelihood theory developed in Fang and Anderson (1990) assumes that the samples $\mathbf{x}_1,\ldots,\mathbf{x}_n$ have the same mean vector $\boldsymbol{\mu}$ and scale matrix $\boldsymbol{\Sigma}$ and the joint distribution of all the samples is elliptically symmetric. In fact, in this case, the maximum likelihood estimators of $\boldsymbol{\mu}$ and $\boldsymbol{\Sigma}$ are essentially the same as those in the multivariate normal case [see Fang and Anderson (1990, Theorem 1, p. 205)].

Several authors have discussed statistical inference for certain elliptical distributions. For example, Lange *et al.* (1989) used multivariate t-distribution and maximum likelihood method to analyze certain regression and repeated measures data, and Lindsey (1999) used multivariate power exponential distribution to analyze certain repeated measures data. In each case, numerical algorithms were used to find the estimates of the parameters. See Naik *et al.* (2002) for a discussion of likelihood based inference for AR(1) and MA(1) models under elliptical distributions. In the following we discuss estimation of parameters using maximum likelihood methods when an *iid* sample from (7.1) is available.

Suppose $\mathbf{x}_1,\ldots,\mathbf{x}_n$ is a random sample from Kotz-type distribution in (7.1). Then the log-likelihood function is given by

$$\ln L(\boldsymbol{\mu},\boldsymbol{\Sigma}) = n\,\ln c - \frac{n}{2}\ln|\boldsymbol{\Sigma}| - \sum_{i=1}^{n}\sqrt{(\mathbf{x}_i-\boldsymbol{\mu})'\boldsymbol{\Sigma}^{-1}(\mathbf{x}_i-\boldsymbol{\mu})}.$$

The MLEs of $\boldsymbol{\mu}$ and $\boldsymbol{\Sigma}$ are obtained by minimizing

$$\frac{n}{2}\ln|\,\boldsymbol{\Sigma}\,| + \sum_{i=1}^{n}\sqrt{(\mathbf{x}_i-\boldsymbol{\mu})'\boldsymbol{\Sigma}^{-1}(\mathbf{x}_i-\boldsymbol{\mu})} \tag{7.4}$$

simultaneously w.r.t. $\boldsymbol{\mu}$ and $\boldsymbol{\Sigma}$.

When $\boldsymbol{\Sigma} = \mathbf{I}$, the solution to the above problem or the MLE of $\boldsymbol{\mu}$ is the spatial median introduced by Haldane (1948) and for general $\boldsymbol{\Sigma}$ it is generalized spatial median introduced by Rao (1988) and studied by Naik (1993).

7.3.1 Generalized spatial median (GSM)

In this section, we consider the estimation of the location parameter $\boldsymbol{\mu}$. Haldane (1948) defined the spatial median of multivariate data vectors $\mathbf{x}_1, \ldots, \mathbf{x}_n$ as a point (vector) $\hat{\boldsymbol{\mu}} \in \Re^p$ that minimizes

$$\sum_{i=1}^{n} \|\mathbf{x}_i - \boldsymbol{\mu}\| = \sum_{i=1}^{n} \sqrt{(\mathbf{x}_i - \boldsymbol{\mu})'(\mathbf{x}_i - \boldsymbol{\mu})}$$

with respect to $\boldsymbol{\mu}$. For $p > 1$, the vector $\hat{\boldsymbol{\mu}}$ is unique except when all the mass of the distribution is concentrated on a line [Haldane (1948) and Ducharme and Milasevic (1987)] and is invariant under orthogonal transformation, but not under affine transformation [Brown (1983) and Ducharme and Milasevic (1987)].

Rao (1988) defined two generalized spatial medians that are invariant under affine transformation as:

(i) a vector $\hat{\boldsymbol{\mu}}$ that minimizes

$$\sum_{i=1}^{n} \sqrt{(\mathbf{x}_i - \boldsymbol{\mu})'\mathbf{S}^{-1}(\mathbf{x}_i - \boldsymbol{\mu})}$$

with respect to $\boldsymbol{\mu}$, where $\mathbf{S}$ is the usual sample variance covariance matrix, and

(ii) a vector $\hat{\boldsymbol{\mu}}$ that minimizes

$$\frac{n}{2} \ln |\boldsymbol{\Sigma}| + \sum_{i=1}^{n} \sqrt{(\mathbf{x}_i - \boldsymbol{\mu})'\boldsymbol{\Sigma}^{-1}(\mathbf{x}_i - \boldsymbol{\mu})}$$

simultaneously with respect to $\boldsymbol{\mu}$ and $\boldsymbol{\Sigma}$.

Thus, we note that the MLE of $\boldsymbol{\mu}$ under the assumption of a Kotz-type distribution in (7.1) for $\mathbf{x}_1, \ldots, \mathbf{x}_n$ is same as the generalized spatial median defined by Rao (1988).

7.3.2 Computation of GSM and $\widehat{\Sigma}$

Let $\mathbf{x}_1, \ldots, \mathbf{x}_n$ be a random sample from (7.1). Then, the GSM of $\boldsymbol{\mu}$ that minimizes (7.4) can be computed in two stages as follows (see Naik, 1993).

Suppose $\mathbf{\Sigma}$ is known or set to an initial value and $\mathbf{\Sigma} = \mathbf{G}\mathbf{G}'$, for a nonsingular $\mathbf{G}$. Then the generalized spatial median $\hat{\boldsymbol{\mu}}$ that minimizes

$$\sum_{i=1}^{n} \sqrt{(\mathbf{x}_i - \boldsymbol{\mu})'\mathbf{\Sigma}^{-1}(\mathbf{x}_i - \boldsymbol{\mu})}$$

w.r.t. $\boldsymbol{\mu}$ can be obtained as $\hat{\boldsymbol{\mu}} = \mathbf{G}\hat{\boldsymbol{\nu}}$, where $\hat{\boldsymbol{\nu}}$ is the spatial median that minimizes $\sum_{i=1}^{n} \sqrt{(\mathbf{y}_i - \boldsymbol{\nu})'(\mathbf{y}_i - \boldsymbol{\nu})}$ w.r.t. $\boldsymbol{\nu}$. Here, $\mathbf{y}_i = \mathbf{G}^{-1}\mathbf{x}_i$ and $\boldsymbol{\nu} = \mathrm{G}^{-1}\boldsymbol{\mu}$. Spatial median can be computed using an algorithm given in Gower (1974).

Next using $\hat{\boldsymbol{\mu}}$, the maximum likelihood estimate of $\mathbf{\Sigma}$ is obtained as the matrix $\widehat{\mathbf{\Sigma}}$ that minimizes (7.4) with respect to $\mathbf{\Sigma}$ as a solution to the non-linear equation given by

$$\mathbf{\Sigma} = \frac{1}{n} \sum_{i=1}^{n} \frac{(\mathbf{x}_i - \hat{\boldsymbol{\mu}})(\mathbf{x}_i - \hat{\boldsymbol{\mu}})'}{\sqrt{(\mathbf{x}_i - \hat{\boldsymbol{\mu}})'\mathbf{\Sigma}^{-1}(\mathbf{x}_i - \hat{\boldsymbol{\mu}})}}.$$

Solving these equations generally requires computational algorithms. Now the two steps are iterated until a certain convergence criteria are met and thus the maximum likelihood estimates of both $\boldsymbol{\mu}$ and $\mathbf{\Sigma}$ are obtained.

While Naik and Patwardhan (1991) have successfully implemented this algorithm for a bivariate version of the distribution in (7.1), Naik (1993) studied the case when $\mathbf{\Sigma}$ has an equi-correlation structure. It is shown in these works that the maximum likelihood estimates are unique and easy to obtain. However, with the current level of computational advances, it is much easier and efficient to use nonlinear optimization methods to obtain maximum likelihood estimates of all the parameters. We have adopted SAS' IML procedure for writing the computer programs. Using the Newton-Raphson method, the optimization yields unique estimates in the feasible regions under most covariance structures.

7.3.3 The asymptotic distribution of GSM

Using the same arguments and derivations as in Huber (1967, 1981), Ducharme and Milasevic (1987), and Naik (1993), the asymptotic distribution of the maximum likelihood estimate, $\hat{\boldsymbol{\mu}}$ (which is also the generalized spatial median), can be summarized in the following theorem.

Theorem 7.3.1 (Asymptotic Distribution of GSM) *Let $\boldsymbol{x}_1, \ldots, \boldsymbol{x}_n$ be a random sample from p-variate ($p > 1$) Kotz-type distribution (7.1) with parameters $\boldsymbol{\mu}$ and $\mathbf{\Sigma}$, and $\hat{\boldsymbol{\mu}}$ be the maximum likelihood estimate of $\boldsymbol{\mu}$. Then*

$$\sqrt{n}(\hat{\boldsymbol{\mu}} - \boldsymbol{\mu}) \xrightarrow{\mathcal{D}} N(\boldsymbol{0},\ \mathbf{\Sigma}\boldsymbol{A}^{-1}\boldsymbol{B}\boldsymbol{A}^{-1}\mathbf{\Sigma}),$$

where

$$\boldsymbol{B} = E\left[\frac{(\boldsymbol{X}-\boldsymbol{\mu})(\boldsymbol{X}-\boldsymbol{\mu})'}{(\boldsymbol{X}-\boldsymbol{\mu})'\boldsymbol{\Sigma}^{-1}(\boldsymbol{X}-\boldsymbol{\mu})}\right] \quad \text{and}$$

$$\boldsymbol{A} = E\left[\frac{1}{\sqrt{(\boldsymbol{X}-\boldsymbol{\mu})'\boldsymbol{\Sigma}^{-1}(\boldsymbol{X}-\boldsymbol{\mu})}}\left(\boldsymbol{\Sigma}-\frac{(\boldsymbol{X}-\boldsymbol{\mu})(\boldsymbol{X}-\boldsymbol{\mu})'}{(\boldsymbol{X}-\boldsymbol{\mu})'\boldsymbol{\Sigma}^{-1}(\boldsymbol{X}-\boldsymbol{\mu})}\right)\right].$$

Further, $\boldsymbol{B}$ and $\boldsymbol{A}$ can be estimated by

$$\widehat{\boldsymbol{B}} = \frac{1}{n}\sum_{i=1}^{n}\frac{(\boldsymbol{x}_i-\hat{\boldsymbol{\mu}})(\boldsymbol{x}_i-\hat{\boldsymbol{\mu}})'}{(\boldsymbol{x}_i-\hat{\boldsymbol{\mu}})'\widehat{\boldsymbol{\Sigma}}^{-1}(\boldsymbol{x}_i-\hat{\boldsymbol{\mu}})},$$

$$\widehat{\boldsymbol{A}} = \frac{1}{n}\sum_{i=1}^{n}\left\{\frac{1}{\sqrt{(\boldsymbol{x}_i-\hat{\boldsymbol{\mu}})'\widehat{\boldsymbol{\Sigma}}^{-1}(\boldsymbol{x}_i-\hat{\boldsymbol{\mu}})}}\left[\widehat{\boldsymbol{\Sigma}}-\frac{(\boldsymbol{x}_i-\hat{\boldsymbol{\mu}})(\boldsymbol{x}_i-\hat{\boldsymbol{\mu}})'}{(\boldsymbol{x}_i-\hat{\boldsymbol{\mu}})'\widehat{\boldsymbol{\Sigma}}^{-1}(\boldsymbol{x}_i-\hat{\boldsymbol{\mu}})}\right]\right\},$$

where $\widehat{\boldsymbol{\Sigma}}$ is the maximum likelihood estimate of $\boldsymbol{\Sigma}$.

Using Theorem 7.3.1, we can perform statistical inference on $\boldsymbol{\mu}$. For example, a test for $H_0 : \boldsymbol{\mu} = \boldsymbol{\mu}_0$ can be performed using

$$TS = n(\hat{\boldsymbol{\mu}}-\boldsymbol{\mu}_0)'\widehat{\boldsymbol{\Omega}}^{-1}(\hat{\boldsymbol{\mu}}-\boldsymbol{\mu}_0) \sim \chi^2_p, \tag{7.5}$$

where $\widehat{\boldsymbol{\Omega}} = \widehat{\boldsymbol{\Sigma}}\widehat{\mathbf{A}}^{-1}\widehat{\mathbf{B}}\widehat{\mathbf{A}}^{-1}\widehat{\boldsymbol{\Sigma}}$.

Further, the following corollary to Theorem 7.3.1 provides simultaneous confidence intervals for a set of m linear functions, $\mathbf{a}_i'\boldsymbol{\mu}$, $i = 1, \ldots, m$, of $\boldsymbol{\mu}$.

Corollary 7.3.1 (Simultaneous Confidence Intervals) *Using Theorem 7.3.1, the $100(1-\alpha)\%$ Bonferroni simultaneous confidence intervals for m linear functions of μ_i's, are given by*

$$\left(\boldsymbol{a}_i'\hat{\boldsymbol{\mu}} - z_{\alpha/(2m)}\sqrt{\frac{\boldsymbol{a}_i'\widehat{\boldsymbol{\Omega}}\boldsymbol{a}_i}{n}},\ \boldsymbol{a}_i'\hat{\boldsymbol{\mu}} + z_{\alpha/(2m)}\sqrt{\frac{\boldsymbol{a}_i'\widehat{\boldsymbol{\Omega}}\boldsymbol{a}_i}{n}}\right),\ i = 1, \ldots, m, \tag{7.6}$$

where $\boldsymbol{a}_i'$s are vectors of known constants and $z_{\alpha/(2m)}$ is the upper $100(1-\alpha/(2m))$th percentile of a standard normal distribution.

7.4 An Example

Using a multivariate data set from the following example, we illustrate the computation of the maximum likelihood estimates and perform some statistical inference. All the computations are done using programs written in SAS/IML software.

The data measuring cork boring of tress given in Rao (1988) consists of the weights of cork boring in four directions (north, east, south, and west) for 28 trees in a block of plantations. Khattree and Naik (1999) have done an extensive analysis of these data and the data can also be found in that book.

The sample statistics, namely, sample mean, $\bar{\mathbf{x}}$, covariance matrix, $\mathbf{S}_n$, and correlation matrix, $\mathbf{R}$, for these data are:

$$\bar{\mathbf{x}} = \left(50.54,\ 46.18,\ 49.68,\ 45.18\right)',$$

$$\mathbf{S}_n^{\{*\}} = \begin{pmatrix} 280.03 & 215.76 & 278.14 & 218.19 \\ 0.89 & 212.08 & 220.88 & 165.25 \\ 0.90 & 0.83 & 337.50 & 250.27 \\ 0.88 & 0.77 & 0.92 & 217.93 \end{pmatrix}.$$

* Elements of $\mathbf{R}$ are on the lower diagonal.

Next, the ML estimates of $\boldsymbol{\mu}$ and $\boldsymbol{\Sigma}$ using the optimization algorithm are:

$$\hat{\boldsymbol{\mu}} = \left(46.62,\ 42.72,\ 46.01,\ 42.09\right)',$$

$$\widehat{\boldsymbol{\Sigma}}^{\{*\}} = \begin{pmatrix} 60.81 & 46.24 & 62.18 & 49.36 \\ 0.89 & 44.71 & 50.24 & 36.99 \\ 0.90 & 0.85 & 78.35 & 58.41 \\ 0.88 & 0.77 & 0.92 & 51.39 \end{pmatrix}, \text{ and}$$

$$\widehat{\boldsymbol{\Omega}} = \begin{pmatrix} 318.55 & 246.67 & 356.88 & 275.39 \\ 246.67 & 238.98 & 304.91 & 210.48 \\ 356.88 & 304.91 & 492.34 & 351.96 \\ 275.39 & 210.48 & 351.96 & 301.44 \end{pmatrix}.$$

In the following, we construct simultaneous confidence intervals for the contrasts,

$\theta_1 = \mu_{north} - \mu_{east} + \mu_{south} - \mu_{west}$,
$\theta_2 = \mu_{north} - \mu_{south}$, and
$\theta_3 = \mu_{east} - \mu_{west}$.

These contrasts will let us check whether or not the bark deposit is uniform in all the four directions. Estimates of these contrasts are easily determined as $\hat{\theta}_1 = 7.82$, $\hat{\theta}_2 = 0.61$, and $\hat{\theta}_3 = 0.63$. The asymptotic standard errors of these estimates are given by: $\text{SE}(\hat{\theta}_1) = 2.14$, $\text{SE}(\hat{\theta}_2) = 1.86$, and $\text{SE}(\hat{\theta}_3) = 2.07$.

Using Corollary 7.3.1, the 95% Bonferroni simultaneous confidence intervals for θ_1, θ_2, and θ_3 are respectively given by

$$(2.70,\ 12.95),\ (-3.84,\ 5.07) \text{ and } (-4.31,\ 5.58).$$

It may be noted that the only significant contrast is the difference of the mean bark deposits in the directions of north and south and the east and west directions. Neither the contrast of deposits between the south and north directions nor that in the east and west directions is significant.

We have used this example to illustrate the computation of the maximum likelihood estimates and various other quantities of interest under Kotz-type distribution in (7.1). Computation of the estimates is easily done using programs written in SAS/IML software. Multivariate analysis of variance (MANOVA) and problems of discriminant analysis, assuming (7.1) as the underlying probability distribution, are discussed in detail in Plungpongpun (2003).

References

1. Baringhaus, L., and Henze, N. (1992). Limit distributions for Mardia's measure of multivariate skewness, *Annals of Statistics*, **20**, 1889–1902.

2. Brown, B. M. (1983). Statistical use of the spatial median, *Journal of the Royal Statistical Society, Series B*, **45**, 25–30.

3. Devroye, L. (1986). *Non-Uniform Random Variate Generation*, Springer-Verlag, New York.

4. Ducharme, G. R., and Milasevic, P. (1987). Spatial median and directional data, *Biometrika*, **74**, 212–215.

5. Fang, K. T., and Anderson, T. W. (1990). *Statistical Inference in Elliptically Contoured and Related Distributions*, Allerton Press, New York.

6. Fang, K. T. and Kotz, S., and Ng, K. W. (1990). *Symmetric Multivariate and Related Distributions*, Chapman and Hall, London.

7. Gómez, E. Gómez-Villegas, M. A., and Marín, J. M. (1998). A multivariate generalization of the power exponential family of distributions, *Communications in Statistics—Theory and Methods*, **27**, 589–600.

8. Gower, J. S. (1974). The mediancentre, *Applied Statistics*, **23**, 466–470.

9. Haldane, J. B. S. (1948). Note on the median of a multivariate distribution, *Biometrika* , **35**, 414–415.

10. Henze, N. (1994). On Mardia's kurtosis test for multivariate normality, *Communications in Statistics—Theory and Methods*, **23**, 1031–1045.

11. Huber, P. J. (1967). The behavior of maximum likelihood estimates under nonstandard conditions, In *Proceedings of Fifth Berkeley Symposium on Mathematical Statistics and Probability, Vol. 1*, pp. 221–233, University of California Press, Berkeley, CA.

12. Huber, P. J. (1981). *Robust Statistics*, John Wiley & Sons, New York.

13. Johnson, R. A., and Wichern, D. W. (1998). *Applied Multivariate Statistical Analysis*, Prentice Hall, New Jersey.

14. Kano, Y. (1994). Consistency property of elliptical probability density functions, *Journal of Multivariate Analysis*, **51**, 139–147.

15. Kariya, T. K., and Sinha, B. K. (1989). *Robustness of Statistical Tests*, Academic Press, San Diego, CA.

16. Khattree, R., and Naik, D. N. (1999). *Applied Multivariate Statistics with SAS Software*, John Wiley & Sons and SAS Institute, New York.

17. Kotz, S. (1975). Multivariate distributions at a cross-road, In *Statistical Distributions in Scientific Work* (Eds., G. P. Patil, S. Kotz, and J. K. Ord), pp. 247–270, D. Reidel, The Netherlands.

18. Kotz, S., Kozubowski, T. J., and Podgórski, K. (2001). *The Laplace Distribution and Generalizations*, Birkhäuser, Boston.

19. Koutras, M. (1986). On the generalized noncentral chi-squared distribution induced by an elliptical gamma law, *Biometrika*, **73**, 528–532.

20. Lange, K. L., Little, R. J. A., and Taylor, J. M. G. (1989). Robust statistical modeling using the t distribution, *Journal of the American Statistical Association*, **84**, 881–896.

21. Lindsey, J. K. (1999). Multivariate elliptically contoured distributions for repeated measurements, *Biometrics*, **55**, 1277–1280.

22. Mardia, K. V. (1970). Measures of multivariate skewness and kurtosis with applications, *Biometrika*, **57**, 519–530.

23. Muirhead, R. J. (1982). *Aspects of Multivariate Statistical Theory*, John Wiley & Sons, New York.

24. Nadarajah, S. (2003). The Kotz-type distribution with applications, *Statistics*, **37**, 341–358.

25. Naik, D. N. (1993). Multivariate medians: A review, In *Probability and Statistics* (Eds., S. K. Basu and B. K. Sinha), pp. 80–90, Narosa Publishing House, New Delhi.

26. Naik, D., Khattree, R., and Shults, J. (2002). A note on likelihood based inference for AR(1) and MA(1) processes under certain robust alternatives to multivariate normality, *Journal of Statistical Theory and Applications*, **1**, 57–62.

27. Naik, D. N., and Patwardhan, G. R. (1991). A Note on testing for correlation in a certain bivariate distribution, *Journal of Quantitative Economics*, **7**, 295–302.

28. Plungpongpun, K. (2003). Analysis of multivariate data using Kotz type distribution, Ph.D. Thesis, Department of Mathematics and Statistics, Old Dominion University, Norfolk, VA.

29. Rao, C. R. (1988). Methodology based on the L_1-norm in statistical inference, *Sankhyā, Series A*, **50**, 289–313.

30. Simoni, S. de (1968). Su una estensione dello schema delle curve normali di ordina r alle variabili doppie, *Statistica (Bologna)*, **28**, 151–170.

8

Range of Correlation Matrices for Dependent Random Variables with Given Marginal Distributions

Harry Joe

University of British Columbia, Vancouver, BC, Canada

Abstract: Let $X_1, \ldots, X_d$ be d ($d \geq 3$) dependent random variables with finite variances such that $X_j \sim F_j$. Results on the set $S_d(F_1, \ldots, F_d)$ of possible correlation matrices with given margins are obtained; this set is relevant for simulating dependent random variables with given marginal distributions and a given correlation matrix. When $F_1 = \cdots = F_d = F$, we let $S_d(F)$ denote the set of possible correlation matrices. Of interest is the set of F for which $S_d(F)$ is the same as the set of all non-negative definite correlation matrices; using a construction with conditional distributions, we show that this property holds only if F is a (location-scale shift of a) margin of a $(d-1)$-dimensional spherical distribution.

Keywords and phrases: Spherically symmetric, elliptically contoured, copula, partial correlation, Fréchet bounds

8.1 Introduction

This article is concerned with the range of correlation matrices when the univariate margins are specified. This is of interest when simulating random variables with given univariate distributions. In general, one does not get the entire set of non-negative definite correlation matrices. For example, for the singular correlation matrices, there must be linear dependencies in the random variables.

To state the problem more precisely, we introduce the following notation. Let $X_1, \ldots, X_d$ be dependent random variables such that $X_j \sim F_j$, where $F_1, \ldots, F_d$ are univariate distributions with finite variances $\sigma_1^2, \ldots, \sigma_d^2$. The correlation of X_j and X_k is denoted as ρ_{jk}. Let $S_d(F_1, \ldots, F_d)$ be the set of possible correlation matrices $R = (\rho_{jk})$; $S_d(F_1, \ldots, F_d)$ is relevant for simulating dependent random variables with given distributions $F_1, \ldots, F_d$ and a given correlation matrix. When $F_1 = \cdots = F_d = F$ and F has finite variance, the set of correlation matrices is denoted as $S_d(F)$. Also, let S_d^* be the set of all non-negative definite correlation matrices.

Further notation that will be used are the following: F_S is the marginal distribution of $(X_j; j \in S)$; $F_{S|T}$ is the conditional distribution of $(X_j; j \in S)$ given $(X_i; i \in T)$; the Fréchet class with given compatible margins $F_{S_1}, \ldots, F_{S_m}$ is denoted as $\mathcal{F}(F_{S_1}, \ldots, F_{S_m})$. For example, $\mathcal{F}(F_{12}, F_{23})$ denotes the Fréchet class with bivariate margins F_{12}, F_{23} (and univariate margins F_1, F_2, F_3).

Of interest is the set A_d of univariate distributions F for which $S_d(F) = S_d^*$. We show that A_d contains the univariate margins of $(d-1)$-dimensional spherically symmetric or spherical distributions, and their location-scale transforms. One consequence is that all correlation matrices are possible for copulas up to dimension $d = 4$, but not dimensions $d > 4$. Copulas [Sklar (1959)] are multivariate distributions with uniform (0,1) margins and are a convenient way to separate univariate margins from the dependence structure in a multivariate distribution.

It is not possible to characterize $S_d(F_1, \ldots, F_d)$ in general. However, simulations with the multivariate normal copula or multivariate t_ν copula should get close to the whole range of possible correlation matrices given $F_1, \ldots, F_d$. Also the multivariate normal copula may be the easiest approach to generate a distribution with given correlation matrix and given margins, as illustrated below.

Let Φ be the univariate standard normal cumulative distribution function (cdf) and let $\Phi_d(\cdot; \Lambda)$ be the d-variate normal cdf with correlation matrix Λ. Let $S_{d\Phi}(F_1, \ldots, F_d)$ be the set of possible correlation matrices with the multivariate normal copula, $\Phi_d(\Phi^{-1}(u_1,), \ldots, \Phi^{-1}(u_d); \Lambda)$. Of course, $S_{d\Phi}(F_1, \ldots, F_d)$ is a subset of $S_d(F_1, \ldots, F_d)$. Consider a correlation matrix $R = (\rho_{jk})$. The following describes how to check if $R \in S_{d\Phi}(F_1, \ldots, F_d)$. For the (j,k) bivariate margin, suppose the bivariate normal copula with correlation parameter λ_{jk}, $F_{jk} = \Phi_2(\Phi^{-1}(F_j), \Phi^{-1}(F_k); \lambda_{jk})$ leads to the correlation ρ_{jk} for F_j, F_k; note that from Hoeffding's identity [Hoeffding (1935)], ρ_{jk} must be between the correlations from the bivariate Fréchet lower and upper bounds: $\max\{0, F_j + F_k - 1\}$ and $\min\{F_j, F_k\}$, respectively. If the matrix $\Lambda = (\lambda_{jk})$ with diagonal elements of 1 is non-negative definite, then $R \in S_{d\Phi}(F_1, \ldots, F_d)$.

Given a feasible ρ_{jk}, then λ_{jk} can be solved for numerically using Hoeffding's identity:

$$\int_{-\infty}^{\infty}\int_{-\infty}^{\infty}\Big\{\Phi_2(\Phi^{-1}(F_j(x_j)), \Phi^{-1}(F_k(x_k)); \lambda_{jk}) - F_j(x_j)F_k(x_k)\Big\}dx_j dx_k = \rho_{jk}\sigma_j\sigma_k. \tag{8.1}$$

The bivariate normal cdf Φ_2 can be numerically computed up to 15 decimal place precision using the code in Donnelly (1973), and a two-dimensional numerical integration method [see, Davis and Rabinowitz (1984)] can be used in (8.1). A similar idea works for the bivariate t_ν distribution.

References relevant to the topic of this article are Cuadras (1992), Song (1997), Emrich and Piedmonte (1991), and the references therein. The above approach with the multivariate normal distribution is used by Emrich and Piedmonte (1991) for Bernoulli margins. Cuadras (1992) generates multivariate distributions with linear regressions and given correlations/margins but does not mention the range of possible correlation matrices. Song (1997) generates multivariate distributions with given correlations, margins in the exponential dispersion family and a specific construction.

The outline of the remainder of this article is as follows. Section 8.2 contains a summary of known results when $S_d(F) = S_d^*$. Section 8.3 has the approach of obtaining bounds sequentially for diagonals of the correlation matrix. Section 8.4 has the new results concerning conditions for $S_d(F) = S_d^*$ for $d \geq 3$.

8.2 Known Results on a Range of Correlations

In this section, we state known results for the case $F_1 = \cdots = F_d = F$. Note that $S_d(F) = S_d(F^*)$ if F^* is obtained as a location-scale transform of F.

The obvious results for $d = 2$ are the following:

1. If F is symmetric (about c), then all bivariate correlation matrices are possible. There exists $X_1 \sim F$, $X_2 \sim F$ such that $\mathrm{Corr}\,(X_1, X_2) = \rho$ for any $\rho \in [-1, 1]$. The two extreme cases come from the Fréchet lower and upper bounds [stochastic representations $X_2 = 2c - X_1$, $X_2 = X_1$].
2. If F is not symmetric, then correlation of -1 cannot be achieved, because correlation for the Fréchet lower bound [stochastic representation $X_2 = F^{-1}(1 - F(X_1))$ when F is continuous], the correlation is strictly greater than -1.

The above implies that $S_2(F) = S_2^*$ if and only if F is symmetric. Hence for $d \geq 3$, $S_d(F)$ cannot be equal to S_d^* unless F is symmetric. So the next question is: for which symmetric F does $S_d(F) = S_d^*$ for $d \geq 3$? Note also that $S_2(F_1, F_2)$ does not have the full range of correlations if F_1, F_2 are not in the same location-scale family; for example, the correlation of 1 is achievable only if the Fréchet upper bound corresponds to a linear function.

In the case where F has finite variance and is a margin of a spherical or elliptical distribution of dimension d, then $S_d(F) = S_d^*$. We introduce some notation for these univariate distributions as they come up in the new results in Section 8.4. Let $\mathcal{M}_d$ ($d \geq 2$) be the set of possible univariate margins (with finite variances) of spherical distributions in dimension d.

Properties of $\mathcal{M}_d$ for different d are the following:

- $\mathcal{M}_2 \supset M_3 \supset \cdots \supset \mathcal{M}_\infty$.
- $\mathcal{M}_\infty$ is set of scale mixtures of normal random variables with mean 0 and finite variance.
- $\mathcal{M}_3$ is set of all symmetric distributions decreasing on $[0, \infty)$ with finite variance.
- $S_d(F) = S_d^*$ for $F \in \mathcal{M}_d$ (all correlation matrices are possible if F is the margin of d-variate spherical distribution).

Because the $U(-1,1)$ distribution is in $\mathcal{M}_3 \backslash \mathcal{M}_4$, then all three-dimensional correlation matrices are possible for $U(-1,1)$ and $U(0,1)$ margins. Hence, all correlation matrices are possible for trivariate copulas. If all correlation matrices are possible for d-variate copulas for dimensions $d \geq 4$, some of them would have to be attained from nonelliptical distributions.

Some linear properties of elliptical distributions used in later sections are mentioned below. Let $(X_1, \ldots, X_d)$ be elliptically distributed.

(a) The conditional expectation of X_j given $(X_i : i \in S)$, $S \subset \{1, \ldots, d\} \backslash \{j\}$ is linear in X_i for $i \in S$.

(b) The conditional covariance matrix of (X_{j_1}, X_{j_2}) given $(X_i : i \in S)$, $S \subset \{1, \ldots, d\} \backslash \{j_1, j_2\}$ is constant over values of $(x_i; i \in S)$.

These are well-known properties of the multivariate normal distribution which extend to elliptical families. For the properties of elliptical distributions mentioned above, see Kelker (1970), Fang, Kotz and Ng (1990), and Section 4.9 of Joe (1997).

8.3 Conditional Approach

One approach to define the inequalities for the set $S_d(F_1, \ldots, F_d)$ is something analogous to the inequalities from partial correlations. Before outlining the sequence of conditional distributions, we review partial correlations for multivariate normal distributions.

8.3.1 Multivariate normal and partial correlations

The bounds for a positive definite correlation matrix can be specified one diagonal at a time. By allowing for equalities at the boundaries, singular correlation matrices can be obtained. Let

$$R = \begin{pmatrix} 1 & \rho_{12} & \rho_{13} & \rho_{14} & \cdots & \rho_{1d} \\ \rho_{12} & 1 & \rho_{23} & \rho_{24} & \cdots & \rho_{2d} \\ \vdots & \vdots & \vdots & \vdots & \ddots & \vdots \\ \rho_{d-2,1} & \rho_{d-2,2} & \rho_{d-2,3} & \rho_{d-2,4} & \cdots & \rho_{d-2,d} \\ \rho_{d-1,1} & \rho_{d-1,2} & \rho_{d-1,3} & \rho_{d-1,4} & \cdots & \rho_{d-1,d} \\ \rho_{d1} & \rho_{d2} & \rho_{d3} & \rho_{d4} & \cdots & 1 \end{pmatrix}$$

be a correlation matrix. The first diagonal consist of $\rho_{j,j+1}$ for $j = 1, \ldots, d-1$. For multivariate normal distributions, these are independently free to vary in $(-1, 1)$. The second diagonal consist of $\rho_{j,j+2}$ for $j = 1, \ldots, d-2$, the mth diagonal consist of $\rho_{j,j+m}$ for $j = 1, \ldots, d-m$, for $m = 1, \ldots, d-1$, and so the last diagonal consists of ρ_{1d}.

For $m \geq 2$, there are bounds on $\rho_{j,j+m}$ which depend on $\{\rho_{s,t} : j \leq s < t \leq j+m,\ (s,t) \neq (j, j+m)\}$. For normal random variables, using the partial correlation $\rho_{j,j+m|j+1,\ldots,j+m-1}$ for the conditional correlation of X_j, X_{j+m} given $X_{j+1}, \ldots, X_{j+m-1}$, the bound for $\rho_{j,j+m}$ can be obtained from $-1 < \rho_{j,j+m|j+1,\ldots,j+m-1} < 1$. Suppose the submatrix $R[j : j+m]$, consisting of rows and columns $j, \ldots, j+m$ of R, is decomposed as

$$R[j : j+m] = \begin{pmatrix} 1 & \mathbf{r}_1^T(j,m) & \rho_{j,j+m} \\ \mathbf{r}_1(j,m) & R_2(j,m) & \mathbf{r}_3(j,m) \\ \rho_{j+m,j} & \mathbf{r}_3^T(j,m) & 1 \end{pmatrix}, \tag{8.2}$$

where $\mathbf{r}_1^T(j,m) = (\rho_{j,j+1}, \ldots, \rho_{j,j+m-1})$, $\mathbf{r}_3^T(j,m) = (\rho_{j+m,j+1}, \ldots, \rho_{j+m,j+m-1})$, and $R_2(j,m)$ consists of the middle $m-2$ rows and columns of $R[j : j+m]$. Assuming that $R_2(j,m)$ is nonsingular, the partial correlation $\rho_{j,j+m|j+1,\ldots,j+m-1}$ is

$$\frac{\rho_{j,j+m} - \mathbf{r}_1^T(j,m)(R_2(j,m))^{-1}\mathbf{r}_3(j,m)}{\Big\{1 - \mathbf{r}_1(j,m)^T(R_2(j,m))^{-1}\mathbf{r}_1(j,m)\Big\}^{1/2}\Big\{1 - \mathbf{r}_3^T(j,m)(R_2(j,m))^{-1}\mathbf{r}_3(j,m)\Big\}^{1/2}}.$$

This leads to the inequality on $\rho_{j,j+m}$:

$$\begin{aligned} &\mathbf{r}_1^T(j,m)(R_2(j,m))^{-1}\mathbf{r}_3(j,m) - D_{jm} \\ &\quad < \rho_{j,j+m} < \mathbf{r}_1^T(j,m)(R_2(j,m))^{-1}\mathbf{r}_3(j,m) + D_{jm}, \end{aligned} \tag{8.3}$$

where

$$D_{jm}^2 = \Big\{1 - \mathbf{r}_1(j,m)^T(R_2(j,m))^{-1}\mathbf{r}_1(j,m)\Big\}\Big\{1 - \mathbf{r}_3^T(j,m)(R_2(j,m))^{-1}\mathbf{r}_3(j,m)\Big\}.$$

Note that a different set of inequalities obtain when the indices $1, \ldots, d$ of the random variables are permuted.

8.3.2 General case

More generally, following constructions in Section 4.5 of Joe (1997), we can consider inequalities based on conditional distributions, one diagonal at a time. The first diagonal consist of the bivariate marginal distributions $F_{j,j+1}$, $j = 1,\ldots,d-1$. The second diagonal consist of the bivariate conditional distributions $F_{j,j+2|j+1}$, $j = 1,\ldots,d-2$, and for $m = 2,\ldots,d-2$, the mth diagonal consist of the bivariate conditional distributions $F_{j,j+m|j+1,\ldots,j+m-1}$, $j = 1,\ldots,d-m$, and finally the last $(d-1)$th diagonal consists of $F_{1,d|2,\ldots,d-1}$. Each conditional bivariate distribution satisfies Fréchet bounds, which are given below.

Consider the Fréchet class $\mathcal{F}(F_{j\cdots j+m-1}, F_{j+1\cdots j+m})$, where $F_{j\cdots j+m-1}$ and $F_{j+1\cdots j+m}$ have a common $F_{j+1\cdots j+m-1}$ $(m-1)$-variate margin. Members of this class have the form

$$F_{j\cdots j+m}(\mathbf{x}) = \int_{-\infty}^{x_{j+1}} \cdots \int_{-\infty}^{x_{j+m-1}} F_{j,j+m|j+1\cdots j+m-1}(x_j, x_{j+m}|\mathbf{z})\, dF_{j+1\cdots j+m-1}(\mathbf{z}), \tag{8.4}$$

where $\mathbf{z} = (z_{j+1},\ldots,z_{j+m-1})$ and $\mathbf{x} = (x_j,\ldots,x_{j+m})$. Note that the conditional distribution $F_{j,j+m|j+1\cdots j+m-1}(x_j, x_{j+m}|\mathbf{z})$ is a bivariate distribution for every $\mathbf{z}$, and is bounded by the Fréchet lower and upper bounds:

$$\begin{aligned}
&\max\{0,\, F_{j|j+1\cdots j+m-1}(x_j|\mathbf{z}) + F_{j+m|j+1\cdots j+m-1}(x_{j+m}|\mathbf{z}) - 1\} \\
&\quad \le F_{j,j+m|j+1\cdots j+m-1}(x_j, x_{j+m}|\mathbf{z}) \\
&\quad \le \min\{F_{j|j+1\cdots j+m-1}(x_j|\mathbf{z}),\, F_{j+m|j+1\cdots j+m-1}(x_{j+m}|\mathbf{z})\}.
\end{aligned}$$

By integrating over $\mathbf{z}$,

$$\begin{aligned}
&F_{j,j+m,L}(x_j, x_{j+m}) \\
&\quad = \int_{(-\infty,\infty)^{m-1}} \max\{0,\, F_{j|j+1\cdots j+m-1}(x_j|\mathbf{z}) \\
&\qquad + F_{j+m|j+1\cdots j+m-1}(x_{j+m}|\mathbf{z}) - 1\}\, dF_{j+1\cdots j+m-1}(\mathbf{z}) \\
&\quad \le F_{j,j+m}(x_j, x_{j+m}) \le \int_{(-\infty,\infty)^{m-1}} \min\{F_{j|j+1\cdots j+m-1}(x_j|\mathbf{z}), \\
&\qquad F_{j+m|j+1\cdots j+m-1}(x_{j+m}|\mathbf{z})\}\, dF_{j+1\cdots j+m-1}(\mathbf{z}) \\
&\quad = F_{j,j+m,U}(x_j, x_{j+m}).
\end{aligned} \tag{8.5}$$

If $F_{j\cdots j+m-1}, F_{j+1\cdots j+m}$ are given, the correlation of $F_{j,j+m}$ is bounded by the correlations of the two conditional Fréchet bound distributions in the inequalities in (8.5). Using Hoeffding's (1940) identity, the correlation of $F_{j,j+m}$ can be written as

$$\left\{\int_{-\infty}^{\infty}\int_{-\infty}^{\infty} [F_{j,j+m}(x_j, x_{j+m}) - F_j(x_j)F_{j+m}(x_{j+m})]dx_j dx_{j+m}\right\} \Big/ (\sigma_j\sigma_{j+m}).$$

But if only the correlations $\{\rho_{s,t} : j \le s < t \le j+m,\ (s,t) \ne (j, j+m)\}$ are given, there are in general an infinite number of choices of $F_{j\cdots j+m-1}, F_{j+1\cdots j+m}$ leading to the given correlations. This is why inequalities for $R \in \mathcal{S}_d(F_1, \ldots, F_d)$ are difficult to obtain in general.

To see the effect of the choice of bivariate distributions with given correlations, we specialize to the $d = 3$ case, for which it is easier to do some numerical comparisons.

For $d = 3$, the two-dimensional numerical integrations needed to obtain the correlations are easily computable. Given feasible correlations ρ_{12}, ρ_{23} for $\mathcal{F}(F_1, F_2, F_3)$, one can find dependence parameters δ_{12}, δ_{23} within a parametric family $C(\cdot; \delta)$ [for example, one of the families B1–B7 in Section 5.1 of Joe (1997)] leading to the specified correlations ρ_{12}, ρ_{23} and then compute bounds on ρ_{13} based on (8.4) and (8.5).

For a specific numerical example to illustrate the above, we take F_j, $j = 1, 2, 3$, to be the exponential distributions with mean 1, and set $\rho_{12} = 0.1$, $\rho_{23} = 0.5$.

copula	LB ρ_{13}	UB ρ_{13}
B1	-0.524	0.879
B2	-0.498	0.873
B3	-0.488	0.862
B4	-0.418	0.817
B5	-0.553	0.881
B6	-0.552	0.887
B7	-0.551	0.887

The bounds on ρ_{13} are summarized in the above table when copula families B1-B7 in Section 5.1 of Joe (1997) are used for F_{12}, F_{23}. B1 is the bivariate normal copula; the other copula families interpolate independence and Fréchet upper bound, and some of them extend to the Fréchet lower bound. All are one-parameter families, and have reflection symmetry $[c(u_1, u_2) = c(1-u_1, 1-u_2)$ for the copula density$]$, or upper or lower tail dependence. Computations were obtained through Monte Carlo simulation and numerical integration, and sometimes it is faster to get 3-digit accuracy using Monte Carlo simulation. When the random variables are all continuous, the simulation algorithm for the conditional Fréchet bounds is the following:

1. Generate $X_2 \sim F_2$, let x_2 be the realization.
2. Generate $X_1 \sim F_{1|2}(\cdot|x_2)$, let x_1 be the realization.
3. Let $x_3 = F_{3|2}^{-1}(F_{1|2}(x_1|x_2))$ for the conditional Fréchet upper bound, and let $x_3 = F_{3|2}^{-1}(1 - F_{1|2}(x_1|x_2))$ for the conditional Fréchet lower bound.

With $F_1 = F_2 = F_3 = F$ being the exponential distribution and $\rho_{12} = 0.1$, $\rho_{23} = 0.5$, some comparisons from the above are:

- In $S_{3\Phi}(F)$, $-0.524 \le \rho_{13} \le 0.879$ from the bivariate normal B1 copula with the conditional Fréchet lower and upper bounds. Note that this is the same as solving (8.1) to get $\lambda_{12} = 0.1195$, $\lambda_{23} = 0.5466$, leading to bounds on λ_{13} as $(-0.7661, 0.8967)$ from $\lambda_{12}\lambda_{23} \pm \sqrt{(1-\lambda_{12}^2)(1-\lambda_{23}^2)}$; with exponential margins, the bivariate normal copula with correlations -0.7661 and 0.8967, the correlations of the exponential random variables are -0.524 and 0.879, respectively.
- In $S_3(F)$, ρ_{13} extends beyond $[-0.553, 0.887]$. To get the complete range, one would need to optimize over all $F_{123} \in \mathcal{F}(F, F, F)$ with (1,2) correlation 0.1 and (2,3) correlation 0.5.
- In S_3^*, from $-1 \le \rho_{13|2} \le 1$, the range for ρ_{13} is $[-0.812, 0.912]$.

The above example may suggest that in some cases, $S_{d\Phi}(F_1, \ldots, F_d)$ comes close to $S_d(F_1, \ldots, F_d)$.

8.4 Characterization of F for $S_d(F) = S_d^*$

In this section, we use the conditional approach of the preceding section to assess when it is possible for $S_d(F) = S_d^*$. The main result is that for $d \ge 3$ and F symmetric (about 0), a necessary and sufficient condition for $S_d(F) = S_d^*$ is $F \in \mathcal{M}_{d-1}$. This means that for F that is a marginal distribution of a spherical distribution in dimension $d-1$, one can attain all possible correlation matrices up to dimension d, and this implies that some nonelliptical distributions are needed for achieve S_d^* for $F \in \mathcal{M}_{d-1} \backslash \mathcal{M}_d$. Because the uniform $(-1, 1)$ distribution is in $\mathcal{M}_3 \backslash \mathcal{M}_4$, all correlation matrices are possible for 4-variate copulas, but not copulas of dimensions greater than 4.

To check if the full range of correlation matrices is possible for $S_d(F)$, where F is symmetric, the following procedure can be used:

- For $d = 3$, given ρ_{12}, ρ_{23}, show that there exist $F_{12}, F_{23} \in \mathcal{F}(F, F)$ with respective correlations ρ_{12}, ρ_{23} for which the conditional Fréchet bounds will have correlations equaling the bounds for ρ_{13} in S_3^*.
- For $d = 4$, given arbitrary $\rho_{12}, \rho_{13}, \rho_{23}, \rho_{24}, \rho_{34}$ such that

$$\begin{pmatrix} 1 & \rho_{12} & \rho_{13} \\ \rho_{12} & 1 & \rho_{23} \\ \rho_{13} & \rho_{23} & 1 \end{pmatrix} \quad \text{and} \quad \begin{pmatrix} 1 & \rho_{23} & \rho_{24} \\ \rho_{23} & 1 & \rho_{34} \\ \rho_{24} & \rho_{34} & 1 \end{pmatrix}$$

  are positive definite matrices in S_3^*, show that there exist compatible $F_{123}, F_{234} \in \mathcal{F}(F, F, F)$ with the specified correlations for which the conditional Fréchet bounds will have correlations equaling the bounds for ρ_{14} in S_4^*.

- The above idea extends to $d \geq 5$.

8.4.1 $d = 3$

We first show the main ideas with $d = 3$.

Consider the Fréchet class $\mathcal{F}(F_{12}, F_{23})$ when $F_1 = F_2 = F_3 = F$ with variance σ^2. Suppose the correlations for F_{12}, F_{32} are ρ_{12}, ρ_{23}. As a special case of (8.4), members of this class have the form

$$F_{123}(x_1, x_2, x_3) = \int_{-\infty}^{x_2} F_{13|2}(x_1, x_3|z)\, dF_2(z). \tag{8.6}$$

Given F_{12}, F_{32}, the most negatively (positively) dependent distribution in (8.6) obtains when $F_{13|2}(\cdot|z)$ corresponds to the Fréchet lower (upper) bound for all z.

From S_3^*, correlation matrices of the form

$$\begin{pmatrix} 1 & \rho_{12} & \rho_{13} \\ \rho_{12} & 1 & \rho_{23} \\ \rho_{13} & \rho_{23} & 1 \end{pmatrix}$$

must satisfy

$$\rho_{12}\rho_{23} - \sqrt{(1-\rho_{12}^2)(1-\rho_{23}^2)} \leq \rho_{13} \leq \rho_{12}\rho_{23} + \sqrt{(1-\rho_{12}^2)(1-\rho_{23}^2)}. \tag{8.7}$$

From (8.6), one gets

$$\begin{aligned} &\mathrm{Cov}\,[\mathrm{E}\,(X_1|X_2), \mathrm{E}\,(X_3|X_2)] - \mathrm{E}\,[\{\mathrm{Var}\,(X_1|X_2)\mathrm{Var}\,(X_3|X_2)\}^{1/2}] \\ &\quad \leq \mathrm{Cov}\,(X_1, X_3) = \mathrm{Cov}\,[\mathrm{E}\,(X_1|X_2), \mathrm{E}\,(X_3|X_2)] + \mathrm{E}\,[\mathrm{Cov}\,(X_1, X_3|X_2)] \\ &\quad \leq \mathrm{Cov}\,[\mathrm{E}\,(X_1|X_2), \mathrm{E}\,(X_3|X_2)] + \mathrm{E}\,[\{\mathrm{Var}\,(X_1|X_2)\mathrm{Var}\,(X_3|X_2)\}^{1/2}] \end{aligned} \tag{8.8}$$

with equality only if X_1, X_3 are linearly related given X_2; that is $X_1 + c(X_2)X_3 = b(X_2)$ for functions b, c with $c > 0$ for the lower bound, and $X_1 - c(X_2)X_3 = b(X_2)$ for functions b, c with $c > 0$ for the upper bound.

Suppose the following stochastic representation holds for X_1, X_2, X_3:

$$X_1 = \rho_{12}X_2 + \epsilon_1, \quad X_3 = \rho_{23}X_2 + \epsilon_3, \tag{8.9}$$

where $\mathrm{E}\,(\epsilon_j|X_2) = 0$ for $j = 1, 3$ (so that ϵ_1, ϵ_3 are each uncorrelated with X_2), and ϵ_1, ϵ_3 can be chosen to be positively or negatively linearly related. Then, $\mathrm{E}\,(X_1|X_2) = \rho_{12}X_2$, $\mathrm{E}\,(X_3|X_2) = \rho_{23}X_2$, $\mathrm{Corr}\,(X_1, X_2) = \rho_{12}$, $\mathrm{Corr}\,(X_2, X_3) = \rho_{23}$, $\mathrm{Var}\,(\epsilon_1) = (1-\rho_{12}^2)\sigma^2$, $\mathrm{Var}\,(\epsilon_3) = (1-\rho_{23}^2)\sigma^2$, and $\mathrm{Cov}\,(X_1, X_3|X_2) = \mathrm{Cov}\,(\epsilon_1, \epsilon_3)$.

For the upper bound of (8.8), take ϵ_1, ϵ_3 to be positively linearly related with correlation 1, so that $\mathrm{Cov}(\epsilon_1, \epsilon_3) = \sigma^2[(1-\rho_{12}^2)(1-\rho_{23}^2)]^{1/2}$, and the last term in (8.8) becomes

$$\sigma^2\rho_{12}\rho_{23} + \sigma^2\{(1-\rho_{12}^2)(1-\rho_{23}^2)\}^{1/2},$$

leading to the correlation upper bound in (8.7). Similarly for the lower bound of (8.8), take ϵ_1, ϵ_3 to be negatively linearly related with correlation -1, so that $\mathrm{Cov}(\epsilon_1, \epsilon_3) = -\sigma^2\{(1-\rho_{12}^2)(1-\rho_{23}^2)\}^{1/2}$, and the first term in (8.8) becomes

$$\sigma^2\rho_{12}\rho_{23} - \sigma^2\{(1-\rho_{12}^2)(1-\rho_{23}^2)\}^{1/2},$$

leading to the correlation lower bound in (8.7).

The stochastic representation in (8.9) is possible for any $F \in \mathcal{M}_2$ when (X_1, X_2) and (X_3, X_2) have elliptical distributions; in this case, $\epsilon_1 = (1-\rho_{12})^{1/2}Z_1$ and $\epsilon_3 = (1-\rho_{23}^2)^{1/2}Z_3$ are chosen so that (X_2, Z_1) and (X_2, Z_3) have spherical distributions and $Z_j \sim F$ for $j = 1, 3$. For a value of ρ_{13} between the upper and lower bounds in (8.7), perhaps the simplest distribution leading to ρ_{13} is an appropriate convex combination of the conditional Fréchet upper and lower bound distributions.

Hence, we have shown that $S_3(F) = S_3^*$ for $F \in \mathcal{M}_2$, that is, $F \in \mathcal{M}_2$ is a sufficient condition for $S_3(F) = S_3^*$.

We next proceed to prove necessity. To show that $S_3(F) \neq S^*(F)$ for F symmetric and $F \notin \mathcal{M}_2$, we only have to pick some ρ_{12}, ρ_{23} so that one of the bounds in (8.7) is not reached.

To get some necessary conditions for $S_3(F) = S_3^*$, we take the case $\rho_{12} = \rho_{23} = \rho \neq 0$ to get some simpler inequalities for ρ_{13}. If (X_1, X_2, X_3) with $X_j \sim F$, then we assume that $(X_1, X_2) \overset{d}{=} (X_3, X_2)$ or $F_{12} = F_{32}$ without loss of generality, because otherwise we can always convert (X_1, X_2, X_3) to

$$(X_1', X_2', X_3') = \begin{cases} (X_1, X_2, X_3) & \text{with probability } 1/2, \\ (X_3, X_2, X_1) & \text{with probability } 1/2, \end{cases}$$

so that $X_j' \sim F$ and the correlations are $\rho_{12}' = \rho_{23}' = \rho$ and $\rho_{13}' = \rho_{13}$, and the distribution of (X_1', X_2') and (X_3', X_2') are both $(F_{12} + F_{32})/2$.

We will use the following lemma.

Lemma 8.4.1 *Let X_1, X_2 be random variables with correlation ρ. Then*

$$\mathrm{Var}\,(\mathrm{E}\,(X_1|X_2)) \geq \rho^2\mathrm{Var}\,(X_1), \tag{8.10}$$

with equality only if $\mathrm{E}\,(X_1|X_2)$ *is linear in* X_2.

PROOF. Inequality (8.10) is the same as

$$\mathrm{Var}\,(\mathrm{E}\,(X_1|X_2))\ \mathrm{Var}\,(X_2) \geq \{\mathrm{Cov}\,(X_1, X_2)\}^2.$$

Assume X_1, X_2 have been standardized so that $\mathrm{E}(X_j) = 0$, $j = 1, 2$. Then it is the same as

$$\mathrm{E}\left[\{\mathrm{E}(X_1|X_2)\}^2\right] \mathrm{E}(X_2^2) \geq \{\mathrm{E}(X_1 X_2)\}^2.$$

Let $g(X_2) = \mathrm{E}(X_1|X_2)$. Then

$$\{\mathrm{E}(X_1 X_2)\}^2 = \{\mathrm{E}(g(X_2)X_2)\}^2 \leq \mathrm{E}\{g^2(X_2)\} \ \mathrm{E}(X^2)$$

by the Cauchy-Schwarz inequality. Equality holds only if $g(X_2) = \mathrm{E}(X_1|X_2)$ is proportional to X_2. ■

With the assumption of $F_{12} = F_{32}$, the upper bound on ρ_{13} of 1 can be attained by taking $X_1 = X_3$ (with probability 1). So we consider the lower bound on ρ_{13}. With (8.6),

$$\begin{aligned} \mathrm{Cov}(X_1, X_3) &= \mathrm{Cov}\left[\mathrm{E}(X_1|X_2), \mathrm{E}(X_3|X_2)\right] + \mathrm{E}\left[\mathrm{Cov}(X_1, X_3|X_2)\right] \\ &= \mathrm{Var}(\mathrm{E}(X_1|X_2)) + \mathrm{E}\left[\mathrm{Cov}(X_1, X_3|X_2)\right] \\ &\geq \mathrm{Var}(\mathrm{E}(X_1|X_2)) - \mathrm{E}\left[\mathrm{Var}(X_1|X_2)\right]. \end{aligned} \tag{8.11}$$

Equality holds for the left part of (8.7) only if X_1, X_3 are negatively linearly related given $X_2 = z$ for all z or if there exists a function $b(z)$ such that $X_1 + X_3 = b(X_2)$ for the conditional Fréchet lower bound.

Because

$$\mathrm{Var}(X_1) = \mathrm{E}\left[\mathrm{Var}(X_1 \mid X_2)\right] + \mathrm{Var}(\mathrm{E}(X_1|X_2)),$$

(8.11) can be written as

$$\mathrm{Cov}(X_1, X_3) \geq 2\mathrm{Var}(\mathrm{E}(X_1|X_2)) - \mathrm{Var}(X_1) = 2\mathrm{Var}(\mathrm{E}(X_1|X_2)) - \sigma^2.$$

From Lemma 8.4.1,

$$\mathrm{Cov}(X_1, X_3) \geq 2\rho^2\sigma^2 - \sigma^2 = \sigma^2(2\rho^2 - 1)$$

or $\rho_{13} \geq 2\rho^2 - 1$, with equality only if $\mathrm{E}(X_1|X_2)$ is linear in X_2. Note that $\rho_{13} \geq 2\rho^2 - 1$ is the inequality from (8.7) with $\rho_{12} = \rho_{23} = \rho$.

Hence, from the above, for $F_{12} = F_{32}$, the lower bound for ρ_{13} is achievable only if two conditions hold:

(a) there is a function b such that $X_1 + X_3 = b(X_2)$,

(b) $\mathrm{E}(X_1|X_2)$ $(= \mathrm{E}(X_3|X_2))$ is linear in X_2.

For both conditions to hold, we must have

$$\mathrm{E}(X_1 + X_3|X_2) = 2\mathrm{E}(X_1|X_2) = b(X_2)$$

is linear in X_2. From (a) and the assumption that F is symmetric (about 0), there is a constant c such that $X_1+X_3=cX_2$. Furthermore, for the conditional Fréchet lower bound, this must mean

$$X_1=\rho X_2+\epsilon, \quad X_3=\rho X_2-\epsilon,$$

with ϵ a symmetric random variable about 0 satisfying $\mathrm{E}(\epsilon|X_2)=0$, $\mathrm{Var}(\epsilon)=(1-\rho^2)\sigma^2$. Hence (8.9), is a necessary condition for $\rho_{12}=\rho_{23}=\rho$.

The above lead to the following lemma.

Lemma 8.4.2 *Let F be symmetric with variance σ^2, and $X\sim F$. Then, a necessary condition for $S_3(F)$ to equal S_3^* is that, for all $-1<\rho<1$, there is a symmetric random variable $\epsilon(\rho)$ satisfying $\mathrm{E}(\epsilon(\rho)|X)=0$ and $\mathrm{Var}(\epsilon(\rho))=(1-\rho^2)\sigma^2$ such that X has the stochastic representation*

$$X\stackrel{d}{=}\rho X+\epsilon(\rho).$$

To complete the characterization that for F symmetric about 0, $S_3(F)=S_3^*$ if and only if $F\in\mathcal{M}_2$, we go back to consider ρ_{12},ρ_{23} different. From Lemma 8.4.2, linearity of X_1, X_3 in X_2 is needed, so we consider the stochastic representation in (8.9):

$$X_1=\rho_{12}X_2+\epsilon_1, \quad X_3=\rho_{23}X_2+\epsilon_3,$$

where $\epsilon_1=\epsilon(\rho_{12})$, $\epsilon_3=\epsilon(\rho_{23})$. In order for the extreme correlations in (8.7) to be attainable, the conditional Fréchet lower and upper bounds must correspond to conditional correlations of ±1. Hence ϵ_1,ϵ_3 must be able to be linearly related. This means the family of random variables $\epsilon(\rho)$ must be related with stochastic representation $\epsilon(\rho)=c(\rho)\epsilon_0$. If ϵ_0 is taken as $\epsilon(0)$ with variance σ^2, then $c(\rho)=\sqrt{1-\rho^2}$. As $\rho\to0$, we get that $\epsilon_0\sim F$.

We can now write $X_1\stackrel{d}{=}\rho X_2+\sqrt{1-\rho^2}\,\epsilon_0$ for all $\rho\in(-1,1)$. Let $\phi_{\epsilon_0X}(t_1,t_2)=\mathrm{E}\left[\exp\{i(t_1\epsilon_0+t_2X_2)\}\right]$ be the characteristic function of (ϵ_0,X_2), and let ϕ_{X_1} be the characteristic function of X_1. The stochastic representation for X_1 means that

$$\phi_{X_1}(t)=\phi_{\epsilon_0X}\left(t\sqrt{1-\rho^2}\,,t\rho\right)$$

for all $\rho\in(-1,1)$. Because F is symmetric, this means that there is a function ψ such that $\phi_{\epsilon_0X}(t_1,t_2)=\psi(t_1^2+t_2^2)$, that is, (ϵ_0,X_2) has bivariate spherical distribution and $F\in\mathcal{M}_2$.

8.4.2 $d>3$

The above ideas extend to any dimension $d\geq4$. For example with $d=4$, with the conditional Fréchet bounds given F_{123},F_{423}, the bounds on ρ_{14} depend on $\rho_{12},\rho_{23},\rho_{13},\rho_{34},\rho_{24}$. For $d=4$, consider the Fréchet class $\mathcal{F}(F_{123},F_{423})$ where

F_{123}, F_{423} have a common F_{23} bivariate margin, and $F_1 = F_2 = F_3 = F_4 = F$. As a special case of (8.4), members of this class have the form

$$F_{1234}(x_1, x_2, x_3, x_4) = \int_{-\infty}^{x_2} \int_{-\infty}^{x_3} F_{14|23}(x_1, x_4|z_2, z_3)\, dF_{23}(z_2, z_3).$$

The conditional Fréchet lower and upper bounds lead to the smallest and largest value of ρ_{14}. The bounds for ρ_{14} based on (8.3) can be attained if $F \in \mathcal{M}_3$ which include the $U(-1, 1)$ distribution.

The algorithmic details, which include an extension of the stochastic representation in (8.9), for simulation of four $U(-1, 1)$ dependent random variables with ρ_{14} at the upper or lower bound are given below. For simulation with other $F \in M_3$, replace "uniform on the three-dimensional sphere" with the spherical distribution with margin F:

- Input $\rho_{12}, \rho_{13}, \rho_{23}, \rho_{42}, \rho_{43}$. Check that $(\rho_{12}, \rho_{13}, \rho_{23})$ leads to a positive definite matrix; same for $(\rho_{42}, \rho_{43}, \rho_{23})$;
- The bounds for ρ_{14} given $\rho_{12}, \rho_{13}, \rho_{23}, \rho_{42}, \rho_{43}$:
  $t_1 = 1 - (\rho_{12}\rho_{12} + \rho_{13}\rho_{13} - 2\rho_{12}\rho_{13}\rho_{23})/(1 - \rho_{23}^2)$;
  $t_2 = 1 - (\rho_{42}\rho_{42} + \rho_{43}\rho_{43} - 2\rho_{42}\rho_{43}\rho_{23})/(1 - \rho_{23}^2)$;
  $t_3 = ((\rho_{12} - \rho_{13}\rho_{23})\rho_{42} + (\rho_{13} - \rho_{12}\rho_{23})\rho_{43})/(1 - \rho_{23}^2)$;
  $\rho_{14u} = \sqrt{t_1 t_2} + t_3$; $\rho_{14l} = -\sqrt{t_1 t_2} + t_3$;
- Initialization from Cholesky decompositions:
  $a_{23} = (1 - \rho_{23}^2)^{1/2}$; $a_{13} = (\rho_{13} - \rho_{12}\rho_{23})/a_{23}$; $a_{43} = (\rho_{43} - \rho_{42}\rho_{23})/a_{23}$;
  $a_{11} = (1 - \rho_{12}^2 - a_{13}^2)^{1/2}$; $a_{44} = (1 - \rho_{42}^2 - a_{43}^2)^{1/2}$;
- Repeat for simulation: Generate (z_1, z_2, z_3) uniform on three-dimensional sphere of radius 1:
  $x_2 = z_2$;
  $x_3 = \rho_{23} z_2 + a_{23} z_3$;
  $x_1 = \rho_{12} z_2 + a_{13} z_3 + a_{11} z_1$;
  $x_{4u} = \rho_{42} z_2 + a_{43} z_3 + a_{44} z_1$; [for ρ_{14u} (conditional correlation $= 1\ \forall z_2, z_3$)]
  $x_{4l} = \rho_{42} z_2 + a_{43} z_3 - a_{44} z_1$; [for ρ_{14l} (conditional correlation $= -1\ \forall z_2, z_3$)]
- To get uniform(0,1) random variables, let $u_j = (x_j + 1)/2$, $j = 1, 2, 3, 4l, 4u$.

The above algorithm extends to dimensions $d > 4$. Using the matrix notation of (8.2), let $R[1 : d-1]$ and $R[2 : d]$ be positive definite correlation submatrices, and let A be a lower triangular matrix in the Cholesky decomposition of $R[2 : d-1]$, i.e., $R[2 : d-1] = AA^T$. Let $\mathbf{a}_d^T = (a_{d2}, \ldots, a_{d,d-1}, a_{dd})$ be the last row in the Cholesky decomposition $R[2 : d] = \binom{A}{\mathbf{a}_d^T}(A^T\ \mathbf{a}_d)$, and let $\mathbf{a}_1^T = (a_{12}, \ldots, a_{1,d-1}, a_{11})$ be the last row in the Cholesky decomposition

$$\begin{pmatrix} R[2 : d-1] & \mathbf{r} \\ \mathbf{r}^T & 1 \end{pmatrix} = \begin{pmatrix} A \\ \mathbf{a}_1^T \end{pmatrix} \begin{pmatrix} A^T & \mathbf{a}_1 \end{pmatrix},$$

where $\mathbf{r}^T = (\rho_{12}, \ldots, \rho_{1,d-1})$.

The algorithm for generating random variables at the conditional Fréchet upper or lower bound is the following:

- Generate $(z_1, \ldots, z_{d-1})$ spherical with $z_j \sim F \in \mathcal{M}_{d-1}$.
- Let $(x_2, \ldots, x_{d-1})^T = A(z_2, \ldots, z_{d-1})^T$, let $x_1 = a_{12}z_2 + \cdots + a_{1,d-1}z_{d-1} + a_{11}z_1$, and $x_d = a_{d2}z_2 + \cdots + a_{d,d-1}z_{d-1} \pm a_{dd}z_1$, with the sign determining the conditional Fréchet upper or lower bound.

As before, for a value of ρ_{1d} between the upper and lower bounds, an appropriate convex combination of the conditional Fréchet upper and lower bound distributions can be used.

The above shows that for $F \in \mathcal{M}_{d-1}$, one can achieve the lower/upper bound for ρ_{1d} and any value in between the bounds given arbitrary correlation matrices for $(X_1, \ldots, X_{d-1})$ and $(X_2, \ldots, X_d)$ [recall from Section 8.2, that any $(d-1)$-dimensional correlation matrix is possible for $F \in \mathcal{M}_{d-1}$].

To prove our general result in $d \geq 4$ dimensions, we extend the lemmas in the previous subsection. To get some necessary conditions for $S_d(F) = S_d^*$, we take the case of arbitrary $R[2:d-1]$, and $\rho_{1j} = \rho_{dj}$, $j = 2, \ldots, d-1$, to get some simpler inequalities for ρ_{1d}. Using an argument like that for $d = 3$, then we can assume without loss of generality that- $(X_1, X_2, \ldots, X_{d-1}) \stackrel{d}{=} (X_d, X_2, \ldots, X_{d-1})$ or $F_{12\cdots d-1} = F_{d2\cdots d-1}$.

The extension of Lemma 8.4.1 is as follows.

Lemma 8.4.3 *Let $X_1, X_2, \ldots, X_{d-1}$ be random variables with variance σ^2 and correlation matrix* $R[1:d-1] = \begin{pmatrix} 1 & \mathbf{r}_1^T \\ \mathbf{r}_1 & R_2 \end{pmatrix}$, *where R_2 is nonsingular and* $\mathbf{r}_1^T = (\rho_{12}, \ldots, \rho_{1,d-1})$. *Let $\mathbf{Z} = (X_2, \ldots, X_{d-1})^T$. Then*

$$\mathrm{Var}\,(\mathrm{E}\,(X_1|\mathbf{Z})) \geq \sigma^2 \mathbf{r}_1^T R_2^{-1} \mathbf{r}_1$$

with equality only if $\mathrm{E}\,(X_1|\mathbf{Z})$ *is linear in* $\mathbf{Z}$.

PROOF. The conditional expectation $g(\mathbf{Z}) = \mathrm{E}\,(X_1|\mathbf{Z})$ is the function of $\mathbf{Z}$ that minimizes $\mathrm{E}\,\{[X_1 - h(\mathbf{Z})]^2\}$ over real-valued functions $h(\mathbf{Z})$. Therefore, for any linear function h,

$$\mathrm{E}\,\{[X_1 - g(\mathbf{Z})]^2\} \leq \mathrm{E}\,\{[X_1 - h(\mathbf{Z})]^2\}. \tag{8.12}$$

From regression theory, the linear function that minimizes the right-hand side of (8.12) is $\mathbf{r}_1^T R_2^{-1}\mathbf{Z}$; that is,

$$\mathrm{E}\,\{[X_1 - g(\mathbf{Z})]^2\} \leq \mathrm{E}\,\{[X_1 - \mathbf{r}_1^T R_2^{-1}\mathbf{Z})]^2\}. \tag{8.13}$$

Assume that $X_1, \ldots, X_{d-1}$ have been standardized so that $\mathrm{E}\,(X_j) = 0$, $j = 1, \ldots, d-1$. Then, (8.13) simplifies to

$$\mathrm{E}\,(X_1^2) - \mathrm{E}\,\{g^2(\mathbf{Z})\} \leq \mathrm{E}\,(X_1^2) - 2\mathrm{E}\,\{X_1\mathbf{r}_1^T R_2^{-1}\mathbf{Z}\} + \mathrm{E}\,\{(\mathbf{r}_1^T R_2^{-1}\mathbf{Z})^2\}$$

or

$$\mathrm{Var}\,(\mathrm{E}\,(X_1|\mathbf{Z})) - \mathrm{E}\,\{g^2(\mathbf{Z})\} \geq 2\mathrm{Cov}\,(X_1, \mathbf{r}_1^T R_2^{-1}\mathbf{Z})) - \sigma^2 \mathbf{r}_1^T R_2^{-1}\mathbf{r}_1 = \sigma^2 \mathbf{r}_1^T R_2^{-1}\mathbf{r}_1$$

■

Let $\mathbf{Z} = (X_2, \ldots, X_{d-1})^T$. From (8.3) with $j = 1$ and $m = d-1$, with the assumption of $F_{12\cdots d-1} = F_{d2\cdots d-1}$,

$$\begin{aligned} \mathrm{Cov}\,(X_1, X_d) &= \mathrm{Cov}\,[\mathrm{E}\,(X_1|\mathbf{Z}), \mathrm{E}\,(X_d|\mathbf{Z})] + \mathrm{E}\,[\mathrm{Cov}\,(X_1, X_d|\mathbf{Z})] \\ &\geq \mathrm{Var}\,(\mathrm{E}\,(X_1|\mathbf{Z})) - \mathrm{E}\,[\mathrm{Var}\,(X_1|\mathbf{Z})] = 2\mathrm{Var}\,(\mathrm{E}\,(X_1|\mathbf{Z})) - \sigma^2. \end{aligned} \tag{8.14}$$

Equality holds in (8.14) only if X_1, X_d are negatively linearly related given $\mathbf{Z} = \mathbf{z}$ for all $\mathbf{z}$ or if there exists a function $b(\mathbf{z})$ such that $X_1 + X_d = b(\mathbf{Z})$ for the conditional Fréchet lower bound. Using Lemma 8.4.3,

$$\mathrm{Cov}\,(X_1, X_d) \geq \sigma^2\{2\mathbf{r}_1^T R_2^{-1}\mathbf{r}_1 - 1\}$$

or $\rho_{1d} \geq 2\mathbf{r}_1^T R_2^{-1}\mathbf{r}_1 - 1$, with equality only if $\mathrm{E}\,(X_1|\mathbf{Z})$ is linear in $\mathbf{Z}$. Note that this inequality is also the inequality from (8.3) [with $\mathbf{r}_1 = \mathbf{r}_1(1, d-1) = \mathbf{r}_3(1, d-1)$ and $R_2 = R_2(1, d-1)$].

Hence, from the above, for $F_{12\cdots d-1} = F_{d2\cdots d-1}$, the lower bound for ρ_{1d} is achievable only if two conditions hold:

(a) there is a function b such that $X_1 + X_d = b(\mathbf{Z})$,

(b) $\mathrm{E}\,(X_1|\mathbf{Z})$ $(= \mathrm{E}\,(X_d|\mathbf{Z}))$ is linear in $\mathbf{Z}$.

For both conditions to hold, we must have

$$\mathrm{E}\,(X_1 + X_d|\mathbf{Z}) = 2\mathrm{E}\,(X_1|\mathbf{Z}) = b(\mathbf{Z})$$

is linear in $\mathbf{Z}$. From (a) and the assumption that F is symmetric, there is a vector $\mathbf{c}$ such that $X_1 + X_d = \mathbf{c}^T\mathbf{Z}$. Furthermore, for the conditional Fréchet lower bound, this must mean

$$X_1 = \mathbf{r}_1^T R_2^{-1}\mathbf{Z} + \epsilon, \quad X_d = \mathbf{r}_1^T R_2^{-1}\mathbf{Z} - \epsilon,$$

with ϵ a symmetric random variable about 0 satisfying $\mathrm{E}\,(\epsilon|\mathbf{Z}) = 0$, $\mathrm{Var}\,(\epsilon) = (1 - \mathbf{r}_1^T R_2^{-1}\mathbf{r}_1)\sigma^2$.

The above lead to the following lemma.

Lemma 8.4.4 *Let F be symmetric with variance σ^2 such that $S_{d-1}(F) = S_{d-1}^*$. Suppose $X_j \sim F$, for $j = 1, \ldots, d-1$. Then a necessary condition for $S_d(F)$ to equal S_d^* is that for all nonsingular matrices* $R[1:d-1] = \begin{pmatrix} 1 & \mathbf{r}_1^T \\ \mathbf{r}_1 & R_2 \end{pmatrix}$, *where R_2 is nonsingular and $\mathbf{r}_1^T = (\rho_{12}, \ldots, \rho_{1,d-1})$, there is (a) a random vector $(X_1, \ldots, X_{d-1})^T$ with the given correlation matrix $R[1:d-1]$, and (b) a symmetric random variable $\epsilon(\mathbf{r}_1, R_2)$ satisfying $\mathrm{E}\,(\epsilon(\mathbf{r}_1, R_2)|X_2, \ldots, X_{d-1}) = 0$ and $\mathrm{Var}\,[\epsilon(\mathbf{r}_1, R_2)] = (1 - \mathbf{r}_1^T R_2^{-1}\mathbf{r}_1)\sigma^2$, such that the stochastic representation*

$$X_1 \stackrel{d}{=} \mathbf{r}_1^T R_2^{-1}(X_2, \ldots, X_{d-1})^T + \epsilon(\mathbf{r}_1, R_2)$$

holds.

To complete the characterization that for F symmetric about 0, $S_d(F) = S_d^*$ if and only if $F \in \mathcal{M}_{d-1}$, we consider ρ_{1j}, ρ_{dj} to be different for $j = 2, \ldots, d-1$. From Lemma 8.4.4, linearity of X_1, X_d in $\mathbf{Z} = (X_2, \ldots, X_{d-1})^T$ is needed, so we consider the stochastic representation

$$X_1 = \mathbf{r}_1^T R_2^{-1}\mathbf{Z} + \epsilon_1, \quad X_3 = \mathbf{r}_3^T R_2^{-1}\mathbf{Z} + \epsilon_3,$$

where $\epsilon_1 = \epsilon(\mathbf{r}_1, R_2)$, $\epsilon_3 = \epsilon(\mathbf{r}_3, R_2)$, $\mathbf{r}_3^T = (\rho_{d2}, \ldots, \rho_{d,d-1})$, and $\begin{pmatrix} 1 & \mathbf{r}_1^T \\ \mathbf{r}_1 & R_2 \end{pmatrix}$, $\begin{pmatrix} 1 & \mathbf{r}_3^T \\ \mathbf{r}_3 & R_2 \end{pmatrix}$ are correlation matrices. In order for the extreme correlations in (8.3) to be attainable, the conditional Fréchet lower and upper bounds must correspond to conditional correlations of ± 1. Hence, ϵ_1, ϵ_3 must be able to be linearly related, and the family $\epsilon(\mathbf{r}_1, R_2)$ must be related with stochastic representation $\epsilon(\mathbf{r}_1, R_2) = \epsilon_0\sqrt{1 - \mathbf{r}_1^T R_2^{-1}\mathbf{r}_1}$, where ϵ_0 has variance σ^2. As $\mathbf{r}_1 \to \mathbf{0}$, we get that $\epsilon_0 \sim F$.

We can now write $X_1 \stackrel{d}{=} \mathbf{r}_1^T R_2^{-1}\mathbf{Z} + \epsilon_0\sqrt{1 - \mathbf{r}_1^T R_2^{-1}\mathbf{r}_1}$ for $\mathbf{r}_1$ and R_2 (corresponding to a proper nonsingular correlation matrix). Let $\phi_{\epsilon_0\mathbf{Z}}(t_1, \mathbf{t}_2) = \mathrm{E}\,[\exp\{i(t_1\epsilon_0 + t_2X_2 + \cdots t_{d-1}X_{d-1})\}]$ be the characteristic function of $(\epsilon_0, \mathbf{Z})$, and let ϕ_{X_1} be the characteristic function of X_1. The stochastic representation for X_1 means that

$$\phi_{X_1}(t) = \phi_{\epsilon_0\mathbf{Z}}\left(t\sqrt{1 - \mathbf{r}_1^T R_2^{-1}\mathbf{r}_1}\,, tR_2^{-1}\mathbf{r}_1\right)$$

for all $\mathbf{r}_1, R_2$. Because F is symmetric, this means that there is a function ψ such that $\phi_{\epsilon_0\mathbf{Z}}(t_1, \mathbf{t}_2) = \psi(t_1^2 + \mathbf{t}_2^T R_2\mathbf{t}_2)$. Letting $t_1 = 0$, the marginal distribution of $\mathbf{Z}$ is $\psi(\mathbf{t}_2^T R_2\mathbf{t}_2)$ so that $\mathbf{Z}$ is elliptical. Write $\mathbf{Z} = A_2\mathbf{Z}_0$, where $A_2A_2^T = R_2$ and $\mathbf{Z}_0$ has a spherical distribution. Let $\phi_{\epsilon_0\mathbf{Z}_0}$ be the characteristic function of $(\epsilon_0, \mathbf{Z}_0)$. Then

$$\phi_{\epsilon_0\mathbf{Z}}(t_1, \mathbf{u}_2) = \phi_{\epsilon_0\mathbf{Z}_0}(t_1, A_2^T\mathbf{u}_2) = \psi(t_1^2 + \mathbf{u}_2^T A_2A_2^T\mathbf{u}_2),$$

so that

$$\phi_{\epsilon_0\mathbf{Z}_0}(t_1, \mathbf{t}_2) = \psi(t_1^2 + \mathbf{t}_2^T\mathbf{t}_2).$$

That is, $(\epsilon_0, \mathbf{Z}_0)$ has $(d-1)$-dimensional spherical distribution and $F \in \mathcal{M}_{d-1}$.

8.5 Discussion

Problem 4.17 of my book [Joe (1997)] suggests that $S_d(F) = S_d^*$ if and only if F is a location-scale transform univariate margin of a spherical distribution in d dimensions. At the time of writing the book, I thought the converse was intuitively obvious, because linear expectation properties and finite variances are associated only with elliptical distributions.

Because of queries on the problem, I have in this article filled in the details, which turn out to be quite intricate, and the statement of the problem must be qualified a little to be correct. For the proof of necessary conditions for $S_d(F) = S_d^*$ to hold, linear properties of expectation play an important role.

Acknowledgements. This research was supported with an NSERC Canada grant.

References

1. Caudras, C. M. (1992). Probability distributions with given multivariate margins and given dependence structure, *Journal of Multivariate Analysis*, **42**, 51–66.

2. Davis, P. J., and Rabinowitz, P. (1984). *Methods of Numerical Integration*, Second edition, Academic Press, Orlando.

3. Donnelly, T. G. (1973). Algorithm 462: Bivariate normal distribution, *Communications of the Association for Computing Machinery*, **16**, 638.

4. Emrich, L. J., and Piedmonte, M. R. (1991). A method for generating high-dimensional multivariate binary variates, *The American Statistician*, **45**, 302–304.

5. Fang, K.-T., Kotz, S., and Ng, K.-W. (1990). *Symmetric Multivariate and Related Distributions*, Chapman & Hall, London.

6. Hoeffding, W. (1940). Maßstabinvariante Korrelationstheorie, *Schriftenreihe des Mathematischen Instituts der Universität Berlin*, **5**, 181–233.

7. Joe, H. (1997). *Multivariate Models and Dependence Concepts*, Chapman & Hall, London.

8. Kelker, D. (1970). Distribution theory of spherical distributions and a location–scale parameter generalization, *Sankhyā, Series A*, **32**, 419–430.

9. Sklar, A. (1959). Fonctions de répartition à n dimensions et leurs marges. *Publ. Inst. Statist. Univ. Paris*, **8**, 229–231.

10. Song, P. X. (1997). Generating dependent random numbers with given correlations and margins from exponential dispersion models, *Journal of Statistical Computation and Simulation*, **56**, 317–335.

9

Multifractional Probabilistic Laws

M. D. Ruiz-Medina and J. M. Angulo

University of Granada, Granada, Spain

Abstract: In this paper, we apply the theory of pseudodifferential operators and Sobolev spaces to characterize fractional and multifractional probability densities. In the fractional case, local regularity properties of the probability density function are given in terms of fractional moment conditions satisfied by the characteristic function. Conversely, the parameter defining the order of the fractional Sobolev space where the characteristic function lies provides the index of stability in relation to fractional moment conditions of the probability density. The extension to the multifractional case leads to the introduction of new probabilistic models considering the theory of pseudodifferential operators and fractional Sobolev spaces of variables order.

Keywords and phrases: Bessel distribution, fractional pseudodifferential operators, Laplace distribution, multifractional pseudodifferential operators

9.1 Introduction

Fractional differential calculus allows the definition of functions with fractional regularity/singularity orders that interpolate the classical integer-order differentiable functions. The classical theory of integer-order differential equations is then extended to the theory of fractional diffusions (anomalous diffusions). In particular, the Gaussian kernel is associated with second-order diffusion theory, the heat kernel; see, for example, Gnedenko and Kolmogorov (1954). Stable laws are associated with fractional derivatives in space, that is, fractional diffusion or anomalous diffusion equations; see, for example, Feller (1971) and Samorodnitski and Taqqu (1994). The fractional order of differentiation defines the stability index of the probabilistic law. While the classical diffusion equation represents Brownian motion, anomalous diffusion equations govern fractional Brownian motion and Lévy motion; see Seshadri and West (1982),

Gorenflo and Mainardi (1998), and Meerschaert *et al.* (1999). The definition of fractional probability densities is not necessarily restricted to parabolic equations. Elliptic equations, particularly fractional pseudodifferential elliptic equations, also define important models in probability theory. The symmetric Bessel distribution [Donoghue (1969)], the Linnik distribution, and the generalized Linnik distribution are examples of fractional probability densities, given by fractional elliptic pseudodifferential equations; see Erdogan and Ostrovskii (1998) and Kemp (2003). These equations characterize the local regularity properties of the functions of the fractional Sobolev space where such densities lie.

In this paper, we consider the characterization of Bessel, Linnik, and generalized Linnik distributions in terms of fractional pseudodifferential equations. We then formulate a multifractional version of symmetric Bessel, Linnik, and generalized Linnik distributions, based on the theory of pseudodifferential operators and fractional Sobolev spaces of variable order. We analyze the local regularity/singularity properties of the characteristic function to define fractional moment laws. This analysis is extended to the multifractional case, providing a framework for the introduction of probabilistic laws with heterogeneous heavy tails. Possible extensions in relation to the definition of multistable and multifractal distributions are also discussed.

9.2 Preliminaries

We first introduce basic elements related to the theory of pseudodifferential operators and fractional Sobolev spaces of variable order. Such operators and spaces will be considered in the characterization of fractional and multifractional probability densities in Sections 9.3, 9.4, and 9.5.

Let δ and ρ be real numbers, with $0 \leq \delta < \rho \leq 1$, and let σ be a real-valued function in $\mathcal{B}^{\infty}(\mathbb{R}^n)$, the space of all C^{∞}-functions on $\mathbb{R}^n$ whose derivatives of all orders are bounded. We say that a function $p(\mathbf{x}, \boldsymbol{\xi}) \in \mathcal{B}^{\infty}\left(\mathbb{R}^n_{\mathbf{x}} \times \mathbb{R}^n_{\boldsymbol{\xi}}\right)$ belongs to $\mathcal{S}^{\sigma}_{\rho,\delta}$ if and only if for any multi-indices α and β there exists some positive constant $C_{\alpha,\beta}$ such that

$$|D^{\alpha}_{\boldsymbol{\xi}} D^{\beta}_{\mathbf{x}} p(\mathbf{x}, \boldsymbol{\xi})| \leq C_{\alpha,\beta} \langle \boldsymbol{\xi} \rangle^{\sigma(\mathbf{x}) - \rho|\alpha| + \delta|\beta|}, \tag{9.1}$$

where $D^{\alpha}_{\boldsymbol{\xi}}$ and $D^{\beta}_{\mathbf{x}}$, respectively, denote the derivatives with respect to $\boldsymbol{\xi}$ and $\mathbf{x}$, and $\langle \boldsymbol{\xi} \rangle = (1 + |\boldsymbol{\xi}|^2)^{1/2}$. The following seminorm is considered for the elements of $\mathcal{S}^{\sigma}_{\rho,\delta}$:

$$|p|^{(\sigma)}_l = \max_{|\alpha+\beta| \leq l} \sup_{(\mathbf{x},\boldsymbol{\xi}) \in \mathbb{R}^n \times \mathbb{R}^n} \left\{ |D^{\alpha}_{\boldsymbol{\xi}} D^{\beta}_{\mathbf{x}} p(\mathbf{x}, \boldsymbol{\xi})| \langle \boldsymbol{\xi} \rangle^{-\sigma(\mathbf{x}) + \rho|\alpha| - \delta|\beta|} \right\}.$$

Definition 9.2.1 [Kikuchi and Negoro (1995, 1997)] For $u \in \mathcal{S}(\mathbb{R}^n)$, the set of rapidly decreasing Schwartz functions, and $p \in S^{\sigma}_{\rho,\delta}$, let $P : \mathcal{S}(\mathbb{R}^n) \longrightarrow \mathcal{S}(\mathbb{R}^n)$ be defined as

$$Pu(\mathbf{x}) = (2\pi)^{-n} \int_{\mathbb{R}^n} e^{i\mathbf{x}\boldsymbol{\xi}} p(\mathbf{x}, \boldsymbol{\xi}) \hat{u}(\boldsymbol{\xi}) d\boldsymbol{\xi}, \tag{9.2}$$

where $\hat{u}(\boldsymbol{\xi}) = \int_{\mathbb{R}^n} e^{-i\mathbf{x}\boldsymbol{\xi}} u(\mathbf{x}) d\mathbf{x}$ is the Fourier transform of u. We refer to $P = p(\mathbf{x}, D_{\mathbf{x}})$ as a pseudodifferential operator of variable order with symbol $p \in S^{\sigma}_{\rho,\delta}$. The set of all pseudodifferential operators with symbol p of the class $S^{\sigma}_{\rho,\delta}$ is denoted by $\boldsymbol{S}^{\sigma}_{\rho,\delta}$.

A pseudodifferential operator $P \in \boldsymbol{S}^{\sigma}_{\rho,\delta}$ is elliptic if there exist constants $c > 0$ and $M > 0$ such that

$$|p(\mathbf{x}, \boldsymbol{\xi})| \geq c \langle \boldsymbol{\xi} \rangle^{\sigma(\mathbf{x})}, \quad |\boldsymbol{\xi}| \geq M. \tag{9.3}$$

Furthermore, $Q \in \boldsymbol{S}^{\infty}_{\rho,\delta} = \bigcup_{m\in\mathbb{R}} \boldsymbol{S}^{m}_{\rho,\delta}$ is said to be a left (resp. right) parametric of P if there exists $R_L \in \boldsymbol{S}^{-\infty}_{\rho,\delta} = \bigcap_{m\in\mathbb{R}} \boldsymbol{S}^{m}_{\rho,\delta}$ (resp. $R_R \in \boldsymbol{S}^{-\infty}_{\rho,\delta} = \bigcap_{m\in\mathbb{R}} \boldsymbol{S}^{m}_{\rho,\delta}$) such that

$$QP = I + R_L \quad (\text{resp.} \quad PQ = I + R_R),$$

where I denotes the identity operator. A pseudodifferential operator Q is a parametric of P if and only if Q is simultaneously a left and right parametric of P.

Definition 9.2.2 Let σ be a real-valued function in $\mathcal{B}^{\infty}(\mathbb{R}^n)$. The Sobolev space of variable order σ on $\mathbb{R}^n$ is defined as

$$H^{\sigma(\cdot)}(\mathbb{R}^n) = \left\{ u \in H^{-\infty} = \bigcup_{s\in\mathbb{R}} H^s(\mathbb{R}^n) : \langle D.\rangle^{\sigma(\cdot)} u \in L^2(\mathbb{R}^n) \right\}, \tag{9.4}$$

where

$$\langle D_{\mathbf{x}} \rangle^{\sigma(\mathbf{x})} u = \int_{\mathbb{R}^n} (2\pi)^{-n} \exp(i\mathbf{x}\boldsymbol{\xi}) \langle \boldsymbol{\xi} \rangle^{\sigma(\mathbf{x})} \hat{u}(\boldsymbol{\xi}) d\boldsymbol{\xi} \tag{9.5}$$

with $\langle \boldsymbol{\xi} \rangle = (1 + |\boldsymbol{\xi}|^2)^{1/2}$, as before, and

$$H^s(\mathbb{R}^n) = \left\{ u \in \mathcal{S}'(\mathbb{R}^n) : \langle D_{\mathbf{x}} \rangle^s u \in L^2(\mathbb{R}^n) \right\}.$$

Proposition 9.2.1 [Kikuchi and Negoro (1997)] *The above-introduced fractional Sobolev spaces of variable order satisfy the following properties:*

(i) If $u \in H^{\sigma(\cdot)}(\mathbb{R}^n)$, then, for $P \in \boldsymbol{S}^{\sigma}_{\rho,\delta}$, $Pu \in L^2(\mathbb{R}^n)$.

(ii) Let σ_1 and σ_2 be functions in $\mathcal{B}^{\infty}(\mathbb{R}^n)$, with $\sigma_1(\mathbf{x}) \geq \sigma_2(\mathbf{x})$, for each $\mathbf{x} \in \mathbb{R}^n$. Then, $H^{\sigma_1(\cdot)}(\mathbb{R}^n) \subset H^{\sigma_2(\cdot)}(\mathbb{R}^n)$. In particular, $H^{\sigma(\cdot)}(\mathbb{R}^n) \subset H^{\underline{\sigma}(\cdot)}(\mathbb{R}^n)$.

(iii) $H^{\sigma(\cdot)}(\mathbb{R}^n)$ *is a Hilbert space with the inner product*

$$\begin{aligned}\langle u, v\rangle_{H^{\sigma(\cdot)}(\mathbb{R}^n)} &= \int_{\mathbb{R}^n} \left(\langle D_{\mathbf{x}}\rangle^{\sigma(\mathbf{x})} u\right)(\mathbf{x})\overline{\left(\langle D_{\mathbf{x}}\rangle^{\sigma(\mathbf{x})} v\right)(\mathbf{x})}d\mathbf{x} \\ &+ \int_{\mathbb{R}^n} \left(\langle D_{\mathbf{x}}\rangle^{\underline{\sigma}} u\right)(\mathbf{x})\overline{\left(\langle D_{\mathbf{x}}\rangle^{\underline{\sigma}} v\right)(\mathbf{x})}d\mathbf{x}, \end{aligned} \tag{9.6}$$

where $\underline{\sigma} = \inf_{\mathbf{x}\in\mathbb{R}^n} \sigma(\mathbf{x})$. *Moreover,* $\mathcal{S}(\mathbb{R}^n)$ *is dense in* $H^{\sigma(\cdot)}(\mathbb{R}^n)$.

(iv) Let σ *and* τ *be functions in* $\mathcal{B}^\infty(\mathbb{R}^n)$. *Suppose that* $P \in \boldsymbol{S}^{\sigma}_{\rho,\delta}$. *Then, there exist some constant* $C > 0$ *independent of* P *and some positive integer* l *depending only on* $\sigma, \tau, \rho, \delta$, *and* n *such that*

$$\|Pu\|_{H^{\tau(\cdot)}(\mathbb{R}^n)} \leq C|p|_l^{(\sigma)}\|u\|_{H^{\sigma(\cdot)+\tau(\cdot)}(\mathbb{R}^n)},$$

for $u \in H^{\sigma(\cdot)+\tau(\cdot)}(\mathbb{R}^n)$, *which provides the continuity of* P *from* $H^{\sigma(\cdot)+\tau(\cdot)}(\mathbb{R}^n)$ *into* $H^{\tau(\cdot)}(\mathbb{R}^n)$.

Theorem 9.2.1 [Kikuchi and Negoro (1997) *Let* $P \in \boldsymbol{S}^{\sigma}_{\rho,\delta}$ *be elliptic. Then,*

$$H^{\sigma(\cdot)}(\mathbb{R}^n) = \left\{u \in H^{-\infty}(\mathbb{R}^n) : \; Pu \in L^2(\mathbb{R}^n)\right\} \tag{9.7}$$

as a set. Moreover, the norm $\|u\|_{H^{\sigma(\cdot)}(\mathbb{R}^n)}$ *is equivalent to the norm*

$$\|u\|_{H^{\sigma(\cdot),P}(\mathbb{R}^n)} = \left(\|Pu\|^2_{L^2(\mathbb{R}^n)} + \|u\|^2_{H^{\underline{\sigma}}(\mathbb{R}^n)}\right)^{1/2}. \tag{9.8}$$

The following results on embeddings and lifting properties for fractional Sobolev spaces of variable order on $L^p(\mathbb{R}^n)$ hold (see Jacob and Leopold, 1993).

Theorem 9.2.2 *Let* $1 < p < \infty$ *and* $j \in \mathbb{N}$, *and let* $\sigma(\mathbf{x}) = s + \psi(\mathbf{x})$, *with* $\psi \in \mathcal{S}(\mathbb{R}^n)$, *satisfying* $0 < m' \leq \sigma(\mathbf{x}) \leq m \leq 2$, *for all* $\mathbf{x} \in \mathbb{R}^n$. *Then, the following assertions hold:*

(i) The space

$$H_p^{j,\sigma(\cdot)}(\mathbb{R}^n) = \left\{f \in \mathcal{S}'(\mathbb{R}^n) : \langle D_{\mathbf{x}}\rangle^{j\sigma(\mathbf{x})} f \in L^2(\mathbb{R}^n)\right\}$$

is a Banach space and $C_0^\infty(\mathbb{R}^n)$ *is dense in this space.*

(ii) For $m'j > n/p$, *the embedding of* $H_p^{j,\sigma(\cdot)}(\mathbb{R}^n)$ *into* $C^\infty(\mathbb{R}^n)$ *is continuous.*

9.3 Fractional Differential Characterization

In this section, the symmetric Bessel, Linnik, and generalized Linnik distributions are characterized as the fundamental solutions (Green functions) of fractional pseudodifferential models. We apply the spectral theory of self-adjoint operators on a Hilbert space [see, e.g., Dautray and Lions (1985)], and, in particular, of fractional pseudodifferential operators on fractional Sobolev spaces [see, e.g., Triebel (1997)].

The characteristic function $\hat{f}_X$ of the symmetric Bessel distribution is given by

$$\hat{f}_X(\lambda) = E[\exp\{i\lambda X\}] = \frac{1}{(1+\lambda^2)^{\alpha}}, \quad 0 < \alpha < 1. \tag{9.9}$$

Equation (9.9) provides the Fourier transform of the Bessel potential kernel [see Stein (1970)]. From Dautray and Lions (1985, pp. 119–126 and p. 140), the probability density f_X then satisfies the fractional pseudodifferential equation

$$(I-\Delta)^{\alpha} f_X(x) = \left(I - \frac{d^2}{dx^2}\right)^{\alpha} f_X(x) = \delta(x), \quad x \in \mathbb{R}, \tag{9.10}$$

where $-\Delta$ represents the negative Laplacian operator on $\mathbb{R}$, and δ denotes the Dirac-delta distribution. That is, f_X is the fundamental solution to Eq. (9.10). Thus, the probability density f_X belongs to the fractional Sobolev space $H^{2\alpha}(\mathbb{R})$. Local regularity and asymptotic properties of f_X are then given as follows:

(i) For $\alpha > 1/4$, f_X is Hölder continuous. For $\alpha < 1/4$, f_X is square-integrable, and its local properties must be analyzed in terms of suitable test function systems [see embedding theorems between fractional Besov spaces, Triebel (1978)].

(ii) The fractional heavy tail behaviour of f_X follows from the well-known asymptotic properties of the Bessel potential kernel. We then have

$$f_X(z) = \mathcal{O}\left(|x|^{-1-\alpha}\right), \quad |x| \longrightarrow \infty;$$

see Donoghue (1969) and Stein (1970).

Let Y be a random variable with Linnik distribution. Then, Y has its characteristic function as

$$\hat{f}_Y(\lambda) = E[\exp\{i\lambda Y\}] = \frac{1}{1+|\lambda|^{\beta}}, \quad 0 < \beta < 1.$$

The probability density f_Y satisfies the fractional pseudodifferential equation

$$\left(I + \left(-i\frac{d}{dy}\right)^{\beta}\right) f_Y(y) = \delta(y), \quad y \in \mathbb{R}; \tag{9.11}$$

see Dautray and Lions (1985, pp. 119–126 and p. 140). Thus, f_Y belongs to the fractional Sobolev space $H^{\beta}(\mathbb{R})$, and f_Y is Hölder continuous for $\beta > 1/2$, and square-integrable for $\beta < 1/2$, having square-integrable weak-sense fractional derivatives up to order β.

For a random variable Z with generalized Linnik distribution, the characteristic function $\hat{f}_Z$ of Z is given by

$$\hat{f}_Z(z) = E[\exp\{i\lambda Z\}] = \frac{1}{(1+|\lambda|^{\beta})^{\nu}}, \quad 0 < \beta\nu < 1;$$

see Erdogan and Ostrovskii (1998). The probability density f_Z is then the fundamental solution of the fractional pseudodifferential equation

$$\left(I + \left(-i\frac{d}{dz}\right)^{\beta}\right)^{\nu} f_Z(z) = \delta(z), \quad z \in \mathbb{R}. \tag{9.12}$$

Remark 9.3.1 Note that in reliability theory the failure rate function is defined in terms of a differential model. Here, we adopt this framework to characterize the probability density. Such characterization is specially useful in engineering systems.

9.4 Multifractional Versions

In the examples analyzed in the previous section, the fractional parameter α characterizes the symmetric Bessel distribution, and the fractional parameters β and ν characterize the Linnik and generalized Linnik distributions. In this section, we formulate the multifractional versions of these distributions, given by a functional parameter defining their heterogeneous local regularity properties. The conditions assumed on the functional parameter are given in Section 9.2, because we are considering the space $\mathcal{S}^{\sigma}_{\rho,\delta}$ in the formulation of the characteristic function.

Let $\sigma(\cdot) = \alpha(\cdot)$ be a real-valued function in $\mathcal{B}^{\infty}(\mathbb{R})$ satisfying

$$\begin{aligned} \overline{\alpha} &= \sup_{x\in\mathbb{R}} \alpha(x) < 1, \\ \underline{\alpha} &= \inf_{x\in\mathbb{R}} \alpha(x) > 0. \end{aligned} \tag{9.13}$$

The following multifractional version of the symmetric Bessel distribution is considered:

$$\begin{aligned} \hat{f}_{\widetilde{X}}(\lambda) &= E[\exp\{i\lambda\widetilde{X}\}] = \frac{1}{(1+\lambda^2)^{\alpha(\cdot)}}, \\ f_{\widetilde{X}}(\widetilde{x}) &= \int_{\mathbb{R}} \exp\{-i\lambda\widetilde{x}\}\frac{1}{(1+\lambda^2)^{\alpha(\widetilde{x})}}\,d\lambda. \end{aligned} \tag{9.14}$$

Equivalently, $f_{\widetilde{X}}$ is the fundamental solution of the multifractional pseudo-differential equation

$$(I-\Delta)^{\alpha(\widetilde{x})} f_{\widetilde{X}}(\widetilde{x}) = \left(I - \frac{d^2}{d\widetilde{x}^2}\right)^{\alpha(\widetilde{x})} f_X(\widetilde{x}) = \delta(\widetilde{x}), \quad \widetilde{x} \in \mathbb{R}. \tag{9.15}$$

Function $f_{\widetilde{X}}$ belongs to the fractional space $H^{2\alpha(\cdot)}(\mathbb{R})$ defined in Eq. (9.4). Under condition (9.13), $f_{\widetilde{X}}$ is a probability density. Properties of this function can be formulated in terms of upper and lower bounds, based on the associated symmetric Bessel distributions

$$f_{X_1}(x_1) = \int_{\mathbb{R}} \exp\{-i\lambda x_1\} \frac{1}{(1+\lambda^2)^{\underline{\alpha}}} d\lambda \tag{9.16}$$

$$f_{X_2}(x_2) = \int_{\mathbb{R}} \exp\{-i\lambda x_2\} \frac{1}{(1+\lambda^2)^{\overline{\alpha}}} d\lambda. \tag{9.17}$$

Note that the asymptotic behavior of $f_{\widetilde{X}}$ is now characterized in terms of the functional parameter $\alpha(\cdot)$. The probability of extreme values is then respectively upper and lower bounded by the fractional power laws $|x|^{-1-\underline{\alpha}}$ and $|x|^{-1-\overline{\alpha}}$. Thus, $f_{\widetilde{X}}$ presents a heterogeneous band-limited heavy tail behavior.

Let now $\beta(\cdot)$ and $\nu(\cdot)$ be real-valued functions in $\mathcal{B}^{\infty}(\mathbb{R})$ satisfying

$$\overline{\beta\nu} = \sup_{x\in\mathbb{R}} \beta(x)\nu(x) < 1,$$

$$\underline{\beta\nu} = \inf_{x\in\mathbb{R}} \beta(x)\nu(x) > 0. \tag{9.18}$$

The functional parameter versions of the Linnik and generalized Linnik distributions can then be formulated as follows:

$$\hat{f}_{\widetilde{Y}}(\lambda) = E[\exp\{i\lambda\widetilde{Y}\}] = \frac{1}{1+|\lambda|^{\beta(\cdot)}},$$

$$f_{\widetilde{Y}}(\widetilde{y}) = \int_{\mathbb{R}} \exp\{-i\lambda\widetilde{y}\} \frac{1}{1+|\lambda|^{\beta(\widetilde{y})}} d\lambda, \tag{9.19}$$

$$\hat{f}_{\widetilde{Z}}(\lambda) = E[\exp\{i\lambda\widetilde{Z}\}] = \frac{1}{\left(1+|\lambda|^{\beta(\cdot)}\right)^{\nu(\cdot)}},$$

$$f_{\widetilde{Z}}(\widetilde{z}) = \int_{\mathbb{R}} \exp\{-i\lambda\widetilde{z}\} \frac{1}{\left(1+|\lambda|^{\beta(\widetilde{z})}\right)^{\nu(\widetilde{z})}} d\lambda. \tag{9.20}$$

Functions $f_{\widetilde{Y}}$ and $f_{\widetilde{Z}}$, respectively, are the fundamental solutions to the following multifractional pseudodifferential equations:

$$\left(I + \left(-i\frac{d}{d\widetilde{y}}\right)^{\beta(\widetilde{y})}\right) f_Y(\widetilde{y}) = \delta(\widetilde{y}), \quad \widetilde{y} \in \mathbb{R}, \tag{9.21}$$

$$\left(I + \left(-i\frac{d}{d\widetilde{z}}\right)^{\beta(\widetilde{z})}\right)^{\nu(\widetilde{z})} f_Z(\widetilde{z}) = \delta(\widetilde{z}), \quad \widetilde{z} \in \mathbb{R}. \tag{9.22}$$

They belong to the fractional Sobolev spaces $H^{\beta(\cdot)}(\mathbb{R})$ and $H^{\beta(\cdot)\nu(\cdot)}(\mathbb{R})$. From condition (9.18), $f_{\widetilde{Y}}$ and $f_{\widetilde{Z}}$ are probability densities. Indeed, similar to the multifractional symmetric Bessel model, the local regularity/singularity properties of the multifractional Linnik and generalized Linnik distributions can be formulated in terms of the local properties of ordinary Linnik distributions with parameters $\overline{\beta}$ and $\underline{\beta}$, and generalized Linnik distributions with parameters $(\overline{\beta}, \overline{\nu})$ and $(\underline{\beta}, \underline{\nu})$.

9.5 Fractional and Multifractional Moment Laws

In this section, we study the local regularity properties of characteristic functions that belong to an element of the continuous scale of fractional Sobolev spaces. From this study, we infer the existence of fractional moments of the associated probability distribution. The multifractional formulation of this characteristic function family, using the theory of fractional Sobolev spaces and pseudodifferential operators of variable order, allows the introduction of heterogeneous heavy-tail probabilistic laws.

Fractional Sobolev spaces of fixed order can be considered as particular cases of fractional Sobolev spaces of variable order (see Definition 9.2.2). Specifically, for $s \in \mathbb{R}$, $H^s(\mathbb{R})$ is the space of tempered distributions u such that

$$(1+ \mid \xi \mid^2)^{s/2}\hat{u}(\xi) \in L^2(\mathbb{R}), \quad \xi \in \mathbb{R}. \tag{9.23}$$

In this space the following inner product is considered:

$$\langle u, v \rangle_{H^s(\mathbb{R})} = \int_{\mathbb{R}} (1+ \mid \xi \mid^2)^s \hat{u}(\xi)\hat{v}(\xi) d\xi, \tag{9.24}$$

with associated norm

$$\| u \|_{H^s(\mathbb{R})} = \left(\int_{\mathbb{R}} (1+ \mid \xi \mid^2)^s \mid \hat{u}(\xi) \mid^2 d\xi \right)^{1/2},$$

where $\hat{}$ stands for the Fourier transform.

Because operator $(-\Delta)^{s/2}$ defines an equivalent norm in the space $H^s(\mathbb{R})$, that is, $(-\Delta)^{s/2}$ is bounded and elliptic in such space [see, for example, Stein (1970) and Triebel (1978)], then, for any function $u \in H^s(\mathbb{R})$,

$$\int_{\mathbb{R}} |\hat{u}(\xi)|^2 |\xi|^{2s} d\xi < \infty. \tag{9.25}$$

From Eq. (9.25),

$$|\hat{u}(\xi)| = \mathcal{O}\left(|\xi|^{-1-s-\varepsilon}\right), \quad |\xi| \longrightarrow \infty, \tag{9.26}$$

for a certain $\varepsilon > 0$.

Let X be a random variable with characteristic function $\hat{f}_X \in H^s(\mathbb{R})$, with $s > 0$, that is,

$$(I - \Delta)^{s/2}\hat{f}_X \in L^2(\mathbb{R}), \quad s > 0.$$

From Eq. (9.25), the probability density f_X satisfies

$$\int_{\mathbb{R}} |f_X(x)|^2 |x|^{2s} d\xi < \infty. \tag{9.27}$$

Hence, f_X has finite fractional moments up to order s. From Eq. (9.26), depending on the range considered for the parameter $s+\varepsilon$, f_X can be a heavy-tail distribution. For example, X has infinite variance for $s + \varepsilon \in (0, 2)$, and has infinite first-order moment for $s + \varepsilon \in (0, 1)$.

9.5.1 Multifractional moment laws

The asymptotic properties of probability densities with associated characteristic function in a fractional Sobolev space of variable order are now studied.

Let $\hat{f}_{\widetilde{X}}$ be the characteristic function of a random variable $\widetilde{X}$. Assume that $\hat{f}_{\widetilde{X}} \in H^{\sigma(\cdot)}(\mathbb{R})$. Then, $\hat{f}_{\widetilde{X}}$ satisfies

$$\langle D.\rangle^{\sigma(\cdot)}\hat{f}_{\widetilde{X}} = (I - \Delta)^{\sigma(\cdot)/2}\hat{f}_{\widetilde{X}} \in L^2(\mathbb{R}),$$

that is,

$$\int_{\mathbb{R}} \exp\{i\widetilde{x}\xi\} f_{\widetilde{X}}(\widetilde{x})(1 + |\widetilde{x}|^2)^{\sigma(\xi)/2} d\widetilde{x}, \quad \xi \in \mathbb{R}, \tag{9.28}$$

belongs to $L^2(\mathbb{R})$. Furthermore, for $0 < \underline{\sigma} < \sigma(\widetilde{x}) < \overline{\sigma}$, with $\widetilde{x} \in \mathbb{R}$,

$$\begin{aligned}
C_1\left[(-\Delta)^{\underline{\sigma}/2}\hat{f}_{\widetilde{X}}\right](\xi) &= C_1\int_{\mathbb{R}} \exp\{i\widetilde{x}\xi\}\, |f_{\widetilde{X}}(\widetilde{x})||\widetilde{x}|^{\underline{\sigma}} d\widetilde{x} \\
&\leq \int_{\mathbb{R}} \exp\{i\widetilde{x}\xi\}\, |f_{\widetilde{X}}(\widetilde{x})|(1 + |\widetilde{x}|^2)^{\underline{\sigma}/2} d\widetilde{x} \\
&\leq \int_{\mathbb{R}} \exp\{i\widetilde{x}\xi\}\, |f_{\widetilde{X}}(\widetilde{x})|(1 + |\widetilde{x}|^2)^{\sigma(\xi)/2} d\widetilde{x} \\
&\leq \int_{\mathbb{R}} \exp\{i\widetilde{x}\xi\}\, |f_{\widetilde{X}}(\widetilde{x})|(1 + |\widetilde{x}|^2)^{\overline{\sigma}/2} d\widetilde{x} \\
&\leq C_2\int_{\mathbb{R}} \exp\{i\widetilde{x}\xi\}\, |f_{\widetilde{X}}(\widetilde{x})||\widetilde{x}|^{\overline{\sigma}} d\widetilde{x} \\
&= C_2\left[(-\Delta)^{\overline{\sigma}/2}\hat{f}_{\widetilde{X}}\right](\xi).
\end{aligned} \tag{9.29}$$

Thus, the asymptotic behavior of $f_{\widetilde{X}}$ is upper and lower bounded by the asymptotic behavior of the probability densities whose characteristic functions are in the spaces $H^{\overline{\sigma}}(\mathbb{R})$ and $H^{\underline{\sigma}}(\mathbb{R})$, respectively. In that sense, we can say that $f_{\widetilde{X}}$ presents an heterogeneous heavy-tail behavior according to Eq. (9.26), with $s = \underline{\sigma}$ and $s = \overline{\sigma}$, considering suitable ranges of such parameters.

Remark 9.5.1 Because heavy-tail probability distributions are in the domain of attraction of stable probability laws, multistable probability distributions can be introduced considering the above framework and the generalized central limit theorem.

9.6 Conclusion

The theory of pseudodifferential operators and fractional Sobolev spaces is considered to characterize the local and asymptotic behavior of fractional probability densities and characteristic functions. The extended theory on pseudodifferential operators and fractional Sobolev spaces of variable order allows the characterization of new probabilistic models with multifractional parameters defining their local regularity and asymptotic properties; see Ruiz-Medina *et al.* (2004) for the Gaussian random field case. Some extensions of the stable probability laws can be formulated in this context from the application of the generalized central limit theorem. Additionally, multifractal probabilistic models can also be constructed using cascade algorithms with log-multistable random weights. The last two aspects will be undertaken in a subsequent paper by the authors.

Acknowledgements. This work was supported in part by project BFM2002-01836 of the DGI, Spain.

References

1. Dautray, R., and Lions, J. L. (1985). *Mathematical Analysis and Numerical Methods for Science and Technology, Spectral Theory and Applications*, Vol. 3, Springer-Verlag, New York.

2. Donoghue, W. J. (1969). *Distributions and Fourier Transforms*, Academic Press, New York.

3. Erdogan, M. B., and Ostrovskii, I. V. (1998). Analytic and asymptotic properties of generalized Linnik probability density, *Journal of Mathematical Analysis and Applications*, **217**, 555–578.

4. Feller, W. (1971). *An Introduction to Probability Theory and Its Applications,* Vol. II, John Wiley & Sons, New York.

5. Gnedenko, B. V., and Kolmogorov, A. N. (1954). *Limit Distributions for Sums of Random Variables*, Addison-Wesley, Reading, MA.

6. Gorenflo, R., and Mainardi, F. (1998). Fractional calculus and stable probability distributions, *Archive of Mechanics,* **50**, 377–388.

7. Jacob, N., and Leopold, H.-G. (1993). Pseudodifferential operators with variable order of differentiation generating Feller semigroup, *Integral Equation Operator Theory*, **17**, 544–553.

8. Kemp, F. (2003). The Laplace distribution and generalizations: a revisit with applications to communications, economics, engineering, and finance, *Journal of the Royal Statistical Society, Series D*, **52**, 698–699.

9. Kikuchi, K., and Negoro, A. (1995). Pseudodifferential operators with variable order of differentiation, *Reports of Liberal Arts and Science Faculty, Shizuoka University*, **31**, 19–27.

10. Kikuchi, K., and Negoro, A. (1997). On Markov processes generated by pseudodifferential operator of variable order, *Osaka Journal of Mathematics*, **34**, 319–335.

11. Meerschaert, M. M., Benson, D. A., and Baümer, B. (1999). Multidimensional advection and fractional dispersion, *Physical Review E*, **59**, 5026–5028.

12. Ruiz-Medina, M. D., Anh, V. V., and Angulo, J .M. (2004). Fractional generalized random fields of variable order, *Stochastic Analysis and Its Applications*, **22**, 775–799.

13. Samorodnitsky, G., and Taqqu, M. S. (1994). *Stable Non-Gaussian Random Processes: Stochastic Models with Infinite Variance*, Chapman & Hall, New York.

14. Seshadri, V., and West, B.J. (1982). Fractal dimensionality of Lévy processes, *Proceedings of the National Academy of Sciences,* **79**, 4501–4505.

15. Stein, E. M. (1970). *Singular Integrals and Differential Properties of Functions*, Princeton University Press, Princeton, NJ.

16. Triebel, H. (1978). *Interpolation Theory, Function Spaces, Differential Operators*, North-Holland Publishing Co., Amsterdam.

17. Triebel, H. (1997). *Fractals and Spectra*, Birkhäuser, Boston.

PART III
Order Statistics and Applications

10

Topics in the History of Order Statistics

H. A. David

Iowa State University, Ames, IA, USA

Abstract: The term "order statistics" was introduced only in 1942, by Wilks. However, the subject is much older, astronomers having long been interested in estimates of location beyond the sample mean. By early in the nineteenth century measures considered included the median, symmetrically trimmed means, the midrange, and related functions of order statistics. In 1818, Laplace obtained (essentially) the distribution of the rth-order statistic in random samples and also derived a condition on the parent density under which the median is asymptotically more efficient than the mean. Other topics considered are of more recent origin: extreme-value theory and the estimation of location and scale parameters by order statistics.

Keywords and phrases: Measures of location, distribution theory, extreme-value theory, estimation of parameters

10.1 Introduction

Before we get into the history of order statistics a word on the term "order statistics" is needed. The history is much older than the term which was introduced in Wilks (1942, p. 401) for the ordered variates in a sample. Wilks was concerned with setting tolerance limits. For example, he asked for the probability that at least N_0 of N measurements on a second random sample will lie between the smallest and the largest value of a first sample of n taken from the same population. This is a nonparametric use of order statistics. In an extensive review paper, actually entitled "Order Statistics," Wilks (1948) deals with both parametric and nonparametric procedures. He includes nonparametric tests based on the ordered observations, such as Friedman's (1937) two-way analysis of variance rank test, but because only the ranks are *required*, such tests are no longer regarded as part of the subject of order statistics. This is in contrast to nonparametric tolerance limits and nonparametric confidence intervals, which do require the order statistics.

An invaluable aid to writing on the history of order statistics is the extraordinary compilation of lightly annotated abstracts of relevant papers up to 1949 prepared by Harter (1978). He has extended his abstracts up to 1969, with a supplement up to 1992, in Harter (1983–1993). Nevertheless, writing a coherent account requires more than a selection from these abstracts. Harter himself has written a historical article (1988). Particularly valuable, as will be seen, are parts of a paper of wider scope by Stigler (1973). I have endeavored to avoid undue repetition and have selected from advances made more than 50 years ago.

As far back as the second century b.c., the Greek astronomer Hipparchus noticed variation in the length of the year. He estimated this as at most 3/4 day, "apparently by taking half the range of his observations" [Plackett (1958)]. This early date is an extreme outlier in the history of order statistics, but outliers in observational data have long drawn astronomers and others to pay special attention to extreme observations. The main concern was the effect of outliers on the estimation of location. One long-standing common sense practice was to take the average of the observations only after eliminating an equal number of the largest and smallest values, that is, to calculate a symmetrically trimmed mean [Anonymous (1821)]. Another approach, also still with us, was to develop what were inevitably debatable rules for the rejection of outliers and to apply these before calculating the mean. We turn now to a more detailed account of measures of location. For measures of dispersion, not confined to order statistics, see David (1998). We take this opportunity to make a slight correction to that paper (p. 375): The exact pdf of the range in random samples was first derived by Craig (1932).

10.2 Early Measures of Location

Given a set of comparable observations $x_1, \ldots, x_n$, or the corresponding order statistics $x_{(1)} \leq \cdots \leq x_{(n)}$, astronomers, geodesists, and others have long searched for the best estimate of the mean μ. From an early time the sample mean, $\bar{x}$, was a natural choice [see, e.g., Plackett (1958)]. The method of least squares [Legendre (1805)] when applied to a random sample leads to $\bar{x}$, that is, $\sum_{i=1}^{n}(x_i - \mu)^2$ is minimized for $\mu = \bar{x}$. Gauss (1809) in fact asked the question: Which distribution makes $\bar{x}$ the most probable estimator of μ? He showed that among symmetric, unimodal, and differentiable pdf's the normal is the one for which $\bar{x}$ is the maximum likelihood estimate. Actually, following Laplace, Gauss used not the likelihood but the posterior density $f(\mu|\bar{x})$ with a uniform prior, which is conceptually different but otherwise equivalent. With this result Gauss gave a great boost to both the use of $\bar{x}$ and the normal distribution, discovered by de Moivre in 1733 as an approximation to the binomial distribution. However, the frequent presence of outliers had made astronomers dubious about relying on $\bar{x}$. Stigler (1986) quotes (in translation from the French) from a 1772 review by Jean Bernoulli III: "The problem of finding the true mean among several observations, which is rarely the arithmetic mean, is of considerable interest to astronomers." Bernoulli goes on to mention Boscovich, Lambert, Daniel Bernoulli, and De La Grange [Lagrange]. Boscovich's method for dealing with linear regression, nearly 50 years before the method of least squares, led in the special case of a random sample to the median, m (for n odd), i.e., $\sum |x_i - \mu|$ is minimized for $\mu = m$.

The median is obviously stabler than the mean. With the help of his central limit theorem Laplace (1818) was able to make an asymptotic comparison of the two, finding the median asymptotically more efficient than the mean if

$$f(0) > 1/(2\sigma), \tag{10.1}$$

where the parent density $f(x)$ is symmetric about zero, with variance σ^2. The inequality (10.1) holds for what has come to be known as the Laplace distribution $f(x) = \frac{1}{2}e^{-|x|}$, $-\infty < x < \infty$, but does not hold for the normal. Moreover, Laplace showed in the normal case that no linear combination of mean and median could improve on the mean alone.

There were other possibilities besides mean and median. Anonymous (1821) in an interesting, wide-ranging discussion (in French) on the choice of location estimate mentions $\frac{1}{2}(x_{(1)} + x_{(n)})$, $\frac{1}{4}(x_{(1)} + x_{(2)} + x_{(n-1)} + x_{(n)})$, etc. and more importantly the trimmed means $(x_{(2)} + \cdots + x_{(n-1)})/(n-2)$, $(x_{(3)} + \cdots + x_{(n-2)})/(n-4)$, etc., He refers to "certain provinces in France, where in order to determine the mean income from a landed property it is customary to consider this income over a period of 20 consecutive years, to subtract the largest and the smallest income, and then to take (1/18)th of the sum of the others." Cournot (1843, p. 142) even cites a law of May 15, 1818, on the transfer of property that assesses value by the average market price over 10 of the preceding 14 years, after elimination of the two highest and two lowest values.

Median. The notion of the median, according to, for example, Hald (1990, p. 108), goes back to the brothers Huygens in 1669, the motivation being Graunt's 1662 life table. In addition to the residual expectation of life, the median residual life time is also featured. For example, from a continuous graph of the estimated survival probability as a function of age, one can easily determine that a person of age 36 (one of the 16% to have reached this age!) has probability $\frac{1}{2}$ of living another 16 years, the median residual life time.

As a more ordinary estimator of location the median makes its first appearance in a surprising way, as a special case of Boscovich's 1757 method of dealing with linear regression, nearly 50 years before Legendre proposed the method of least squares. To determine a and b in the line fitted to points $(x_i, y_i), i = 1, \ldots, n$, Boscovich required

$$\sum (y_i - a - bx_i) = 0$$

and

$$\sum |y_i - a - bx_i| = \text{minimum}.$$

The first condition, sometimes given up by later writers, ensured that the fitted line $y = a + bx$ passed through the centroid. The second condition, minimizing the sum of the absolute deviations, was already natural in its day, but Boscovich had a means of implementing it. His ingenious geometric method is described by Hald (1998, p. 99). Laplace recognized its importance and provided an algebraic proof in 1789. We present here a later version of the proof [Laplace (1818)] which leads explicitly to the median and some of its properties. Considering without essential loss of generality regression through the origin, Laplace minimizes $\sum_{i=1}^{n} |y_i - bx_i|$ as follows: If x_i is negative, change its sign and that of y_i. Then renumber the observations so that $y_1/x_1, y_2/x_2, \ldots$ form a decreasing sequence. Boscovich's choice of b is y_r/x_r, where r is determined by

$$x_1 + \cdots + x_{r-1} < x_r + \cdots + x_n \text{ and } x_1 + \cdots + x_r > x_{r+1} + \cdots + x_n. \quad (10.2)$$

To see that $b_r = y_r/x_r$ minimizes the sum of the absolute deviations, write $e_i = y_i - bx_i$. Then, in view of the renumbering, $e_1, \ldots, e_{r-1}$ will be positive and $e_{r+1}, \ldots, e_n$ will be negative. If b is increased by the infinitesimal quantity δb, the sum of the positive deviations will decrease by

$$(x_1 + \cdots + x_{r-1})\delta b$$

but the sum of the absolute values of the negative deviations will increase by

$$(x_{r+1} + \cdots + x_n)\delta b$$

and e_r will become $-x_r \delta b$. The sum of the absolute values of all the deviations will therefore be increased by

$$(x_r + \cdots + x_n - x_1 - \cdots - x_{r-1})\delta b.$$

By (10.2) this quantity is positive. Likewise, if b_r is decreased by $\delta\beta$, the sum of the absolute deviations will be increased by the positive quantity

$$(x_1 + \cdots + x_r - x_{r+1} - \cdots - x_n)\delta b.$$

Thus in both cases the sum of the absolute deviations is increased. The resulting b_r may be called a weighted median, the ordinary median being the special case $x_1 = \cdots = x_n = 1$ (n odd).

Midrange. For distributions of finite range it has long been realized that the sample mean may not be the best estimator of location and that more weight should be placed on the extreme observations. For example, see Hald (1998, p. 85) for a discussion of Daniel Bernoulli's 1778 approach to estimating μ in the semicircular distribution

$$f(x) = [a^2 - (x-\mu)^2]^{\frac{1}{2}}, \qquad \mu - a \le x \le \mu + a.$$

Fisher (1922), in the course of a famous paper (p. 347) gives an interesting argument for distributions depending only on a location parameter and whose pdf's make a finite angle with the x-axis. Taking the lower endpoint, ℓ, let $f(x) = kx^\alpha$ in the neighborhood of ℓ, where x is the distance from ℓ. Then

$$F(x) = \frac{k}{\alpha+1} x^{\alpha+1}$$

and

$$\begin{aligned} Pr(X_{(1)} > x) &= (1-F)^n \\ &\doteq e^{-Fn}, \end{aligned}$$

the approximation holding when n is large and F correspondingly small. Equating this to e^{-c}, where c is a constant, we have

$$\frac{k}{\alpha+1} x^{\alpha+1} = \frac{c}{n}.$$

This means that if we use $X_{(1)}$ to estimate ℓ, the error x is proportional to $n^{-1/(\alpha+1)}$. For $\alpha < 1$ this quantity decreases more rapidly than $n^{-1/2}$ and, in large samples is therefore superior to the mean as a basis for a location estimator [provided $x_{(1)}$ is not an outlier!]

Interestingly, in the same year Dodd (1922) compared mean, median, and midrange for symmetric distributions. One of his conclusions is that the midrange may be better than the mean if the pdf meets the x-axis at right angles. He establishes the superiority of the midrange M' over both mean and median for a uniform distribution by comparing the densities at the population mean of the three statistics. Dodd also derives the pdf of M' for any distribution having a density as

$$f_{M'}(m) = 2n(n-1)\int_m^\infty [F(y) - F(2m-y)]^{n-2} f(y) f(2m-y)dy,$$

a result overlooked by Gumbel (1958, p. 108) when obtaining the simpler cdf of M' which may be written as

$$F_{M'}(m) = n\int_{-\infty}^m [F(2m-x) - F(x)]^{n-1} f(x)dx.$$

10.3 Distribution Theory

The first derivation of the distribution of $X_{(r)}$ in a random sample $X_1, \ldots, X_n$ from a population with cdf $F(x)$ and pdf $f(x)$ may be ascribed to Laplace (1818). Being incidental to an examination of Boscovich's 1757 method of estimation, later known as L_1-estimation, Laplace's result was largely overlooked until pointed out by Stigler (1973) and explicitly by Hald (1998, p. 448). See also David and Edwards (2001) for a translation, with commentary, of the relevant section of Laplace (1818).

What is truly surprising is that this first derivation occurs in the course of a more general study of the distribution of the rth-order statistic among $X_1/c_1, \ldots, X_n/c_n$, where the c_i are positive constants. Stripped of its specific context, Laplace's reasoning is a generalization of what is now a very familiar argument: If $X_r = x$ is to make X_r/c_r the rth largest among the X_i/c_i, then $r-1$ of the X_i must satisfy $X_i/c_i < x/c_r, n-r$ must satisfy $X_i/c_i > x/c_r$, so that the combined probability is proportional to

$$f(x)\Pi_{i=1}^{r-1}F(c_ix/c_r)\Pi_{i=r+1}^{n}[1-F(c_ix/c_r)].$$

If $c_1 = \cdots = c_n = 1$, this gives, in modern terms, the pdf of $X_{(r)}$ as proportional to

$$g(x) = F^{r-1}(x)[1-F(x)]^{n-r}f(x). \tag{10.3}$$

Laplace assumes $f(x) = f(-x)$, but he uses this symmetry assumption only later when obtaining asymptotic results. We present his asymptotic approach, applying it however not to the special situation considered by him, but to the "near-median" case when $|r - \frac{1}{2}n| < \frac{a}{n}$, where a is a constant. With x assumed small, we have to order x^2

$$F(x) = \frac{1}{2} + xf(0) + \frac{1}{2}x^2 f'(0), \tag{10.4}$$

the last term vanishing by the symmetry assumption. Also

$$f(x) = f(0) + \frac{1}{2}x^2 f''(0).$$

Then to order x^2 we have

$$\begin{aligned}\log g(x) &= -2(n-2r+1)[xf(0) - \log 2] \\ &\quad -2(n-1)[x^2 f^2(0)] + \log[f(0) + \frac{1}{2}x^2 f''(0)].\end{aligned}$$

Assuming x to be of order $1/\sqrt{n}$, we see that only the second term is important. Thus asymptotically

$$X_{(r)} \stackrel{d}{=} N(0, \frac{1}{4f^2(0)n}) \quad |r - \frac{1}{2}n| < \frac{a}{n}.$$

The constant of proportionality in (10.3) was of no importance to Laplace. Pearson (1902) in the course of arriving at the formula

$$E(X_{(r+1)} - X_{(r)}) = \binom{n}{r} \int_{-\infty}^{\infty} F^r(x)[1 - F(x)]^{n-r} dx \quad r = 1, \ldots, n-1$$

goes through the arguments needed for the derivation of $f_{X_{(r)}}(x)$ without writing down the result which is perhaps first given in von Bortkiewicz (1922). Strictly, this author's formula applies to the case $F(x) = Pr\{|X| \leq x\}$, where $X \stackrel{d}{=} N(0,1)$, but his argument holds for any distribution having a density function. The formula may be regarded as well known only with its appearance in *Biometrika* [Irwin (1925)]. However, it is interesting to note that a brilliant, long overlooked paper by Daniell (1920) [see Stigler (1973)] begins by obtaining mathematically the result (in present notation)

$$E(X_{(r)}) = \frac{n!}{(r-1)!(n-r)!} \int_0^1 F^{-1}(u) u^{r-1} (1-u)^{n-r} du$$

as well as the corresponding result for $E(X_{(r)} X_{(s)})$.

10.4 Extreme-Value Theory

One of the oldest nontrivial results in order statistics arose from the following question considered by Nicolas Bernoulli in his 1709 Ph.D. dissertation:

> Given that b individuals die in a time span of a years, during which the probability of death is constant, what is the number of years the last survivor can expect to live?

Bernoulli reduces this to finding the expected value of the maximum of b independent variates uniform in $(0, a)$. After giving a combinatorial argument he offers a second solution by what he calls a geometric approach. If the abscissa x denotes time to death of the longest living and the ordinate y is proportional to the probability of $b-1$ deaths before time x, then the desired expectation is the x-coordinate of the "center of gravity" of the area under the curve $y = x^{b-1}$, namely,

$$\frac{\int_0^a x \cdot x^{b-1} dx}{\int_0^a x^{b-1} dx} = \frac{ab}{b+1}.$$

We note that the argument would have been more general if Bernoulli had taken y to be proportional to the pdf of x. Only in the uniform case does his approach work. Bernoulli goes on to consider the life expectancy of the longer living of two individuals of different ages. The original (in Latin) may be found in Jakob (James) Bernoulli (1975, p. 296) with a detailed commentary (in German) by K. Kohli (p. 545). See also Hald (1990, p. 114).

Now we fast forward to von Mises (1923) who, triggered by von Bortkiewicz (1922), pioneered the asymptotic theory of extremes of iid variates. Under the conditions $E|X| < \infty$ and, for fixed positive c,

$$\lim_{x\to\infty} \frac{1 - F(x+c)}{1 - F(x)} = 0, \tag{a}$$

he shows that

$$\lim_{n\to\infty} \frac{E(X_{(n)})}{F^{-1}(1 - \frac{1}{n})} = 1. \tag{b}$$

This gives a convenient asymptotic approximation for $E(X_{(n)})$. The condition (a) is typical of the tail-behavior assumptions made in subsequent extreme-value theory work. In particular, (a) is satisfied when X is normal, in which case von Mises proves the result, stronger than (b), that

$$\lim_{n\to\infty} \left[E(X_{(n)}) - F^{-1}(1 - \frac{1}{n})\right] = 0.$$

See David and Edwards (2001) for a translation from the German, with commentary, of von Mises's paper.

Explicit results for the asymptotic distribution of the normalized maximum in samples from a variety of initial distributions are given by Dodd (1923). An elegant clarifying breakthrough is achieved by Fisher and Tippett (1928). They point out that if a limiting distribution, $\Lambda(x)$, of the maximum exists, then the distribution of the largest in a sample of n drawn from $\Lambda(x)$ must be "similar" to $\Lambda(x)$, that is, differing only in location and scale. This gives the functional equation

$$\Lambda^n(x) = \Lambda(a_n x + b_n), \qquad a_n > 0, -\infty < b_n < \infty.$$

If $a_n \neq 1$, then $x = a_n x + b_n$ when $x = b_n/(1 - a_n)$. At this point $\Lambda^n = \Lambda$, that is, $\Lambda = 0$ or 1. Consequently the solutions fall into three classes or types:

1. $a_n = 1$ $\quad\Lambda_1^n(x) = \Lambda_1^n(x + b_n)$
2. $\Lambda = 0$ when $x = 0$ $\quad\Lambda_2^n(x) = \Lambda_2(a_n x)$
3. $\Lambda = 1$ when $x = 0$ $\quad\Lambda_3^n(x) = \Lambda_3(a_n x)$

The authors then show that

$$\begin{aligned}
\Lambda_1(x) &= e^{\,c^{-x}}, && -\infty < x < \infty \\
\Lambda_2(x) &= 0, && x \le 0 \\
&= e^{-x^{-\alpha}}, && x > 0, \alpha > 0 \\
\Lambda_3(x) &= e^{-(-x)^{\alpha}}, && x \le 0, \alpha > 0 \\
&= 1, && x > 0.
\end{aligned}$$

Actually $\Lambda_2(x)$ had essentially also been obtained, and with a wider range of validity, by Fréchet (1927) in whose honor it is sometimes named. Wilks (1948, p. 430) notes that in spite of the different dates the two papers appeared "almost simultaneously." Fréchet was influenced by Lévy's (1925, Chapter 3) notion of the "stability" (in distribution) of the sums of independent normal and Cauchy variates. He points out that the cdf of the maximum is the product of the component cdf's, just as the characteristic function of the sum is the product of the component characteristic functions. Thus it is natural for Fréchet to allow for differences in scale among the component variates. He restricts himself to non-negative variates $X_1, \ldots, X_n$ with measures of scale $\sigma_1, \ldots, \sigma_n$ (not necessarily standard deviations) of the same order of magnitude.

Fréchet now solves the functional equation

$$\Lambda_2(x/\sigma) = \Lambda_2(x/\sigma_1) \cdots \Lambda_2(x/\sigma_n),$$

for both Λ_2 and σ. He shows that

$$F_{X_{(n)}/\sigma}(x) \to e^{-x^{-\alpha}} \text{as } n \to \infty, \quad x \ge 0,$$

where $\sigma^\alpha = \sigma_1^\alpha + \cdots + \sigma_n^\alpha$.

Juncosa (1949), examining the asymptotic behavior of the minimum of independent nonidentically distributed variates, shows that many more than the three limiting forms become possible when identity of component distributions is given up. Although citing Fréchet's paper, he makes no reference to the above result.

It is interesting to note that none of the 1920s authors above—von Mises, Dodd, Fréchet, or Fisher/Tippett—refers to the work of the others. However, Tippett (1925) in his important finite-sample paper on the extremes and the range in normal samples compares some of his exact calculations with approximations suggested in von Bortkiewicz (1922) and Dodd (1923).

The next major development was by von Mises (1936) who provided convenient sufficient conditions on the initial distribution leading to the three types. Necessary and sufficient conditions on the initial distribution were given in a masterly paper by Gnedenko (1943), with further improvements for type 1 above by de Haan (1970). Gumbel (1958) continues to be a useful review, especially on applications. For a recent summary see, for example, David and Nagaraja

(2003, Section 10.5) and for an extended account, see Galambos (1987). See also the end-of-chapter surveys of the literature in Galambos (1987) and Reiss (1989).

10.5 Estimation of Location and Scale Parameters by Linear Functions of Order Statistics

Selected order statistics, such as the median, the upper and lower quartiles, and the extremes have long been used in an ad hoc way to estimate location and scale parameters. A unified approach is possible by the method of maximum likelihood applied to ordered samples or subsets thereof. But this is often laborious and the estimators do not necessarily have good small-sample properties. It was not until 1952 that E.H. Lloyd, in a very influential paper, showed how the method of least squares could be used to estimate the parameters μ and σ (not confined to denote mean and standard deviation) in distributions with pdf of the form

$$f(x;\mu,\sigma)=\frac{1}{\sigma}g\left(\frac{x-\mu}{\sigma}\right).$$

If $X_i, i=1,\ldots,n$, are independent, with pdf $f(x;\mu,\sigma)$, then $Y_i=(X_i-\mu)/\sigma$ has pdf $g(y)$, not depending on μ and σ. The transformation also takes $X_{(r)}$ into $Y_{(r)}, r=1,\ldots,n$.

Let

$$E(Y_{(r)})=\alpha_r \quad \text{and} \quad \operatorname{cov}(Y_{(r)},X_{(s)})=\sigma^2\beta_{rs} \qquad s=1,\ldots n.$$

Then

$$E(X_{(r)})=\mu+\sigma\alpha_r \quad \text{and} \quad \operatorname{cov}(X_{(r)},X_{(s)})=\sigma^2\beta_{rs}.$$

For given $g(y)$, the α_r and β_{rs} can be computed once and for all. Then the $X_{(r)}$ have expectations that are linear functions of μ and σ, with known coefficients, and covariances (including variances) that are known up to the scale factor σ^2. Lloyd realized that consequently Gauss's least-squares theory [see, e.g., Plackett (1949)], generalized by Aitken (1935) to cover nondiagonal covariance matrices, results in estimators

$$\mu^*=\sum_{i=1}^{n}\gamma_i X_{(i)} \quad \text{and} \quad \sigma^*=\sum_{i=1}^{n}\delta_i X_{(i)},$$

that have minimum variance in the class of linear functions of the $X_{(i)}$. Again, the γ_i and δ_i can be tabulated once and for all, making the estimation immediate.

Apparently independently, Gupta (1952) also obtained these results. He introduced the terms type I and type II censoring and made the important observation that type II censoring, for example, terminating a life test at the time of the rth failure $x_{(r)}$, can be treated in the same way by simply using α_i and β_{ij} for just $i = 1, \ldots, r$ and $j = 1, \ldots, r$. Type II censoring at each end can obviously be treated similarly.

For the corresponding asymptotic theory we refer the reader to an excellent review [Stigler (1973)] that includes coverage of the remarkably modern paper by Daniell (1920).

10.6 Tables

Two-decimal tables of the expectations of order statistics from standard normal samples for $n \leq 50$ are given in Fisher and Yates (1938) (and subsequent editions). The entries are called scores for ordinal (or ranked) data and are recommended for data that can be ranked but not measured, as in psychological preference tests.

The first systematic table of means, variances, and covariances of order statistics is given in Hastings *et al.* (1947). This is truly a pioneering paper. The authors write:

> It would be very helpful to have (1) at least the first two moments (including product moments) of the order statistics, and (2) tables of the percentage points of their distributions, for samples of sizes from 1 to some moderately large value such as 100 and for a large representative family of distributions. This is a large order and will require much computation

Hastings *et al.* deal for $n \leq 10$ with the uniform, normal, and a specially devised long-tailed distribution given by representing X as $X = (1-U)^{-1/10} - U^{-1/10}$, where U is uniform over $[0, 1]$. The covariances in the normal case could be computed to just 2D (decimal places), five places being provided elsewhere. Comparisons with asymptotic approximations are also made.

In the normal case Godwin (1949) gives also the covariances for $n \leq 10$ to 5D and obtains all first two moments and product moments for $n \leq 6$ in terms of elementary functions. Other authors were also involved but the real breakthrough came with the advent of the high-speed computer. Teichroew (1956) tabulates all first two raw moments and product moments for $n \leq 20$ to 10D. Sarhan and Greenberg (1956) use these, following Gupta (1952), to obtain to 8D the coefficients of the best linear estimators of μ and σ for singly or doubly censored type II samples. Since then, numerous such tables for location-scale distributions have appeared. For a listing see the Appendix, Section 8.5, of David and Nagaraja (2003). The construction of these convenient tables involves, among other operations, inversion of the covariance matrix of the relevant order statistics. A listing of tables of covariance matrices, which have of course also other uses, is given in Appendix Section 3.2. It should be noted that the range, W_n, in normal samples received earlier attention in the remarkable 5D tables of $E(W_n)$ by Tippett (1925) for $n = 2(1)1000$!

References

1. Aitken, A. C. (1935). On least squares and linear combinations of observations, *Proceedings of the Roual Society of Edinburgh,* **55**, 42–47.

2. Anonymous (1821). Dissertation sur la recherche du milieu le plus probable entre les résultats de plusieurs observations ou expériences, *Ann. Math. Pures Appliquées* **12**, 181-204.

3. Bernoulli, J. (1975). *Die Werke von Jakob Bernoulli*, Vol. 3, Birkhäuser, Basel.

4. Boscovich, R. J. (1757). De litteraria expenditione per pontificiam ditionem, et synopsis amplioris operist, *Bononiensi Scientiarum et Artum Instituto atque Academia Commentarii*, **4**, 353–396.

5. Cournot, A. A. (1843). *Exposition de la théorie des chances et des probabilités*, Hachette, Paris. Reprinted in 1984 with changed pagination and extensive notes by B. Bru. Vrin, Paris.

6. Craig, A. T. (1932). On the distribution of certain statistics, *American Journal of Mathematics,* **54**, 353–366.

7. Daniell, P. J. (1920). Observations weighted according to order, *American Journal of Mathematics,* **42**, 222–236.

8. David, H. A. (1998). Early sample measures of variability, *Statistical Science*, **13**, 368-377.

9. David, H. A., and Edwards, A. W. F. (2001). *Annotated Readings in the History of Statistics*, Springer-Verlag, New York.

10. David, H. A. and Nagaraja, H. N. (2003). *Order Statistics*, Third edition, John Wiley & Sons, Hoboken, NJ.

11. de Haan, L. (1970). *On Regular Variation and Its Application to the Weak Convergence of Sample Extremes*, Mathematical Centre Tracts **32**, Mathematisch Centrum, Amsterdam.

12. de Moivre, A. (1733). Approximatio ad summam terminorum binomii $(a+b)^n$ in seriem expansi, *Printed for private circulation.*

13. Dodd, E. L. (1922). Functions of measurements under general laws of error, *Skandinavsik Aktuarietidskrift*, **5**, 133-158.

14. Dodd, E. L. (1923). The greatest and least variate under general laws of error, *Transactions of the American Mathematical Society*, **25**, 525–539.

15. Fisher, R. A. (1922). On the mathematical foundations of theoretical statistics, *Philosophical Transactions of the Royal Society of London, Series A*, **222**, 309–368. Reprinted in Fisher (1950).

16. Fisher, R. A. (1950). *Contributions to Mathematical Statistics.* John Wiley & Sons, New York.

17. Fisher, R. A., and Tippett, L. H. C. (1928). Limiting forms of the frequency distribution of the largest or smallest member of a sample, *Proceedings of the Cambridge Philosophical Society*, **24**, 180–190. Reprinted in Fisher (1950).

18. Fisher, R. A. and Yates, F. (1938). *Statistical Tables for Biological, Agricultural and Medical Research*, Oliver and Boyd, London.

19. Fréchet, M. (1927). Sur la loi de probabilité de l'écart maximum, *Ann. Soc. Polonaise de Mathématique*, **6**, 93–116.

20. Friedman, M. (1937). The use of ranks to avoid the assumption of normality implicit in the analysis of variance, *Journal of the American Statistical Association*, **32**, 675–701.

21. Galambos, J. (1987). *The Asymptotic Theory of Extreme Order Statistics*, Second edition, Krieger, Malabar, FL.

22. Gauss, C.F. (1809). *Theoria Motus Corporum Coelestium.*, Perthes and Besser, Hamburg. English translation by C. H. Davis. Little Brown, Boston, 1857; Dover, New York, 1963.

23. Gnedenko, B. (1943). Sur la distribution limite du terme maximum d'une série aléatoire, *Annals of Mathematics*, **44**, 423–453.

24. Godwin, H. J. (1949). Some low moments of order statistics, *Annals of Mathematical Statistics*, **20**, 279–285.

25. Gumbel, E. J. (1958). *Statistics of Extremes*, Columbia University Press, New York.

26. Gupta, A. K. (1952). Estimation of the mean and standard deviation of a normal population from censored samples, *Biometrika*, **39**, 260–273.

27. Hald, A. (1990). *A History of Probability and Statistics and Their Applications Before 1750*, John Wiley & Sons, New York.

28. Hald, A. (1998). *A History of Mathematical Statistics from 1750 to 1930*, John Wiley & Sons, New York.

29. Harter, H. L. (1978). *A Chronological Annotated Bibliography of Order Statistics*, **1**. Pre-1950, U.S. Government Printing Office, Washington, DC.

30. Harter, H. L. (1983-1993). *A Chronological Annotated Bibliography of Order Statistics*, **1-8**. American Sciences Press, Columbus, OH.

31. Harter, H. L. (1988). History and role of order statistics, *Communications in Statistics—Theory and Methods*, **17**, 2091–2108.

32. Hastings, C., Jr., Mosteller, F., Tukey, J. W., and Winsor, C. P. (1947). Low moments for small samples: A comparative study of order statistics, *Annals of Mathematical Statistics*, **18**, 413–426.

33. Irwin, J. O. (1925). Theory of Francis Galton's individual difference problem, *Biometrika*, **17**, 100–128.

34. Juncosa, M. L. (1949). The asymptotic behavior of the minimum in a sequence of random variables, *Duke Mathematical Journal*, **16**, 609–618.

35. Laplace, P. S. (1774). Mémoire sur la probabilité des causes par les évènemens, *Mem. Acad. Roy. Sci.* **6**, 621–656. English translation, with introduction, by S. Stigler in *Statistical Science*, **1**, 359–378 (1986).

36. Laplace, P. S. (1818). *Théorie analytique des probabilités, deuxième supplément*, Section 2. Reprinted in *Oeuvres de Laplace* **7**. Imprimerie Royale, Paris (1847).

37. Legendre, A. M. (1805). *Nouvelles méthodes pour la détermination des orbites des comètes*, Appendix. Courcier, Paris.

38. Lévy, P. (1925). *Calcul des probabilités*, Gauthier-Villars, Paris.

39. Lloyd, E. H. (1952). Least-squares estimation of location and scale parameters using order statistics, *Biometrika* **39**, 88–95.

40. Pearson, E. S., and Kendall, M. G. (1970). *Studies in the History of Statistics and Probability*, Hafner, Darien, CT.

41. Pearson, K. (1902). Note on Francis Galton's difference problem, *Biometrika*, **1**, 390–399.

42. Plackett, R. L. (1949). A historical note on the method of least squares, *Biometrika*, **36**, 458–460.

43. Plackett, R. L. (1958). Studies in the history of probability and statistics. VII. The principle of the arithmetic mean, *Biometrika*, **45**, 130–135. Reprinted in Pearson and Kendall (1970).

44. Reiss, R.-D. (1989). *Approximate Distributions of Order Statistics*, Springer-Verlag, New York.

45. Sarhan, A. E., and Greenberg, B. G. (1956). Estimation of location and scale parameters by order statistics from singly and doubly censored samples. Part I. The normal distribution up to samples of size 10, *Annals of Mathematical Statistics*, **27**, 427–451. *Correction*: **40**, 325.

46. Stigler, S. M. (1973). Simon Newcomb, Percy Daniell, and the history of robust estimation 1885–1920, *Journal of the American Statistical Association*, **68**, 872–879.

47. Stigler, S. M. (1986). Laplace's 1774 memoir on inverse probability, *Statistical Science*, **1**, 359–378.

48. Teichroew, D. (1956). Tables of expected values of order statistics and products of order statistics for samples of size twenty and less from the normal distribution, *Annals of Mathematical Statistics*, **27**, 410–426.

49. Tippett, L. H. C. (1925). On the extreme individuals and the range of samples taken from a normal population, *Biometrika*, **17**, 364–387.

50. von Bortkiewicz, L. (1922). Variationsbreite und mittlerer Fehler, *Sitzungsberichte der Berliner Math. Gesellschaft*, **21**, 3–11.

51. von Mises, R. (1923). Über die Variationsbreite einer Beobachtungsreihe, *Sitzungsberichte der Berliner Math. Gesellschaft*, **22**, 3–8. Reproduced in von Mises (1964).

52. von Mises, R. (1936). La distribution de la plus grande de n valeurs, *Rev. Math. Union Interbalkanique 1*, 141–160. Reproduced in von Mises (1964).

53. von Mises, R. (1964). *Selected Papers of Richard von Mises, Vol. 2*, American Mathematical Society, Providence, RI.

54. Wilks, S. S. (1942). Statistical prediction with special reference to the problem of tolerance limits, *Annals of Mathematical Statistics*, **13**, 400–409.

55. Wilks, S. S. (1948). Order statistics, *Bulletin of the American Mathematical Society*, **5**, 6–50.

11

Order Statistics from Independent Exponential Random Variables and the Sum of the Top Order Statistics

H. N. Nagaraja

The Ohio State University, Columbus, OH, USA

Abstract: Let $X_{(1)} < \cdots < X_{(n)}$ be the order statistics from n independent nonidentically distributed exponential random variables. We investigate the dependence structure of these order statistics, and provide a distributional identity that facilitates their simulation and the study of their moment properties. Next, we consider the partial sum $T_i = \sum_{j=i+1}^{n} X_{(j)}$, $0 \leq i \leq n-1$. We obtain an explicit expression for the cdf of T_i, exploiting the memoryless property of the exponential distribution. We do this for the identically distributed case as well, and compare the properties of T_i under the two settings.

Keywords and phrases: Markov property, equal in distribution, simulation, mixtures, selection differential

11.1 Introduction

Let $X_1, \ldots, X_n$ be independent nonidentically distributed (inid) random variables (rvs), where X_j is $\text{Exp}(\lambda_j)$, $j = 1, \ldots, n$; that is, the pdf of X_j is given by

$$f_j(x) = \lambda_j e^{-\lambda_j x}, \ x \geq 0,$$

and the λ_j are possibly distinct. Let $X_{(1)} < \cdots < X_{(n)}$ be the order statistics from this sample. We investigate their dependence structure and provide a distributional identity that facilitates their simulation and investigation of distributional and moment properties. This is done in Section 11.2.

The work in Section 11.3 is motivated by a personal communication from

Dr. Yang-Seok Choi who was interested in the distribution of

$$T_i = \sum_{j=i+1}^{n} X_{(j)}, \ 0 \le i \le n-1. \tag{11.1}$$

There we obtain an explicit expression for the cdf of T_i. We also consider the independent identically distributed (iid) case and relate T_i to a rv known as *selection differential* in the genetics literature. We then compare the properties of T_i under the iid and inid models.

11.2 Distributional Representations and Basic Applications

We begin with a discussion of the stochastic structure of and distributional representations for the vector of order statistics $(X_{(1)}, \ldots, X_{(n)})$. When the λ_j are identical and equal to, say λ, it is known that (see, e.g., David and Nagaraja, 2003, p. 18)

$$(X_{(i)}, i = 1, \ldots, n) \stackrel{d}{=} \frac{1}{\lambda}\left(\sum_{j=1}^{i} \frac{Z_j}{n-j+1}, i = 1, \ldots, n\right), \tag{11.2}$$

where the Z_j are iid standard exponential (i.e., Exp(1)) rvs. This is known as Rényi's representation [Rényi (1953)].

Let $\mathbf{X} = (X_{(1)}, \ldots, X_{(n)})'$ and $\mathbf{Z} = (Z_1, \ldots, Z_n)'$, and define a vector $\boldsymbol{\alpha}_i = (\alpha_1, \ldots, \alpha_i, 0, \ldots, 0)'$ where $\alpha_j = 1/\{\lambda(n-j+1)\}$, $1 \le i, j \le n$. Then, $X_{(i)} \stackrel{d}{=} \boldsymbol{\alpha}_i{}'\mathbf{Z}$ and (11.2) can be expressed as

$$\mathbf{X} \stackrel{d}{=} \mathrm{C}\mathbf{Z}, \tag{11.3}$$

where C is the $n \times n$ matrix of constants whose ith row is $\boldsymbol{\alpha}_i{}'$. This relation is helpful in simulating all or a subset of order statistics from a random sample of size n from an Exp(λ) parent.

When the λ_j are not identical, representations for the exponential order statistics do exist. Nevzorov (1984) shows that [see also Nevzorova and Nevzorov (1999)] the joint distribution of order statistics can be expressed as a mixture distribution with $n!$ components where the various component vectors are chosen with probability p_l of picking certain permutation of the λ_j for ordering the observed rvs. To be precise, Nevzorov shows that the cdf of $X_{(i)}$, the ith component of $\mathbf{X}$, can be expressed as a mixture cdf given by

$$F_{(i)}(x) = \sum_{l=1}^{n!} p_l F_l(x), \tag{11.4}$$

where

$$p_l = \frac{\lambda_1 \cdots \lambda_n}{(\lambda_{d(1)} + \cdots + \lambda_{d(n)})(\lambda_{d(2)} + \cdots + \lambda_{d(n)}) \cdots \lambda_{d(n)}} \tag{11.5}$$

and F_l is the cdf of the rv

$$\frac{Z_1}{(\lambda_{d(1)} + \cdots + \lambda_{d(n)})} + \cdots + \frac{Z_i}{(\lambda_{d(i)} + \cdots + \lambda_{d(n)})}, \quad 1 \le i \le n,$$

and the mixture includes all $n!$ vectors corresponding to the $n!$ permutations $(d(1), d(2), \ldots, d(n))$ of integers $1, 2, \ldots, n$.

Tikhov (1991) gave another, simpler, form of the above representation by introducing *antiranks* $D(1), \ldots, D(n)$ defined by

$$\{D(i) = m\} = \{X_{(i)} = X_m\}, \quad 1 \le i, \ m \le n. \tag{11.6}$$

With these random subscripts, one can write the distributional equality

$$X_{(i)} \stackrel{d}{=} \frac{Z_1}{(\lambda_{D(1)} + \cdots + \lambda_{D(n)})} + \cdots + \frac{Z_i}{(\lambda_{D(i)} + \cdots + \lambda_{D(n)})}, 1 \le i \le n, \tag{11.7}$$

where the Z_j are iid standard exponentials and are independent of the antirank vector $(D(1), \ldots, D(n))$. The form in (11.3) also holds in this case, with a modification that lets the elements of C to be rvs. Let us define a random vector $\mathbf{a}_i = (A_1, \ldots, A_i, 0, \ldots, 0)', 1 \le i \le n$, where

$$A_j = (\lambda_{D(j)} + \cdots + \lambda_{D(n)})^{-1}, 1 \le j \le n. \tag{11.8}$$

Then the following distributional equality holds:

$$\mathbf{X} \stackrel{d}{=} \mathbf{AZ}, \tag{11.9}$$

where $\mathbf{A}$ is an $n \times n$ random matrix whose ith row is $\mathbf{a}_i'$. The elements of $\mathbf{A}$ are independent of the vector $\mathbf{Z}$ whose components themselves are iid standard exponential rvs. The elements of $\mathbf{A}$ are functions of $A_1, \ldots, A_n$ that are dependent and depend on the distribution of $(D(1), \ldots, D(n))$, given by the p_l in (11.5).

11.2.1 Remarks

1. The joint distribution of $(D(1), \ldots, D(n))$, given in (11.5), can be used to simulate this vector. We now describe how it can be done easily and more efficiently in a sequential manner. We start with $D(1)$; it is a discrete rv with support $\Omega_0 = \{1, 2, \ldots, n\}$ and $P(D(1) = i) = \lambda_i / (\sum_{j \in \Omega_0} \lambda_j)$. Once $D(1)$ is selected from this distribution, $D(2)$ is chosen from $\Omega_1 = \{1, 2, \ldots, n\} - \{D(1)\}$ using the probability distribution given by $P(D(2) = i) = \lambda_i/(\sum_{j\in\Omega_1} \lambda_j)$. In general, for $1 \le k \le n-1$, after $D(1), \ldots, D(k)$ are chosen, $D(k+1)$ is chosen from

$$\Omega_k = \{1, 2, \ldots, n\} - \{D(1), D(2), \ldots, D(k)\}$$

using the probabilities

$$P(D(k+1) = i) = \lambda_i \bigg/ \left(\sum_{j\in\Omega_k} \lambda_j\right), \quad i \in \Omega_k, 1 \le k \le n-1.$$

2. The representation in (11.9) can be used to simulate exponential order statistics or functions of these order statistics. If the quantity of interest is a function of the first i order statistics, one need to simulate only $D(1), \ldots, D(i)$ and these choices will determine the sum $\sum_{k=i+1}^{n} \lambda_{D(k)}$ that is needed to evaluate the observed values of $A_j, j \le i$. Also, we need to simulate only $Z_k, 1 \le k \le i$.

3. The representation for the cdf of $X_{(i)}$ given in (11.4) and the distributional identity for the rv $X_{(i)}$ given in (11.7) have different purposes and applications. The former can be used to determine probabilities associated with $X_{(i)}$ assuming that the explicit form for F_l is available, whereas the latter gives a handy framework for simulation. There is a distinction between (11.4) and an equality in distribution ($\stackrel{d}{=}$) relation obtained by replacing the cdfs with the associated rvs in that equation. Tikhov's (1991, p. 630) interpretation of Nevzorov's result makes this improper leap.

11.2.2 Applications

Moments

We can use the distributional equality in (11.7) to obtain expressions for the moments of order statistics. Because

$$X_{(i)} \stackrel{d}{=} \sum_{j=1}^{i} A_j Z_j,$$

A_j and Z_j are independent, and the Z_j are iid standard exponential, it follows that

$$E(X_{(i)}) = \sum_{j=1}^{i} E(A_j)$$

and

$$Var(X_{(i)}) = E(X_{(i)}^2) - \{E(X_{(i)})\}^2 = \sum_{j=1}^{i} E(A_j^2) + Var\left(\sum_{j=1}^{i} A_j\right),$$

upon simplification. Further, for $1 \leq i < k \leq n$,

$$\begin{aligned} Cov(X_{(i)}, X_{(k)}) &= Var(X_{(i)}) + \sum_{j=1}^{i} \sum_{l=i+1}^{k} Cov(A_j, A_l) \\ &= \sum_{j=1}^{i} E(A_j^2) + Cov\left(\sum_{j=1}^{i} A_j, \sum_{l=i+1}^{k} A_l\right). \end{aligned} \tag{11.10}$$

In the iid case, the A_j's are all constants and $A_j = 1/\{\lambda(n-j+1)\}$, and the classical results follow immediately.

Spacings of order statistics

The relation in (11.7) can also be used to study the distributional representations for spacings. For example,

$$X_{(i)} - X_{(i-1)} \stackrel{d}{=} A_i Z_i, \quad 2 \leq i \leq n,$$

and hence for $2 \leq i \leq n-1$,

$$Cov(X_{(i)} - X_{(i-1)}, X_{(i+1)} - X_{(i)}) = Cov(A_i Z_i, A_{i+1} Z_{i+1}) = Cov(A_i, A_{i+1}).$$

In the iid case, it is wellknown that the spacings are independent and thus are uncorrelated. It appears that the covariance is zero if and only if the λ_i are identical. Such a conjecture is also made in Khaledi and Kochar (2000) and a proof is given of the claim for $n = 3$. (They actually prove a stronger result.) The case where $n > 3$ appears to be open.

Other linear functions

For a vector $\boldsymbol{\beta} = (\beta_1, \ldots, \beta_n)'$, one can simulate $\boldsymbol{\beta}'\mathbf{X}$ values as $\boldsymbol{\beta}'\mathbf{AZ}$ using (11.9). For example, the T_i in (11.1) can be simulated as the sum

$$T_i = (n-i)\sum_{j=1}^{i} A_j Z_j + \sum_{j=i+1}^{n} (n-j+1) A_j Z_j. \tag{11.11}$$

In the iid case, T_i is related to the *selection differential*, given by

$$D_k = \frac{1}{\sigma}\left(\frac{1}{k}\sum_{j=n-k+1}^{n} X_{(j)} - \mu\right), \tag{11.12}$$

where μ and σ are the mean and standard deviation of the parent population. For the Exp(λ) parent, both these moments are $1/\lambda$. The rv D_k is used to measure the improvement due to selection where the top values in the sample are selected and for small k $(= n - i)$, it provides a good test for checking for outliers at the upper end of the sample.

Another linear function is the *total time on test* given by

$$\sum_{j=1}^{i} X_{(j)} + (n-i)X_{(i)},$$

and serves as the best estimator of $1/\lambda$ based on type II right censored sample in the iid case.

11.3 Sum of the Top Order Statistics

The following classical result (see, e.g., David and Nagaraja, 2003, pp. 137–138) is helpful in our pursuit of the cdf of the sum T_i.

Lemma 11.3.1. *Suppose Z_r, $r = 1, \ldots, m$, are independent standard exponential random variables and c_r's are distinct positive numbers. Then*

$$P\left(\sum_{r=1}^{m}\left(\frac{Z_r}{c_r}\right) > z\right) = \sum_{r=1}^{m} w_r e^{-c_r z}, \quad z > 0,$$

where

$$w_r = 1\Big/\prod_{s\neq r}\left(1 - \frac{c_r}{c_s}\right),$$

and the probability is 1 if $z < 0$.

Now recall the representation (11.11) for T_i where the joint distribution of the A_j is as described in Section 11.2 and the Z_j are iid standard exponential rvs.

11.3.1 The IID case

When the X_i are identically distributed each being standard exponential, say, A_j would be a constant $1/c_j$ where $c_j \equiv n - j + 1$. In that case,

$$T_i \stackrel{d}{=} (n-i)\sum_{j=1}^{i}\left(\frac{1}{c_j}\right) Z_j + W_i, \quad 0 \le i < n-1, \tag{11.13}$$

where W_i is the sum of $(n-i)$ standard exponential rvs, and is a gamma$(n-i, 1)$ rv with pdf

$$f_i(w) = \frac{1}{(n-i-1)!} e^{-w} w^{n-i-1}, w > 0.$$

Thus, T_0 is a gamma$(n, 1)$ rv. Also, because $T_{n-1} = X_{(n)}$,

$$P(T_{n-1} > t) = 1 - (1 - e^{-t})^n, \ t > 0.$$

For $0 < i < n-1$, one can use Lemma 11.3.1 and conditioning argument in the representation (11.13) to obtain an explicit expression for the survival function of T_i as follows:

$$\begin{aligned} P(T_i > t) &= P\left((n-i)\sum_{j=1}^{i}\frac{1}{c_j}Z_j + Y_i > t\right) \\ &= \int_{y=0}^{t}\left(\sum_{j=1}^{i}\frac{1}{c_j}Z_j > \frac{1}{c_{i+1}}(t-y)\right) f_i(y)dy + P(Y_i > t) \\ &= \sum_{j=1}^{i} w_j \exp\{-c_j t/c_{i+1}\} \frac{1}{(n-i-1)!}\int_0^t \exp(d_j y) y^{n-i-1} dy \\ &\quad + \sum_{k=0}^{n-i-1} e^{-t}\frac{t^k}{k!}. \end{aligned} \tag{11.14}$$

Here, $c_j = n - j + 1$,

$$d_j = \frac{c_j}{c_{i+1}} - 1 = \frac{i+1-j}{n-i} > 0.$$

The w_j are obtained using Lemma 11.3.1, and have alternating signs. They are given by

$$w_j = \prod_{\substack{k=1 \\ \neq j}}^{i} \frac{n-k+1}{j-k} = \frac{1}{n-j+1}\frac{n!}{(n-i)!}\frac{(-1)^{i-j}}{(j-1)!(i-j)!}.$$

The pdf of T_i can be obtained by differentiating (11.14). Upon some simplification the pdf can be expressed as

$$f_{T_i}(t) = \sum_{j=1}^{i} w_j \frac{c_j}{c_{i+1}} \exp\left\{-\frac{c_j}{c_{i+1}} t\right\} \frac{1}{(n-i-1)!} \int_0^t \exp(d_j y) y^{n-i-1} dy,$$

or as

$$\begin{aligned} f_{T_i}(t) &= n\binom{n-1}{i-1} \sum_{j=1}^{i} \binom{i-1}{j-1} (-1)^{i-j} \exp\left\{-\frac{n-j+1}{n-i} t\right\} \\ &\times \frac{1}{(n-i)!} \int_0^t \exp\left(\frac{i+1-j}{n-i} y\right) y^{n-i-1} dy. \end{aligned}$$

Nagaraja (1981) has obtained a similar expression for the pdf of $T_i/(n-i)$ in his study of the selection differential D_k in (11.12) arising from a random sample from an exponential distribution. From Nagaraja (1982), one can obtain the asymptotic distribution of $T_i - (n-i)\log(n)$ if n approaches infinity such that $k = n - i$ is held fixed. Becasue the exponential distribution is in the domain of attraction of the Gumbel distribution, the cdf of $T_i - k\log(n)$ converges to the following cdf for $k \geq 2$:

$$\frac{k^{k-1}}{(k-2)!} \sum_{j=0}^{k-1} \frac{\exp\{-(jx/k)\}}{j!} \int_0^{\infty} \exp\left\{-\exp\left(y - \frac{x}{k}\right)\right\} \exp\{-y(k-j)\} y^{k-2} dy.$$

Andrews (1996) has studied the finite-sample moment and distributional properties of the selection differential D_k for the exponential and uniform parents. From his work, one can obtain explicit expressions for the first four moments of $T_i = (n-i)(\mu + \sigma D_{n-i})$ in the iid case. He also discusses asymptotes for the moments of D_k when $k \approx np$, $0 < p < 1$, and the rate of convergence of the finite-sample moments.

11.3.2 The non-IID case

Let us assume that the λ_j are all distinct. As in the iid case, we dispose of the special situations first. When $i = 0$,

$$T_0 = \sum_{j=1}^{n} X_{(j)} \equiv \sum_{j=1}^{n} X_j \stackrel{d}{=} \sum_{j=1}^{n} Z_j/\lambda_j.$$

Hence, Lemma 11.3.1 can be used directly to obtain an explicit expression for $P(T_0 > t)$.

When $i = n-1$, $T_i = X_{(n)}$ and hence

$$P(T_{n-1} > t) = 1 - \prod_{j=1}^{n} \left(1 - e^{-\lambda_j t}\right). \tag{11.15}$$

As we see below, for $1 \leq i < n-1$, the expression for $P(T_i > t)$ is more involved.

For a given $j, 1 \leq j \leq n$, let $S(j)$ be a set with $(i-1)$ elements taken from $\{1, 2, \ldots, n\} - \{j\}$. There are $\binom{n-1}{i-1}$ different choices for $S(j)$. For each such choice, let $\overline{S}(j) = \{1, 2, \ldots, n\} - \{j\} - S(j)$.

Theorem 11.3.1. *Let T_i be given by (11.1) with $1 \leq i < n-1$. Then, for $t > 0$, $P(T_i > t)$ can be expressed as*

$$\begin{aligned}&\sum_{j=1}^{n} \lambda_j \sum_{S(j)} \sum_{k \in \overline{S}(j)} w_k(\overline{S}(j)) e^{-\lambda_k t} \\ &\quad \times \int_0^{t/(n-i)} \prod_{m \in S(j)} (1 - e^{-\lambda_m x}) \exp\left\{ - \left[\lambda_j + \sum_{r \in \overline{S}(j)} \lambda_r - (n-i)\lambda_k \right] x \right\} dx \\ &\quad + \sum_{j=1}^{n} \lambda_j \sum_{S(j)} \int_{t/(n-i)}^{\infty} \prod_{m \in S(j)} (1 - e^{-\lambda_m x}) \exp\left\{ -(\lambda_j + \sum_{r \in \overline{S}(j)} \lambda_r) x \right\} dx, \end{aligned} \tag{11.16}$$

where

$$w_k(\overline{S}(j)) = \frac{1}{\prod_{l \neq k \in \overline{S}(j)} \left(1 - \frac{\lambda_k}{\lambda_l}\right)}.$$

Proof. The joint pdf of $X_{(1)}, \ldots, X_{(n)}$ is the sum of $n!$ terms where each term has the form

$$\prod_{k=1}^{n} \lambda_{r(k)} e^{-\lambda_{r(k)} x_k}, \quad 0 < x_1 < \cdots < x_n,$$

where $(r(1), \ldots, r(n))$ is a permutation of $(1, \ldots, n)$. Then

$$P(T_i > t) = \sum_{n!} \int \cdots \int_{\substack{0 < x_1 < \cdots < x_n < \infty \\ x_{i+1} + \cdots + x_n > t}} \prod_{k=1}^{n} \lambda_{r(k)} e^{-\lambda_{r(k)} x_k} dx_k. \tag{11.17}$$

We split and group the $n!$ terms using the following procedure:

(a) We fix $X_{(i)} = x$ and its parameter λ_j, $j = 1, \ldots, n$.

(b) Given j, we fix the parameters associated with $X_{(1)}, \ldots, X_{(i-1)}$. There are

$$(n-1) \cdots (n-i+1) = \binom{n-1}{i-1} (i-1)!$$

such distinct ways of choosing their parameters.

(c) The remaining parameters associated with $X_{(i+1)}, \ldots, X_{(n)}$ can be ordered in $(n-i)!$ ways.

Let $S^o(j)$ be a typical (ordered) set in (b) and $\overline{S}^o(j)$ be a typical ordered set in (c). The expression for $P(T_i > t)$ given in (11.17) above can be written as

$$\sum_{j=1}^{n} \sum_{S^o(j)} \sum_{\overline{S}^o(j)} \int_{x=0}^{\infty} \lambda_j e^{-\lambda_j x} \left\{ \int \cdots \int_{0<x_1<\cdots<x_{i-1}<x} \prod_{k=1}^{i-1} \lambda_{r(k)} e^{-\lambda_{r(k)} x_k} dx_k \right\} \cdot \left\{ \int \cdots \int_{\substack{0<x<x_{i+1}<\cdots<x_n<\infty \\ x_{i+1}+\cdots+x_n>t}} \prod_{k=i+1}^{n} \lambda_{r(k)} e^{-\lambda_{r(k)} x_k} dx_k \right\} dx. \tag{11.18}$$

For every unordered set $S(j)$ that leads to $S^o(j)$,

$$\sum_{S^o(j);\ S(j)\text{ fixed}} \left\{ \int \cdots \int_{0<x_1<\cdots<x_{i-1}<x} \prod_{k=1}^{i-1} \lambda_{r(k)} e^{-\lambda_{r(k)} x_k} dx_k \right\}$$

can be seen as

$$P(\max_{k\in S(j)} X_k < x) = \prod_{k\in S(j)} (1-e^{-\lambda_k x}), \quad x>0. \tag{11.19}$$

Further, in (11.18), for every unordered set $\overline{S}(j)$ that leads to $\overline{S}^o(j)$,

$$\sum_{\overline{S}^o(j);\ \overline{S}(j)\text{ fixed}} \left\{ \int \cdots \int_{\substack{x<x_{i+1}<\cdots<x_n<\infty \\ x_{i+1}+\cdots+x_n>t}} \prod_{k=i+1}^{n} \lambda_{r(k)} e^{-\lambda_{r(k)} x_k} dx_k \right\}$$

can be expressed as

$$\sum_{\overline{S}^o(j);\overline{S}(j)\text{fixed}} e^{-x\sum_{r\in\overline{S}(j)} \lambda_r} \cdot \left\{ \int \cdots \int_{\substack{0<y_{i+1}<\cdots<y_n<\infty \\ y_{i+1}+\cdots+y_n>t-(n-i)x}} \prod_{k=i+1}^{n} \lambda_{r(k)} e^{-\lambda_{r(k)} y_k} dy_k \right\}, \tag{11.20}$$

by taking $y_k = x_k - x$, $k = i+1, \ldots, n$. The multiple integral in (11.20), when summed over $\overline{S}^o(j)$ for a fixed $\overline{S}(j)$, represents

$$P(Y_{(1)} + \cdots + Y_{(n-i)} > t-(n-i)x)$$

where $Y_{(1)}, \ldots, Y_{(n-i)}$ are the sample order statistics generated from $(n-i)$ independent exponential rvs having $\exp(\lambda_r)$distribution, $r \in \overline{S}(j)$. Thus, the above expression is nothing but

$$P\left(\sum_{r\in\overline{S}(j)} Y_r > t-(n-i)x\right) = P\left(\sum_{r\in\overline{S}(j)} \tfrac{1}{\lambda_r} Z_r > t-(n-i)x\right), \tag{11.21}$$

where the Z_r are iid standard exponential rvs. Thus, in view of Lemma 11.3.1, for a fixed x and $\overline{S}(j)$, the expression in (11.21) reduces to

$$\sum_{r \in \overline{S}(j)} w_r(\overline{S}(j)) e^{-\lambda_r \{t-(n-i)x\}}$$

if $x < t/(n-i)$, where the $w_r(\overline{S}(j))$ are as given in the theorem. The expression in (11.21) is clearly 1 if $x \geq t/(n-i)$.

Combining the above with (11.19) and (11.20), and recalling (11.18), we are led to the expression for $P(T_i > t)$ given in (11.16). ■

Notes

1. The first summation in (11.16) above has $n \times \binom{n-1}{i-1} \times (n-i)$ distinct terms and the second summation has $n \times \binom{n-1}{i-1}$ terms.

2. The form given by (11.16) holds when $i = n-1$ as well. In that case $\overline{S}(j)$ has only one element, $w_k(\overline{S}(j)) = 1$, and $\sum_{r \in \overline{S}(j)} \lambda_r - (n-i)\lambda_k = 0$ in the above expression. However, the expression given by (11.12) is much easier to work with.

3. If some of the λ_r's coincide, one could use limiting argument to obtain the relevant expression for $P(T_i > t)$. The extreme setup of this type is the iid case.

4. The distribution of the random variable T_i is helpful in finding probabilities of interest in the performance analysis of multiple antenna systems. See for example, Choi *et al.* (2003). There, the inid case is of interest.

11.3.3 The IID case vs. the INID case

It would be interesting to study the changes in the distributional properties of T_i as one moves from the iid case to the inid case. Of course, the additional complications that arise in the expression for the cdf in the inid case are evident in the above discussion. The question of interest could be in terms of stochastic comparisons. For example, how do the cdf of T_i in the inid case compare with the one in the iid case?

Proschan and Sethuraman (1976) obtained a majorization result for order statistics from heterogeneous populations with proportional hazard functions. They showed that if the vector $\boldsymbol{\lambda} = (\lambda_1, \ldots, \lambda_n)'$ majorizes $\boldsymbol{\nu} = (\nu_1, \ldots, \nu_n)'$, X_i is $\exp(\lambda_i)$, Y_i is $\exp(\nu_i)$, and they are all mutually independent, then $(X_{(1)}, \ldots, X_{(n)})$ is stochastically larger than $(Y_{(1)}, \ldots, Y_{(n)})$. Without loss of generality, we can take $\lambda_1 \geq \cdots \geq \lambda_n$ and $\nu_1 \geq \cdots \geq \nu_n$, then the first vector majorizes the second if $\sum_{j=1}^{i} \lambda_j \geq \sum_{j=1}^{i} \nu_j$ for $1 \leq i < n$, and equality holds when $i = n$. This means any monotonically increasing function of order statistics is stochastically larger with parameter vector $\boldsymbol{\lambda}$ than with $\boldsymbol{\nu}$, and in particular, this property holds for T_i. The iid case corresponds to the vector $(\lambda, \ldots, \lambda)'$ and is majorized by any $\boldsymbol{\lambda}$ with at least two distinct components. Thus, T_i will have a larger mean under heterogeneity than under homogeneity when the sum of the hazard rates remains the same. But, then one has to keep in mind that

$$\begin{aligned} E(X_1 + \cdots + X_n) \equiv E(T_0) &= \frac{n}{\lambda} \quad \text{(iid case)} \\ &= \sum_{i=1}^{n} \frac{1}{\lambda_i} \quad \text{(inid case)}. \end{aligned}$$

When $\sum \lambda_i = n\lambda$, from the "arithmetic mean-harmonic mean inequality," it is clear that the mean of the sample average $(= T_0/n)$ in the iid case is itself (much) smaller than its mean in the inid case. Thus, a similar result for T_i when $i > 0$ is hardly surprising given that components of T_i tend to be those X_j with larger means or smaller hazard rates.

References

1. Andrews, D. M. (1996). Moments of the selection differential from exponential and uniform parents, In *Statistical Theory and Applications: Papers in Honor of Herbert A. David* (Eds. H. N. Nagaraja, P. K. Sen, and D. F. Morrison), pp. 67–80, Springer-Verlag, New York.

2. Choi, Y.-S., Nagaraja, H. N., and Alamouti, S. M. (2003). Performance analysis and comparisons of antenna and beam selection/combining diversity, *Submitted for publication.*

3. David, H. A., and Nagaraja, H. N. (2003). *Order Statistics*, Third edition, John Wiley & Sons, New York.

4. Khaledi, B.-E., and Kochar, S. (2000). Dependence among spacings, *Probability in the Engineering and Information Sciences*, **14**, 461–472.

5. Nagaraja, H. N. (1981). Some finite sample results for the selection differential, *Annals of the Institute of Statistical Mathematics*, **33**, 437–448.

6. Nagaraja, H. N. (1982). Some nondegenerate limit laws for the selection differential, *Annals of Statistics*, **10**, 1306–1310.

7. Nevzorov, V. B. (1984). Representations of order statistics, based on exponential variables with different scaling parameters, *Zapiksi Nauchnykh Seminarov Leningradskogo Otdeleniya Matematicheskogo Instituta imeni V. A. Steklova Akademii Nauk SSSR (LOMI)*, **136**, 162–164; English translation (1986). *Journal of Soviet Mathematics*, **33**, 797–798.

8. Nevzorova, L., and Nevzorov, V. (1999). Ordered random variables, *Acta Applicandae Mathematicae*, **58**, 217-229.

9. Proschan, F., and Sethuraman, J. (1976). Stochastic comparisons of order statistics from heterogeneous populations, with applications in reliability, *Journal of Multivariate Analysis*, **6**, 608–616.

10. Rényi, A. (1953). On the theory of order statistics, *Acta Mathematica Academiae Scientiarum Hungaricae*, **4**, 191–231.

11. Tikhov, M. (1991). Reducing of test duration for censored samples, *Theory of Probability and Applications*, **36**, 604–607.

12

Fisher Information and Tukey's Linear Sensitivity Measure Based on Ordered Ranked Set Samples

N. Balakrishnan and T. Li

McMaster University, Hamilton, ON, Canada

Abstract: Stokes (1995) derived the Fisher information and discussed the maximum likelihood estimation (MLE) of the parameters of a location-scale family $F\left(\frac{x-\mu}{\sigma}\right)$ based on the ranked set sample (RSS). She found that a RSS provided more information about both μ and σ than a simple random sample (SRS) of the same size. We also focus here on the location-scale family. We use the idea of order statistics from independent and nonidentical random variables (INID) to propose an ordered ranked set sample (ORSS) and develop the Fisher information and the maximum likelihood estimation based on such an ORSS. We use logistic, normal, and one-parameter exponential distributions as examples and conclude that in all these three cases, the ORSS does not provide as much Fisher information as the RSS, and consequently the MLEs based on the ORSS (MLE-ORSS) are not as efficient as the MLEs based on the RSS (MLE-RSS). In addition to the MLEs, we are also interested in best linear unbiased estimators (BLUE). For this purpose, we apply another measure of information, viz., Tukey's linear sensitivity. Tukey (1965) proposed linear sensitivity to measure information contained in an ordered sample. We use logistic, normal, one- and two-parameter exponential, two-parameter uniform, and right triangular distributions as examples and show that in all these cases except the one-parameter exponential, in terms of linear sensitivity, the ORSS has more information than the RSS, and consequently the BLUEs based on the ORSS (BLUE-ORSS) are more efficient than the BLUEs based on the RSS (BLUE-RSS). In the case of one-parameter exponential, the ORSS has only slightly less information than the RSS with the relative efficiency being very close to 1.

Keywords and phrases: Ranked set samples, ordered ranked set samples, Fisher information, linear sensitivity measure, best linear unbiased estimators

12.1 Introduction

The basic procedure for obtaining a ranked set sample is as follows. First, we draw a random sample of size n from the population and order it (without actual measurement, e.g., visually). Then, the smallest observation is measured and the remaining are not measured. Next, another sample of size n is drawn and ordered, and only the second smallest observation is measured. This procedure is continued until the largest observation of the n-th sample of size n is measured. This process is called as a *one-cycle ranked set sample* of size n. If we replicate the above procedure m times, we obtain a ranked set sample of total size $N = mn$. The data thus observed is denoted by $\boldsymbol{X}_{\text{RSS}} = \{X_{1(1)}, X_{2(1)}, \ldots, X_{m(1)}, \ldots, X_{1(n)}, X_{2(n)}, \ldots, X_{m(n)}\}$. We use the following figure to describe this observational process:

$$
\begin{array}{cccccc}
 & & \underline{\text{Cycle 1}} & & & \\
\underline{X_{1:n}} & X_{2:n} & \cdots & X_{n:n} & \longrightarrow & X_{1(1)} \\
X_{1:n} & \underline{X_{2:n}} & \cdots & X_{n:n} & \longrightarrow & X_{1(2)} \\
\vdots & \vdots & \vdots & \vdots & \longrightarrow & \vdots \\
X_{1:n} & X_{2:n} & \cdots & \underline{X_{n:n}} & \longrightarrow & X_{1(n)} \\
\vdots & \vdots & \vdots & \vdots & \vdots & \vdots \\
 & & & & & \\
 & & \underline{\text{Cycle } m} & & & \\
\underline{X_{1:n}} & X_{2:n} & \cdots & X_{n:n} & \longrightarrow & X_{m(1)} \\
X_{1:n} & \underline{X_{2:n}} & \cdots & X_{n:n} & \longrightarrow & X_{m(2)} \\
\vdots & \vdots & \vdots & \vdots & \longrightarrow & \vdots \\
X_{1:n} & X_{2:n} & \cdots & \underline{X_{n:n}} & \longrightarrow & X_{m(n)}
\end{array}
$$

The ranked set sampling was first proposed by McIntyre (1952) in order to find a more efficient method to estimate the average yield of pasture. Since then, numerous parametric and nonparametric procedures based on ranked set samples have been developed in the literature. In the parametric case, Stokes (1995) examined both maximum likelihood estimates (MLE) and best linear unbiased estimates (BLUE) for location-scale distributions based on RSS. The BLUE based on RSS have been further discussed by Chuiv and Sinha (1998), Barnett and Barreto (2001), Hossain and Muttlak (2000), Zheng and Al-Saleh (2003), and Bhoj and Ahsanullah (1996). For some other parametric aspects of RSS, we refer the readers to Kim and Arnold (1999), Perron and Sinha (2004), Stokes (1980b), Barreto and Barnett (1999), and Chen (2000). In the nonparametric case, the estimation of the population mean and variance based on RSS and the properties of these estimators have been investigated. We refer the readers to Takahasi and Wakimoto (1968), Dell and Clutter (1972), and Stokes (1977, 1980a). The estimation of the parent cdf and the pdf based on RSS have been discussed by Stokes and Sager (1988), Chen (1999), and Kvam and Tiwari (1999).

This chapter is motivated by the work of Stokes (1995) who derived the Fisher information and discussed the maximum likelihood estimation of the parameters from a location-scale family $F\left(\frac{x-\mu}{\sigma}\right)$ based on the RSS. She found that the RSS provided more information about both μ and σ than a SRS of the same size. We also focus here on the location-scale family. For the purpose of computational simplicity, we discuss one-cycle ranked set sample of size n, which is denoted by $\boldsymbol{X}_{\text{RSS}} = \left\{X_{(1)}, X_{(2)}, \ldots, X_{(n)}\right\}$. In Section 12.2, we present the likelihood function based on the ORSS, the score equations, and the Fisher information. Then, we compare this information measure to that of the RSS. Next, we use the Newton-Raphson method to compare the MLE-RSS and the MLE-ORSS. We consider three examples, viz., logistic, normal, and one-parameter exponential distributions, and find that in all these three cases, the ORSS does not provide as much Fisher information as the RSS, and consequently the MLE-ORSS are not as efficient as the MLE-RSS. In addition to the MLEs, we are also interested in the BLUEs. Hence, we discuss in Section 12.3 another measure of information, viz., Tukey's linear sensitivity. We use logistic, normal, one- and two-parameter exponential, two-parameter uniform, and right triangular distributions as examples and show that in all these cases except the one-parameter exponential, in terms of linear sensitivity, the ORSS has more information than the RSS, and consequently the BLUE-ORSS are more efficient than the BLUE-RSS. In the case of one-parameter exponential distribution, the ORSS has only slightly less information than the RSS with the relative efficiency being very close to 1.

12.2 Maximum Likelihood Estimation Based on the ORSS

Let $\boldsymbol{X}_{\text{RSS}} = \left\{X_{(1)}, X_{(2)}, \ldots, X_{(n)}\right\}$ be the RSS from a location-scale population with pdf $\frac{1}{\sigma} f\left(\frac{x-\mu}{\sigma}\right)$ and cdf $F\left(\frac{x-\mu}{\sigma}\right)$. It is then evident that if the ranking of the RSS is perfect, the pdf of $X_{(r)}$ is

$$\begin{aligned} f_{r:n}(x_{(r)}) &= \frac{n!}{(r-1)!(n-r)!}\frac{1}{\sigma}\left[F\left(\frac{x_{(r)}-\mu}{\sigma}\right)\right]^{r-1} f\left(\frac{x_{(r)}-\mu}{\sigma}\right) \\ &\quad \times\left[1-F\left(\frac{x_{(r)}-\mu}{\sigma}\right)\right]^{n-r}, \quad -\infty < x_{(r)} < \infty. \quad (12.1) \end{aligned}$$

Because the likelihood function of the $\boldsymbol{X}_{\rm RSS}$ is simply the product of $f_{1:n}(x_{(1)}), \ldots,$ $f_{n:n}(x_{(n)})$ from (12.1) due to the independence of $X_{(1)}, X_{(2)}, \ldots, X_{(n)}$, we have the log-likelihood function of the $\boldsymbol{X}_{\rm RSS}$ to be

$$\begin{aligned} l &= C - n\ln\sigma + \sum_{r=1}^{n} \ln f(z_{(r)}) + \sum_{r=1}^{n} (r-1)\ln F(z_{(r)}) \\ &\quad + \sum_{r=1}^{n} (n-r)\ln[1-F(z_{(r)})], \qquad -\infty < z_{(1)}, \ldots, z_{(n)} < \infty, \end{aligned} \tag{12.2}$$

where C is a constant and $z_{(r)} = \frac{x_{(r)}-\mu}{\sigma}$.

Therefore, the MLE-RSS, denoted by $(\hat{\mu}_{\rm MLE}, \hat{\sigma}_{\rm MLE})$, is the solution of the equations

$$\begin{cases} \sum_{r=1}^{n} \frac{f'(z_{(r)})}{f(z_{(r)})} + \sum_{r=1}^{n} (r-1)\frac{f(z_{(r)})}{F(z_{(r)})} - \sum_{r=1}^{n} (n-r)\frac{f(z_{(r)})}{1-F(z_{(r)})} = 0, \\ n + \sum_{r=1}^{n} \frac{z_{(r)}f'(z_{(r)})}{f(z_{(r)})} + \sum_{r=1}^{n} (r-1)\frac{z_{(r)}f(z_{(r)})}{F(z_{(r)})} \\ \qquad\qquad - \sum_{r=1}^{n} (n-r)\frac{z_{(r)}f(z_{(r)})}{1-F(z_{(r)})} = 0. \end{cases} \tag{12.3}$$

Stokes (1995) also derived the Fisher information in RSS, from (12.2), as

$$\begin{aligned} I_{11} &= \mathrm{E}\left\{-\frac{\partial^2 l}{\partial \mu^2}\right\} \\ &= \frac{n}{\sigma^2}\mathrm{E}\left\{\frac{f'(Z)}{f(Z)}\right\}^2 + \frac{n(n-1)}{\sigma^2}\mathrm{E}\left\{\frac{f^2(Z)}{F(Z)[1-F(Z)]}\right\}, \end{aligned} \tag{12.4}$$

$$\begin{aligned} I_{12} &= \mathrm{E}\left\{-\frac{\partial^2 l}{\partial \mu \partial \sigma}\right\} \\ &= \frac{n}{\sigma^2}\mathrm{E}\left\{Z\left[\frac{f'(Z)}{f(Z)}\right]^2\right\} + \frac{n(n-1)}{\sigma^2}\mathrm{E}\left\{\frac{Zf^2(Z)}{F(Z)[1-F(Z)]}\right\}, \end{aligned} \tag{12.5}$$

$$\begin{aligned} I_{22} &= \mathrm{E}\left\{-\frac{\partial^2 l}{\partial \sigma^2}\right\} \\ &= \frac{n}{\sigma^2}\mathrm{E}\left\{\left[\frac{Zf'(Z)}{f(Z)}\right]^2 - 1\right\} + \frac{n(n-1)}{\sigma^2}\mathrm{E}\left\{\frac{[Zf(Z)]^2}{F(Z)[1-F(Z)]}\right\}, \end{aligned} \tag{12.6}$$

where Z is a random variate with the standard density $f(z)$.

Now ordering $\boldsymbol{X}_{\text{RSS}}$ in an increasing order of magnitude, we get $\boldsymbol{X}_{\text{ORSS}} = \{X_{1:n}^{\text{ORSS}} \leq X_{2:n}^{\text{ORSS}} \leq \cdots \leq X_{n:n}^{\text{ORSS}}\}$, which is called the *ordered ranked set sample* (ORSS). Using the results of order statistics from INID random variables [see David and Nagaraja (2003)], the density function of $X_{r:n}^{\text{ORSS}}$ $(1 \leq r \leq n)$ is given by

$$\begin{aligned} & f_{r:n}^{\text{ORSS}}(x_r) \\ &= \frac{1}{(r-1)!(n-r)!} \sum_{P} \left\{ \prod_{k=1}^{r-1} [F_{i_k:n}(x_r)]\, f_{i_r:n}(x_r) \prod_{k=r+1}^{n} [1 - F_{i_k:n}(x_r)] \right\}, \\ & \qquad -\infty < x_r < \infty, \qquad (12.7) \end{aligned}$$

where $\sum\limits_{P}$ denotes the summation over all $n!$ permutations $(i_1, i_2, \ldots, i_n)$ of $(1, 2, \ldots, n)$. The likelihood function of $\boldsymbol{X}_{\text{ORSS}}$ can then be written as

$$\begin{aligned} L^* &= \sum_{P} \left[\prod_{k=1}^{n} f_{i_k:n}^{\text{ORSS}}(x_k) \right] \\ &= \frac{1}{\sigma^n} \sum_{P} \prod_{k=1}^{n} \left[\frac{n!}{(i_k-1)!(n-i_k)!} \left[F(z_k)\right]^{i_k-1} \left[1 - F(z_k)\right]^{n-i_k} f(z_k) \right] \\ &= \frac{1}{\sigma^n} \prod_{k=1}^{n} \left[\frac{f(z_k)}{\mathrm{B}(k, n-k+1)} \right] \left\{ \sum_{P} \prod_{k=1}^{n} \left[\left[F(z_k)\right]^{i_k-1} \left[1 - F(z_k)\right]^{n-i_k} \right] \right\}, \\ & \qquad -\infty < z_1 < \cdots < z_n < \infty, \end{aligned}$$

where $z_k = \frac{x_k - \mu}{\sigma}$, and $B(a, b) = \frac{(a-1)!(b-1)!}{(a+b-1)!}$ is the complete beta function. The log-likelihood function is then

$$\begin{aligned} l^* &= D - n \ln \sigma + \sum_{k=1}^{n} \ln f(z_k) + \ln \left\{ \sum_{P} \prod_{k=1}^{n} \left[(F(z_k))^{i_k-1} (1 - F(z_k))^{n-i_k} \right] \right\}, \\ & \qquad -\infty < z_1 < \cdots < z_n < \infty, \end{aligned}$$

where D is a constant. The MLE-ORSS, denoted by $(\mu_{\text{MLE}}^*, \sigma_{\text{MLE}}^*)$, is the solution of the equations

$$\begin{cases} \sum_{k=1}^{n} \frac{f'(z_k)}{f(z_k)} + \frac{a_1}{b} = 0, \\ \\ n + \sum_{k=1}^{n} \left(z_k \frac{f'(z_k)}{f(z_k)} \right) + \frac{a_2}{b} = 0, \end{cases} \qquad (12.8)$$

where

$$\begin{aligned} a_1 &= \sum_P \left\{ \prod_{s=1}^{n} \left[(F(z_s))^{i_s-1}(1-F(z_s))^{n-i_s} \right] \right. \\ &\qquad\qquad \left. \times \sum_{k=1}^{n} \left[\left(\frac{i_k-1}{F(z_k)} - \frac{n-i_k}{1-F(z_k)} \right) f(z_k) \right] \right\}, \\ a_2 &= \sum_P \left\{ \prod_{s=1}^{n} \left[(F(z_s))^{i_s-1}(1-F(z_s))^{n-i_s} \right] \right. \\ &\qquad\qquad \left. \times \sum_{k=1}^{n} \left[\left(\frac{i_k-1}{F(z_k)} - \frac{n-i_k}{1-F(z_k)} \right) z_k f(z_k) \right] \right\}, \\ b &= \sum_P \prod_{s=1}^{n} \left[(F(z_s))^{i_s-1}(1-F(z_s))^{n-i_s} \right]. \end{aligned}$$

We are also interested in the Fisher information in ORSS, because it will allow us to compare the relative efficiency of MLE-ORSS with respect to MLE-RSS. The Fisher information in ORSS can be derived as follows:

$$\begin{aligned} I_{11}^* &= \mathrm{E}\left\{ -\frac{\partial^2 l^*}{\partial \mu^2} \right\} \\ &= \frac{n}{\sigma^2}\mathrm{E}\left\{ \frac{f'(Z)}{f(Z)} \right\}^2 + \frac{1}{\sigma^2}\mathrm{E}\left\{ \frac{a_3}{b} \right\} + \frac{1}{\sigma^2}\mathrm{E}\left\{ \left(\frac{a_1}{b}\right)^2 - \frac{a_4}{b} \right\}, \end{aligned} \tag{12.9}$$

$$\begin{aligned} I_{12}^* &= \mathrm{E}\left\{ -\frac{\partial^2 l^*}{\partial \mu \partial \sigma} \right\} \\ &= \frac{n}{\sigma^2}\mathrm{E}\left\{ Z\left[\frac{f'(Z)}{f(Z)} \right]^2 \right\} + \frac{1}{\sigma^2}\mathrm{E}\left\{ \frac{a_5}{b} \right\} + \frac{1}{\sigma^2}\mathrm{E}\left\{ \frac{a_1 a_2}{b^2} - \frac{a_6}{b} \right\}, \end{aligned} \tag{12.10}$$

$$\begin{aligned} I_{22}^* &= \mathrm{E}\left\{ -\frac{\partial^2 l^*}{\partial \sigma^2} \right\} \\ &= \frac{n}{\sigma^2}\mathrm{E}\left\{ \left[\frac{Zf'(Z)}{f(Z)} \right]^2 - 1 \right\} + \frac{1}{\sigma^2}\mathrm{E}\left\{ \frac{a_7}{b} \right\} + \frac{1}{\sigma^2}\mathrm{E}\left\{ \left(\frac{a_2}{b}\right)^2 - \frac{a_8}{b} \right\}, \end{aligned} \tag{12.11}$$

where

$$\begin{aligned}
a_3 &= \sum_P \left\{ \coprod_{s=1}^{n} \left[(F(Z_s))^{i_s-1} (1-F(Z_s))^{n-i_s} \right] \right. \\
&\qquad \times \left[\sum_{k=1}^{n} \left(\left(\frac{i_k-1}{(F(Z_k))^2} + \frac{n-i_k}{(1-F(Z_k))^2} \right) f^2(Z_k) \right) \right. \\
&\qquad \left. \left. - \sum_{k=1}^{n} \left(\left(\frac{i_k-1}{F(Z_k)} - \frac{n-i_k}{1-F(Z_k)} \right) f'(Z_k) \right) \right] \right\},
\end{aligned}$$

$$\begin{aligned}
a_4 &= \sum_P \left\{ \prod_{s=1}^{n} [(F(Z_s))^{i_s-1}(1-F(Z_s))^{n-i_s}] \right. \\
&\qquad \left. \times \left[\sum_{k=1}^{n} \left(\frac{i_k-1}{F(Z_k)} - \frac{n-i_k}{1-F(Z_k)} \right) f(Z_k) \right]^2 \right\},
\end{aligned}$$

$$\begin{aligned}
a_5 &= \sum_P \left\{ \prod_{s=1}^{n} [(F(Z_s))^{i_s-1}(1-F(Z_s))^{n-i_s}] \right. \\
&\qquad \times \left[\sum_{k=1}^{n} \left(\left(\frac{i_k-1}{(F(Z_k))^2} + \frac{n-i_k}{(1-F(Z_k))^2} \right) Z_k f^2(Z_k) \right) \right. \\
&\qquad \left. \left. - \sum_{k=1}^{n} \left(\left(\frac{i_k-1}{F(Z_k)} - \frac{n-i_k}{1-F(Z_k)} \right) Z_k f'(Z_k) \right) \right] \right\},
\end{aligned}$$

$$\begin{aligned}
a_6 &= \sum_P \left\{ \prod_{s=1}^{n} [(F(Z_s))^{i_s-1}(1-F(Z_s))^{n-i_s}] \right. \\
&\qquad \times \sum_{k=1}^{n} \left(\left(\frac{i_k-1}{F(Z_k)} - \frac{n-i_k}{1-F(Z_k)} \right) f(Z_k) \right) \\
&\qquad \left. \times \sum_{k=1}^{n} \left(\left(\frac{i_k-1}{F(Z_k)} - \frac{n-i_k}{1-F(Z_k)} \right) Z_k f(Z_k) \right) \right\},
\end{aligned}$$

$$\begin{aligned}
a_7 &= \sum_P \left\{ \prod_{s=1}^{n} [(F(Z_s))^{i_s-1}(1-F(Z_s))^{n-i_s}] \right. \\
&\qquad \times \left[\sum_{k=1}^{n} \left(\left(\frac{i_k-1}{(F(Z_k))^2} + \frac{n-i_k}{(1-F(Z_k))^2} \right) Z_k^2 f^2(Z_k) \right) \right. \\
&\qquad \left. \left. - \sum_{k=1}^{n} \left(\left(\frac{i_k-1}{F(Z_k)} - \frac{n-i_k}{1-F(Z_k)} \right) Z_k^2 f'(Z_k) \right) \right] \right\},
\end{aligned}$$

$$
\begin{aligned}
a_8 &= \sum_P \left\{ \prod_{s=1}^{n} \left[(F(Z_s))^{i_s - 1} (1 - F(Z_s))^{n - i_s} \right] \right. \\
&\qquad \left. \times \left[\sum_{k=1}^{n} \left(\left(\frac{i_k - 1}{F(Z_k)} - \frac{n - i_k}{1 - F(Z_k)} \right) f(Z_k) \right) \right]^2 \right\}.
\end{aligned}
$$

From Eqs. (12.9)-(12.11), we can see that the first terms in these equations are actually the corresponding terms in the Fisher information in an ordered simple random sample [see Stokes (1995)]. Moreover, we can prove that the third term in Eqs. (12.9) and (12.11) are always less than zero (see Appendix). The complexity of the other terms in Eqs. (12.9)-(12.11) makes it difficult for us to determine a relation between the Fisher information in RSS and ORSS. But, from our study of logistic, normal, and one-parameter exponential distributions, we observe that the RSS has larger Fisher information than the ORSS in these three cases. Stokes (1995) noted that, for the RSS, the term I_{12} in (12.5) is zero for symmetric distributions, but this may not be true for I_{12}^* in (12.10) in the case of ORSS in general, even though it is true in the case of the logistic distribution.

12.2.1 Logistic distribution

Let $\boldsymbol{X}_{\text{RSS}}$ and $\boldsymbol{X}_{\text{ORSS}}$ be from the logistic population with pdf

$$
f(x) = \frac{1}{\sigma} \frac{e^{-(x-\mu)/\sigma}}{(1 + e^{-(x-\mu)/\sigma})^2}, \qquad -\infty < x < \infty.
$$

From Eqs. (12.4)-(12.6) and Eqs. (12.9)-(12.11), we can derive the Fisher information in RSS as

$$
\begin{aligned}
I_{11} &= \frac{n(n+1)}{6\sigma^2}, \qquad I_{12} = 0, \\
I_{22} &= \frac{n}{\sigma^2} \left\{ \mathrm{E}[Z^2(1 - 2F(Z))]^2 - 1 \right\} + \frac{n(n-1)}{\sigma^2} \mathrm{E}[Z^2 F(Z)(1 - F(Z))],
\end{aligned}
$$

and the Fisher information in ORSS as

$$
\begin{aligned}
I_{11}^* &= \frac{n(n+1)}{6\sigma^2}, \qquad I_{12}^* = 0, \\
I_{22}^* &= \frac{n}{\sigma^2} \left\{ \mathrm{E}[Z^2(1 - 2F(Z))]^2 - 1 \right\} + \frac{n(n-1)}{\sigma^2} \mathrm{E}[Z^2 F(Z)(1 - F(Z))] \\
&\qquad + \frac{1}{\sigma^2} \mathrm{E} \left\{ \left(\frac{a_2}{b} \right)^2 - \frac{a_8}{b} \right\}.
\end{aligned}
$$

It is easy to see that $I_{11}^* = I_{11}$, $I_{12}^* = I_{12}$, but $I_{22}^* \leq I_{22}$, because we know that $\mathrm{E}\left\{ \left(\frac{a_2}{b}\right)^2 - \frac{a_8}{b} \right\} \leq 0$ (see Appendix). Table 12.1 presents the Fisher information about σ in RSS and ORSS, respectively, which are based on Monte

Table 12.1: Comparison of Fisher information between RSS and ORSS from the logistic distribution

n	2	3	4	5	6	7	8	9	10
$\frac{1}{\sigma^2}I_{22}$	3.29440	5.58191	8.29451	11.45899	15.03105	19.08059	23.50773	28.33388	33.67061
$\frac{1}{\sigma^2}I_{22}^*$	3.04032	5.01068	7.36663	10.16772	13.35357	17.01754	21.04628	25.54500	30.92350

Table 12.2: Bias and MSE of MLEs based on RSS from the logistic distribution

n	Bias($\hat{\mu}_{\text{MLE}}$)	MSE($\hat{\mu}_{\text{MLE}}$)	Bias($\hat{\sigma}_{\text{MLE}}$)	MSE($\hat{\sigma}_{\text{MLE}}$)
2	−0.00123	1.14555	−2.12526	5.97207
3	0.00103	0.54891	−0.16112	0.19406
4	−0.00017	0.31869	−0.10695	0.12830
5	0.00110	0.20932	−0.07887	0.09209
6	0.00002	0.14808	−0.06070	0.06933
7	0.00019	0.11038	−0.04831	0.05477
8	0.00015	0.08548	−0.03993	0.04421
9	−0.00136	0.06875	−0.03254	0.03603
10	0.00169	0.05538	−0.02809	0.03043

Carlo simulations. It is clear to see that the Fisher information in the ORSS is moderately less than that in the RSS.

By using the Newton-Raphson method to solve Eqs. (12.3) and (12.8), we obtain the MLE-RSS and the MLE-ORSS, which we shall denote by $(\hat{\mu}_{\text{MLE}}, \hat{\sigma}_{\text{MLE}})$ and $(\mu^*_{\text{MLE}}, \sigma^*_{\text{MLE}})$, respectively. Tables 12.2 and 12.3 present the bias and mean square error of the MLE-RSS and MLE-ORSS determined from 10,000 Monte Carlo simulations (with $\mu = 0$ and $\sigma = 1$). When $n = 2$, the MLE-RSS as well as the MLE-ORSS are far removed from the true value of μ and σ, but σ^*_{MLE} is better than $\hat{\sigma}_{\text{MLE}}$. When $n \geq 3$, the efficiency of the MLE-ORSS of μ is almost the same as the MLE-RSS of μ, while the relative efficiency of σ^*_{MLE} with respect to $\hat{\sigma}_{\text{MLE}}$ is around 90%.

12.2.2 Normal distribution

Let $\boldsymbol{X}_{\text{RSS}}$ and $\boldsymbol{X}_{\text{ORSS}}$ be from the normal population with density function $f(x) = \frac{1}{\sqrt{2\pi}\sigma}e^{-\frac{(x-\mu)^2}{2\sigma^2}}$, $-\infty < x < \infty$. Table 12.4 presents the Fisher information in RSS and ORSS. It is clear that both I^*_{11} and I^*_{22} are less than I_{11} and I_{22}, respectively. Tables 12.5 and 12.6 present the bias and MSE of the MLE-RSS and the MLE-ORSS, respectively. When $n = 2$, the Newton-Raphson method does not often converge based on either RSS or ORSS. When $n = 3$, the

Table 12.3: Bias and MSE of MLEs based on ORSS from the logistic distribution and relative efficiencies

n	Bias(μ^*_{MLE})	MSE(μ^*_{MLE})	RE($\mu^*_{\text{MLE}}, \hat{\mu}_{\text{MLE}}$)	Bias(σ^*_{MLE})	MSE(σ^*_{MLE})	RE($\sigma^*_{\text{MLE}}, \hat{\sigma}_{\text{MLE}}$)
2	−0.00123	1.14555	1.00000	−1.67503	3.06856	1.94621
3	0.00108	0.54980	0.99838	−0.18265	0.21666	0.89570
4	−0.00018	0.31940	0.99780	−0.12261	0.14504	0.88461
5	0.00108	0.20972	0.99812	−0.09068	0.10460	0.88042
6	0.00000	0.14836	0.99809	−0.07018	0.07855	0.88265
7	0.00020	0.11055	0.99853	−0.05562	0.06178	0.88653
8	0.00010	0.08560	0.99866	−0.04604	0.04962	0.89091
9	−0.00137	0.06882	0.99904	−0.03726	0.04032	0.89374
10	0.00178	0.05545	0.99868	−0.03164	0.03368	0.90347

Table 12.4: Comparison of Fisher information between RSS and ORSS from the normal distribution

n	$\sigma^2 I_{11}$	$\sigma^2 I_{12}$	$\sigma^2 I_{22}$	$\sigma^2 I^*_{11}$	$\sigma^2 I^*_{12}$	$\sigma^2 I^*_{22}$
2	2.96123	0.00039	4.53074	2.95681	0.00044	4.20724
3	5.88288	−0.00561	7.62492	5.86996	−0.00493	6.88766
4	9.76617	−0.00149	11.23962	9.74324	−0.00198	10.04306
5	14.61027	−0.00437	15.40385	14.57601	−0.00446	13.71449
6	20.41955	0.00434	20.07481	20.37240	0.00335	17.86093
7	27.18308	0.00227	25.29647	27.12498	0.00216	22.56148
8	34.90793	−0.06191	31.15110	34.83470	−0.06409	27.85410
9	43.59402	−0.07638	37.63034	43.49940	−0.07851	33.69233
10	53.24776	−0.02337	44.40295	53.14073	−0.04098	39.92216

Newton-Raphson method still often does not converge based on RSS, but it converges based on ORSS. We can also see that the relative efficiency of μ^*_{MLE} with respect to $\hat{\mu}_{\text{MLE}}$ is very close to 1, while the relative efficiency of σ^*_{MLE} with respect to $\hat{\sigma}_{\text{MLE}}$ is around 90%.

12.2.3 One-parameter exponential distribution

Let $\boldsymbol{X}_{\text{RSS}}$ and $\boldsymbol{X}_{\text{ORSS}}$ be from an exponential population with density function $f(x) = \frac{1}{\sigma}\exp\left(-\frac{x}{\sigma}\right)$, $x > 0$, $\sigma > 0$. Table 12.7 presents the Fisher information about σ in RSS and ORSS, and it is clear from this table that the Fisher information in ORSS is slightly less than in RSS. The bias and MSE of the MLE from RSS and ORSS are presented in Table 12.8. The relative efficiency of σ^*_{MLE} with respect to $\hat{\sigma}_{\text{MLE}}$ is about 98%.

Table 12.5: Bias and MSE of MLEs based on RSS from the normal distribution

n	Bias($\hat{\mu}_{\text{MLE}}$)	MSE($\hat{\mu}_{\text{MLE}}$)	Bias($\hat{\sigma}_{\text{MLE}}$)	MSE($\hat{\sigma}_{\text{MLE}}$)
3	—	—	—	—
4	0.00141	0.10676	−0.12910	0.10160
5	−0.00396	0.06799	−0.09069	0.07030
6	−0.00262	0.04912	−0.07149	0.05349
7	−0.00073	0.03804	−0.05576	0.04285
8	−0.00170	0.02834	−0.04439	0.03381
9	−0.00020	0.02309	0.03441	0.02877
10	−0.00424	0.01931	−0.03675	0.02501

Table 12.6: Bias and MSE of MLEs based on ORSS from the normal distribution and relative efficiencies

n	Bias(μ^*_{MLE})	MSE(μ^*_{MLE})	RE($\mu^*_{\text{MLE}}, \hat{\mu}_{\text{MLE}}$)	Bias(σ^*_{MLE})	MSE(σ^*_{MLE})	RE($\sigma^*_{\text{MLE}}, \hat{\sigma}_{\text{MLE}}$)
3	−0.00054	0.17397	—	−0.22907	0.18255	—
4	0.00125	0.10706	0.99721	−0.14710	0.11498	0.88360
5	−0.00400	0.06814	0.99772	−0.10249	0.08032	0.87518
6	−0.00243	0.04935	0.99527	−0.08175	0.06104	0.87636
7	−0.00076	0.03815	0.99706	−0.06212	0.04820	0.88904
8	−0.00161	0.02840	0.99786	−0.04924	0.03838	0.88108
9	−0.00028	0.02316	0.99661	0.03868	0.03214	0.89535
10	−0.00418	0.01936	0.99729	−0.04068	0.02745	0.91093

12.2.4 Conclusions

From the above three examples, we see that even though the ORSS does not have as much Fisher information as the RSS, the relative efficiencies are very high, especially for the location parameter μ for normal and logistic distributions. The Newton-Raphson method to obtain the MLE-RSS and the MLE-ORSS does not often converge, or converges to a value away from the true value when $n = 2$ or 3. In this case, the MLE-ORSS seems to be better than the MLE-RSS for normal and logistic distributions in terms of both convergence and mean square error. In the case of the one-parameter exponential distribution, the Fisher information in ORSS is only slightly less than in RSS, and the relative efficiency of the MLE-ORSS compared to the MLE-RSS is nearly 98%.

Table 12.7: Comparison of Fisher information between RSS and ORSS from the one-parameter exponential distribution

n	2	3	4	5	6	7	8	9	10
$\sigma^2 I_{22}$	2.79955	5.41277	8.84513	13.08436	18.10901	23.95813	30.59065	38.15936	46.23967
$\sigma^2 I_{22}^*$	2.76277	5.32411	8.70176	12.88173	17.84192	23.61300	30.18296	37.64389	45.66347

Table 12.8: Bias and MSE of the MLE based on RSS and ORSS from the one-parameter exponential distribution, and relative efficiency

n	Bias($\hat{\sigma}_{\text{MLE}}$)	MSE($\hat{\sigma}_{\text{MLE}}$)	Bias(σ^*_{MLE})	MSE(σ^*_{MLE})	RE($\sigma^*_{\text{MLE}}, \hat{\sigma}_{\text{MLE}}$)
2	0.01834	0.36682	0.02226	0.37486	0.97856
3	0.01415	0.19200	0.01685	0.19625	0.97837
4	0.01022	0.11638	0.01198	0.11881	0.97960
5	0.00731	0.07811	0.00851	0.07955	0.98190
6	0.00670	0.05614	0.00754	0.05712	0.98280
7	0.00540	0.04242	0.00604	0.04307	0.98494
8	0.00045	0.03290	0.00088	0.03335	0.98629
9	0.00428	0.02649	0.00447	0.02694	0.98345
10	0.00403	0.02133	0.00442	0.02158	0.98840

12.3 Tukey's Linear Sensitivity Measure Based on ORSS

Fisher information revealed that the RSS is more efficient than the ORSS in the three examples discussed in the last section. Moreover, we also noted that the MLE-RSS is in general more efficient than the MLE-ORSS in these three cases. How about other estimators, such as the best linear unbiased estimators based on RSS (BLUE-RSS) and ORSS (BLUE-ORSS)? Tukey's linear sensitivity measure naturally comes in to play in this context.

Tukey (1965) proposed linear sensitivity as a measure of information in an ordered sample. Nagaraja (1994) showed that the linear sensitivity of an ordered sample is actually the inverse of the variance of BLUE based on this ordered sample. This definition of linear sensitivity and its connection to the BLUE was extended to the multiparameter version by Chandrasekar and Balakrishnan (2002). In this section, we will examine the linear sensitivity in an ORSS and compare it with that in a RSS.

Let $\boldsymbol{X}_{\text{RSS}} = \{X_{(1)}, X_{(2)}, \ldots, X_{(n)}\}$ and $\boldsymbol{X}_{\text{ORSS}} = \{X_{1:n}^{\text{ORSS}} \leq X_{2:n}^{\text{ORSS}} \leq \cdots$

$\le X_{n:n}^{\text{ORSS}}\}$ be the RSS and ORSS from a location-scale population with pdf $\frac{1}{\sigma} f\left(\frac{x-\mu}{\sigma}\right)$ and cdf $F\left(\frac{x-\mu}{\sigma}\right)$. The pdf of $X_{r:n}^{\text{ORSS}}$ is as given in Eq. (12.7). Furthermore, the joint pdf of $X_{r:n}^{\text{ORSS}}$ and $X_{s:n}^{\text{ORSS}}$ $(1 \le r < s \le n)$ can be shown to be

$$
\begin{aligned}
f_{r,s:n}^{\text{ORSS}}(x_r, x_s) &= \sum_{P} \sum_{k_1=j_1}^{n} \cdots \sum_{k_{s-1}=j_{s-1}}^{n} \sum_{k_s=0}^{j_s-1} \cdots \sum_{k_n=0}^{j_n-1} \sum_{l_1=0}^{n-k_1} \cdots \sum_{l_{r-1}=0}^{n-k_{r-1}} \\
&\quad \times \sum_{l_{r+1}=k_{r+1}+1-j_{r+1}}^{k_{r+1}} \cdots \sum_{l_{s-1}=k_{s-1}+1-j_{s-1}}^{k_{s-1}-1} \sum_{l_{s+1}=0}^{k_{s+1}} \cdots \sum_{l_n=0}^{k_n} W_{r,s}^{*} f_{\tilde{r},\tilde{s}:n^2}(x_r, x_s), \\
&\qquad -\infty < x_r < x_s < \infty,
\end{aligned} \tag{12.12}
$$

where

$$
\begin{aligned}
W_{r,s}^{*} &= W_{j,k,l} \frac{(\tilde{r}-1)!(\tilde{s}-\tilde{r}-1)!(n^2-\tilde{s})!}{(r-1)!(s-r-1)!(n-s)!(n^2)!}, \\
W_{j,k,l} &= \left\{\prod_{a=1}^{r-1} \binom{n}{k_a}\binom{n-k_a}{l_a}\right\} \cdot \left\{j_r \binom{n}{j_r}\binom{n-j_r}{k_r-j_r}\right\} \\
&\quad \times \left\{\prod_{a=r+1}^{s-1} \binom{n}{k_a}\binom{k_a}{l_a}\right\} \left\{j_s \binom{n}{j_s}\binom{j_s-1}{k_s}\right\} \left\{\prod_{a=s+1}^{n} \binom{n}{k_a}\binom{k_a}{l_a}\right\},
\end{aligned}
$$

$$
\begin{aligned}
\tilde{r} &= \sum_{\substack{a=1 \\ a\ne r,s}}^{n} k_a + j_r + j_s - \sum_{\substack{a=r+1 \\ a\ne s}}^{n} l_a - k_s - 1, \\
\tilde{s} &= \sum_{\substack{a=1 \\ a\ne r,s}}^{n} k_a + j_s + \sum_{a=1}^{r-1} l_a.
\end{aligned}
$$

From Eqs. (12.7) and (12.12), the mean vector and the variance-covariance matrix of ORSS can be computed with which the BLUE-ORSS can be obtained using the general formula of BLUEs which was first derived by Lloyd (1952). Specifically, with $\boldsymbol{X}_{\text{ORSS}} = (X_{1:n}^{\text{ORSS}}, \ldots, X_{n:n}^{\text{ORSS}})'$ denoting the ordered ranked set sample from a location-scale family with location parameter μ and scale parameter σ (> 0) and $\boldsymbol{Y} = \left(\frac{X_{1:n}^{\text{ORSS}}-\mu}{\sigma}, \ldots, \frac{X_{n:n}^{\text{ORSS}}-\mu}{\sigma}\right)'$ denoting the corresponding standard random vector, the BLUE of $(\mu, \sigma)'$ is given by

$$
\begin{pmatrix} \mu^* \\ \sigma^* \end{pmatrix} = \left(\boldsymbol{B}'\boldsymbol{\Sigma}^{-1}\boldsymbol{B}\right)^{-1} \boldsymbol{B}'\boldsymbol{\Sigma}^{-1}\boldsymbol{X}_{\text{ORSS}},
$$

and its variance-covariance matrix is

$$
\text{Var}\begin{pmatrix} \mu^* \\ \sigma^* \end{pmatrix} = \sigma^2 \left(\boldsymbol{B}'\boldsymbol{\Sigma}^{-1}\boldsymbol{B}\right)^{-1},
$$

Table 12.9: Comparison of linear sensitivity of RSS and ORSS from the logistic distribution

n	$\sigma^2 \boldsymbol{S}_{11}$	$\sigma^2 \boldsymbol{S}_{22}$	$\sigma^2 \boldsymbol{S}_{11}^*$	$\sigma^2 \boldsymbol{S}_{22}^*$	$ARE(\mu_{\text{BLUE}}^*, \hat{\mu}_{\text{BLUE}})$	$ARE(\sigma_{\text{BLUE}}^*, \hat{\sigma}_{\text{BLUE}})$
2	0.87341	0.87341	0.87341	1.73650	1.00000	1.98818
3	1.75573	2.20602	1.82447	3.60392	1.03915	1.63367
4	2.96026	3.96609	3.11446	5.85548	1.05209	1.47638
5	4.49111	6.14675	4.74096	8.51207	1.05563	1.38481
6	6.35048	8.74662	6.70318	11.58308	1.05554	1.32429
7	8.53967	11.76594	9.01309	15.10464	1.05544	1.28376
8	11.05950	15.20538	11.62819	19.01391	1.05142	1.25047
9	13.91056	19.06575	14.60118	23.33060	1.04965	1.22369
10	17.09325	23.34781	17.90163	28.07809	1.04729	1.20260

where $\boldsymbol{B} = (\boldsymbol{1}\ \boldsymbol{\mu})$, $\boldsymbol{1} = (1, 1, \ldots, 1)'_{1\times n}$, and $\boldsymbol{\mu}$ and $\boldsymbol{\Sigma}$ are the mean vector and the variance-covariance matrix of $\boldsymbol{Y}$, respectively. Similarly, if $\boldsymbol{X}_{\text{ORSS}}$ denotes an ordered ranked set sample from a scale family with scale parameter σ (> 0) and $\boldsymbol{Y} = \boldsymbol{X}_{\text{ORSS}}/\sigma$ denotes the corresponding standard random vector, the BLUE of σ is given by

$$\sigma^* = \boldsymbol{\mu}'\boldsymbol{\Sigma}^{-1}\boldsymbol{X}_{\text{ORSS}}/\left(\boldsymbol{\mu}'\boldsymbol{\Sigma}^{-1}\boldsymbol{\mu}\right)$$

and its variance is

$$\text{Var}(\sigma^*) = \sigma^2/\left(\boldsymbol{\mu}'\boldsymbol{\Sigma}^{-1}\boldsymbol{\mu}\right).$$

From the above formulas, linear sensitivities of RSS and ORSS, denoted by $\boldsymbol{S}$ and $\boldsymbol{S}^*$, respectively, can be computed. Note that the means and variances of RSS are exactly the same as the means and variances of the usual order statistics which have been computed rather extensively; see, for example, Tietjen *et al.* (1977) and Balakrishnan (1992) for tables for normal and logistic distributions, respectively.

Tables 12.9–12.14 present the linear sensitivity of RSS and ORSS from the logistic, normal, one- and two-parameter exponential, two-parameter uniform, and right triangular distributions, respectively. Bhoj and Ahsanullah (1996) discussed the estimation of parameters of the generalized geometric distribution using RSS. They used the two-parameter uniform distribution with pdf $f(x) = \frac{1}{2\sqrt{3}\sigma}$, $\mu - \sqrt{3}\sigma \le x \le \mu + \sqrt{3}\sigma$, and right triangular distribution with pdf $f(x) = \frac{1}{9\sigma}\left(\frac{x-\mu}{\sigma} + 2\sqrt{2}\right)$, $\mu - 2\sqrt{2}\sigma \le x \le \mu + \sqrt{2}\sigma$, as specific examples and showed that when sample size is small $(n \le 5)$, the BLUE-RSS is not as efficient as BLUE based on the usual order statistics (BLUE-OS). Here, we want to compare the BLUE-ORSS to the BLUE-RSS.

It is clear from these tables that ORSS possesses more linear sensitivity than the RSS for both location and scale parameters of logistic, normal, two-

Table 12.10: Comparison of linear sensitivity of RSS and ORSS from the normal distribution

n	$\sigma^2 \boldsymbol{S}_{11}$	$\sigma^2 \boldsymbol{S}_{22}$	$\sigma^2 \boldsymbol{S}^*_{11}$	$\sigma^2 \boldsymbol{S}^*_{22}$	$ARE(\mu^*_{\text{BLUE}}, \hat{\mu}_{\text{BLUE}})$	$ARE(\sigma^*_{\text{BLUE}}, \hat{\sigma}_{\text{BLUE}})$
2	2.93388	0.93388	2.93388	2.00532	1.00000	2.14729
3	5.80363	2.56028	5.81448	4.52789	1.00187	1.76851
4	9.61593	4.79934	9.65043	7.56371	1.00359	1.57599
5	14.37543	7.61739	14.44306	11.11572	1.00470	1.45926
6	20.08524	10.99780	21.06979	16.91149	1.04902	1.53772
7	26.74753	14.93140	26.89108	19.81793	1.00537	1.32726
8	34.36385	19.41280	34.64788	24.97151	1.00827	1.28634
9	42.93534	24.43864	43.27344	30.67988	1.00787	1.25538
10	52.46290	30.00675	52.90568	36.86042	1.00844	1.22840

Table 12.11: Comparison of linear sensitivity of RSS and ORSS from the one-parameter exponential distribution

n	2	3	4	5	6	7	8	9	10
$\sigma^2 \boldsymbol{S}$	2.80000	5.39246	8.77927	12.96275	17.94482	23.72697	30.31037	37.69593	45.88441
$\sigma^2 \boldsymbol{S}^*$	2.76213	5.30766	8.64837	12.78917	17.73261	23.49164	30.09246	37.43908	45.60811
ARE	0.98648	0.98427	0.98509	0.98661	0.98817	0.99008	0.99281	0.99319	0.99398

Table 12.12: Comparison of linear sensitivity of RSS and ORSS from the two-parameter exponential distribution

n	$\sigma^2 \boldsymbol{S}_{11}$	$\sigma^2 \boldsymbol{S}_{22}$	$\sigma^2 \boldsymbol{S}^*_{11}$	$\sigma^2 \boldsymbol{S}^*_{22}$	$ARE(\mu^*_{\text{BLUE}}, \hat{\mu}_{\text{BLUE}})$	$ARE(\sigma^*_{\text{BLUE}}, \hat{\sigma}_{\text{BLUE}})$
2	1.14286	0.66667	3.02314	1.19512	2.64524	1.79268
3	4.29170	1.85085	11.18056	2.93206	2.60516	1.58417
4	10.06989	3.56145	25.35888	5.26941	2.51829	1.47957
5	18.96039	5.81344	46.14243	8.22941	2.43362	1.41558
6	31.35595	8.61967	109.53242	11.82492	3.49319	1.37185
7	47.58677	11.99067	109.53242	16.07860	2.30174	1.34093
8	67.93731	15.93509	152.51807	21.01255	2.24498	1.31863
9	92.65721	20.46016	204.07229	26.55999	2.20244	1.29813
10	121.96878	25.57201	263.42567	32.75320	2.15978	1.28082

Table 12.13: Comparison of linear sensitivity of RSS and ORSS from the two-parameter uniform distribution

n	$\sigma^2 \boldsymbol{S}_{11}$	$\sigma^2 \boldsymbol{S}_{22}$	$\sigma^2 \boldsymbol{S}^*_{11}$	$\sigma^2 \boldsymbol{S}^*_{22}$	$ARE(\mu^*_{\text{BLUE}}, \hat{\mu}_{\text{BLUE}})$	$ARE(\sigma^*_{\text{BLUE}}, \hat{\sigma}_{\text{BLUE}})$
2	3.00000	1.00000	3.00000	2.57143	1.50000	1.28571
3	6.11111	3.33333	6.62196	7.60041	1.98659	1.52008
4	10.41667	7.25000	11.94679	15.69508	2.38936	1.74390
5	15.98333	12.95000	19.04924	27.19859	2.72132	1.94276
6	22.86667	20.60000	28.00786	42.31571	3.00084	2.11579
7	31.11429	30.34286	38.81823	61.15990	3.23485	2.26518
8	40.76786	42.30357	51.61002	83.99617	3.44067	2.39989
9	51.86442	56.59325	66.52622	110.87540	3.62870	2.51990
10	64.43730	73.31190	83.46182	141.96166	3.79372	2.62892

Table 12.14: Comparison of linear sensitivity of RSS and ORSS from the right triangular distribution

n	$\sigma^2 \boldsymbol{S}_{11}$	$\sigma^2 \boldsymbol{S}_{22}$	$\sigma^2 \boldsymbol{S}^*_{11}$	$\sigma^2 \boldsymbol{S}^*_{22}$	$ARE(\mu^*_{\text{BLUE}}, \hat{\mu}_{\text{BLUE}})$	$ARE(\sigma^*_{\text{BLUE}}, \hat{\sigma}_{\text{BLUE}})$
2	2.94118	0.94118	2.94118	2.19635	1.00000	2.33363
3	5.82492	2.93670	5.98326	5.79730	1.02718	1.97409
4	9.64718	6.03895	10.03109	10.87361	1.03979	1.80058
5	14.40398	10.28052	15.05528	17.40289	1.04522	1.69280
6	20.09201	15.68473	21.04194	25.36716	1.04728	1.61732
7	26.70861	22.26964	27.96545	34.75701	1.04706	1.56074
8	34.25156	30.04991	35.88645	45.57950	1.04773	1.51679
9	42.71906	39.03785	44.68710	57.76560	1.04607	1.47973
10	52.10957	49.24401	54.48783	71.39798	1.04564	1.44988

parameter exponential, two-parameter uniform, and right triangular distributions. But, for the one-parameter exponential distribution, just as in the case of Fisher information, ORSS possesses slightly less linear sensitivity than the RSS. Because the ORSS has more linear sensitivity than the RSS in all cases except the one-parameter exponential, the BLUE-ORSS turns out to be naturally more efficient than the BLUE-RSS in all these cases.

Appendix

Result 1: The third term in Eq. (12.9) is nonpositive, viz.,

$$\mathrm{E}\left\{\left(\frac{a_1}{b}\right)^2 - \frac{a_4}{b}\right\} \leq 0.$$

PROOF. Let

$$p_{\boldsymbol{p}_i} = \frac{\prod_{k=1}^{n} (F(z_k))^{i_k-1}(1-F(z_k))^{n-i_k}}{\sum_{P}\prod_{k=1}^{n} [(F(z_k))^{i_k-1}(1-F(z_k))^{n-i_k}]},$$

and

$$q_{\boldsymbol{p}_i} = \sum_{k=1}^{n}\left[\left(\frac{i_k-1}{F(z_k)} - \frac{n-i_k}{1-F(z_k)}\right) f(z_k)\right],$$

where $\boldsymbol{p}_i = (i_1, i_2, \ldots, i_n) \in P(1, 2, \cdots, n)$ and $P(1, 2, \ldots, n)$ denotes the group of $n!$ permutations of $(1, 2, \ldots, n)$.

It is evident that $p_{\boldsymbol{p}_i} \geq 0$ and $\sum_{i=1}^{n!} p_{\boldsymbol{p}_i} = 1$. Hence, $\left(\frac{a_1}{b}\right)^2 - \frac{a_4}{b}$ can be written as

$$\left(\frac{a_1}{b}\right)^2 - \frac{a_4}{b} = \left(\sum_{i=1}^{n!} p_{\boldsymbol{p}_i} q_{\boldsymbol{p}_i}\right)^2 - \sum_{i=1}^{n!} p_{\boldsymbol{p}_i} q_{\boldsymbol{p}_i}^2 = -\sum_{i=1}^{n!}\sum_{j>i}^{n!} p_{\boldsymbol{p}_i} p_{\boldsymbol{p}_j} (q_{\boldsymbol{p}_i} - q_{\boldsymbol{p}_j})^2 \leq 0.$$

Therefore, $\mathrm{E}\left\{\left(\frac{a_1}{b}\right)^2 - \frac{a_4}{b}\right\} \leq 0.$ ■

Result 2: The third term in Eq. (12.11) is nonpositive, viz.,

$$\mathrm{E}\left\{\left(\frac{a_2}{b}\right)^2 - \frac{a_8}{b}\right\} \leq 0.$$

PROOF. Following the above notations and setting

$$v_{\boldsymbol{p}_i} = \sum_{k=1}^{n}\left[\left(\frac{i_k-1}{F(z_k)} - \frac{n-i_k}{1-F(z_k)}\right) z_k f(z_k)\right],$$

$\left(\frac{a_2}{b}\right)^2 - \frac{a_8}{b}$ can be actually written as

$$\left(\frac{a_2}{b}\right)^2 - \frac{a_8}{b} = \left(\sum_{i=1}^{n!} p_{\boldsymbol{p}_i} v_{\boldsymbol{p}_i}\right)^2 - \sum_{i=1}^{n!} p_{\boldsymbol{p}_i} v_{\boldsymbol{p}_i}^2 = -\sum_{i=1}^{n!} \sum_{j>i}^{n!} p_{\boldsymbol{p}_i} p_{\boldsymbol{p}_j} (v_{\boldsymbol{p}_i} - v_{\boldsymbol{p}_j})^2 \leq 0.$$

Therefore, $\mathrm{E}\left\{\left(\frac{a_2}{b}\right)^2 - \frac{a_8}{b}\right\} \leq 0.$ ■

References

1. Balakrishnan, N. (Ed.) (1992). *Handbook of the Logistic Distribution*, Marcel Dekker, New York.

2. Barnett, V., and Barreto, M. C. M. (2001). Estimators for a Poisson parameter using ranked set sampling, *Journal of Applied Statistics*, **28**, 929–941.

3. Barreto, M. C. M., and Barnett, V. (1999). Best linear unbiased estimators for the simple linear regression model using ranked set sampling, *Environmental and Ecological Statistics*, **6**, 119–133.

4. Bhoj, D. S., and Ahsanullah, M. (1996). Estimation of parameters of the generalized geometric distribution using ranked set sampling, *Biometrics*, **52**, 685–694.

5. Chandrasekar, B., and Balakrishnan, N. (2002). On a multiparameter version of Tukey's linear sensitivity measure and its properties, *Annals of the Institute of Statistical Mathematics*, **54**, 796–805.

6. Chen, Z. (1999). Density estimation using ranked set sampling data, *Environmental and Ecological Statistics*, **6**, 135–146.

7. Chen, Z. (2000). The efficiency of ranked-set sampling relative to simple random sampling under multiparameter families, *Statistica Sinica*, **10**, 247–263.

8. Chuiv, N. N., and Sinha, B. K. (1998). On some aspects of ranked set sampling in parametric estimation, In *Handbook of Statistics* (Eds., N. Balakrishnan and C. R. Rao) Vol. 17, pp. 337–377, Elsevier, Amsterdam.

9. David, H. A., and Nagaraja, H. N. (2003). *Order Statistics*, Third edition, John Wiley & Sons, New York.

10. Dell, T. R., and Clutter, J. L. (1972). Ranked set sampling theory with order statistics background, *Biometrics*, **28**, 545–555.

11. Hossain, S. S., and Muttlak, H. A. (2000). MVLUE of population parameters based on ranked set sampling, *Applied Mathematics and Computation*, **108**, 167–176.

12. Kim, Y., and Arnold, B. C. (1999). Parameter estimation under generalized ranked set sampling, *Statistics & Probability Letters*, **42**, 353–360.

13. Kvam, P. H., and Tiwari, R. C. (1999). Bayes estimation of a distribution function using ranked set samples, *Environmental and Ecological Statistics*, **6**, 11–22.

14. Lloyd, E. H. (1952). Least squares estimation of location and scale parameters using order statistics, *Biometrika*, **39**, 88–95.

15. McIntyre, G. A. (1952). A method for unbiased selective sampling using ranked sets, *Australian Journal of Agricultural Research*, **3**, 385–390.

16. Nagaraja, H. N. (1994). Tukey's linear sensitivity and order statistics, *Annals of the Institute of Statistical Mathematics*, **46**, 757–768.

17. Perron, F., and Sinha, B. K. (2004). Estimation of variance based on a ranked set sample, *Journal of Statistical Planning and Inference*, **120**, 21–28.

18. Stokes, S. L. (1977). Ranked set sampling with concomitant variables, *Communications in Statistics—Theory and Methods*, **6**, 1207–1211.

19. Stokes, S. L. (1980a). Estimation of variance using judgement ordered ranked set samples, *Biometrics*, **36**, 35–42.

20. Stokes, S. L. (1980b). Inferences on the correlation coefficient in bivariate normal populations from ranked set samples, *Journal of the American Statistical Association*, **75**, 989–995.

21. Stokes, S. L. (1995). Parametric ranked set sampling, *Annals of the Institute of Statistical Mathematics*, **47**, 465–482.

22. Stokes, S. L., and Sager, T. W. (1988). Characterization of a ranked-set sample with application to estimating distribution functions, *Journal of the American Statistical Association*, **83**, 35–42.

23. Takahasi, K., and Wakimoto, K. (1968). On unbiased estimates of the population mean based on the sample stratified by means of ordering, *Annals of the Institute of Statistical Mathematics*, **20**, 1–31.

24. Tietjen, G. L., Kahaner, D. K., and Beckman, R. J. (1977). Variances and covariances of the normal order statistics for sample sizes 2 to 50, In *Selected Tables in Mathematical Statistics*, Vol. 5, American Mathematical Society, Providence, RI.

25. Tukey, J. W. (1965). Which part of the sample contains the information? *Proceedings of the National Academy of Sciences of the USA*, **53**, 127–134.

26. Zheng, G., and Al-Saleh, M. F. (2003). Improving the best linear unbiased estimator for the scale parameter of symmetric distribution by using the absolute value of ranked set samples, *Journal of Applied Statistics*, **30**, 253–265.

13

Information Measures for Pareto Distributions and Order Statistics

Majid Asadi, Nader Ebrahimi, G. G. Hamedani, and Ehsan S. Soofi

University of Isfahan, Isfahan, Iran
Northern Illinois University, DeKalb, IL, USA
Marquette University, Milwaukee, WI, USA
University of Wisconsin-Milwaukee, Milwaukee, WI, USA

Abstract: This paper consists of three sections. The first section gives an overview of the basic information functions, their interpretations, and dynamic information measures that have been recently developed for lifetime distributions. The second section summarizes the information features of univariate Pareto distributions, tabulates transformations of a Pareto random variable under which information measures of numerous distributions can be obtained, and gives a few characterizations of the generalized Pareto distribution. The final section summarizes information measures for order statistics and tabulates the expressions for Shannon entropies of order statistics for numerous distributions.

Keywords and phrases: Characterization, entropy, hazard rate, Kullback-Leibler, reliability, Rényi, residual life, Shannon

13.1 Introduction

Professor Barry Arnold has made significant contributions to the distribution theory and statistics. Two examples of his contributions to the field are the theory and applications of Pareto distributions [Arnold (1983)] and order statistics [Arnold *et al.* (1992)]. The Pareto distributions provide models for many applications in social, natural, and physical sciences, and are related to numerous other families of distributions. Order statistics have applications in a wide range of problems in many fields, provide numerous characterizations of probability distributions, and serve as building blocks for some statistical methodologies including robust statistical estimation and detection of outliers, goodness-of-fit tests, entropy estimation, and analysis of censored samples. In this paper, we summarize information properties of Pareto distributions and order statistics. We only discuss univariate Pareto distributions and refer to Darbellay and Vajda (2000) for the multivariate case.

13.2 Information Measures

Two probability distributions F_1 and F_2 with continuous densities f_j, $j = 1, 2$, on the support $\mathcal{S}$ are under consideration as models for a random prospect X. The fundamental information measure for comparing the two distributions is the Kullback-Leibler discrimination information,

$$\begin{aligned} K(f_1 : f_2) &= \int_{\mathcal{S}} f_1(x) \log \frac{f_1(x)}{f_2(x)} dx \\ &= E_1 \log \frac{f_1(X)}{f_2(X)}, \end{aligned} \tag{13.1}$$

where f_1 is absolutely continuous with respect to f_2 and E_1 denotes the expectation with respect to f_1. $K(f_1 : f_2) \geq 0$, where equality holds if and only if $f_1(x) = f_2(x)$ almost everywhere. But $K(f_1 : f_2)$ is not symmetric, so it is not a distance function. It is a measure of directed divergence between f_1 and f_2, where f_2 is the *reference distribution*. It is also referred to as cross-entropy and relative entropy.

The term *information* reflects two aspects of (13.1). First, $K(f_1 : f_2)$ generalizes two measures of information, entropy and mutual information, developed by Shannon (1948) for communication theory. Second, the statistical interpretation of information stems from the foundation of $K(f_1 : f_2)$ in probabilistic inference via Bayes theorem [Kullback and Leibler (1951) and Kullback (1959)]. The log-ratio in (13.1) is the difference between the logarithms of the posterior and prior odds in favor of F_1, referred to as the *weight of evidence* for F_1 provided by an observation x [Good (1950)]. Thus, $K(f_1 : f_2)$ is the expected information in favor of F_1 provided by X for discriminating between the two models.

The discrimination information $K(f_1 : f_2)$ quantifies *loss* or *gain* of information per natures of F_1 and F_2. When an F_j is an ideal distribution (e.g., the true data-generating distribution), $K(f_1 : f_2)$ measures loss of information in using the other distribution instead of the ideal one. In this case, $K(f_1 : f_2)$ is also referred to as *entropy loss*; see Soofi (1997) for references. When F_1 and F_2 reflect two states of knowledge (e.g., prior and posterior distributions), then $K(f_1 : f_2)$ measures the gain (loss) of information in using the distribution that is reflective of more (less) knowledge instead of the alternative. In this case, $K(f_1 : f_2)$ is also referred to as a *utility function* [Bernardo (1979)].

The properties of (13.1) is studied extensively by Kullback (1959). Two properties of interest in this study are invariance and decomposition.

(a) If $Y = \phi(X)$ is a one-to-one transformation, then $K(f_{1_Y} : f_{2_Y}) = K(f_{1_X} : f_{2_X})$, where f_{j_Y} denotes the distributions induced by ϕ on f_{j_X}, $j = 1, 2$.

(b) Let $\mathcal{E} = \{\mathcal{E}_1, \ldots, \mathcal{E}_n\}$ be a partition of the support $\mathcal{S}$. Then,

$$K(f_1 : f_2) = K(P_1 : P_2; \mathcal{E}) + \sum_{i=1}^{n} P_1(\mathcal{E}_i) K(f_1 : f_2; \mathcal{E}_i), \tag{13.2}$$

where $K(P_1 : P_2; \mathcal{E}) = \sum_{i=1}^{n} P_1(\mathcal{E}_i) \log \frac{P_1(\mathcal{E}_i)}{P_2(\mathcal{E}_i)}$ and $P_j(\mathcal{E}_i) = \int_{\mathcal{E}_i} f_j(x)dx$, $j = 1, 2$. This property is obtained by combining two expressions in Kullback (1959).

13.2.1 Shannon entropy

Shannon entropy [Shannon (1948)] is defined by

$$H(X) \equiv H(f) = -\int_{\mathcal{S}} f(x) \log f(x) dx. \tag{13.3}$$

Shannon entropy measures lack of uniformity (concentration of probabilities) under f. With a less concentrated distribution, it is more difficult to predict an outcome. The negative entropy $-H(f) = E_f[\log f(X)]$ is the average log-height of the density. It is the discrimination function between F and the uniform distribution and is a measure of informativeness of F about the prediction of its outcomes [Zellner (1971)].

The entropy is not invariant under nonsingular transformations of X. If $Y = \phi(X)$ is a one-to-one transformation, then

$$H(Y) = H(X) - E\left[\log\left|\frac{d}{dY}\phi^{-1}(Y)\right|\right]. \tag{13.4}$$

Decomposition of entropy over the partition $\mathcal{E}$ of the support $\mathcal{S}$ is given by

$$H(f) = H(P_f; \mathcal{E}) + \sum_{i=1}^{n} P_f(\mathcal{E}_i) H(f; \mathcal{E}_i), \tag{13.5}$$

where $H(P_f; \mathcal{E})$ is the entropy of the multinomial distribution implied by F on the partition.

13.2.2 Rényi information measures

The information divergence of order α [Rényi (1961)] between two distributions is defined by

$$K_\alpha(f_1 : f_2) = \frac{1}{\alpha - 1} \log \int_{\mathcal{S}} f_1^\alpha(x) f_2^{1-\alpha}(x) dx, \tag{13.6}$$

where $\alpha \neq 1$.

The following representations provide some insights about the role of α in (13.6):

$$\begin{aligned} K_\alpha(f_1 : f_2) &= \frac{1}{\alpha - 1} \log E_2 \left[\frac{f_1(X)}{f_2(X)}\right]^\alpha, \qquad \alpha > 1 \\ &= \frac{1}{\alpha - 1} \log E_2 \left[\frac{f_1(X)}{f_2(X)}\right]^\alpha, \qquad \alpha < 1 \\ &= \frac{1}{\alpha - 1} \log E_1 \left[\frac{f_2(X)}{f_1(X)}\right]^{1-\alpha}, \quad \alpha < 1 \end{aligned}$$

That is, for $\alpha > 1$, $K_\alpha(f_1 : f_2)$ is log of the expected odd in favor of F_1, given F_2, where the magnitude of α is the weight given to the odd ratio. However, for $\alpha < 1$, pending on the weight, α or $1 - \alpha$, $K_\alpha(f_1 : f_2)$ can be interpreted as the log of the expected odd in favor of F_j, given F_k, $k \neq j = 1, 2$. A useful case is when $\alpha = \frac{1}{2}$, where $K_{1/2}(f_1 : f_2)$ is symmetric in f_1 and f_2.

It is well known that $\lim_{\alpha \to 1} K_\alpha(f_1 : f_2) = K(f_1 : f_2) \equiv K_1(f_1 : f_2)$. Like (13.1), $K_\alpha(f_1 : f_2)$ is non-negative and invariant under one-to-one transformations of X.

The entropy of order α of a distribution (Rényi 1961) is defined as

$$H_\alpha(f) = \frac{1}{1 - \alpha} \log \int_{\mathcal{S}} f^\alpha(x) dx, \tag{13.7}$$

where $\alpha > 0$, $\alpha \neq 1$.

It is well known that $\lim_{\alpha \to 1} H_\alpha(f) = H(f)$. Like (13.3), $H_\alpha(f)$ is not invariant under one-to-one transformations of X. However, there is no useful formula like (13.4) for $H_\alpha(f)$.

Rényi entropy expressions for univariate distributions are given in Song (2001) and Nadarajah and Zografos (2003).

13.2.3 Dynamic information

Frequently, in reliability one has information about the current age of the system under consideration. In such cases, the age must be taken into account when measuring information. Ebrahimi and Kirmani (1996a,b) considered the discrimination information between two residual distributions that take age t into account. In this case, the set of interest is the *residual lifetime*

$$\mathcal{E}_t = \mathcal{S}_t = \{x : x > t\}.$$

The discrimination information function between two residual life distributions $F_j(x;t) = P_j(X - t \leq x | X > t)$ implied by two lifetime distributions $F_j(x)$, $j = 1, 2$, is given by

$$K(f_1 : f_2; t) \equiv K[f_1(x;t) : f_2(x;t)] = \int_t^\infty f_1(x;t) \log \frac{f_1(x;t)}{f_2(x;t)} dx,$$

where $f_j(x;t) = \frac{f_j(x)}{\bar{F}_j(t)}$, $j = 1, 2$, denote the conditional densities and $\bar{F}_j(t) = P_j(\mathcal{S}_t) = 1 - F_j(t)$, $j = 1, 2$. It is clear that for $t_0 = \inf\{x : \bar{F}(x) = 1\}$, $K(f_1 : f_2; t_0) = K(f_1 : f_2)$. For each t, $t \geq 0$, $K(f_1 : f_2; t)$ possesses all the properties of the discrimination information function (13.1). If we consider t as an index ranging over $\mathcal{S}_t$, then $K(f_1 : f_2; t)$ provides a dynamic discrimination information function indexed by t for measuring the discrepancy between the residual life distributions $F_j(x;t)$, $j = 1, 2$.

The entropy of residual life distribution is defined similarly as

$$H(X;t) \equiv H(f;t) = -\int_t^\infty \frac{f(x)}{\bar{F}(t)} \log \frac{f(x)}{\bar{F}(t)} dx;$$

[Ebrahimi (1996)]. It is clear that for $t_0 = \inf\{x : \bar{F}(x) = 1\}$, $H(f; t_0) = H(f)$.

Another set of interest that leads to dynamic information measures is the past lifetime of the individual

$$\mathcal{S}_{[t]} = \{x : x \leq t\}.$$

The discrimination information function between two past lifetime distributions implied by two lifetime distributions F_1 and F_2 is given by

$$K(f_1 : f_2; [t]) \equiv \int_0^t \frac{f_1(x)}{F_1(t)} \log \frac{f_1(x)/F_1(t)}{f_2(x)/F_2(t)} dx,$$

where $\frac{f_j(x)}{F_j(t)}$, $j = 1, 2$ are the conditional densities. It is clear that for $t^* = \inf\{x : F(x) = 1\}$, $K(f_1 : f_2; [t^*]) = K(f_1, f_2)$.

By (13.2), for partition $\mathcal{E} = \{\mathcal{E}_t, \mathcal{E}_{[t]}\}$, we have the following dynamic information decomposition:

$$K(f_1 : f_2) = K(P_1 : P_2; t) + \bar{F}_1(t) K(f_1 : f_2; t) + F_1(t) K(f_1 : f_2; [t]),$$

where

$$K(P_1 : P_2; t) = F_1(t) \log \frac{F_1(t)}{F_2(t)} + \bar{F}_1(t) \log \frac{\bar{F}_1(t)}{\bar{F}_2(t)};$$

[Di Crescenzo and Longobardi (2004)]

The entropy of the past lifetime distribution is defined similarly as

$$H(X;[t]) = -\int_t^\infty \frac{f(x)}{F(t)} \log \frac{f(x)}{F(t)} dx.$$

The entropy decomposition (13.5) gives

$$H(f) = H(P_f;t) + \bar{F}(t)H(f;t) + F(t)H(f;[t]),$$

where

$$H(P_f;t) = -F(t)\log F(t) - \bar{F}(t)\log \bar{F}(t)$$

is the entropy of the Bernoulli distribution implied by F on the partition [Di Crescenzo and Longobardi (2002)].

The Rényi measures for the residual lifetime distributions $K_\alpha(f_1 : f_2; t)$ and $H_\alpha(f;t)$, and for the past lifetime distributions $K_\alpha(f_1 : f_2; [t])$ and $H_\alpha(f;[t])$ are defined similarly.

13.2.4 Maximum entropy and maximum dynamic entropy

Laplace's principle of insufficient reason assigns probability uniformly in the absence of any constraint on the probabilities. The maximum entropy (ME) principle extends this idea to producing probability models closest to uniform, which are most noncommittal to information other than that explicitly taken into account via some moment constraints [Jaynes (1957, 1982)].

The ME method subject to moment conditions seeks a distribution function F^* with the density that maximizes $H(f)$ in a class of all distributions with given moments

$$\Omega_{\boldsymbol{\theta}} = \{f : E_f[T_j(X)] = \theta_j,\ j = 0, 1, \ldots, J\},$$

where $T_j(X)$ are integrable functions with respect to the density, $T_0(X) = \theta_0 = 1$, and $\boldsymbol{\theta} = (\theta_1, \ldots, \theta_J)$ is a vector of moments.

Recently, Asadi *et al.* (2004) proposed a maximum dynamic entropy (MDE) procedure that develops lifetime models when the information is given in terms of differential inequality constraints describing the growth of the hazard rate $\lambda_F(t)$. The MDE model in a set of distributions $\Omega_F = \{f\}$ is the distribution with density f^* such that

$$H(f;t) \le H(f^*;t) \quad \forall t \ge 0.$$

That is, $f^*(x;t)$ retains its ME property among all the residual lifetime distributions induced by all members of Ω_F.

Like the Shannon entropy, $H_\alpha(f)$ is concave for all $\alpha > 0$. However, the Rényi entropy does not share the nice ME property of the Shannon entropy.

Consequently, ME of order α subject to moment constraints has not been developed. Golan and Perloff (2002) have used Rényi entropy in the context of an ME estimation where the support of distribution is a finite number of points. Asadi *et al.* (2005) have shown that developing MDEα models subject to differential inequality constraints is feasible.

13.3 Information Properties of Pareto Distributions

Consider Pareto distribution with survival function

$$\bar{F}_\beta(x) = (x+1)^{-\beta}, \; x > 0, \quad \beta > 0.$$

We denote this distribution by $\mathcal{P}_\beta$.

The Kullback-Leibler information function between two Pareto distributions $\mathcal{P}_{\beta_j}$, $j = 1, 2$, is given by

$$K(\mathcal{P}_{\beta_1} : \mathcal{P}_{\beta_2}) = \rho - \log \rho - 1, \tag{13.8}$$

where $\rho = \frac{\beta_2}{\beta_1}$.

Shannon entropy of $\mathcal{P}_\beta$ is

$$H(\mathcal{P}_\beta) = 1 + \frac{1}{\beta} - \log \beta.$$

Because $H(\mathcal{P}_\beta)$ is a decreasing function of β, the distributions are ordered by Shannon entropy within the $\mathcal{P}_\beta$ family. Also, let X be distributed as $\mathcal{P}_\beta$. Then, X has a decreasing failure rate and by a result of Ebrahimi *et al.* (2004) any non negative random variable Z stochastically dominated by X has a smaller entropy.

Rényi information divergence between two Pareto distributions $\mathcal{P}(\beta_j)$, $j = 1, 2$ with densities f_j, $j = 1, 2$ is given by

$$K_\alpha(f_1 : f_2) = \frac{1}{1-\alpha} \log \left(\alpha \rho^{\alpha-1} + (1-\alpha)\rho^\alpha \right), \qquad \alpha + (1-\alpha)\rho > 0. \tag{13.9}$$

Rényi entropy of $\mathcal{P}(\beta)$ is

$$H_\alpha(\mathcal{P}_\beta) = \frac{1}{1-\alpha} \log \frac{\beta^\alpha}{\alpha(\beta+1) - 1}, \qquad \alpha > \frac{1}{\beta+1}.$$

Numerous distributions can be obtained as distributions of one-to-one transformations of a random variable X distributed as $\mathcal{P}_\beta$. Therefore, their information functions can be derived and studied via the information functions of $\mathcal{P}_\beta$.

Table 13.1 lists several families of distributions and the transformations under which they can be obtained from $\mathcal{P}_\beta$. The Kullback-Leibler and Rényi information functions between two members of a family, obtained from $\mathcal{P}_{\beta_j}$, $j=1,2$, are the same as (13.8) and (13.9), where β_j, $j=1,2$, are determined by the parameters of the transformed models. Shannon entropy of these distributions are related to $H(\mathcal{P}_\beta)$ by (13.4), where the expectation is taken with respect to $\mathcal{P}_\beta$. These distributions do not include location parameter because it does not affect Shannon entropy.

Two Pareto distributions $\mathcal{P}_{\beta_j}$, $j=1,2$, are a proportional hazard. Thus, $K(\mathcal{P}_{\beta_1}:\mathcal{P}_{\beta_2};t)=K(\mathcal{P}_{\beta_1}:\mathcal{P}_{\beta_2})$ and $K_\alpha(\mathcal{P}_{\beta_1}:\mathcal{P}_{\beta_2};t)=K_\alpha(\mathcal{P}_{\beta_1}:\mathcal{P}_{\beta_2})$, [Ebrahimi and Kirmani (1996a) and Asadi *et al.* (2005)]. Other dynamic measures for $\mathcal{P}(\beta)$ are as follows.

$$H(\mathcal{P}_\beta;t)=H(\mathcal{P}_\beta)+\log(1+t),$$

$$H(\mathcal{P}_\beta;[t])=H(\mathcal{P}_\beta)+\log F_\beta(t)+\frac{F_\beta(t)}{\bar{F}_\beta(t)}\log\bar{F}_{\beta+1}(t),$$

$$K(\mathcal{P}_{\beta_1}:\mathcal{P}_{\beta_2};[t])=K(\mathcal{P}_{\beta_1}:\mathcal{P}_{\beta_2})-\log\frac{F_{\beta_1}(t)}{F_{\beta_2}(t)}-\frac{\bar{F}_{\beta_1}(t)}{F_{\beta_1}(t)}\log\frac{\bar{F}_{\beta_1}(t)}{\bar{F}_{\beta_2}(t)},$$

$$H_\alpha(\mathcal{P}_\beta;t)=H_\alpha(\mathcal{P}_\beta)+\frac{1}{1-\alpha}\log\bar{F}_{\beta_\alpha}(t),\qquad \alpha>\frac{1}{\beta+1},$$

$$H_\alpha(\mathcal{P}_\beta;[t])=H_\alpha(\mathcal{P}_\beta)+\frac{1}{1-\alpha}\log\frac{F_{\beta_\alpha}(t)}{[F_\beta(t)]^\alpha},\qquad \alpha>\frac{1}{\beta+1},$$

$$K_\alpha(\mathcal{P}_{\beta_1}:\mathcal{P}_{\beta_2};[t])=K_\alpha(\mathcal{P}_{\beta_1}:\mathcal{P}_{\beta_2})+\frac{1}{1-\alpha}\log\frac{F_{\beta_{1,2}}(t)}{[F_{\beta_1}(t)]^\alpha\,[F_{\beta_2}(t)]^{1-\alpha}},$$

where $\beta_\alpha=\alpha(\beta+1)-1$ and $\beta_{1,2}=\alpha\beta_1+(1-\alpha)\beta_2$.

We note that $H(\mathcal{P}_\beta;t)$ is an increasing function of t and is a decreasing function of the shape parameter β. Also because the density of $\mathcal{P}_\beta$ is strictly decreasing over $\mathcal{S}_t$, by a result of Asadi *et al.* (2005), $H_\alpha(\mathcal{P}_\beta;t)$, $\alpha>1$, uniquely determines F_β.

13.3.1 Characterizations of generalized Pareto

The generalized Pareto (GP) family shown in Table 13.1 includes the Pareto type II when $\delta>1$, the power distribution when $-1<\delta<0$, and the exponential distribution when $\lambda\to 0$. We present a few information characterizations of GP.

(a) Let X be a non-negative random variable with a hazard function $r_X(t)=\frac{f_X(t)}{\bar{F}_X(t)}$ and mean residual life function $m_X(t)=E(X-t|X>t)$.

Table 13.1: Distributions related to Pareto distribution $\mathcal{P}_\beta$ by transformation

Distribution, Support	Survival Function	Transformation
Burr type III, $\Re^+$	$\bar{F}_Y(y) = 1 - \left(1 + y^{-\nu}\right)^{-\delta}$	$y = \left[(1+x)^{\beta/\delta} - 1\right]^{-1/\nu}$
Burr type IX, $\Re$	$\bar{F}_Y(y) = 2\left[\lambda\,(1+e^y)^{\underline{\delta}}\,\lambda + 2\right]^{-1}$	$y = \log\left(\left\{\frac{2}{\lambda}\left[(1+x)^{\underline{\beta}}\,1\right] + 1\right\}^{1/\delta} - 1\right)$
Burr type X, $\Re^+$	$\bar{F}_Y(y) = 1 - \left(1 - e^{-y^2}\right)^{\delta}$	$y = \left[-\log\left(1 - (1+x)^{-\beta/\delta}\right)\right]^{1/2}$
Compound extreme value, $\Re$	$\bar{F}_Y(y) = 1 - \left(1 + \frac{\lambda}{\delta}e^{-y/\lambda}\right)^{-1}$	$y = \lambda \log\left(\frac{\delta}{\lambda}\left[(1+x)^{\beta} - 1\right]\right)$
Exponential, $\Re^+$	$\bar{F}_Y(y) = e^{-\lambda y}$	$y = \frac{\beta}{\lambda}\log(1+x)$
Extreme value type I, $\Re^+$	$\bar{F}_Y(y) = \exp\left[\frac{\lambda^2}{\delta}\left(1 - e^{\lambda y/\delta}\right)\right]$	$y = \frac{\lambda}{\delta}\log\left(1 + \frac{\beta\delta}{\lambda^2}\log(1+x)\right)$
Extreme value type II, $\Re^+$	$\bar{F}_Y(y) = 1 - \exp\left[-(\lambda y)^{-\delta}\right]$	$y = -\frac{1}{\lambda}\log\left[\beta\log(1+x)\right]^{-1/\delta}$
F, $\Re^+$	$\bar{F}_Y(y) = \delta^{\delta/2}(\delta + 2y)^{-\delta/2}$	$y = \frac{\delta}{2}\left[(1+x)^{2\beta/\delta} - 1\right]$
Generalized logistic, $\Re$ (Dubey)	$\bar{F}_Y(y) = \delta^{\delta}\left(\delta + \lambda e^{y/\lambda}\right)^{-\delta}$	$y = \lambda\log\left(\frac{\delta}{\lambda}\left[(1+x)^{\beta/\delta} - 1\right]\right)$
Generalized Pareto, $\Re^+$	$\bar{F}_Y(y) = \left(1 + \frac{\delta}{\lambda}y\right)^{-1/\delta - 1}$	$y = \frac{\lambda}{\delta}\left[(1+x)^{\beta\delta/(\delta+1)} - 1\right]$
Half-Cauchy, $\Re^+$	$\bar{F}_Y(y) = 1 - \frac{2}{\pi}\arctan y$	$y = \tan\left[\frac{\pi}{2}(1+x)^{-\beta}\right]$
Half-logistic, $\Re^+$	$\bar{F}_Y(y) = (\delta+1)\left(\delta + e^{\lambda y}\right)^{-1}$	$y = \frac{1}{\lambda}\log\left((\delta+1)(1+x)^{\beta} - \delta\right)$
Linear failure rate, $\Re^+$	$\bar{F}_Y(y) = e^{-(\delta y + \lambda y^2/2)}$	$y = \frac{\delta}{\lambda}\left[\left(1 + \frac{2\beta\lambda}{\delta^2}\log(1+x)\right)^{1/2} - 1\right]$
Logistic, $\Re$ (Burr type II, $\lambda = 1$)	$\bar{F}_Y(y) = 1 - \left(1 + e^{-\lambda y}\right)^{-\delta}$	$y = -\frac{1}{\lambda}\log\left((1+x)^{\beta/\delta} - 1\right)$
Pareto type I, $[\kappa, \infty)$	$\bar{F}_Y(y) = \kappa^{\delta}y^{-\delta}$	$y = \kappa(1+x)^{\beta/\delta}$
Pareto type IV, $\Re^+$ (type II, $\nu = 1$, type III, $\delta = 1$) (Burr type XII)	$\bar{F}_Y(y) = \kappa^{\delta}\left(\kappa + y^{\nu}\right)^{-\delta}$	$y = \kappa^{1/\nu}\left[(1+x)^{\beta/\delta} - 1\right]^{1/\nu}$
Truncated exponential, $[0, \kappa]$	$\bar{F}_Y(y) = \frac{e^{-\lambda y} - e^{-\lambda\kappa}}{1 - e^{-\lambda\kappa}}$	$y = -\frac{1}{\lambda}\log\left(1 - \frac{1 - e^{-\lambda\kappa}}{(1+x)^{\beta}}\right)$
Weibull, $\Re^+$	$\bar{F}_Y(y) = \exp\left(-\lambda y^{\delta}\right)$	$y = \left[\frac{\beta}{\lambda}\log(1+x)\right]^{1/\delta}$

Then

$$H_\alpha(X;t) = a - \log r_X(t), \quad \forall \alpha > 0,$$

where a is a constant, if and only if F_X is GP. Also,

$$H_\alpha(X;t) = b + \log m_X(t), \quad \forall \alpha > 0,$$

where b is a constant, if and only if F_X is GP.

(b) Let X_1 and X_2 be two continuous non-negative random variables with density functions f_j, $j = 1, 2$, and proportional hazard functions $r_1(t) = cr_2(t)$. If

$$H(f_1;t) = d + H(f_2;t), \tag{13.10}$$

where c is a constant, then F_j, $j = 1, 2$, are members of the GP family. The converse holds for $|\lambda| = \delta$ as well as for the case of $\lambda = 0$, that is, X_j, $j = 1, 2$, are both exponential random variables. In (13.10), $d = -\log c$, $d < -\log c$ and $d > -\log c$ imply the exponential, Pareto, and the power distributions, respectively.

(c) Let X_1 and X_2 be defined as above. Then, F_1 and F_2 are members of the GP family if

$$H_\alpha(f_1;t) = d_\alpha + H_\alpha(f_2;t).$$

(d) Let $X_1, \ldots, X_n$ be a sample from distribution F_X and $Y = \min(X_1, \ldots, X_n)$. Then

$$H(Y;t) = k + H(X;t),$$

where k is a constant, if and only if F_X is GP.

The proofs for (a) are given in Asadi and Ebrahimi (2000) for the case of Shannon entropy ($\alpha = 1$), and in Asadi *et al.* (2005) for the general case. The proofs of (c) and (d) are simple and follow from (b). The proof for (b) is as follows. Note that $r_1(x) = cr_2(x)$ is equivalent to $\bar{F}_1(x) = \bar{F}_2^c(x)$. Using this and (13.10), we obtain

$$1 - \frac{c\log c}{\bar{F}_2^c(x)} \int_x^\infty f_2(u)\bar{F}_2^{c-1}(u) r_2(u)du = d + 1 - \frac{1}{\bar{F}_2(x)} \int_x^\infty f_2(u)\log r_2(u)du.$$

Equivalently,

$$d\bar{F}_2^c(x) - \bar{F}_2^{c-1}(x) \int_x^\infty f_2(u)\log r_2(u)du = -c\log c \int_x^\infty \bar{F}_2^{c-1}(u) f_2(u) r_2(u)du.$$

Differentiating both sides with respect to x, after some simplification, we obtain

$$H(f_2,t) = 1 - \frac{cd + c\log c}{c-1} - \log r_2(t),$$

which implies that F_2 (and hence F_1) is a member of GP.

13.3.2 ME, MED, and MDEα characterizations of Pareto

Consider the class of distributions with moment constraints

$$\Omega_\theta = \{f : E_f[\log(a + bX^c)] = \theta_{abc}\}.$$

It can be shown that the Pareto distributions with the survival functions shown in Table 13.1 are the ME model in Ω_θ for various values of a, b, and c. The GP is the ME when $a = c = 1$, $b = \delta/\lambda$, and $\mathcal{S} = \Re^+$. Pareto type I is ME when $a = 0, b = c = 1$ and $\mathcal{S} = [\kappa, \infty)$. Pareto II is the ME when $a = \kappa$, $b = c = 1$, and $\mathcal{S} = \Re^+$. Pareto III is the ME when $a = \kappa$, $b = 1$, $c = \nu$, and $\mathcal{S} = \Re^+$.

The MDE characterizations of Pareto II, GP, the minimum of an exponential and a Pareto, and mixture of two Paretos in the classes of distributions with differential inequalities describing the growth rate of their hazard functions are given in Asadi *et al.* (2004). These characterizations are obtained based on the monotonicity of the densities of these distributions. We should note that the ME characterization for the minimum of exponential and Pareto can be formulated but for the mixture of two Paretos, no ME characterization is available. The MDEα characterizations of these distributions are given in Asadi *et al.* (2004) based on results for decreasing failure rate (DFR) distributions.

13.4 Information Properties of Order Statistics

The information properties of order statistics have been studied by a few authors. Wong and Chen (1990) showed that the difference between the average entropy of order statistics and the entropy of data distribution is a constant. They also showed that for symmetric distributions, the entropy of order statistics is symmetric about the median. Park (1995) showed some recurrence relations for the entropy of order statistics and Park (1996) provided similar results in terms of the Fisher information. Ebrahimi *et al.* (2004) provided several results on the entropy of order statistics and showed that the Kullback-Leibler functions involving order statistics are distribution-free. This section summarizes some of these results and presents entropies of order statistics for numerous distributions.

Let $X_1, \ldots, X_n$ be independent and identically distributed observations from a distribution F_X, where F_X is differentiable with a density f_X which is positive in an interval and zero elsewhere. Denote their order statistics by $Y_1 < \cdots < Y_n$. It is well known that the distribution $F_i(y) = P(Y_i \leq y)$, $i = 1, \ldots, n$, has density

$$f_i(y) = \frac{\Gamma(n+1)}{\Gamma(n-i+1)\Gamma(i)}[F_X(y)]^{i-1}[1 - F_X(y)]^{n-i} f_X(y),$$

where for a positive integer z, $\Gamma(z) = (z-1)!$ is the gamma function.

The probability integral transformation of the random variable, $U = F_X(X)$, is pivotal in developing information results for order statistics [Ebrahimi *et al.* (2004)]. The distribution of U is uniform over the unit interval. The order statistics of a sample from uniform distribution $U_1, \ldots, U_n$ are denoted by $W_1 < \cdots < W_n$ and W_i, $i = 1, \ldots, n$ has beta distribution with density

$$g_i(w) = \frac{1}{B(i, n-i+1)} w^{i-1}(1-w)^{n-i}, \quad 0 \leq w \leq 1, \tag{13.11}$$

where $B(z_1, z_2) = \Gamma(z_1)\Gamma(z_2)/\Gamma(z_1+z_2)$.

The entropy of the beta distribution is

$$\begin{aligned} H_n(W_i) &= \log B(i, n-i+1) - (i-1)[\psi(i) - \psi(n+1)] \\ &\quad -(n-i)[\psi(n-i+1) - \psi(n+1)], \end{aligned}$$

where $\psi(z) = \frac{d \log \Gamma(z)}{dz}$ is the digamma function.

Noting that $W_i = F_X(Y_i)$ and $Y_i = F_X^{-1}(W_i)$ $i = 1, \ldots, n$, are one-to-one transformations, Ebrahimi *et al.* (2004) found the following information functions for order statistics.

(a) The Kullback-Leibler discrimination information measures between the distributions of X and its order statistics Y_i are given by:

$$\begin{aligned} K_n(f_i : f_X) &= -H_n(W_i), \\ K_n(f_X : f_i) &= \log B(i, n-i+1) + n - 1. \end{aligned}$$

According to both measures, the information discrepancy between the distribution of order statistics and f_X decreases up to the median and then increases. Thus, amongst the order statistics, the median has the closest distribution to the data distribution.

(b) The discrimination information between distributions of ith and jth order statistics is given by

$$K_n(f_i : f_j) = \log \frac{\Gamma(j)\Gamma(n-j+1)}{\Gamma(i)\Gamma(n-i+1)} - (i-j)[\psi(i) - \psi(n-i)] - \frac{i-j}{n-i}.$$

Special cases of interest are

$$\begin{aligned} K_n(f_{i+1} : f_i) &= \frac{1}{i} + H_n(W_i) - H_n(W_{i+1}) \\ K_n(f_i : f_{i+1}) &= \frac{1}{n-i} - H_n(W_i) + H_n(W_{i+1}). \end{aligned}$$

Both measures are decreasing for $i \leq (n+1)/2$ and increasing for $i \geq (n+1)/2$. Therefore, the distributions of the consecutive order statistics become closer to each other as they approach the median from either extremes.

(c) The degree of dependency among $Y_1, \ldots, Y_n$ is measured by the mutual information between consecutive order statistics, defined by

$$\begin{aligned} M_n(Y_i, Y_{i+1}) &\equiv K_n(f_{i,i+1} : f_i f_{i+1}) \\ &= M_n(W_i, W_{i+1}) - \log \binom{n}{i} + n\psi(n) - i\psi(i) \\ &\quad -(n-i)\psi(n-i) - 1, \end{aligned}$$

where $f_{i,i+1}$ is the joint density of (Y_i, Y_{i+1}),

$$\begin{aligned} f_{i,i+1}(y_i, y_{i+1}) &= \frac{\Gamma(n+1)}{\Gamma(n-i)\Gamma(i)}[F_X(y_i)]^{i-1} \\ &\quad \times[1-F_X(y_{i+1})]^{n-i-1} f_X(y_i) f_X(y_{i+1}), \quad \text{for } y_i < y_{i+1}, \\ &= 0, \qquad \text{otherwise.} \end{aligned}$$

For a given n, $M_n(Y_i, Y_{i+1})$ is symmetric in i and $n-i$, increases in i for $i < n/2$, and decreases in i for $i > n/2$. $M_n(Y_i, Y_{i+1})$ is also increasing in n.

(d) By (13.4), the entropies of order statistics can be computed

$$H(Y_i) = H_n(W_i) - E_{g_i}\left[\log f_X\left(F_X^{-1}(W_i)\right)\right], \tag{13.12}$$

where $H_n(W_i)$ is the entropy and E_{g_i} is an expectation of the beta distribution. Thus, entropies of order statistics can be derived in terms of the entropy of beta distributions and various beta expectations $E_{g_i}[\cdot]$.

Ebrahimi *et al.* (2004) showed an application of (13.12) for the exponential distribution. Table 13.2 lists several distributions and the entropies of their order statistics obtained via (13.12). The distributions do not include scale and location parameters. Adjustment can simply be made using (13.4) which for $Y^* = \lambda Y + \mu$ gives $H(Y^*) = H(Y) - \log \lambda$. In Table 13.2,

$$a_i = E_{g_i}[\log(W_i)] = \psi(i) - \psi(n+1) \tag{13.13}$$

$$b_i = E_{g_i}[\log(1-W_i)] = \psi(n-i+1) - \psi(n+1) \tag{13.14}$$

$$c_i = E_{g_i}[\log(1+W_i)] = \frac{n!}{(i-1)!}\sum_{k=1}^{\infty}(-1)^{k+1}\frac{(i+k-1)!}{k[(n+k)!]}. \tag{13.15}$$

Other beta expectations $E_{g_i}[\cdot]$ in Table 13.2 can be computed numerically.

The properties of Rényi and dynamic information measures of order statistics are currently under investigation by the authors. Here, we close by reporting that because Rényi information divergence is invariant under one-to-one transformations, the Rényi entropy of beta distribution plays the same role as that

Table 13.2: Entropy of order statistics for several distributions

Distribution, Support	Survival Function	Entropy of Order Statistics*
Beta, $[0,1]$	$\bar{F}_X(x) = x^\delta$	$H_n(W_i) - \log\delta - \left(1-\delta^{-1}\right)a_i$
Beta, $[0,1]$	$\bar{F}_X(x) = (1-x)^\delta$	$H_n(W_i) - \log\delta - \left(1-\delta^{-1}\right)b_i$
Bradford distribution, $[0,1]$	$\bar{F}_X(x) = 1 - \dfrac{\log(1+x)}{\log 2}$	$H_n(W_i) - \log\log 2 + (\log 2)a_i$
Burr type III, $\Re^+$	$\bar{F}_X(x) = 1 - \left(1+x^{-1}\right)^{-1}$	$H_n(W_i) - 2b_i$
Compound extreme value, $\Re$ (Logistic, $\delta = 1$)	$\bar{F}_X(x) = 1 - \left(1+\delta^{-1}e^{-x}\right)^{-1}$	$H_n(W_i) - a_i - b_i$
Exponential, $\Re^+$	$\bar{F}_X(x) = e^{-x}$	$H_n(W_i) - b_i$
Extreme value type I, $\Re^+$	$\bar{F}_X(x) = \exp\left[\frac{1}{\delta}\left(1-e^{\delta x}\right)\right]$	$H_n(W_i) - b_i - E_{g_i}\left[\log\left(1-\log(1-W_i)^\delta\right)\right]$
F, $\Re^+$	$\bar{F}_X(x) = \delta^{\delta/2}(\delta+2x)^{-\delta/2}$	$H_n(W_i) - \left(1+2\delta^{-1}\right)b_i$
Generalized logistic, $\Re$ (Dubey)	$\bar{F}_X(x) = \delta^\delta\,(\delta+e^x)^{-\delta}$	$H_n(W_i) - \log\delta - \left(1+\delta^{-1}\right)b_i$ $-E_{g_i}\left[\log\left((1-W_i)^{-1/\delta}-1\right)\right]$
Generalized Pareto, $\Re^+$	$\bar{F}_X(x) = (1+\delta x)^{-1/\delta-1}$	$H_n(W_i) - \log(1+\delta) - \dfrac{1+2\delta}{1+\delta}b_i$
Half-Cauchy, $\Re^+$	$\bar{F}_X(x) = 1 - \dfrac{2}{\pi}\arctan x$	$H_n(W_i) + \log\dfrac{\pi}{2}$ $+E_{g_i}\left[\log\left(1+\tan^2\dfrac{\pi W_i}{2}\right)\right]$
Half-logistic, $\Re^+$ (Half-Burr II, $\nu = 1$)	$\bar{F}_X(x) = (\nu+1)\,(\nu+e^x)^{-1}$	$H_n(W_i) + \log 2 - b_i - c_i$
Linear failure rate, $\Re^+$	$\bar{F}_X(x) = e^{-(\delta x+\lambda x^2/2)}$	$H_n(W_i) - a_i - E_{g_i}\left[\log\left(\delta^2 - 2\lambda\log W_i\right)\right]$
Pareto type II, $\Re^+$ (type I, $z = x+1$)	$\bar{F}_X(x) = (1+x)^{-\delta}$	$H_n(W_i) - \log\delta - \left(1+\delta^{-1}\right)b_i$
Pareto type III, $\Re^+$ (Burr XII)	$\bar{F}_X(x) = \left(1+x^\delta\right)^{-1}$	$H_n(W_i) - \log\delta - \left(1-\delta^{-1}\right)a_i$ $-\left(1+\delta^{-1}\right)b_i$
Weibull, $\Re^+$	$\bar{F}_X(x) = e^{-x^\delta}$	$H_n(W_i) - \log\delta - b_i$ $-\left(1-\delta^{-1}\right)E_{g_i}\left[\log\log(1-W_i)^{-1}\right]$

* a_i, b_i, and c_i are defined in (13.13), (13.14), and (13.15), respectively.

seen for Shannon entropy. For example, Rényi information measures between the distributions of X and its order statistics Y_i are given by

$$K_{\alpha,n}(f_i : f_X) \quad = \quad -H_{\alpha,n}(W_i)$$

$$K_{\alpha,n}(f_X : f_i) \quad = \quad \frac{\alpha}{1-\alpha} H_{1-\alpha,n}(W_i),$$

where

$$H_{\alpha,n}(W_i) = \frac{1}{1-\alpha} \log \frac{B([i-1]\alpha+1, [n-i]\alpha+1)}{B^{\alpha}(i, n-i+1)}$$

is Rényi entropy of beta distribution with density (13.11).

Acknowledgement. The research of M. Asadi was supported by the University of Isfahan grant 831125.

References

1. Arnold, B. C. (1983). *Pareto Distributions*, International Co-operative, Publishing House, Baltimore, MD.

2. Arnold, B. C., Balakrishnan, N., and Nagaraja, H. N. (1992). *A First Course in Order Statistics*, John Wiley & Sons, New York.

3. Asadi, M., and Ebrahimi, N. (2000). Residual entropy and its characterizations in terms of hazard function and mean residual life function, *Statistics & Probability Letters*, **49**, 263–269.

4. Asadi, M., Ebrahimi, N., Hamedani, G. G., and Soofi, E. S. (2004). Maximum dynamic entropy models, *Journal of Applied Probability*, **41**, 379–390.

5. Asadi, M., Ebrahimi, N., and Soofi, E. S. (2005). Dynamic generalized information measures, *Statistics & Probability Letters*, **71**, 85–98.

6. Bernardo, J. M. (1979). Expected information as expected utility, *Annals of Statistics*, **7**, 686–690.

7. Darbellay, G. A., and Vajda, I. (2000). Entropy expressions for multivariate continuous distributions, *IEEE Transactions on Information Theory*, **46**, 709–712.

8. Di Crescenzo, A., and Longobardi, M. (2002). Entropy-based measure of uncertainty in past lifetime distributions, *Journal of Applied Probability*, **39**, 434–440.

9. Di Crescenzo, A., and Longobardi, M. (2004). A measure of discrimination between past lifetime distributions, *Statistics & Probability Letters*, **67**, 173–182.

10. Ebrahimi, N. (1996). How to measure uncertainty in the residual lifetime distributions, *Sankhyā, Series A*, **58**, 48–57.

11. Ebrahimi, N., and Kirmani, S. N. U. A. (1996a). A characterization of the proportional hazards model through a measure of discrimination between two residual life distributions, *Biometrika*, **83**, 233–235.

12. Ebrahimi, N., and Kirmani, S. N. U. A. (1996b). A measure of discrimination between two residual lifetime distributions and its applications, *Annals of Institute of Statistical Mathematics*, **48**, 257–265.

13. Ebrahimi, N., Soofi, E. S., and Zahedi, H. (2004). Information properties of order statistics and spacings, *IEEE Transactions on Information Theory*, **50**, 177–183.

14. Golan, A., and Perloff, J. M. (2002). Comparison of maximum entropy and higher-order entropy estimators, *Journal of Econometrics*, **107**, 195–211.

15. Good, I. J. (1950). *Probability and Weighting of Evidence*, Griffin, London.

16. Jaynes, E. T. (1957). Information theory and statistical mechanics, *Physics Review*, **106**, 620–630.

17. Jaynes, E. T. (1982). On the rationale of maximum-entropy methods, *Proceedings of IEEE*, **70**, 939–952.

18. Kullback, S. (1959). *Information Theory and Statistics*, John Wiley & Sons, New York.

19. Kullback, S., and Leibler R. A. (1951). On information and sufficiency, *Annals of Mathematical Statistics*, **22**, 79–86.

20. Nadarajah, S., and Zografos, K. (2003). Formulas for Rényi information and related measures for univariate distributions, *Information Science*, **155**, 119–138.

21. Park, S. (1995). The entropy of consecutive order statistics, *IEEE Transactions on Information Theory*, **41**, 2003–2007.

22. Park, S. (1996). Fisher information on order statistics, *Journal of the American Statistical Association*, **91**, 385–390.

23. Rényi, A. (1961). On measures of entropy and information, *Proceedings of the Fourth Berkeley Symposium*, **1**, 547–561, University of California Press, Berkeley.

24. Shannon, C. E. (1948). A mathematical theory of communication, *Bell System Technical Journal*, **27**, 379–423.

25. Shore, J. E., and Johnson R. W. (1980). Axiomatic derivation of the principle of maximum entropy and principle of minimum cross-entropy, *IEEE Transactions on Information Theory*, **26**, 26–37.

26. Song, K. (2001). Rényi information, loglikelihood and an intrinsic distribution measure, *Journal of Statistical Planning and Inference*, **93**, 51–69.

27. Soofi, E. S. (1997). Information theoretic regression methods, In *Advances in Econometrics: Applying Maximum Entropy to Econometric Problems*, **12** (Eds., T. B. Fomby and R. C. Hill), pp. 25–83, JAI Press, Greenwich, CT.

28. Wong, K.M., and Chen, S. (1990). The entropy of ordered sequences and order statistics, *IEEE Transactions on Information Theory*, **36**, 276–284.

29. Zellner, A. (1971). *An Introduction to Bayesian Inference in Econometrics*, John Wiley & Sons, New York (reprinted in 1996).

14

Confidence Coefficients of Interpolated Nonparametric Sign Intervals for Medians Under No or Weak Shape Assumptions

Olivier Guilbaud

AstraZeneca, Södertälje, Sweden

Abstract: Non-parametric "sign" intervals for a parent median based on order statistics have the important property of being generally valid. With small sample sizes, the available confidence coefficients (CCs) are sparse, however, and it is natural to try to interpolate between adjacent sign intervals to attain intermediate levels. This chapter provides the CC associated with weighted means of adjacent sign intervals over some interesting classes of parent distributions, including: (a) all distributions, (b) all symmetric distributions, and (c) all symmetric and unimodal distributions. The behavior of these CCs as functions of the weight is simple but intuitively quite surprising, with certain discontinuities and intervals of constancy. Some unexpected domination relations among weighted means of adjacent sign intervals follow from these results. The resulting nondominated intervals constitute a considerable extension of the sign intervals, with substantially more confidence-coefficient levels; and they are valid under no or weak shape assumptions about the parent distribution.

Keywords and phases: Confidence interval, general distribution, interpolation, median, nonparametric, order statistic, symmetric distribution, unimodal distribution

14.1 Introduction

Let $X_{(1)} \leq \cdots \leq X_{(n)}$ be the order statistics of a random sample $X_1, \ldots, X_n$ of size $n \geq 2$ from a parent X-distribution with distribution function $F(x) = \Pr[X \leq x]$. No shape assumption is made about this X-distribution; it is only assumed that F belongs to the general class $\mathbf{F}$ of all right-continuous and proper distribution functions. Moreover, let θ_F be the midpoint of the possibly degenerate interval of all medians of F, so θ_F is a uniquely defined median for any $F \in \mathbf{F}$. This article concerns interval estimation of the median θ_F. For convenience, no notational distinction is made between random quantities and the corresponding realizations.

Let r and s be any given integers satisfying $1 \le r < s = n - r + 1$. As is well known, the interval

$$I_{(r)} = [X_{(r)}, X_{(s)}] \tag{14.1}$$

constitutes a nonparametric confidence interval for θ_F. This interval is closely related to the sign test, and therefore is sometimes called a sign interval. The confidence coefficient over $\mathbf{F}$ associated with (14.1), that is, the infimum over $F \in \mathbf{F}$ of the coverage probability of (14.1), is equal to

$$C_{r;n} = 1 - 2\sum_{i=0}^{r-1} \binom{n}{i} 2^{-n}, \tag{14.2}$$

and this infimum is attained for any F that is continuous at θ_F; see, for example, David and Nagaraja (2003, Section 7.1).

The general validity of the sign intervals (14.1) is of considerable practical importance. However, a weakness is that if n is small, the available confidence coefficients (14.2) are sparse. It is then natural to consider weighted means $I_{(r,w)}$ of adjacent sign intervals $I_{(r+1)} \subset I_{(r)}$ of the form

$$I_{(r,w)} = \Big[wX_{(r)} + (1-w)X_{(r+1)}, (1-w)X_{(s-1)} + wX_{(s)}\Big] \tag{14.3}$$

with $0 \le w \le 1$ in attempts to attain, at least approximately, other levels in the range $[C_{r+1;n}, C_{r;n}]$, because as a function of $0 \le w \le 1$, the interval (14.3) is a continuously nondecreasing set that satisfies $I_{(r+1)} \subset I_{(r,w)} \subset I_{(r)}$, and equals $I_{(r+1)}$ for $w = 0$, and $I_{(r)}$ for $w = 1$. In (14.3) and subsequently, it is assumed that $n \ge 3$, $1 \le r < s = n - r + 1$, and $s - r \ge 2$.

Approximations for the coverage probability of (14.3) in case F is continuous and "sufficiently smooth" have been proposed that are of the form

$$w_* C_{r;n} + (1 - w_*) C_{r+1;n}, \tag{14.4}$$

where the weight w_* is a continuous and strictly increasing function of $0 \le w \le 1$ with range [0,1]. The inverse function evaluated at $w_* = (\gamma - C_{r+1;n})/(C_{r;n} - C_{r+1;n})$ then gives the weight w to be used in (14.3) to get a nominal level (14.4) equal to $\gamma \in [C_{r+1;n}, C_{r;n}]$. In particular: (a) Hettmansperger and Sheather (1986) proposed the use of $w_* = w(n-r)/[w(n-r)+(1-w)r]$; (b) Hutson (1999) proposed another nonlinear function of w, $w_* = (2I_{1/2} - 1 - C_{r+1;n})/(C_{r;n} - C_{r+1;n})$ with $I_{1/2}$ defined in terms of the incomplete beta function as $I_{1/2}(r+1-$

$w, n-r+w$); whereas (c) Beran and Hall (1993) gave certain rate-of-convergence ($n \to \infty$) arguments for using $w_* = w$, that is simple linear interpolation.

In contrast to such developments, the present article deals with the following question. What can be said about the confidence coefficient

$$CC(r, w, \mathbf{\Phi}) = \inf_{F \in \mathbf{\Phi}} \Pr[\theta_F \in I_{(r,w)}] \tag{14.5}$$

associated with (14.3) over a given nonempty class $\mathbf{\Phi} \in \mathbf{F}$ of parent distribution functions F of interest? The answer is provided in Sections 14.2–14.4 for the classes $\mathbf{F} \supset \mathbf{F}_S \supset \mathbf{F}_{SU}$ corresponding to all X-distributions, all symmetric X-distributions, and all symmetric and unimodal X-distributions; whereas a partial answer (covering the case $1/2 \leq w \leq 1$) is provided in Section 14.7 for the class $\mathbf{F}_U \subset \mathbf{F}$ corresponding to all unimodal X-distributions. [An $F \in \mathbf{F}$ belongs to $\mathbf{F}_U$ if and only if its graph $\{(x, y); y = F(x)\}$ is convex over $-\infty < x < x_*$ and concave over $x_* < x < \infty$ for some point x_*. Any such x_* is called a mode of F. If $F \in \mathbf{F}_{SU}$, the median θ_F is a mode of F, and F is continuous everywhere except possibly at θ_F. See Dharmadhikari and Joag-dev (1988, Chapter 1) for general properties of an $F \in \mathbf{F}_U$]. For the classes $\mathbf{\Phi}$ considered, the behavior of (14.5) as a function of w is simple but intuitively quite surprising, with certain discontinuities and intervals of constancy. These results constitute the principal achievements of this article.

Some unexpected domination relations among intervals (14.3) follow from these results. Here, briefly, domination means that an interval is entirely contained within another although it has the same confidence coefficient. The nondominated intervals constitute a considerable extension of the sign intervals. These results are considered in Sections 14.5 and 14.6. Some concluding comments and additional results are given in Section 14.7.

14.2 Confidence Coefficient Under No Shape Assumption

The best possible lower bound for the coverage probability of (14.3) valid for all X-distributions is given in Theorem 14.2.1. The result is stated in terms of (14.2) and (14.5).

Theorem 14.2.1 *The confidence coefficient $CC(r, w, \mathbf{F})$ associated with (14.3) equals*

$$\begin{cases} C_{r+1;n}, & \text{if } 0 \leq w < 1/2, \\ (C_{r;n} + C_{r+1;n})/2, & \text{if } 1/2 \leq w < 1, \\ C_{r;n}, & \text{if } w = 1. \end{cases} \tag{14.6}$$

PROOF. This equality follows for $w = 0, 1$ from (14.1)-(14.2) and the fact that there are $F \in \mathbf{F}$ which are continuous at θ_F, and for $w = 1/2$ from Guilbaud (1979, Theorem 2.1). Moreover, it is shown: (a) in Appendix A that there is a (symmetric) X-distribution $D_\varepsilon^{(S)}$ indexed by $\varepsilon > 0$ such that for any given $0 < w < 1/2$, the coverage probability of (14.3) tends to $C_{r+1;n}$ as $\varepsilon \to 0$; and (b) in Appendix B that there is a (unimodal) X-distribution $D_\varepsilon^{(U)}$ indexed by $\varepsilon > 0$ such that for any given $1/2 < w < 1$, the coverage probability of (14.3) tends to $(C_{r;n} + C_{r+1;n})/2$ as $\varepsilon \to 0$. Theorem 14.2.1 then follows from the fact that (14.5) is a nondecreasing function of $0 \le w \le 1$. ■

As a function of $0 \le w \le 1$, the confidence coefficient $CC(r, w, \mathbf{F})$ thus is constant over $[0, 1/2)$ and over $[1/2, 1)$, with jumps at $w = 1/2, 1$. This is intuitively quite surprising in view of how the interval (14.3) behaves as a function of $0 \le w \le 1$. One may wonder whether it makes any difference if one restricts considerations to the subclass $\mathbf{F}_C$ of $\mathbf{F}$ that consists of all continuous distribution functions $F \in \mathbf{F}$. The answer is no in that $CC(r, w, \mathbf{F}_C)$ equals $CC(r, w, \mathbf{F})$ for $0 \le w \le 1$. This follows immediately from the fact that the X-distributions $D_\varepsilon^{(S)}$ and $D_\varepsilon^{(U)}$ referred to in the proof of Theorem 14.2.1 have continuous distribution functions.

14.3 Confidence Coefficient Under Symmetry

The best possible lower bound for the coverage probability of (14.3) valid for all symmetric X-distributions is given in Theorem 14.3.1. The result is stated in terms of (14.2) and (14.5).

Theorem 14.3.1 *The confidence coefficient $CC(r, w, \mathbf{F}_S)$ associated with (14.3) equals*

$$\begin{cases} C_{r+1;n}, & \text{if } 0 \le w < 1/2, \\ C_{r;n-1}, & \text{if } 1/2 \le w < 1, \\ C_{r;n}, & \text{if } w = 1. \end{cases} \tag{14.7}$$

PROOF. This equality follows for $w = 0, 1$ from (14.1)-(14.2) and the fact that there are $F \in \mathbf{F}_S$ which are continuous at θ_F, and for $w = 1/2$ from Guilbaud (1979, end of last paragraph, p. 32). Now: (a) as mentioned in the proof of Theorem 14.2.1, there is a symmetric X-distribution $D_\varepsilon^{(S)}$ such that for any given $0 < w < 1/2$, the coverage probability of (14.3) tends to $C_{r+1;n}$ as $\varepsilon \to 0$; and (b) it is shown in Appendix C that there is a symmetric (and unimodal) X-distribution $D_m^{(SU)}$ indexed by an integer $m \ge 2$ such that for any given $1/2 < w < 1$, the coverage probability of (14.3) tends to $C_{r;n-1}$

as $m \to \infty$. Theorem 14.3.1 then follows from the fact that (14.5) is a non-decreasing function of $0 \le w \le 1$. ■

Thus, as a function of $0 \le w \le 1$, the confidence coefficient $CC(r, w, \mathbf{F}_S)$ has the same kind of intuitively surprising behavior as $CC(r, w, \mathbf{F})$. And again, the restriction to continuous distribution functions makes no difference in that $CC(r, w, \mathbf{F}_C \cap \mathbf{F}_S)$ equals $CC(r, w, \mathbf{F}_S)$ for $0 \le w \le 1$. This follows from the fact that the X-distributions $D_\varepsilon^{(S)}$ and $D_m^{(SU)}$ referred to in the proof of Theorem 14.3.1 have continuous distribution functions. The value of $C_{r;n-1}$ in (14.7) is strictly larger than $(C_{r;n} + C_{r+1;n})/2$ in (14.6), so the restriction to symmetric X-distributions has increased the confidence coefficient, though surprisingly, only for $1/2 \le w < 1$.

For any $F \in \mathbf{F}_C \cap \mathbf{F}_S$, the coverage probability of (14.3) with $w = 1/2$ is equal to $C_{r;n-1}$. This is well known, and shown for example by Noether (1973). Actually, Noether essentially showed also the result in Theorem 14.3.1 for $w = 1/2$, though the "projection" method he used for an F which is not continuous requires that F has at most a finite number of discontinuity points in any bounded interval—a condition that is not satisfied for all $F \in \mathbf{F}_S$.

14.4 Confidence Coefficient Under Symmetry and Unimodality

The best possible lower bound for the coverage probability of (14.3) valid for all symmetric and unimodal X-distributions is given in Theorem 14.4.1. The result is stated in terms of (14.2), (14.5) and the function $\delta_r(w)$ of $0 \le w \le 1/2$ given by

$$\delta_r(w) = 2^{1-n} \binom{n}{r} \sum_{i=1}^{n-r} \binom{n-r}{i} p^i (1-p)^{n-r-i} i/(i+r) \tag{14.8}$$

with $p = w/(1-w)$. It can be verified that (14.8) is a continuous and strictly increasing function of $0 \le w \le 1/2$ such that $\delta_r(0) = 0$ and $\delta_r(1/2) = C_{r;n-1} - C_{r+1;n}$.

Theorem 14.4.1 *The confidence coefficient $CC(r, w, \mathbf{F}_{SU})$ associated with (14.3) equals*

$$\begin{cases} C_{r+1;n} + \delta_r(w), & \text{if } 0 \le w < 1/2, \\ C_{r;n-1}, & \text{if } 1/2 \le w < 1, \\ C_{r;n}, & \text{if } w = 1. \end{cases} \tag{14.9}$$

Proof. This equality follows for $w = 0, 1$ from (14.1) and (14.2) and the fact that there are $F \in \mathbf{F}_{SU}$ which are continuous at θ_F, and is shown in Appendix E for $0 < w < 1/2$. Moreover, as mentioned in the proof of Theorem 14.3.1, there is a symmetric and unimodal X-distribution $D_m^{(SU)}$ such that for any given $1/2 < w < 1$, the coverage probability of (14.3) tends to $C_{r;n-1}$ as $m \to \infty$. Theorem 14.4.1 then follows from the properties of (14.8) just mentioned and the fact that (14.5) is a nondecreasing function of $0 \le w \le 1$. ■

Compared to (14.7), the additional restriction to unimodal X-distributions has increased the confidence coefficient, though surprisingly, now only for $0 < w < 1/2$. And once again, the restriction to continuous distribution functions makes no difference in that $CC(r, w, \mathbf{F}_C \cap \mathbf{F}_{SU})$ equals $CC(r, w, \mathbf{F}_{SU})$ for $0 \le w \le 1$; see Appendix E.

14.5 Domination Relations Among Interval Estimators

The following terminology concerning interval estimators (14.3) of θ_F with a common r is used subsequently.

Definition 14.5.1 An interval $I_{(r,w')}$ is said to $\mathbf{\Phi}$-dominate another interval $I_{(r,w'')}$ if: (a) it is "smaller" in that $0 \le w' < w'' \le 1$; and (b) it nevertheless has the same confidence coefficient over $\mathbf{\Phi}$, that is $CC(r, w', \mathbf{\Phi}) = CC(r, w'', \mathbf{\Phi})$ in terms of (14.5).

The notion of "smaller" used here is quite strong because $0 \le w' < w'' \le 1$ implies that: (i) $I_{(r,w')}$ is a subset of $I_{(r,w'')}$ with probability 1 for any $F \in \mathbf{F}$, and (ii) the endpoints of $I_{(r,w')}$ are strictly between those of $I_{(r,w'')}$ with probability 1 for any $F \in \mathbf{F}_C$.

Now, it is evident from the behavior of (14.6) as a function of $0 \le w \le 1$ that: (a) the interval $I_{(r,0)} \equiv I_{(r+1)}$ $\mathbf{F}$-dominates each interval $I_{(r,w)}$ with $0 < w < 1/2$; and (b) the interval $I_{(r,1/2)}$ $\mathbf{F}$-dominates each interval $I_{(r,w)}$ with $1/2 < w < 1$. Thus, in case one is not willing to assume anything about $F \in \mathbf{F}$ (except possibly that $F \in \mathbf{F}_C$, which as mentioned in Section 14.2 does not change the confidence coefficient), it seems reasonable to restrict considerations to interval estimators (14.3) of θ_F that are not $\mathbf{F}$-dominated by others, that is, to intervals (14.3) with $w \in \{0, 1/2, 1\}$.

The behavior of (14.7) is similar to that of (14.6), so similar conclusions can be drawn. Thus, in case one is willing to assume that $F \in \mathbf{F}_S$, but nothing else (except possibly that $F \in \mathbf{F}_C \cap \mathbf{F}_S$, which as mentioned in Section 14.3 does not change the confidence coefficient), it seems reasonable to restrict considerations to intervals (14.3) with $w = \{0, 1/2, 1\}$.

The behavior of (14.9) implies that: (a) among the intervals $I_{(r,w)}$ with $0 \leq w < 1/2$, no interval $\mathbf{F}_{SU}$-dominates any other; whereas (b) the interval $I_{(r,1/2)}$ $\mathbf{F}_{SU}$-dominates each interval $I_{(r,w)}$ with $1/2 < w < 1$. Thus, in case one is willing to assume that $F \in \mathbf{F}_{SU}$, but nothing else (except possibly that $F \in \mathbf{F}_C \cap \mathbf{F}_{SU}$, which as mentioned in Section 14.4 does not change the confidence coefficient), it seems reasonable to restrict considerations to intervals (14.3) with $w \in [0, 1/2] \cup \{1\}$.

14.6 Nondominated Interval Estimators and Available Confidence Coefficients

Among the interval estimators (14.3) of θ_F, it thus seems reasonable to restrict considerations to the nondominated ones, that is those with

$$
\begin{array}{ll}
w = 0, 1/2, 1, & \text{under no shape assumption,} \\
w = 0, 1/2, 1, & \text{under symmetry,} \\
0 \leq w \leq 1/2 \text{ or } w = 1, & \text{under symmetry and unimodality.}
\end{array}
$$

These nondominated interval estimators constitute a considerable extension of the sign intervals (14.2), with substantially more confidence-coefficient levels; and they are valid under no or weak shape assumptions about the X-distribution.

Table 14.1 provides some numerical details about available confidence coefficients given by Theorems 14.2.1-14.4.1. The last two columns of this table give the w-values $0 \leq w_{90} \leq 1/2$ and $0 \leq w_{95} \leq 1/2$ satisfying

$$
CC(r, w_{90}, \mathbf{F}_{SU}) = 0.90 \qquad \text{and} \qquad CC(r, w_{95}, \mathbf{F}_{SU}) = 0.95 \qquad (14.10)
$$

when such w-values exist, that is, when the levels 0.90 and 0.95 belong to the interval $[C_{r+1;n}, C_{r,n-1}]$; see (14.9). Given any desired level $\gamma \in (C_{r+1;n}, C_{r;n-1})$, the weight $0 < w < 1/2$ satisfying $CC(r, w, \mathbf{F}_{SU}) = \gamma$ can be determined by solving the equation $\delta_r(w) = \gamma - C_{r+1;n}$ numerically through some suitable search method, for example, regula falsi or some improved variant, with starting values $\delta_r(0) = 0$ and $\delta_r(1/2) = C_{r;n-1} - C_{r+1;n}$. Theoretically such a search method always converges to the unique solution, because of the properties of (14.8) mentioned previously.

Consider for example the row corresponding to the sample size $n = 10$ in Table 14.1. This row shows that with this sample size: (a) the sign intervals $I_{(2)} = [X_{(2)}, X_{(9)}]$ and $I_{(3)} = [X_{(3)}, X_{(8)}]$ cover the median θ_F with probability

Table 14.1: Confidence coefficients given by (14.2), (14.6), (14.7), and (14.9) of interval estimators (14.3), as well as w-weights w_{90} and w_{95} given by (14.10)

n	r	$C_{r;n}$	$C_{r+1;n}$	$(C_{r;n} + C_{r+1;n})/2$	$C_{r,n-1}$	w_{90}	w_{95}
6	1	.9688	.7813	.8750	.9375	.306	
7	1	.9844	.8750	.9297	.9688	.081	.309
8	1	.9922	.9297	.9609	.9844		.104
8	2	.9297	.7109	.8203	.8750		
9	2	.9609	.8203	.8906	.9297	.300	
10	2	.9785	.8906	.9346	.9609	.041	.360
11	2	.9883	.9346	.9614	.9785		.107
11	3	.9346	.7734	.8540	.8906		
12	3	.9614	.8540	.9077	.9346	.223	
13	3	.9775	.9077	.9426	.9614		.331
13	4	.9077	.7332	.8204	.8540		
14	3	.9871	.9426	.9648	.9775		.064
14	4	.9426	.8204	.8815	.9077	.433	
15	4	.9648	.8815	.9232	.9426	.109	
16	4	.9787	.9232	.9510	.9648		.254
16	5	.9232	.7899	.8565	.8815		
17	5	.9510	.8565	.9037	.9232	.276	
18	5	.9691	.9037	.9364	.9510		.483
18	6	.9037	.7621	.8329	.8565		
19	5	.9808	.9364	.9586	.9691		.150
19	6	.9364	.8329	.8847	.9037	.462	
20	6	.9586	.8847	.9216	.9364	.111	

≥ 0.9785 and ≥ 0.8906, respectively, for *any* X-distribution; (b) the intermediate "middle" interval $I_{(2,1/2)}$ covers θ_F with probability ≥ 0.9346 for *any* X-distribution, cf. (14.6); (c) this "middle" interval $I_{(2,1/2)}$ covers θ_F with a probability ≥ 0.9609 for *any* symmetric X-distribution, cf. (14.7); and (d) the intervals $I_{(2,w)}$ with $w = w_{90} = 0.041$ and $w = w_{95} = 0.360$ cover θ_F with probability ≥ 0.90 and ≥ 0.95, respectively, for *any* symmetric and unimodal X-distribution, cf. (14.9).

14.7 Concluding Comments and Additional Results

The confidence coefficients given by Theorems 14.2.1-14.4.1 constitute the main results of this discussion. The subsequent results concerning domination relations and nondominated intervals in Sections 14.5 and 14.6 follow naturally from the intuitively surprising behavior of these confidence coefficients as functions of $0 \leq w \leq 1$. The nondominated intervals considered in Section 14.6 are of practical interest in that: (a) they constitute a considerable extension of the sign intervals (14.1); (b) they are valid under no or weak shape assumptions about the parent X-distribution; and (c) they are almost as easily implemented as the sign intervals.

The confidence coefficient (14.5) has been derived in Sections 14.2-14.4 over natural classes of distribution functions, but other classes may of course also be of interest. One such class is the class $\mathbf{F}_U \subset \mathbf{F}$ corresponding to all unimodal X-distributions; see Section 14.1. A partial result concerning this class follows immediately from Theorem 14.2.1 and the fact that the X-distribution $D_\varepsilon^{(U)}$ referred to in its proof is unimodal: *The confidence coefficient $CC(r, w, \mathbf{F}_U)$ associated with (14.3) equals*

$$\begin{cases} C_{r+1;n}, & \text{if } w = 0, \\ (C_{r;n} + C_{r+1;n})/2, & \text{if } 1/2 \leq w < 1, \\ C_{r;n}, & \text{if } w = 1. \end{cases} \tag{14.11}$$

This result is partial in that the case with $0 < w < 1/2$ is not covered. Compared to (14.6), the restriction to unimodal X-distributions thus has not increased the confidence coefficient for $1/2 \leq w < 1$.

The confidence coefficients derived in this discussion can be used to make comparisons versus interpolation methods based on (14.3) and (14.4), including those mentioned in Section 14.1. More precisely, for any such interpolation method, it is possible to derive the confidence coefficient (14.5) corresponding to given values $\gamma \in [C_{r+1;n}, C_{r;n}]$ of the nominal level (14.4), and to study the relation between (14.5) and (14.4) for the classes $\mathbf{\Phi}$ considered. In particular, it is interesting to note from (14.6) and (14.11) that simple linear interpolation, for which Beran and Hall (1993) gave certain rate-of-convergence ($n \to \infty$) arguments, leads to a nominal level (14.4) for the "middle" interval $I_{(r,1/2)}$ that actually equals the confidence coefficient

$$CC(r, 1/2, \mathbf{F}) = CC(r, 1/2, \mathbf{F}_U) = (C_{r;n} + C_{r+1;n})/2 \tag{14.12}$$

of this interval over the classes $\mathbf{F}$ and $\mathbf{F}_U$ - for any sample size $n \geq 3$.

Appendices

Appendix A

The symmetric X-distribution $D_\varepsilon^{(S)}$ referred to in the proofs of Theorems 14.2.1 and 14.3.1 has median $\theta_F = 0$ and density function

$$\begin{cases} 1/(2\varepsilon) - 1, & \text{if } -(1+\varepsilon) < x < -1, \\ \varepsilon, & \text{if } -1 < x < 1, \\ 1/(2\varepsilon) - 1, & \text{if } 1 < x < (1+\varepsilon), \end{cases} \tag{14.13}$$

with $0 < \varepsilon < 1/2$, so suppose this is the actual X-distribution. Let $0 < w < 1/2$ be given. It can be verified that if $0 < \varepsilon < \min(1/2, 1/w - 2)$, then $A \leq \Pr[0 \in I_{r,w})] \leq A + B$ with $A = \Pr[X_{(r+1)} \leq -1, X_{(s-1)} \geq 1]$ and B equal to the probability that at least one of $X_{(r)}, X_{(r+1)}, X_{(s-1)}, X_{(s)} \in (-1, 1)$. As $\varepsilon \to 0 : B \to 0$, so also the difference between $C_{r+1;n} = \Pr[X_{(r+1)} \leq 0 \leq X_{(s-1)}]$ and A tends to 0, and thus $\Pr[0 \in I_{(r,w)}] \to C_{r+1;n}$.

Appendix B

The unimodal X-distribution $D_\varepsilon^{(U)}$ referred to in the proof of Theorem 14.2.1 and in connection with (14.11) has median $\theta_F = 0$ and density function

$$\begin{cases} 1/(2\varepsilon), & \text{if } -\varepsilon < x < 0, \\ \varepsilon, & \text{if } 0 < x < 1/(2\varepsilon), \end{cases} \tag{14.14}$$

with $\varepsilon > 0$, so suppose this is the actual X-distribution. Let $1/2 < w < 1$ be given. Now, $A \equiv \Pr[wX_{(r)} + (1-w)X_{(r+1)} > 0]$ equals the sum of $\Pr[X_{(r)} > 0]$ and $\Pr[X_{(r)} \leq 0, X_{(r+1)} > -cX_{(r)}]$ with $c = w/(1-w) > 1$. The latter term is bounded by $\Pr[X_{(r)} \leq 0, X_{(r+1)} > c\varepsilon]$ and $\Pr[X_{(r)} \leq 0, X_{(r+1)} > 0]$. As $\varepsilon \to 0$: the lower bound tends to the upper, so $A \to \Pr[X_{(r+1)} > 0]$. Similarly, $B \equiv \Pr[(1-w)X_{(s-1)} + wX_{(s)} < 0]$ equals the sum of $\Pr[X_{(s)} < 0]$ and $\Pr[X_{(s)} \geq 0, X_{(s-1)} < -cX_{(s)}]$. The latter term is bounded from above by $\Pr[0 \leq X_{(s)} < \varepsilon/c]$ which as $\varepsilon \to 0$, tends to 0, so $B \to \Pr[X_{(s)} < 0]$. Thus as $\varepsilon \to 0$, $\Pr[0 \in I_{(r,w)}] = 1 - A - B$ tends to $\Pr[X_{(r+1)} \leq 0 \leq X_{(s)}]$, which equals $(C_{r;n} + C_{r+1;n})/2$.

Appendix C: An auxiliary result

Let $\mathbf{F}_{Ci} \subset \mathbf{F}_C$ consist of all $F \in \mathbf{F}_C$ that are strictly increasing in $\{x; 0 < F(x) < 1\}$, and define $\mathbf{F}_{CiSU} = \mathbf{F}_{Ci} \cap \mathbf{F}_{SU}$. Thus if $F \in \mathbf{F}_{CiSU}$, then: (a) the X-distribution is symmetric and unimodal; (b)$F \in \mathbf{F}_{Ci} \subset F_C$; and (c) the inverse function $F^{-1}(u)$ is well defined and continuous for $0 < u < 1$.

Now, suppose $F \in \mathbf{F}_{CiSU}$, and define $F_0 \in \mathbf{F}_{CiSU}$ with median 0 through the translation $F_0(x) = F(x + \theta_F)$. Moreover, let $0 \leq w < 1$ be given, and set $c = w/(1-w)$. It can then be verified through a development similar to that in Guilbaud (1979, Equations (A1.2)-(A1.6)) that

$$C_{r;n} - \Pr[\theta_F \in I_{(r,w)}] = 2n\binom{n-1}{r-1}\int_{u=0}^{1/2} u^{r-1}[K_c(u)]^{n-r}du, \qquad (14.15)$$

with $K_c(u) = F_0(cF_0^{-1}(u))$. Note that $K_c(1/2) = 1/2$, and that for any given $0 < u < 1/2$, $K_c(u)$ is a strictly decreasing function of $0 \leq c < \infty$ such that $K_0(u) = 1/2$, $K_1(u) = u$, and $K_c(u) \to 0$ as $c \to \infty$. It follows immediately: (a) from (14.15) with $w = 1/2$ and the result mentioned at the beginning of the last paragraph of Section 14.3 that $C_{r;n} - C_{r;n-1}$ equals the right member with $K_c(u)$ replaced by u; and (b) from (14.15) with $w = 0$ that $C_{r;n} - C_{r+1;n}$ equals the right member with $K_c(u)$ replaced by $1/2$. These results are used in Appendices D and E.

Appendix D

The symmetric and unimodal X-distribution $D_m^{(SU)}$ referred to in the proofs of Theorems 14.3.1 and 14.4.1 has median $\theta_F = 0$ and distribution function $F \in \mathbf{F}_{CiSU} \subset \mathbf{F}_{SU}$ (subsequently denoted F_m) given for $x \leq 0$ by

$$\begin{cases} 0, & \text{if } x \leq -1, \\ (1 - |x|^{1/m})^m/2, & \text{if } -1 < x \leq 0, \end{cases} \qquad (14.16)$$

and for $x \geq 0$ by symmetry. Suppose this is the actual X-distribution, let $1/2 < w < 1$ be given, set $c = w/(1-w)$, and note that $c > 1$. Then (14.15) holds with $K_c(u) = F_m(cF_m^{-1}(u))$, and it can be verified using (14.16) that for any given $0 < u < 1/2$, $K_c(u) \to u$ as $m \to \infty$. Combining this with the representation of $C_{r;n} - C_{r;n-1}$ mentioned in Appendix C it then follows that as $m \to \infty$: (14.15) tends to $C_{r;n} - C_{r;n-1}$, that is $\Pr[0 \in I_{r,w)}] \to C_{r;n-1}$.

Appendix E

The first step here is to show that $CC(r, w, \mathbf{F}_{CiSU})$ equals (14.9) for $0 < w < 1/2$. Thus, suppose $F \in \mathbf{F}_{CiSU}$, and as in Appendix C, define $F_0 \in \mathbf{F}_{CiSU}$ with median 0 through $F_0(x) = F(x + \theta_F)$. Let $0 < w < 1/2$ be given, set

$c = w/(1-w)$, and note that $0 < c < 1$. Now, for any $0 < u < 1/2$, the slope of the straight line connecting $P_1 = (F_0^{-1}(u), u)$ and $P_2 = (cF_0^{-1}(u), F_0(cF_0^{-1}(u)))$ is $\leq$ the slope of the straight line connecting P_1 and $P_3 = (0, 1/2)$, so

$$F_0(cF_0^{-1}(u)) \leq cu + (1-c)/2, \tag{14.17}$$

where the left member equals $K_c(u)$ in (14.15). Let $K_c^*(u) = cu + (1-c)/2$. The left member of (14.15) is then $\leq$ the right member of (14.15) with $K_c(u)$ replaced by $K_c^*(u)$. Combining this with the representation of $C_{r;n} - C_{r+1;n}$ mentioned in Appendix C it then follows that

$$\begin{aligned}\Pr[\theta_F \in I_{(r,w)}] \quad \geq \quad & C_{r+1;n} + 2n\binom{n-1}{r-1}\int_{u=0}^{1/2} u^{r-1}\left\{[1/2]^{n-r}\right.\\ & \left. -[cu+(1-c)/2]^{n-r}\right\} du. \end{aligned} \tag{14.18}$$

The second term in the right member can be shown to be equal to (14.8) with $p = c$ through straightforward integration (using transformation $u' = 2u$ and obvious binominal expansion). Moreover, equality in (14.17) and (14.18) is attained with the particular $F \in \mathbf{F}_{CiSU}$ corresponding to the uniform X-distribution over $(-1, 1)$, so $CC(r, w, \mathbf{F}_{CiSU}) = C_{r+1;n} + \delta_r(w)$ for $0 < w < 1/2$.

The next step is to use this to show that $CC(r, w, \mathbf{F}_{SU}) \geq C_{r+1;n} + \delta_r(w)$ for $0 < w < 1/2$. Thus suppose $F \in \mathbf{F}_{SU}$, and let $0 < w < 1/2$ be given. For any given $0 < \lambda < 1$, define the auxiliary random sample $X_1', \ldots, X_n'$ with parent distribution function F_λ through

$$X_i' = \begin{cases} X_i & \text{if } |X_i - \theta_F| > \lambda, \\ U_i & \text{if } |X_i - \theta_F| \leq \lambda, \end{cases} \tag{14.19}$$

in terms of a random sample $U_1, \ldots, U_n$ from the uniform distribution over $[\theta_F - \lambda, \theta_F + \lambda]$ that is independent of $X_1, \ldots, X_n$. Note that: (a) X_i' and X_i have common parent median θ_F; (b) the distribution function F_λ of X_i' is linear in $(\theta_F - \lambda, \theta_F + \lambda)$ and equal to F outside this interval; and (c) $F_\lambda \in \mathbf{F}_{CiSU}$, cf. Dharmadhikari and Joag-dev (1988, Chapter 1). With $I'_{(r,w)}$ defined as (14.3) in terms of the order statistics $X'_{(1)}, \ldots, X'_{(n)}$ of (14.19), it follows from the first step that the inequality $\Pr[\theta_F \in I'_{(r,w)}] \geq C_{r+1;n} + \delta_r(w)$ holds. Now, for any given $\varepsilon > 0$, the event $E = [|X_i' - X_i| < \varepsilon, \text{ all } 1 \leq i \leq n]$ is a subset of the event $[|X'_{(i)} - X_{(i)}| < \varepsilon, \text{ all } 1 \leq i \leq n]$, and it can be verified from (14.19) that $\Pr(E) \to 1$ as $\lambda \to 0$; so for all sufficiently small $\lambda > 0$,

$$\Pr[wX_{(r)} + (1-w)X_{(r+1)} - \varepsilon \leq \theta_F \leq (1-w)X_{(s-1)} + wX_{(s)} + \varepsilon] + \varepsilon \tag{14.20}$$

is larger than or equal to $\Pr[\theta_F \in I'_{(r,w)}] \geq C_{r+1;n} + \delta_i(w)$. But the difference between (14.20) and $\Pr[\theta_F \in I_{(r,w)}]$ can be made arbitrarily close to zero by choosing $\varepsilon > 0$ sufficiently small, so the inequality $\Pr[\theta_F \in I_{(r,w)}] \geq C_{r+1;n} + \delta_r(w)$ must hold. Thus $CC(r, w, \mathbf{F}_{SU}) \geq C_{r+1;n} + \delta_r(w)$ for $0 < w < 1/2$.

Finally, note that (14.5) is a non-increasing function of $\boldsymbol{\Phi}$ in that

$$CC(r, w, \boldsymbol{\Phi}_1) \geq CC(r, w, \boldsymbol{\Phi}_2)$$

if $\boldsymbol{\Phi}_1 \subset \boldsymbol{\Phi}_2$. Then, because $\mathbf{F}_{CiSU} \subset \mathbf{F}_C \cap \mathbf{F}_{SU} \subset \mathbf{F}_{SU}$, it follows from the results in the previous two steps that $CC(r, w, \mathbf{F}_{SU})$ and $CC(r, w, \mathbf{F}_C \cap \mathbf{F}_{SU})$ are equal to $CC(r, w, \mathbf{F}_{CiSU}) = C_{r+1;n} + \delta_r(w)$ for $0 < w < 1/2$.

References

1. Beran, R., and Hall, P. (1993). Interpolated nonparametric prediction intervals and confidence intervals, *Journal of the Royal Statistical Society, Series B*, **55**, 643–652.

2. David, H. A., and Nagaraja, H. N. (2003). *Order Statistics*, Third edition, John Wiley & Sons, Hoboken, NJ.

3. Dharmadhikari, S. and Joag-dev, K. (1988). *Unimodality, Convexity, and Applications*, Academic Press, San Diego.

4. Guilbaud, O. (1979). Interval estimation of the median of a general distribution, *Scandinavian Journal of Statistics*, **6**, 29–36.

5. Hettmansperger, T. P., and Sheather, S. J. (1986). Confidence intervals based on interpolated order statistics, *Statistics & Probability Letters*, **4**, 75–79.

6. Hutson, A. D. (1999). Calculating nonparametric confidence intervals for quantiles using fractional order statistics, *Journal of Applied Statistics*, **26**, 343–353.

7. Noether, G. E. (1973). Some simple distribution-free confidence intervals for the center of a symmetric distribution, *Journal of the American Statistical Association*, **68**, 716–719.

15

Small Sample Asymptotics for Higher-Order Spacings

Riccardo Gatto and S. Rao Jammalamadaka

University of Bern, Bern, Switzerland
University of California, Santa Barbara, CA, USA

Abstract: In this chapter, we give conditional representations for families of statistics based on higher-order spacings and spacing frequencies. This allows us to compute accurate approximations to the distribution of such statistics, including tail probabilities and critical values. These results generalize those discussed in Gatto and Jammalamadaka (1999) and are essential in using such statistics in various testing contexts.

Keywords and phrases: Goodness-of-fit tests, nonparametric tests, rank tests, m-step spacings, m-step spacing frequencies, two-sample tests, Dirichlet, gamma, negative binomial distributions

15.1 Introduction

In this article, we provide some conditional representations that allow us to compute accurately the distribution of a large number of test statistics based on higher-order spacings and "spacing frequencies," following the ideas suggested in Gatto and Jammalamadaka (1999). The key point is that many important test statistics including the chi-square goodness-of-fit statistic, can be rewritten as conditional statistics, and the technique we develop here allows for very accurate approximations of their P-values, or in finding the critical values at a given level. Testing problems that were already considered by Gatto and Jammalamadaka (1999) included the two following classes of tests: (i) The class of tests based on simple spacings statistics, that is, based on the gaps between successive values of the ordered sample; and (ii) the class of tests based on the "spacing-frequencies", that is, the frequencies of one sample that fall in between the successive order statistics of the other sample, which includes many rank tests. We generalize (i) to tests based on higher-order spacings, or m-step spacings, which are the gaps between order statistics and ones that are m steps away; and (ii) to tests based on higher-order spacing frequencies, which are the frequencies of one sample that fall in between the order statistic of the other sample that are m steps away. The reason to consider such tests is that they have higher asymptotic local powers, as demonstrated in Rao and Kuo (1984) for higher order spacings, and in Jammalamadaka and Schweitzer (1985) for higher-order spacing frequencies.

For convenience, we first review the "conditional saddlepoint approximation" that has been described in Gatto and Jammalamadaka (1999) which is the main tool for the proposed accurate approximations. The saddlepoint approximation is a well-known method of asymptotic analysis that allows us to approximate efficiently contour integrals of a general type. This method, also called the method of steepest descent, was brought into statistical use by Daniels (1954) and Lugannani and Rice (1980) for approximating the distribution of the sum of independent and identically distributed (i.i.d.) observations. The saddlepoint formula $P_n(t_1 \mid t_2)$ below enables us to find the P-values of a test statistic $T_{1n}(S_1, \ldots, S_n)$ based on the dependent quantities $S_1, \ldots, S_n$ which admit the conditional representation $T_n(S_1, \ldots, S_n) \sim T_{1n}(X_1, \ldots, X_n) \mid T_{2n}(X_1, \ldots, X_n) = t_2$, where "$\sim$" signifies the equivalence in distribution. Consider the independent random variables $X_1, \ldots, X_n$, and a statistic (T_{1n}, T_{2n}), $T_{1n} = T_{1n}(X_1, \ldots, X_n) \in \mathrm{R}$ and $T_{2n} = T_{2n}(X_1, \ldots, X_n) \in \mathrm{R}$, defined by

$$\sum_{i=1}^{n} \begin{pmatrix} \psi_{1i}(X_i, T_{1n}, T_{2n}) \\ \psi_{2i}(X_i, T_{2n}) \end{pmatrix} = 0.$$

The joint cumulant generating function of the sum of score functions ψ_{1i} and ψ_{2i} is given by

$$K_n(\lambda, t) = \sum_{i=1}^{n} \log \mathrm{E}[\exp\{\lambda_1 \psi_{1i}(X_i, t_1, t_2) + \lambda_2 \psi_{2i}(X_i, t_2)\}], \tag{15.1}$$

where $\lambda = (\lambda_1, \lambda_2)$ and $t = (t_1, t_2)$.

Step 1 Find $\alpha \in \mathrm{R}^2$ and $\beta \in \mathrm{R}$, solutions of the equations

$$\frac{\partial}{\partial \lambda} K_n(\lambda, t) = 0, \quad \frac{\partial}{\partial \lambda_2} K_n((0, \lambda_2), t) = 0.$$

Step 2 Define

$$K_n''(\lambda, t) = \frac{\partial^2}{\partial \lambda \partial \lambda^T} K_n(\lambda, t), \;\; K_{2n}''(\lambda_2, t) = \frac{\partial^2}{\partial \lambda_2^2} K_n((0, \lambda_2), t),$$

$$s = \alpha_1 \left| \frac{\det(K_n''(\alpha, t))}{K_{2n}''(\beta, t)} \right|^{\frac{1}{2}}, \;\; r = \mathrm{sgn}(\alpha_1)\{2[K_n((0, \beta), t) - K_n(\alpha, t)]\}^{\frac{1}{2}},$$

and

$$P_n(t_1 \mid t_2) = 1 - \Phi(r) + \phi(r)\left(\frac{1}{s} - \frac{1}{r}\right), \tag{15.2}$$

where $\phi(\cdot)$ and $\Phi(\cdot)$ are the standard normal density and distribution functions, and α_1 is the first element of α. Then, $\forall t_1$, t_2 and as $n \to \infty$,

$$\mathrm{P}[T_{1n} \geq t_1 \mid T_{2n} = t_2] = P_n(t_1 \mid t_2)\{1 + \mathrm{O}(n^{-1})\}. \tag{15.3}$$

Note that there is an asymptotically equivalent version of (15.2) which is given by

$$P_n^*(t_1 \mid t_2) = 1 - \Phi\left(r + \frac{1}{r}\log\left\{\frac{s}{r}\right\}\right), \tag{15.4}$$

and we refer to Example 15.2.2 for a numerical comparison.

The two steps given above allow one to approximate a tail probability or a P-value. If we are interested in quantiles or critical values, see Gatto (2001, Section 1) for an efficient algorithm for inverting this saddlepoint approximation.

15.2 Tests Based on Higher-Order Spacings

Statistics based on spacings play an important role in goodness-of-fit tests and in tests on hazard rates in the context of reliability; see Pyke (1965) for an excellent review. One-step spacings are the gaps between the successive ordered sample values and, more generally, m-step spacings are the gaps between m successive ordered sample values. One-step spacings are also very important with circular data, that is, when data are directions in two dimensions and are represented by angles. Indeed, one-step spacings are maximal invariant under changes of origin and sense of rotation. Except for one or two special cases, the exact distribution of such statistics based on uniform spacings is unknown. For most cases, the asymptotic distribution is known but it can be potentially misleading, especially when the sample size is moderate to small. Gatto and Jammalamadaka (1999, Section 3.1) derived saddlepoint approximations for test statistics based on uniform spacings. In this section, we generalize this result and provide saddlepoint approximations to test statistics based on higher-order or m-step uniform spacings. Tests based on such higher-order spacings are known to be more efficient as shown by Rao and Kuo (1984).

Consider $X_1, \ldots, X_{N-1}$ to be a sample of independent random variables from a given absolute continuous distribution F with support in R. The fundamental problem of goodness-of-fit, is to test $F = F_0$, where F_0 is specified. By the probability integral transform $U_i = F_0(X_i)$, $i = 1, \ldots, N-1$ the goodness-of-fit test is reduced to one of testing if $U_1, \ldots, U_{N-1}$ are uniformly distributed, that is, to test the null hypothesis

$$\mathrm{H}_0 : F(u) = u, \ \forall u \in [0, 1).$$

Let $0 \leq U_{(1)} \leq \cdots \leq U_{(N-1)} \leq 1$, denote the ordered sample. The simple or one-step spacings $D_1, \ldots, D_N$ are the gaps between this ordered sample, viz.,

$$D_i = U_{(i)} - U_{(i-1)}, \ \ i = 1, \ldots, N,$$

where $U_{(0)} \stackrel{\text{def}}{=} 0$ and $U_{(N)} \stackrel{\text{def}}{=} 1$. More generally, the m-step disjoint spacings are the gaps between m successive values of the ordered sample. That is, denoting $\lfloor x \rfloor$ for the greatest integer less than or equal to x, for $M = \lfloor N/m \rfloor$,

$$D_{im}^{(m)} = U_{(im)} - U_{((i-1)m)}, \ \ i = 1, \ldots, M.$$

Let $h(\cdot)$ and $h_i(\cdot)$, $i = 1, \ldots, M$, be real-valued functions that satisfy some weak regularity conditions. Most spacings statistics can then be expressed as

$$T_n^* = \sum_{i=1}^{M} h_i(M D_{im}^{(m)}), \tag{15.5}$$

which is not symmetric in the spacings, or as

$$T_n = \frac{1}{M} \sum_{i=1}^{M} h(M D_{im}^{(m)}), \tag{15.6}$$

which is symmetric in the spacings. Sethuraman and Rao (1970) and Rao and Sethuraman (1975) showed that the class of symmetric tests (15.6) based on one-step spacings cannot discriminate alternatives converging to the null hypothesis at asymptotic rates faster than $N^{-1/4}$, which is a drawback when compared, for example, to the Kolmogorov-Smirnov test. Del Pino (1979) showed that tests based on m-step spacings, $m > 1$, have better asymptotic efficiencies than tests based on one-step spacings. Typical examples of symmetric test statistics (15.6) are obtained with

$$h(x) = \log x, \ |x - 1|, \ x^a,$$

$a > -1/2$ and $\neq 0$ or 1. The first two functions lead to the Rao and the log higher-order test statistics and they will be developed in Examples 15.2.1 and 15.2.2 below. The last function for $a = 2$ leads to the Greenwood higher-order test statistic and will be developed in Example 15.2.3. It has maximum asymptotic relative efficiency among symmetric m-step spacings statistics, is asymptotically more efficient that the one-step Greenwood statistic, and indeed the efficiency grows with m; see Table 2 in Rao and Kuo (1984).

The exact distribution of spacings statistics is unknown in most cases and it is common practice to rely on the limiting normal distribution, which does however not guarantee sufficient accuracy, if we have a sample of small to moderate size, or if we are interested in small tail probabilities. If a higher accuracy is desired, the conditional saddlepoint approximation can be applied with the following conditional representation of the m-step spacings. If $Y_1, \ldots, Y_M$ are independent Gamma(m, b) random variables with density $\{b^m/\Gamma(m)\}y^{m-1}e^{-by}$, $y \geq 0$, then, under H_0 and $\forall b > 0$,

$$(MD_{1\cdot m}^{(m)}, \ldots, MD_{M\cdot m}^{(m)}) \sim \left\{(Y_1, \ldots, Y_M) \mid \sum_{i=1}^{M} Y_i = M\right\}. \tag{15.7}$$

The equivalence in (15.7) is easy to justify; see, for example, Wilks (1962, Section 7.7). Thus $(D_{1\cdot m}^{(m)}, \ldots, D_{(M-1)\cdot m}^{(m)}) \sim$ Dirichlet$(m, \ldots, m; m)$, and these m-spacings admit the conditional Gamma representation (15.7). This conditional representation together with the computational steps given in Section 15.1 allow us to compute a saddlepoint approximation for the distribution of symmetric and asymmetric test statistics based on m-step spacings. The particular case $m = 1$ in (15.7) corresponds to the exponential representation of simple spacings, and using this, Gatto and Jammalamadaka (1999, Section 3.1) developed four examples with one-step spacing statistics: the Rao spacings test, the log spacings test, the Greenwood spacings test, and the locally most powerful spacings test given by $h_i(ND_i) = \Phi^{(-1)}(\frac{i}{N+1})ND_i$. Saddlepoint approximations were computed for these four examples with sample sizes as low as $N = 3$, and they showed a very high accuracy, even for small tail probabilities. By means of this new conditional representation, we provide some further examples for the case of higher-order spacings.

Example 15.2.1 (The Rao higher-order spacings test) In order to apply Steps 1 and 2 of the saddlepoint approximation in Section 15.1, we must determine the joint cumulant generating function of the score functions

$$\begin{aligned}
\psi_{1i}(x, t_1) &= \begin{cases} (1 - x - t_1), \text{ if } x \in [0, 1), \\ (x - 1 - t_1), \text{ if } x \in [1, \infty), \end{cases} \\
\psi_{2i}(x, t_2) &= x - t_2,
\end{aligned}$$

with $\psi_{ji} = \psi_j$, $i = 1, \ldots, n$, $j = 1, 2$. With some algebraic computations, we can see that, for $b = m$ and $t_2 = 1$, this cumulant generating function has the form

$$\begin{aligned}
K_M((\lambda_1, \lambda_2), (t_1, 1)) &= M\Big[m \log m - m \log(m + \lambda_1 - \lambda_2) + \lambda_1(1 - t_1) - \lambda_2 \\
&\quad + \log\Big\{P(m, m + \lambda_1 - \lambda_2) + (\frac{m + \lambda_1 - \lambda_2}{m - \lambda_1 - \lambda_2})^m e^{-2\lambda_1} \\
&\quad [1 - P(m, m - \lambda_1 - \lambda_2)]\Big\}\Big],
\end{aligned}$$

where $P(m, x) = 1 - e^{-x} \sum_{i=1}^{m-1} x^j/j!$, $m = 1, 2, \ldots$, and $x \in \mathrm{R}$. The derivatives of $K_M((\lambda_1, \lambda_2), (t_1, 1))$ with respect to λ_1 and λ_2 can be obtained by automatic symbolic computation (e.g., with *Maple*). The advantage of choosing $b = m$ as scale parameter in the conditional Gamma representation is that the expectation of the sample mean of the Gamma random variables becomes one, and hence the "conditional saddlepoint equation," that is, the second equation in Step 1, has the trivial solution $\beta = 0$. Furthermore, $\beta = 0$ leads to $K''_{2M}(\beta, t) = M\mathrm{Var}(Y_1) = M/m$ and to $K_M((0, \beta), t) = 0$ in the formulas of s and r in Step 2.

Example 15.2.2 (The log higher-order spacings test) The choice of the score function $h(x) = \log x$ in (15.6) was proposed by Darling (1953) and it maximizes Bahadur efficiency; see Zhou and Jammalamadaka (1989). For the case $b = m$ and $t_2 = 1$, the joint cumulant generating function in (15.1) is given by

$$\begin{aligned} &K_M((\lambda_1, \lambda_2), (t_1, 1)) \\ &= M\Big[- \lambda_1 t_1 - \lambda_2 + m \log m - (\lambda_1 + m) \log(m - \lambda_2) + \log \frac{\Gamma(\lambda_1 + m)}{\Gamma(m)}\Big] \end{aligned}$$

provided that $\lambda_1 > -m$ and $\lambda_2 < m$. The second derivatives of $K_M((\lambda_1, \lambda_2), (t_1, 1))$ with respect to λ_1 and λ_2 are the following:

$$\partial^2 K_M((\lambda_1, \lambda_2), (t_1, 1))/(\partial\lambda_1)^2 = \Psi(1, \lambda_1 + m),$$

$$\partial^2 K_M((\lambda_1, \lambda_2), (t_1, 1))/(\partial\lambda_1 \partial\lambda_2) = (m - \lambda_2)^{-1},$$

and

$$\partial^2 K_M((\lambda_1, \lambda_2), (t_1, 1))/(\partial\lambda_2)^2 = \frac{\lambda_1 + m}{(m - \lambda_2)^2},$$

where $\Psi(z) = \Gamma'(z)/\Gamma(z)$ is the digamma function and $\Psi(z, n) = (d/dz)^n \Psi(z)$ is the polygamma function, with $\Re\{z\} > 0$ and $n \in \mathcal{N}$. The first derivatives are not necessary because the saddlepoint equation can be efficiently solved by a minimization routine such as *Matlab*'s routine `fminsearch`. In this example, we consider $N = 6$ and $m = 2$, yielding the very small number of summands or effective sample size $M = 3$. The numerical results are displayed in Figure 15.1 in terms of absolute errors $\mid \mathrm{P_{MC}} - \mathrm{P_{SP}} \mid$ and relative absolute error $\mid \mathrm{P_{MC}} - \mathrm{P_{SP}} \mid / \min\{\mathrm{P_{MC}}, 1 - \mathrm{P_{MC}}\}$, where $\mathrm{P_{MC}}$ and $\mathrm{P_{SP}}$ denote the distribution of the test statistic obtained by the 10^6 Monte Carlo simulated values of the test statistic and by the saddlepoint approximation in the Lugannani and Rice form in (15.2), or in its asymptotic equivalent version in (15.4), sometimes referred to as "Barndorff-Nielsen formula."

From Figure 15.1, we can see that the saddlepoint approximation has a small relative error over the whole domain of the distribution, and therefore is uniformly accurate. The Lugannani and Rice version in (15.2) has all relative errors below 10 %, and it appears substantially more accurate than its asymptotic equivalent formula in (15.4). For this test of uniformity, the small left tail probabilities are the most important. Note that the small increment of relative errors at both ends of the domains is not necessarily due to an inaccuracy of the saddlepoint approximation, because it is based on very few simulated values. (A further analysis based on importance sampling would provide a more reliable comparison.) The domain of the distribution is $(-\infty, 0)$ (all approximated distributions are almost zero at the left of -1), and the density function has a negative skewness.

Matlab programs for the computation of this saddlepoint approximation can be found at the address `http://www.stat.unibe.ch/~gatto`.

Example 15.2.3 (The Greenwood higher-order spacings test) The choice of the score function $h(x) = x^2$ in (15.6) defines the Greenwood test statistic. The joint cumulant generating function (15.1) for $b = m$ and $t_2 = 1$ is given by

$$\begin{aligned} K_M((\lambda_1, \lambda_2), (t_1, 1)) \quad = \quad & M \Bigg[m \log 2 + m \log m - \lambda_1 t_1 - \lambda_2 - \frac{(m-\lambda_2)^2}{4\lambda_1} \\ & -\frac{m}{2} \log(-\lambda_1) + (m-1)\log(m-\lambda_2) \\ & + \log \sum_{j=0}^{m-1} (-1)^{j+m-1} \left(\frac{2}{m-\lambda_2}\right)^j \frac{\Gamma(\frac{j+1}{2}, -\frac{(m-\lambda_2)^2}{4\lambda_1})}{\Gamma(j+1)\Gamma(m-j)} \Bigg] \end{aligned}$$

provided that $\lambda_1 < 0$ and $\lambda_2 < m$, and where $\Gamma(a, x) = \int_x^\infty e^{-t} t^{a-1} dt$ is the incomplete Gamma function.

15.3 Tests Based on Higher-Order Spacing-Frequencies

Consider a first sample of $(N-1)$ independent random variables $X_1, \ldots, X_{N-1}$, with underlying absolute continuous distribution F defined on $A \subset \mathrm{R}$, and a second sample of n independent random variables $Y_1, \ldots, Y_n$, with underlying absolute continuous distribution G, also defined on $A \subset \mathrm{R}$. The general two-sample problem is to test the null hypothesis H_0: $F = G$. Define the random

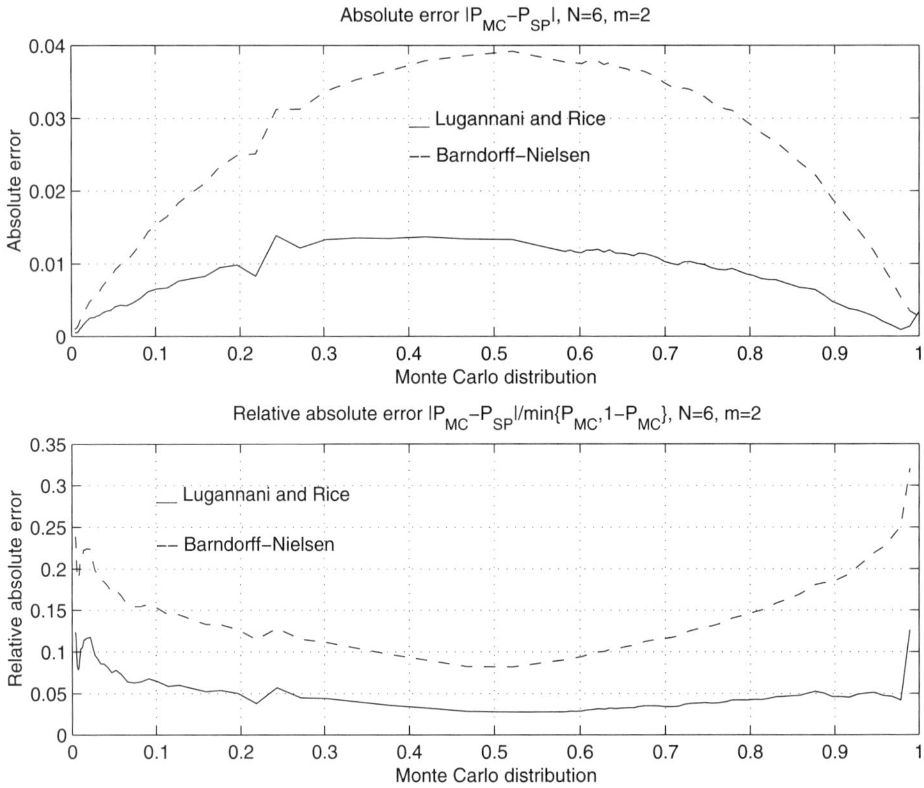

Figure 15.1: Saddlepoint and Monte Carlo approximations to the distribution of the log higher-order spacings statistic, $N = 6$, $m = 2$ and $M = 3$. Upper figure: absolute error $\mid P_{MC} - P_{SP} \mid$. Lower figure: relative absolute error $\mid P_{MC} - P_{SP} \mid / \min\{P_{MC}, 1 - P_{MC}\}$. P_{MC}: Monte Carlo approximation to the distribution. P_{SP}: saddlepoint approximations to the distribution. Solid line: Lugannani and Rice approximation in (15.2). Dashed line: Barndorff-Nielsen approximation in (15.4)

variables

$$S_j = \sum_{i=1}^{n} I\{Y_i \in [X_{(j-1)}, X_{(j)})\}, \;\; j = 1, \ldots, N,$$

where for convenience, we take $X_{(0)} \stackrel{\text{def}}{=} \inf\{A\}$ and $X_{(N)} \stackrel{\text{def}}{=} \sup\{A\}$. The numbers $\{S_1, \ldots, S_N\}$ are called the spacing frequencies becaise they correspond to the frequencies or counts of the $\{Y_i\}$ that fall in between successive $\{X_{(j)}\}$. In fact, if $R(X_{(k)})$ denotes the rank of the kth largest $\{X_j\}$ in the combined sample, $k = 1, \ldots, N$, it is easily seen that $R(X_{(k)}) = \sum_{j=1}^{k}(S_j + 1)$, or, $S_k = R(X_{(k)}) - R(X_{(k-1)}) - 1, \;\; k = 1, \ldots, N$, so that the $\{S_j\}$ are also the "rank differences."

Let $h(\cdot)$ and $h_j(\cdot)$, $j = 1, \ldots, N$, be real-valued functions satisfying certain regularity conditions. Holst and Rao (1980) consider statistics of the form $N^{-1/2} \sum_{j=1}^{N} h_j(S_j)$ and $N^{-1/2} \sum_{j=1}^{N} h(S_j)$ and their asymptotic properties when both N and n tend to infinity; formally, through nondecreasing sequences of positive integers $\{N_\nu\}$ and $\{n_\nu\}$ such that, as $\nu \to \infty$,

$$N_\nu \to \infty, \; n_\nu \to \infty \;\text{ and }\; \frac{N_\nu}{n_\nu} = \rho_\nu \to \rho, \; 0 < \rho < \infty.$$

Specifically, they show that if $V_1, \ldots, V_N$ are independent geometric random variables with probability distribution function

$$\mathrm{P}[V_1 = k] = \{\rho/(\rho + 1)\}^k \cdot 1/(\rho + 1), \;\; k = 0, 1, 2, \ldots, \tag{15.8}$$

then, under H_0,

$$\sum_{j=1}^{N} h_j(S_j) \xrightarrow{\mathcal{D}} \mathcal{N}(\mu, \sigma^{2}), \tag{15.9}$$

where $\mu = \mathrm{E}[\sum_{j=1}^{N} h_j(V_j)]$ and $\sigma^2 = \mathrm{Var}\left(\sum_{j=1}^{N} h_j(V_j) - \beta \sum_{i=j}^{N} V_j\right)$ in which β is the regression coefficient given by

$$\beta = \mathrm{Cov}\left(\sum_{j=1}^{N} h_j(V_j), \sum_{j=1}^{N} V_j\right) \Big/ \mathrm{Var}\left(\sum_{j=1}^{N} V_j\right).$$

As we stated already, the asymptotic efficiencies are improved by considering the corresponding higher-order spacings. Therefore, we now consider the more general case. For $m \geq 1$, denote $M = \lfloor N/m \rfloor$, and define the "nonoverlapping" or disjoint mth order spacing-frequencies

$$S_{k \cdot m}^{(m)} = \sum_{j=0}^{m-1} S_{k \cdot m + j} = \sum_{k=1}^{M} I\{Y_j \in [X_{(k \cdot m - 1)}, X_{(k \cdot m + m - 1)})\}, \;\; k = 1, \ldots, M - 1,$$

where we take $S_k^{(m)} = S_{k-M}^{(m)}$ for $k > M$ circularly, for convenience. Let $h(\cdot)$ and $h_j(\cdot)$, $j = 1, \ldots, N$, be real-valued functions satisfying certain regularity conditions [see Assumption (A), in Jammalamadaka and Schweitzer (1985)], and define the general classes of test statistics

$$T_\nu^* = \sum_{j=1}^{M} h_j(S_{j\cdot m}^{(m)}), \tag{15.10}$$

and

$$T_\nu = \sum_{j=1}^{M} h(S_{j\cdot m}^{(m)}), \tag{15.11}$$

which represent, respectively, the nonsymmetric and the symmetric test statistics based on such higher-order spacing frequencies. Jammalamadaka and Schweitzer (1985) discuss the asymptotic normality of such statistics (and indeed, more general ones based on the "overlapping" mth-order spacing frequencies) both under the null hypothesis, as well as under close alternatives.

The following optimality result has been proved there; see Theorem 3.2 in Jammalamadaka and Schweitzer (1985) for further details. Consider $\{G_N\}$, a smooth sequence of distribution functions converging towards F, as $N \to \infty$. It turns out that the asymptotically most powerful test for the null hypothesis H_0 against the sequence of simple alternatives

$$\mathrm{A}_N : G = G_N$$

is to reject H_0 when

$$\sum_{j=1}^{M} l\left(\frac{j}{M+1}\right) S_{j\cdot m}^{(m)} > c, \tag{15.12}$$

where $l(\cdot)$ is the derivative of $L(u) = \lim_{N\to\infty} N^{\frac{1}{2}}[G_N(F^{(-1)}(u))-u], 0 \leq u \leq 1$. However, such linear combinations of higher-order spacing frequencies in $\{S_{j\cdot m}^{(m)}\}$ are equivalent to linear combinations in one-step spacing frequencies S_j, already discussed in Gatto and Jammalamadaka (1999, Section 4) and need no further elaboration.

However, among the class of symmetric tests, there is reason to consider higher-order spacing frequencies. It is shown there that the sum of squares, leading to the statistic

$$\sum_{j=1}^{M} (S_{j\cdot m}^{(m)})^2, \tag{15.13}$$

is the optimal choice among all such symmetric nonoverlapping statistics. When $m = 1$, this has been introduced by Dixon (1940) and has been shown to be locally most powerful by Holst and Rao (1980) among such tests based on one-step spacing-frequencies.

For the more general statistics based on the mth-order spacing frequencies, consider the independent random variables $\eta_1, \ldots, \eta_M$ with the same negative binomial distribution with parameters m and $\rho/(1+\rho)$, viz.,

$$\mathrm{P}[\eta_1 = j] = \binom{m+j-1}{j} \left(\frac{1}{1+\rho}\right)^j \left(\frac{\rho}{1+\rho}\right)^m, \quad j = 0, 1, \ldots . \tag{15.14}$$

A moment's reflection shows that these negative binomial random variables arise by taking sums of the independent geometric random variables m at a time, corresponding to one-step spacing frequencies. It can be verified that under H_0, the mth-order spacing frequencies have the same distribution as independent negative binomial random variables conditioned to sum up to n, that is, if $\eta_1, \ldots, \eta_M$ are i.i.d. with probability function (15.14), then $\forall p \in (0, 1)$, it can be checked that

$$\{S_1^{(m)}, \ldots, S_M^{(m)}\} \sim \{\eta_1, \ldots, \eta_M\} \mid \sum_{j=1}^{M} \eta_j = n.$$

To illustrate the power of our conditional approach through which accurate saddlepoint approximations can be obtained, we quote a simple result for symmetric statistics based on nonoverlapping mth-order spacing frequencies, which is a consequence of the results of Jammalamadaka and Schweitzer (1985, Theorem 4.2).

Proposition 15.3.1 *Under H_0, if $\eta \sim \eta_1$,*

$$M^{-1/2} \sum_{j=1}^{M} \{h(S_{j \cdot m}^{(m)}) - \mathrm{E}[h(\eta)]\} \xrightarrow{\mathcal{D}} \mathcal{N}(0, \sigma^2), \tag{15.15}$$

where

$$\sigma^2 = \mathrm{Var}(h(\eta)) - \frac{\rho^2}{1+\rho}(\mathrm{Cov}^2(h(\eta), \eta).$$

The same conditioning idea used for obtaining the first-order approximation in (15.15) can be exploited for the construction of our saddlepoint approximation. By defining

$$T_{1\nu}^* = \sum_{j=1}^{M} h_j(\eta_j), \quad T_{1\nu} = \frac{1}{M} \sum_{j=1}^{M} h(\eta_j) \text{ and } T_{2\nu} = \frac{1}{M} \sum_{j=1}^{M} \eta_j,$$

the conditional distributions of $(T_{1\nu}^* \mid T_{2\nu} = 1)$ and $(T_{1\nu} \mid T_{2\nu} = 1)$ can be approximated again by Steps 1 and 2 of Section 15.1 and with the result below. These approximations are also accurate approximations to the distributions of T_ν^* and T_ν in (15.10) and (15.11), respectively. The following results, which can be proved by direct verification from our general results, show how one can find saddlepoint approximations for statistics in (15.12) and (15.13). Numerical evaluations are somewhat straightforward and are omitted.

Proposition 15.3.2 *The joint cumulant generating function on (15.1) for the test statistic (15.12) is given by*

$$\begin{aligned} K_M((\lambda_1,\lambda_2),(t_1,t_2)) &= -\lambda_1 t_1 - M\lambda_2 t_2 + mM\log(1-p) \\ &\quad -m\sum_{j=1}^{M}\log\left[1-p\exp\left\{\lambda_1 l\left(\frac{j}{M+1}\right)+\lambda_2\right\}\right], \end{aligned}$$

where $0<p<1$ *and* $\lambda_1 l(j/(M+1))+\lambda_2<-\log p$*, for* $j=1,\dots,M$*.*

Proposition 15.3.3 *The joint cumulant generating function in (15.1) for the test statistic (15.13) is given by*

$$\begin{aligned} K_M((\lambda_1,\lambda_2),(t_1,t_2)) &= M\Big[-\lambda_1 t_1 - \lambda_2 t_2 + m\log(1-p) \\ &\quad -m\log\{1-pe^{\lambda_2}\}+\kappa(\lambda_1)\Big], \end{aligned}$$

where $\kappa(\lambda_1) = \log \mathrm{E}[e^{\lambda_1 J^2}]$*,* J *is a negative binomial random variable with parameters* m *and* $1-pe^{\lambda_2}$*,* $0<p<1$*,* $\lambda_1<0$ *and* $\lambda_2<-\log p$*.*

15.4 Conclusion

In this discussion, we develop accurate approximations valid for small to moderate sample sizes, for the distributions of statistics based on higher order spacings, and higher-order spacing frequencies, whose exact distributions are unavailable and asymptotics are quite inaccurate.

Acknowledgements. The research of R. Gatto was supported by the Swiss National Science Foundation.

References

1. Daniels, H. E. (1954). Saddlepoint approximations in statistics, *The Annals of Mathematical Statistics*, **25**, 631–650.

2. Darling, D. A. (1953). On a class of problems related to the random division of an interval, *The Annals of Mathematical Statistics*, **24**, 239–253.

3. Del Pino, G. E. (1979). On the asymptotic distributionof k-spacings with applications to goodness-of-fit tests, *The Annals of Statistics*, **7**, 1058–1065.

4. Dixon, W. J. (1940). A criterion for testing the hypothesis that two samples are from the same population, *Annals of Mathematical Statistics*, **11**, 199–204.

5. Gatto, R. (2001). Symbolic computation for approximating the distributions of some families of one and two-sample nonparametric test statistics, *Statistics and Computing*, **11**, 449–455.

6. Gatto, R., and Jammalamadaka, S. R. (1999). A conditional saddlepoint approximation for testing problems, *Journal of the American Statistical Association*, **94**, 533–541.

7. Holst, L., and Rao, J. S. (1980). Asymptotic theory for some families of two-sample nonparametric statistics, *Sankhyā, Series A*, **42**, 19–52.

8. Holst, L., and Rao, J. S. (1981). Asymptotic spacings theory with applications to the two-sample problem, *The Canadian Journal of Statistics*, **9**, 79–89.

9. Lugannani, R., and Rice, S. (1980). Saddle point approximation for the distribution of the sum of independent random variables, *Advances in Applied Probability*, **12**, 475–490.

10. Jammalamadaka, S. R., and Schweitzer, R. L. (1985). On tests for the two-sample problem based on higher order spacing-frequencies, In *Statistical Theory and Data Analysis* (Ed., K. Matusita), pp. 583–618, North-Holland, Amsterdam.

11. Pyke, R. (1965). Spacings, *The Journal of the Royal Statistical Society, Series B*, **27**, 395–449.

12. Rao, J. S. (1976). Some tests based on arc-lengths for the circle, *Sankhyā, Series B*, **4**, 329–338.

13. Rao, J. S., and Kuo, M. (1984). Asymptotic results on the Greenwood statistic and some of its generalizations, *The Journal of the Royal Statistical Society, Series B*, **46**, 228–237.

14. Rao, J. S., and Sethuraman, J. (1975). Weak convergence of empirical distribution functions of random variables subject to perturbations and scale factors, *The Annals of Statistics*, **3**, 299–313.

15. Sethuraman, J., and Rao, J. S. (1970). Pitmann efficiencies of tests based on spacings, In *Nonparametric Techniques in Statistical Inference* (Ed., M. L. Puri), Cambridge University Press, Cambridge.

16. Wilks, S. S. (1962). *Mathematical Statistics*, John Wiley & Sons, New York.

17. Zhou, X., and Jammalamadaka, S. R. (1989). Bahadur efficiencies of spacings test for goodness of fit, *Annals of the Institute of Statistical Mathematics*, **41**, 541–553.

16

Best Bounds on Expectations of L-Statistics from Bounded Samples

Tomasz Rychlik

Polish Academy of Sciences, Toruń, Poland

Abstract: We present two optimal bounds on the expectations of arbitrary L-statistics based on i.i.d. samples with a bounded support expressed in the support length units. One depends on the location of the population mean in the support interval, and the other is general. The results are explicitly described in the special cases of single-order statistics and their differences.

Keywords and phrases: Bounded variable, i.i.d. sample, order statistic, L-statistic, Moriguti inequality

16.1 Introduction

Assume that $X, X_1, \ldots, X_n$ are independent random variables identically distributed on a finite interval $[a, b]$. Let $F(x)$, $F^{-1}(x)$, and

$$\mathbb{E}X = \mu = \int_0^1 F^{-1}(x)\, dx \in (a, b) \tag{16.1}$$

denote the common distribution and quantile functions, and expectation value, respectively. Let $\widetilde{c} = (c_1, \ldots, c_n) \in \mathbb{R}^n$ be an arbitrarily chosen vector of coefficients of a linear combination $\sum_{i=1}^n c_i X_{i:n}$ of order statistics $X_{1:n}, \ldots, X_{n:n}$ based on the sample $X_1, \ldots, X_n$. In this paper, we present two sharp evaluations of $\mathbb{E} \sum_{i=1}^n c_i(X_{i:n} - \mu)/(b - a)$. The first one includes μ. Observe that if μ approaches either of the endpoints a and b, so do the expectations of all order statistics, and the whole expression tends to zero. It is clear that the bound depends, except for the coefficient vector $\widetilde{c}$, on the location of μ in the support interval. This is expressed in terms of the parameter

$$\alpha = \frac{b - \mu}{b - a} \in (0, 1), \tag{16.2}$$

which represents the relative distance of μ from the upper support point in the support length units. A general bound, independent of μ, is derived by maximizing the particular ones with respect to (16.2). We also present one- or two- or three-point distributions which attain the bounds. The results for the general L-statistics are presented in Section 16.2. Special cases of single-order statistics and their differences are studied in Section 16.3.

Optimal evaluations of $\mathbb{E}\sum_{i=1}^{n} c_i(X_{i:n}-\mu)$ in various scale units were presented in the literature. For comprehensive reviews, we refer the reader to Arnold and Balakrishnan (1989, Chapter 3) and Rychlik (2001, Chapter 4). Bounds with the scale parameters $\sigma_p = (\mathbb{E}|X-\mu|^p)^{1/p}$, $1 \le p < \infty$, generated by the pth central absolute moments, including the most popular standard deviation parameter σ_2, were described in Rychlik (1998). Some results for specific L-statistics were known earlier; for instance, for the sample maximum [Hartley and David (1954), Gumbel (1954), Arnold (1985), and Balakrishnan (1993)], sample range [Plackett (1947) and Arnold (1985)], single-order statistics and their differences [Moriguti (1953)], and selection differentials [Nagaraja (1981)]. The dispersion measured by $\mathbb{E}(X-\mu|X>F^{-1}(\gamma))$ for some $\gamma \in (0,1)$ were considered in Balakrishnan and Rychlik (2005), and respective results for single-order statistics are due to Gajek and Okolewski (2000). Using variational methods, Hartley and David (1954) and Rustagi (1957) derived sharp bounds for the sample maximum and range, respectively, in σ_2 units for the i.i.d. samples with a finite support symmetric about the mean.

Similar results were established for other models of ordered statistical data, including the record values, progressively censored, and generalized order statistics. It is worth pointing out that our approach can be easily extended to all these models.

16.2 General Results

Given $\widetilde{c} = (c_1, \ldots, c_n)$, we make use of the following integral representation;

$$\mathbb{E}\sum_{i=1}^{n} c_i(X_{i:n}-\mu) = \int_0^1 [F^{-1}(x)-\mu] f_{\tilde{c}:n}(x)\,dx, \tag{16.3}$$

where

$$f_{\tilde{c}:n}(x) = \sum_{i=1}^{n} c_i f_{i:n}(x) \tag{16.4}$$

is the respective linear combination of density functions

$$f_{i:n}(x) = nB_{i-1,n-1}(x), \quad 1 \le i \le n, \tag{16.5}$$

of order statistics of the i.i.d. standard uniform samples of size n, and

$$B_{l,m}(x) = \binom{m}{l} x^l (1-x)^{m-l}, \quad 0 \leq l \leq m,$$

denote the standard Bernstein polynomials. Functions (16.4) and (16.5) have the antiderivatives

$$F_{\tilde{c}:n}(x) = \sum_{i=1}^{n} c_i F_{i:n}(x), \tag{16.6}$$

and

$$F_{i:n}(x) = \sum_{r=i}^{n} B_{r,n}(x),$$

respectively. Let $\underline{F}_{\tilde{c}:n}(x)$, $0 \leq x \leq 1$, denote the greatest convex minorant of (16.6). Because $F_{\tilde{c}:n}(x)$ is continuously differentiable, so is $\underline{F}_{\tilde{c}:n}(x)$. Denote the respective nondecreasing derivative by $\underline{f}_{\tilde{c}:n}(x)$. Rychlik (2001) showed that $\underline{f}_{\tilde{c}:n}(x)$ is the projection of $f_{\tilde{c}:n}(x)$ onto the family of nondecreasing functions in $L^2([0,1), dx)$. The set

$$\mathcal{A} = \{0 < x < 1 : \underline{F}_{\tilde{c}:n}(x) < F_{\tilde{c}:n}(x)\} \tag{16.7}$$

is open, and, if nonempty, consists of at most countably many disjoint open intervals. Assume that $\mathcal{A} = \bigcup_i (\alpha_i, \beta_i)$ for some $\alpha_i < \beta_i \leq \alpha_{i+1}$. Function $\underline{f}_{\tilde{c}:n}(x)$ is constant on each interval $[\alpha_i, \beta_i]$. On the completion of (16.7), $\underline{f}_{\tilde{c}:n}(x) = f_{\tilde{c}:n}(x)$, and is strictly increasing there. Theorem 1 of Moriguti (1953) implies that

$$\int_0^1 [F^{-1}(x) - \mu] f_{\tilde{c}:n}(x)\, dx \leq \int_0^1 [F^{-1}(x) - \mu] \underline{f}_{\tilde{c}:n}(x)\, dx \tag{16.8}$$

and the equality holds iff

$$\forall i, \quad F^{-1}(x) = \text{const.}, \quad \alpha_i < x < \beta_i. \tag{16.9}$$

The Moriguti inequality is the basic tool for establishing sharp bounds on functionals of ordered statistical data. Most of the results mentioned in Section 16.1 were derived by combining the Moriguti inequality with other ones, including the Schwarz, Hölder, and Steffensen inequalities. Relation (16.8) is also crucial in our study.

Theorem 16.2.1 *Under the above assumptions and notation,*

$$\mathbb{E} \sum_{i=1}^{n} c_i \left(\frac{X_{i:n} - \mu}{b-a} \right) \leq \frac{b-\mu}{b-a} \sum_{i=1}^{n} c_i - \underline{F}_{\tilde{c}:n} \left(\frac{b-\mu}{b-a} \right). \tag{16.10}$$

If $\mathcal{A} = (0,1)$*, then the equality holds if*

$$\mathbb{P}(X = \mu) = 1. \tag{16.11}$$

If $\alpha = \frac{b-\mu}{b-a} \notin \mathcal{A} \neq (0,1)$*, then* (16.10) *becomes the equality if*

$$\begin{aligned} \mathbb{P}(X = a) = \quad & \alpha & = \frac{b-\mu}{b-a}, \\ \mathbb{P}(X = b) = \quad & 1-\alpha & = \frac{\mu-a}{b-a}. \end{aligned} \tag{16.12}$$

If $\alpha = \frac{b-\mu}{b-a} \in (\alpha_i, \beta_i) \subset \mathcal{A} \neq (0,1)$*, then the equality in* (16.10) *is attained if*

$$\begin{aligned} \mathbb{P}(X = a) \quad &= \quad \alpha_i, \\ \mathbb{P}\left(X = \frac{\mu - \alpha_1 a - (1-\beta_1)b}{\beta_i - \alpha_i}\right) \quad &= \quad \beta_1 - \alpha_1, \\ \mathbb{P}(X = b) \quad &= \quad 1 - \beta_i. \end{aligned} \tag{16.13}$$

PROOF. We first prove the inequality. Combining (16.1), (16.3), and (16.8), we obtain

$$\begin{aligned} \mathbb{E}\sum_{i=1}^{n} c_i(X_{i:n} - \mu) \quad \leq \quad & \int_0^{\alpha} [F^{-1}(x) - \mu][\underline{f}_{\tilde{c}:n}(x) - \underline{f}_{\tilde{c}:n}(\alpha)]\,dx \\ + \quad & \int_{\alpha}^{1} [F^{-1}(x) - \mu][\underline{f}_{\tilde{c}:n}(x) - \underline{f}_{\tilde{c}:n}(\alpha)]\,dx. \end{aligned} \tag{16.14}$$

Because $\underline{f}_{\tilde{c}:n}(x)$ is nondecreasing, the latter functions in the integrals are nonpositive and nonnegative, respectively. Function $F^{-1}(x) - \mu$ may range between $a - \mu < 0$ and $b - \mu > 0$. Therefore

$$\begin{aligned} & \int_0^1 [F^{-1}(x) - \mu][\underline{f}_{\tilde{c}:n}(x) - \underline{f}_{\tilde{c}:n}(\alpha)]\,dx \\ \leq \quad & (a-\mu)\int_0^{\alpha} [\underline{f}_{\tilde{c}:n}(x) - \underline{f}_{\tilde{c}:n}(\alpha)]\,dx + (b-\mu)\int_{\alpha}^{1} [\underline{f}_{\tilde{c}:n}(x) - \underline{f}_{\tilde{c}:n}(\alpha)]\,dx \\ = \quad & (b-\mu)\underline{F}_{\tilde{c}:n}(1) - (b-a)\underline{F}_{\tilde{c}:n}(\alpha) = \left[\alpha\sum_{i=1}^{n} c_i - \underline{F}_{\tilde{c}:n}(\alpha)\right](b-a). \end{aligned} \tag{16.15}$$

The last equality follows from

$$\underline{F}_{\tilde{c}:n}(1) = F_{\tilde{c}:n}(1) = \sum_{i=1}^{n} c_i.$$

If $\mathcal{A} = (0,1)$, then $\underline{f}_{\tilde{c}:n}(x)$ is constant on $[0,1]$, and all the integrals in (16.15) amount to zero. The equality in (16.14) holds if $F^{-1}(x) - \mu$ is constant and zero on the whole unit interval, which yields (16.11).

If $\alpha \notin \mathcal{A} \neq (01,)$, it suffices to take

$$F^{-1}(x) = \begin{cases} a, & 0 \le x < \alpha, \\ b, & \alpha \le x < 1, \end{cases} \tag{16.16}$$

which evidently provides the equality in (16.15). The equality in (16.14) is implied by the fact that the only increase point of (16.16) lies beyond (16.7) [cf. (16.9)]. Finally, we observe that it obeys the moment condition

$$\int_0^1 F^{-1}(x)\,dx = \alpha a + (1-\alpha)b = \mu,$$

by definition.

Assume now that $\alpha \in (a_i, b_i) \subset \mathcal{A} \neq (0,1)$, or, equivalently,

$$\mu \in (\beta_i a + (1-\beta_i)b, \alpha_i a + (1-\alpha_i)b). \tag{16.17}$$

The equality in (16.15) holds if

$$F^{-1}(x) = \begin{cases} a, & 0 \le x < \alpha_i, \\ b, & \beta_i \le x < 1. \end{cases} \tag{16.18}$$

For the equality in (16.14), we also need

$$F^{-1}(x) = c, \quad \alpha_i < x < \beta_i. \tag{16.19}$$

The first moment condition is satisfied for

$$c = c(\mu) = \frac{\mu - \alpha_i a - (1-\beta_i)b}{\beta_i - \alpha_i}. \tag{16.20}$$

We immediately check that (16.20) ranges between a and b for μ restricted to (16.17). Therefore, relations (16.18)–(16.20) define a three-point distribution that satisfies the support condition. This completes the proof. ∎

The right-hand side of (16.10) represents the distance between the convex function $\underline{F}_{\tilde{c}:n}(x)$ and the straight line joining its endpoints $\underline{F}_{\tilde{c}:n}(0) = 0$ and $\underline{F}_{\tilde{c}:n}(1) = \sum_{i=1}^n c_i$ at the point $\alpha = \frac{b-\mu}{b-a}$. Hence, it is always non-negative. This vanishes for all $\alpha \in (0,1)$ iff

$$F_{\tilde{c}:n}(x) \ge x \sum_{i=1}^n c_i, \quad 0 \le x \le 1. \tag{16.21}$$

Otherwise, the bound is always positive except for the endpoints. For the majority of L-statistics, the equality conditions presented in Theorem 16.2.1 are necessary and sufficient. Other extreme distributions may occur when $\beta_i = \alpha_{i+1}$ for some i, and $\alpha \in (\alpha_i, \beta_{i+1})$. These are rare cases, though, and we decided not to present all the solutions to the equality problem. Those presented in Theorem 16.2.1 are the simplest ones. Note that in the last case, (16.13) reduces to a two-point distribution if either $\alpha_i = 0$ or $\beta_i = 1$.

A general bound independent of $\mu \in (a,b)$ is derived by maximizing the RHS of (16.10)

$$R_{\tilde{c}}(\alpha) = \alpha \sum_{i=1}^{n} c_i - \underline{F}_{\tilde{c}:n}(\alpha), \quad 0 < \alpha < 1. \tag{16.22}$$

This is a non-negative concave function vanishing at 0 and 1, with the continuous nonincreasing derivative

$$R'_{\tilde{c}}(\alpha) = \sum_{i=1}^{n} c_i - \underline{f}_{\tilde{c}:n}(\alpha). \tag{16.23}$$

Function (16.22) is maximized at any zero of (16.23). Under (16.21), both (16.22) and (16.23) are constant zero. Otherwise, the set of zeros of (16.23) is a possibly degenerate closed interval contained in $(0,1)$. The zero is unique iff it is an interior point of the closed set $[0,1] \setminus \mathcal{A}$. A nondegenerate interval of zeros coincides with an element of at most countable set of closed intervals that form the closure of (16.7). The results are summarized in Theorem 16.2.2.

Theorem 16.2.2 *Put*

$$\mathcal{A}_* = \left\{ 0 < \alpha < 1 : \ \underline{f}_{\tilde{c}:n}(\alpha) = \sum_{i=1}^{n} c_i \right\}. \tag{16.24}$$

Then for every $\alpha_* \in \mathcal{A}_*$, we have

$$\mathbb{E} \sum_{i=1}^{n} c_i \left(\frac{X_{i:n} - \mu}{b-a} \right) \le \alpha_* \sum_{i=1}^{n} c_i - \underline{F}_{\tilde{c}:n}(\alpha_*). \tag{16.25}$$

If $\mathcal{A}_* = (0,1)$, then the RHS of (16.25) is zero, and the equality is attained for the degenerate distributions concentrated at any $\mu \in (a,b)$.

If (16.24) consists of a single point α_*, then the equality in (16.25) holds for (16.12) with $\alpha = \alpha_*$.

If (16.24) is a proper interval and $\alpha_* \in [\alpha_i, \beta_i]$, then the equality in (16.25) is attained by (16.13).

16.3 Special Cases

We first determine the upper bounds for the single-order statistics. Every distribution function $F_{j:n}$ is convex on $(0, \frac{j-1}{n-1})$ and concave on $(\frac{j-1}{n-1}, 1)$. Using standard arguments, we calculate

$$\underline{F}_{1:n}(x) = x, \qquad 0 \le x \le 1,$$

$$\underline{F}_{j:n}(x) = \begin{cases} F_{j:n}(x), & 0 \le x \le \alpha_1, \\ f_{j:n}(\alpha_1)(x-1)+1, & \alpha_1 \le x \le 1, \end{cases} \quad 2 \le j \le n-1, \tag{16.26}$$

where $0 < \alpha_1 < \frac{j-1}{n-1}$ is the unique solution to

$$f_{j:n}(x)(1-x) = 1 - F_{j:n}(x),$$

and

$$\underline{F}_{n:n}(x) = F_{n:n}(x) = x^n, \quad 0 \le x \le 1.$$

Consequently, for the sample minimum, bounds (16.10) and (16.25) amount to zero, which is clear by $\mathbb{E}X_{1:n} \le \mathbb{E}X = \mu$. For $2 \le j \le n-1$, we have two cases. If $a < \mu < \alpha_1 a + (1-\alpha_1)b$, then $\alpha > \alpha_1$, and, applying (16.26), we get

$$\mathbb{E}\left(\frac{X_{j:n}-\mu}{b-a}\right) \le \alpha - f_{j:n}(\alpha_1)(\alpha-1) - 1 = \{f_{j:n}(\alpha_1)-1\}\left(\frac{\mu-a}{b-a}\right). \tag{16.27}$$

The bound is positive, because $\underline{f}_{j:n}(x)$ is a density function on $(0,1)$, and $f_{j:n}(\alpha_1)$ is its maximal value. Here $\alpha \in (\alpha_1, 1) = \mathcal{A}$, and due to (16.13), bound in (16.27) is attained if

$$\begin{aligned} \mathbb{P}(X=a) &= \alpha_1, \\ \mathbb{P}\left(X = \frac{\mu-\alpha_1 a}{1-\alpha_1}\right) &= 1-\alpha_1. \end{aligned}$$

If μ is large enough, that means $\alpha_1 a + (1-\alpha_1)b \le \mu < b$, then $\alpha \le \alpha_1$, and

$$\mathbb{E}\left(\frac{X_{j:n}-\mu}{b-a}\right) \le \frac{b-\mu}{b-a} - F_{j:n}\left(\frac{b-\mu}{b-a}\right). \tag{16.28}$$

Now $\alpha \notin \mathcal{A} = (\alpha_1, 1)$, and we conclude that the equality in (16.28) holds under (16.12). Calculating the general bound (16.25) for nonextreme order statistics, we solve the equation $\underline{f}_{j:n}(x) = 1$. By (16.26), $\underline{f}_{j:n}(x) = f_{j:n}(\min\{x, \alpha_1\})$ is strictly increasing on $(0, \alpha_1)$ from $f_{j:n}(0) = 0$ to $f_{j:n}(\alpha_1) > 1$. Therefore,

$$\mathbb{E}\left(\frac{X_{j:n}-\mu}{b-a}\right) \le \alpha_* - F_{j:n}(\alpha_*),$$

where $0 < \alpha_* < \alpha < \frac{j-1}{n-1}$ is the unique solution to $f_{j:n}(x) = 1$.

In the case of sample maximum, (16.10) takes on the form

$$\mathbb{E}\left(\frac{X_{n:n} - \mu}{b - a}\right) \le \frac{b-\mu}{b-a} - \left(\frac{b-\mu}{b-a}\right)^n,$$

and becomes the equality if (16.12) holds. The bound is maximized with respect to $\alpha = \frac{b-\mu}{b-a}$ at $\alpha_* = n^{-1/(n-1)}$. This implies the general bound

$$\mathbb{E}\left(\frac{X_{n:n} - \mu}{b - a}\right) \le n^{-1/(n-1)} - n^{-n/(n-1)}, \qquad (16.29)$$

with the equality conditions

$$\begin{aligned} \mathbb{P}(X = a) &= n^{-1/(n-1)}, \\ \mathbb{P}(X = b) &= 1 - n^{-1/(n-1)}. \end{aligned}$$

Observe that the right-hand side of (16.29) tends to 1 as $n \to \infty$, which is a trivial deterministic bound for $\frac{X_{n:n}-\mu}{b-a}$.

Now we proceed to the differences of order statistics $X_{k:n} - X_{j:n}$, $1 \le j < k \le n$. We easily check that function

$$F_{j,k:n}(x) = F_{k:n}(x) - F_{j:n}(x), \quad 0 \le x \le 1,$$

vanishes at 0 and 1, and is first concave decreasing (except for the case $j = 1$), then convex decreasing, convex increasing, and ultimately concave increasing (except for $k = n$). It attains its minimum at

$$\alpha_* = \left\{\left[\binom{n-1}{k-1} \Big/ \binom{n-1}{j-1}\right]^{1/(k-j)} + 1\right\}^{-1}. \qquad (16.30)$$

Therefore,

$$\underline{F}_{j,k:n}(x) = \begin{cases} [f_{k:n}(\beta_1) - f_{j:n}(\beta_1)]x, & 0 \le x \le \beta_1, \\ F_{k:n}(x) - F_{j:n}(x), & \beta_1 \le x \le \alpha_2, \\ [f_{k:n}(\alpha_2) - f_{j:n}(\alpha_2)]\,(x-1), & \alpha_2 \le x \le 1, \end{cases} \qquad (16.31)$$

and $\mathcal{A} = (0, \beta_1) \cup (\alpha_2, 1)$, where $0 < \beta_1 < \alpha_*$ uniquely solves

$$[f_{k:n}(x) - f_{j:n}(x)]x = F_{k:n}(x) - F_{j:n}(x)$$

when $j \ge 2$ and $\beta_1 = 0$ for $j = 1$, and $\alpha_* < \alpha_2 < 1$ uniquely solves

$$[f_{k:n}(x) - f_{j:n}(x)](1 - x) = F_{j:n}(x) - F_{k:n}(x)$$

when $k \leq n-1$ and $\alpha_2 = 1$ for $k = n$. This implies that (16.10) has three different forms. If $a < \mu < \alpha_2 a + (1-\alpha_2)b$, then $\alpha > \alpha_2$, and by (16.10) and (16.31),

$$\mathbb{E}\left(\frac{X_{k:n} - X_{j:n}}{b-a}\right) \leq \{f_{k:n}(\alpha_2) - f_{j:n}(\alpha_2)\}\left(\frac{\mu - a}{b-a}\right),$$

and the equality holds if

$$\begin{aligned} \mathbb{P}(X = a) &= \alpha_2, \\ \mathbb{P}\left(X = \frac{\mu - \alpha_2 a}{1-\alpha_2}\right) &= 1 - \alpha_2. \end{aligned}$$

If $\alpha_2 a + (1-\alpha_2)b \leq \mu \leq \beta_1 a + (1-\beta_1)b$, then $\beta_1 \leq \alpha \leq \alpha_2$, and

$$\mathbb{E}\left(\frac{X_{k:n} - X_{j:n}}{b-a}\right) \leq F_{j:n}\left(\frac{b-\mu}{b-a}\right) - F_{k:n}\left(\frac{b-\mu}{b-a}\right) = \sum_{i=j}^{k-1} B_{i,n}\left(\frac{b-\mu}{b-a}\right), \tag{16.32}$$

and the equality holds for (16.12). Eventually, for $\beta_1 a + (1-\beta_1)b < \mu < b$, we have $\alpha < \beta_1$ so that the inequality

$$\mathbb{E}\left(\frac{X_{k:n} - X_{j:n}}{b-a}\right) \leq \{f_{j:n}(\beta_1) - f_{k:n}(\beta_1)\}\left(\frac{b-\mu}{b-a}\right)$$

holds, and the equality conditions are

$$\begin{aligned} \mathbb{P}\left(X = \frac{\mu - (1-\beta_1)b}{\beta_1}\right) &= \beta_1, \\ \mathbb{P}(X = b) &= 1 - \beta_1. \end{aligned}$$

In order to derive the general bound, we solve $\underline{F}'_{j,k:n}(x) = 0$. An analysis carried out above shows that (16.30) is the unique solution that belongs to (β_1, α_2). By (16.32),

$$\mathbb{E}\left(\frac{X_{k:n} - X_{j:n}}{b-a}\right) \leq F_{k:n}(\alpha_*) - F_{j:n}(\alpha_*) = \sum_{i=j}^{k-1} B_{i,n}(\alpha_*).$$

Especially, for spacings we have $\alpha_* = \frac{j}{n}$, and

$$\mathbb{E}\left(\frac{X_{j+1:n} - X_{j:n}}{b-a}\right) \leq B_{j,n}\left(\frac{j}{n}\right). \tag{16.33}$$

By the Stirling approximation, the bound tends to $1/\sqrt{2\pi j}$ if j is fixed, and $1/\sqrt{2\pi(n-j)}$ if $n-j$ is fixed, and n tends to infinity. For the spacings of intermediate and central-order statistics, the right-hand side of (16.33) tends to zero in increasing samples. The maximum point (16.30) has a simple form

$\alpha_* = \frac{1}{2}$ for the symmetric differences of the jth greatest and smallest order statistics. Therefore,

$$\mathbb{E}\left(\frac{X_{n+1-j:n} - X_{j:n}}{b-a}\right) \le \sum_{i=j}^{n-j} B_{j,n}\left(\frac{1}{2}\right), \quad 1 \le j \le \frac{n}{2}.$$

Moreover, it is easy to verify that for the sample range formulae (16.10) and (16.25) take on the forms

$$\mathbb{E}\left(\frac{X_{n:n} - X_{1:n}}{b-a}\right) \le 1 - \left(\frac{\mu - a}{b-a}\right)^n - \left(\frac{b-\mu}{b-a}\right)^n,$$

and

$$\mathbb{E}\left(\frac{X_{n:n} - X_{1:n}}{b-a}\right) \le 1 - \frac{1}{2^{n-1}},$$

respectively.

References

1. Arnold, B.C. (1985). p-Norm bounds on the expectation of the maximum of possibly dependent sample, *Journal of Multivariate Analysis*, **17**, 316–332.

2. Arnold, B. C., and Balakrishnan, N. (1989). *Relations, Bounds, and Approximations for Order Statistics*, Lecture Notes in Statistics, Vol. **53**, Springer-Verlag, New York.

3. Balakrishnan, N. (1993). A simple application of binomial-negative binomial relationship in the derivation of sharp bounds for moments of order statistics based on greatest convex minorants, *Statistics & Probability Letters*, **18**, 301–305.

4. Balakrishnan, N. and Rychlik, T. (2006). Evaluating expectations of L-statistics by the Steffensen inequality, *Metrika* (to appear).

5. Gajek, L., and Okolewski A. (2000). Sharp bounds on moments of generalized order statistics, *Metrika*, **52**, 27–43.

6. Gumbel, E. J. (1954). The maxima of the mean largest value and of the range, *Annals of Mathematical Statistics*, **25**, 76–84.

7. Hartley, H. O., and David, H. A. (1954). Universal bounds for mean range and extreme observation, *Annals of Mathematical Statistics*, **25**, 85–99.

8. Moriguti, S. (1953). A modification of Schwarz's inequality with applications to distributions, *Annals of Mathematical Statistics*, **24**, 107–113.

9. Nagaraja, H. N. (1981). Some finite sample results for the selection differential, *Annals of the Institute of Statistical Mathematics*, **33**, 437–448.

10. Plackett, R. L. (1947). Limits of the ratio of mean range to standard deviation, *Biometrika*, **34**, 120–122.

11. Rustagi, J. S. (1957). On minimizing and maximizing a certain integral with statistical applications, *Annals of Mathematical Statistics*, **28**, 309–328.

12. Rychlik, T. (1998). Bounds on expectations of L-estimates, In *Order Statistics: Theory & Methods* (Eds., N. Balakrishnan and C. R. Rao), Handbook of Statistics, Vol. **16**, pp. 105–145, North-Holland, Amsterdam.

13. Rychlik, T. (2001). *Projecting Statistical Functionals*, Lecture Notes in Statistics, Vol. **160**, Springer-Verlag, New York.

PART IV
Reliability and Applications

17

The Failure Rates of Mixtures

Henry W. Block

University of Pittsburgh, Pittsburgh, PA, USA

Abstract: Mixtures of distributions of lifetimes occur in many settings. In engineering applications, it is often the case that populations are heterogeneous, often with a small number of subpopulations. In survival analysis, selection effects can often occur. The concept of a failure rate in these settings becomes a complicated topic, especially when one attempts to interpret the shape as a function of time. Even if the failure rates of the subpopulations of the mixture have simple geometric or parametric forms, the shape of the mixture is often not transparent.

Recent results, developed by the author (with Joe, Li, Mi, Savits, and Wondmagegnehu) in a series of papers, are presented. These results focus on general results concerning the asymptotic limit and eventual monotonicity of a mixture, and also the overall behavior for mixtures of specific parametric families.

An overall picture is given of different things that influence the behavior of the failure rate of a mixture.

Keywords and phrases: Failure rate, mixture, coherent systems, signature

17.1 Introduction

Mixtures are a common topic in most areas of statistics. They also play a central role in reliability and survival analysis. However, the failure rate of mixed distributions is a source of much confusion. Many questions and anomalies have arisen. We discuss some of these in the following.

A much cited paper is that of Proschan (1963). In this paper, pooled data for airplane air conditioning systems whose lifetimes are known to be exponential exhibit a decreasing failure rate. Because decreasing failure rates are usually associated with systems that improve with age, this was initially thought to be counterintuitive.

A second anomaly, at least to some, was that mixtures of lifetimes with increasing failure rates could be decreasing on certain intervals. Examples of such lifetimes can be found in Vaupel and Yashin (1985) as well as in Barlow and Proschan (1975).

A variant of the above is due to Gurland and Sethuraman (1994, 1995), which gives examples of mixtures of very rapidly increasing failure rates that are eventually decreasing.

In the survival analysis literature [see, e.g., Bretagnolle and Huber-Carol (1988), and other papers cited therein], it is known that if an important random covariate in a Cox model is omitted, the shape of the hazard rate is drastically changed.

A recent paper by Wang, Muller and Capra (1998) (and many articles cited there) mentions that in many biological populations, including humans, lifetimes of organisms at extreme old age exhibit decreasing hazard rate. A natural question to ask is whether this means that some of the individuals in the population are improving or not.

This lack of understanding and general confusion about mixtures is one of the reasons that we attempt to explain the behavior of mixtures. We have been successful in describing the initial and final behaviors. Progress is being made on the intermediate behavior, but a general pattern has not emerged. It is indeed a very challenging problem.

17.2 Notation

In industrial settings, populations of components are rarely homogeneous. There are usually at least two subpopulations. For the purposes of this paper, we consider the case of two. The situation can be described using mixtures of distributions.

Consider two component lifetimes with survival functions $\overline{F}_1$ and $\overline{F}_2$, densities f_1 and f_2, and failure rates λ_1 and λ_2. The mixture has survival function

$$\overline{F}_m(t) = p\overline{F}_1(t) + (1-p)\overline{F}_2(t)$$

where $0 < p < 1$, density

$$f_m(t) = pf_1(t) + (1-p)f_2(t),$$

and failure rate

$$\lambda_m(t) = \frac{pf_1(t) + (1-p)f_2(t)}{p\overline{F}_1(t) + (1-p)\overline{F}_2(t)},$$

which can be rewritten as

$$\lambda_m(t) = p_1(t)\lambda_1(t) + (1 - p_1(t))\lambda_2(t),$$

where

$$p_1(t) = \frac{p\overline{F}_1(t)}{p\overline{F}_1(t) + (1-p)F_2(t)}.$$

Although λ_m is not a simple mixture of λ_1 and λ_2, it nonetheless follows that for every $t \geq 0$, $\lambda_m(t)$ is bounded by $\min(\lambda_1(t), \lambda_2(t))$ and $\max(\lambda_1(t), \lambda_2(t))$.

17.3 Examples

The failure rates of standard distributions in reliability are often monotone (i.e., increasing or decreasing). One result that is known is that mixtures of distributions with decreasing failure rates have decreasing failure rates; see Proschan (1963).

However, if λ_1 and λ_2 are not both decreasing, not much is known about the monotonicity of λ_m. We give examples to illustrate various behaviors.

Example 17.3.1 (IFR Weibulls) We consider two Weibull distributions with increasing failure rates $\lambda_1(t) = 2t$ and $\lambda_2(t) = 3t^2$ and any $0 < p < 1$. It turns out that the mixture of these two distributions has increasing failure rate.

This behavior, however, is not typical as the following example shows.

Example 17.3.2 (IFR with exponential failure rates) Let $\lambda_1(t) = 1 - \exp(-5t)$ and $\lambda_2(t) = 6 - \exp(-5t)$ and any $0 < p < 1$. Here, the mixture has strictly decreasing failure rate.

Gurland and Sethuraman (1994, 1995) believed the behavior of Example 17.3.2 was typical, that is, many mixtures have eventually decreasing failure rates. The following example is similar to examples that these authors studied.

Example 17.3.3 (Exponential and gamma distributions) Consider distributions with densities $f_1(t) = \exp(-t)$ and $f_2(t) = 16t\exp(-4t)$ and any $0 < p < 1$. The failure rate of the mixture starts off increasing and then is eventually decreasing. The shape of this failure rate is called *upside-down bathtub* or *hump-shaped.* Two standard distributions that have a failure rate with a similar shape are the log-normal and the log-logistic.

This behavior is also not typical, as the following example shows.

Example 17.3.4 (Exponential and gamma distributions) Consider densities $f_1(t) = 4\exp(-4t)$ and $f_2(t) = t\exp(-t)$ and any $0 < p < 1$. The failure rate is eventually increasing. This shape is called *bathtub* (BT). Various populations exhibit such behavior [see Klein and Moeschberger (1997, p. 29)], and for industrial populations of this type burn-in is important [see Block and Savits (1997)].

Again, this behavior is not typical as the following example shows.

Example 17.3.5 (IFR Weibulls) Consider distributions with failure rates $\lambda_1(t) = 2t$ and $\lambda_2(t) = 4t^3$ and any $0 < p < 1$. The mixture has a failure rate that is increasing, then decreasing, then increasing. This shape is called *modified bathtub* (MBT). Various industrial lifetimes exhibit this behavior [see Jensen and Petersen (1982)] as do certain human populations [see Klein and Moeschberger (1997, Section 1.9)].

In general, therefore, a multitude of behaviors is possible.

17.4 Asymptotics

Notice in the previous examples that the limit of the mixture as $t \to \infty$ tends to the limit of the lower (i.e., the stronger) failure rate. Furthermore, in Examples 1, 4, and 5 the eventual monotonicity of the mixture is similar to the monotonicity of the stronger failure rate. We state two results. To do this we introduce a more general mixture that has a failure rate

$$\lambda_m(t) = \frac{\int_S f(t,\theta)\ P(d\theta)}{\int_S \overline{F}(t,\theta)\ P(d\theta)}$$

and let $r(t,\theta) = \frac{f(t,\theta)}{\overline{F}(t,\theta)}$.

Theorem 17.4.1 (Block, Mi, and Savits (1993)) *Assume*

(i) $r(t,\theta)$ converges to $a(\theta)$ uniformly on S as $t \to \infty$;

(ii) for any $r(t,\theta)$ that goes to ∞, the rate of convergence is exponentially bounded (i.e., $r(t,\theta) \le \exp(Lt)$ for large t and $L > 0$).

Then $\lambda_m(t)$ has (essentially) the same limit as the strongest $r(t,\theta)$ (i.e., $\lim_{t\to\infty} \lambda_m(t) = ess\ inf_{\theta \in S} a(\theta)$).

Theorem 17.4.2 (Block and Joe (1997)) *Consider the finite mixture*

$$\lambda_m(t) = \frac{\Sigma_{i=1}^n p_i f_i(t)}{\Sigma_{i=1}^n p_i \overline{F}_i(t)}, \quad \Sigma_{i=1}^n p_i = 1,\ 0 < p_i < 1.$$

Then under technical conditions on the first derivatives of $r_i(t) = \frac{f_i(t)}{\overline{F}_i(t)}$, $i = 1, 2, \ldots, n$ (essentially the $r_i'(t)$ behave like ratios of polynomials), the ultimate monotonicity of $\lambda_m(t)$ is the same as that of the strongest component.

Remark 17.4.1 Improved versions of these two theorems can be found in Block, Li, and Savits (2003b).

17.5 Mixtures of Distributions with Linear Failure Rates

Although the asymptotic behavior of the failure rate of a mixture has been studied and the initial behavior is not hard to determine, little is known about the intermediate behavior. Block, Savits, and Wondmagegnehu (2003) determined the overall monotone behavior of the failure rate of a mixture of distributions with a linear failure rate. For two components, the mixture is

$$\lambda_m(t) = \frac{pf_1(t) + (1-p)f_2(t)}{p\overline{F}_1(t) + (1-p)\overline{F}_2(t)}.$$

Consider the two distributions with linear failure rates

$$\lambda_1(t) = c_1 t + d_1 \text{ and } \lambda_2(t) = c_2 t + d_2$$

which are assumed to be increasing, i.e., $c_i > 0$, $i = 1, 2$. Four cases are considered:

(a) parallel rates ($c_1 = c = c_2, d_1 < d_2$);
(b) rates with the same y-intercept ($d_1 = d = d_2, c_1 < c_2$);
(c) noncrossing rates ($c_1 < c_2, d_1 < d_2$, contains a) and b));
(d) crossing rates ($c_1 < c_2, d_1 > d_2$).

The results are as follows.

(a) Parallel Failure Rates: Here $c_1 = c = c_2$ and $d_1 < d_2$. Let $a = d_2 - d_1$ and $\delta = \frac{a}{\sqrt{c}}$. There are two cases:

Case ($0 < \delta < 2$): The mixture failure rate is increasing (IFR).

Case ($2 < \delta$): There exists $\zeta_1 < \zeta_2$ and the following two subcases hold:

Subcase ($0 < p < \zeta_1$): Here, the mixture has a modified bathtub (MBT) failure rate.

Subcase ($\zeta_1 \leq p < \zeta_2$): Here, the failure rate is bathtub-shaped (BT).

Subcase ($\zeta_2 \leq p < 1$): IFR.

(b) Same y-intercept. There exists $0 < \xi < 1$ and there are two cases.

Case ($0 < p < \xi$): MBT.

Case ($\xi \leq p < 1$): IFR.

(c) Noncrossing (includes (a) an (b) above). There are three possibilities: IFR, BT, and MBT.

(d) Crossing

Case ($t < t_0$): Many cases, only three shapes (IFR, BT, MBT).

Case ($t_0 < t$): Many cases only two shapes (IFR, MBT).

These lead to a total of five possible monotonicity behaviors in the crossing cases. There will be at most four changes of monotonicity in these cases. See Block, Savits, and Wondmagegnehu (2003) for details.

17.6 Mixtures of Standard Reliability Distributions

Recently, attempts have been made to study the behavior of mixtures of standard reliability distributions. We summarize some of the recent work.

Weibull [Wondmagegnehu (2002)] Two Weibulls, same shape parameter $\alpha > 1$ with failure rates

$$\lambda_1(t) = \theta_1 \alpha t^{\alpha-1}, \quad \lambda_2(t) = \theta_2 \alpha t^{\alpha-1}.$$

For small p, the only behavior is MBT. For large p, the only behavior is IFR. Jiang and Murthy (1998) have also determined the above by computational methods as well as all other cases. These involve eight different shapes with from 0 to 4 changes of monotonicity.

Exponential and Weibull [Wondmagegnehu, Navarro, and Hernandez (2004)]

$$\lambda_1(t) = \theta_1, \quad \lambda_2(t) = \theta_2 \alpha t^{\alpha-1}$$

Case ($\alpha > 2$): Decreasing, increasing, then decreasing.
Case ($\alpha = 2$): For small p, UBT. For large p, as in the $\alpha > 2$ case.
Case ($1 < \alpha < 2$): Two behaviors are possible.

Gamma All cases for the gamma distributions have been determined by Gupta and Warren (2001) using theoretical, numerical, and graphical techniques. Six different shapes were encountered with from 0 to 4 changes of monotonicity.

Normal There is a long history of the shapes of the densities for mixtures of normal distributions. See Schilling, Watkins, and Watkins (2002) for a summary and also Robertson and Fryer (1969). Recent work on the failure rate appears in Block, Li, and Savits (2004). Failure rates of mixtures of truncated normals can be found in Navarro and Hernandez (2002).

17.7 Preservation Under Mixtures

From Sections 17.2 and 17.4, various examples demonstrate that finite mixtures of distributions with increasing failure rates need not have increasing failure rates. However, for continuous mixtures of distributions with increasing failure rates, there are conditions under which the mixture has an increasing failure rate. This was noted by Lynch (1999). The mixing distribution requires a strong joint property.

Theorem 17.7.1 (Lynch (1999)) *Let $\{\overline{F}(t|\theta)|\theta \geq 0\}$ be a family of survival functions that is logconcave in (t, θ) and an increasing in θ. Also, let M be a distribution with increasing failure rate. Then*

$$\overline{F}(t) = \int \overline{F}(t|\theta)\ dM(\theta)$$

has an increasing failure rate.

Remark 17.7.1 As shown in Block, Li, and Savits (2003a), the logconcave condition is not particularly restrictive. For example, the Weibull distribution with survival function

$$\overline{F}(t|\theta) = \exp(-t^{\alpha}/\theta^{\alpha-1}) \quad \text{for} \quad \theta > 0 \quad \text{and} \quad \alpha > 1$$

is increasing in θ and logconcave in (t, θ). This gives that mixing with respect to Weibulls with increasing failure rates preserves increasing failure rates.

Remark 17.7.2 Similar results for other reliability classes were shown in Block, Li, and Savits (2003a).

17.8 Analytic Tools for Determining the Shape of Mixtures

Puri and Singh (1986), Mi (1996), Block, Savits, and Singh (2002), and Savits (2003) all examine conditions under which:

$$\text{monotonicity of } \frac{N'(t)}{D'(t)} \quad \Rightarrow \quad \text{monotonicity of } \frac{N(t)}{D(t)}.$$

In particular, Puri and Singh (1986) examine functions with no change of monotonicity (i.e., either increasing or decreasing); Mi (1996) and Block, Savits, and Singh (2002) consider one change of monotonicity (i.e., bathtub functions); and Savits (2003) considers multiple changes of monotonicity (i.e., rollercoaster functions). The results obtained in these papers include the marginal cost results of Berg (1986) and Chen and Savits (1992), and also results on the monotonicity of the failure rate obtained by Glaser (1980) and Gupta and Warren (2001). In these latter cases,

$$\text{monotonicity of} \quad \eta(t) = -\frac{f'(t)}{f(t)} \quad \Rightarrow \quad \text{monotonicity of} \quad r(t) = \frac{f(t)}{\overline{F}(t)},$$

where $f(t)$ is the density. The reason this is important for mixtures is that $\eta(t)$ is often much easier to analyze than $r(t)$.

17.9 Coherent Systems

Using some of the techniques described previously, the monotonicity of the failure rate of a coherent system can be determined. In the case where the components are independent, Block, Li, and Savits (2003b) determined conditions for describing the asymptotic behavior of the failure rate of a system in terms of the asymptotic behavior of the failure rate of the components. The conditions involve the min path or the min cut sets of the system. If the components are also identically distributed, Samaniego (1985) showed that the reliability of the system has a representation as a mixture and the mixture coefficients are a probability vector called the *signature*. Recently, Block, Dugas, and Samaniego (2004) showed that the asymptotic failure rate of such a system can be determined by using the signature representation just mentioned. Min path and min cut sets do not appear in this result and this does not follow from the result of Block, Li, and Savits (2003b). We give both results. Results on the eventual monotonicity of the system failure rate follow similarly.

Theorem 17.9.1 (Block, Li, and Savits (2003b)) *Let $r(t)$ be the failure rate of a coherent system with independent components and min path sets $P_1, \ldots, P_p$. Let $r_i(t)$ for $i = 1, \ldots, n$ be the component failure rates and assume that these converge to finite limits $a_1, \ldots, a_n$ respectively. Let $b_i = \Sigma_{k \in P_i} a_i$ for $i = 1, \ldots, p$. If there is a unique smallest b_i, then $r(t)$ converges to this b_i.*

Theorem 17.9.2 (Block, Dugas, and Samaniego (2003)) *Let $r_T(t)$ be the failure rate of a coherent system with iid components each of which has a failure* *rate* $r(t)$ *with limit* r. *Let* $\mathbf{s} = (s_1, \ldots, s_n)$ *be the signature of the system and* $k* = \max\{i | s_i > 0\}$. *Then*

$$\lim_{t \to \infty} r_T(t) = (n - k * +1)r.$$

17.10 Summary of Overall Shape

We give some general rules of thumb concerning the asymptotic behavior of the failure rate when we mix several distributions. First, from one of our basic theorems, the failure rate of the mixture will approach the stronger (i.e., lowest) failure rate so that there is a downward trend. However, if the strongest failure rate is eventually increasing, the mixture will become increasing. If one of the mixture probabilities is close to one, the mixture failure rate will initially behave like that component. If the component with probability close to one becomes the strongest component, then the mixture will eventually behave like that component. If the failure rates cross, then the point of intersection is also a factor. The differences of the y-intercepts and the ratio of the slopes also play a role.

References

1. Barlow, R. E., and Proschan, F. (1975). *Statistical Theory of Reliability*, Holt, Rinehart and Winston, New York.

2. Berg, H., Bienvenu, M., and Cleroux, R. (1986). Age replacement policy with age-dependent minimal repair, *Informatics*, **24**, 26–32.

3. Block, H.W., Dugas, M., and Samaniego, F. J. (2004). Signature-related results on failure rates and lifetimes, submitted for publication.

4. Block, H., and Joe, H. (1997). Tail behavior of the failure rate functions of mixtures, *Lifetime Data Analysis* **3**, 269–288.

5. Block, H. W., Li, Y., and Savits, T. H. (2003a). Preservation of properties under mixtures, *Probability in Engineering and Information Sciences*, **17**, 205–212.

6. Block, H. W., Li, Y., and Savits, T. H. (2003b). Initial and final behavior of failure rate functions for mixtures and systems, *Journal of Applied Probability*, **40**, 721–740.

7. Block, H. W., Li, Y., and Savits, T. H. (2004). Mixtures of normal distributions: modality and failure rate, *Technical Report*, University of Pittsburgh, Pittsburgh, PA.

8. Block, H. W., Mi, J., and Savits, T. H. (1993). Burn–in and mixed populations, *Journal of Applied Probability*, **30**, 692–702.

9. Block, H. W., and Savits, T. H. (1997). Burn-in, *Statistical Science*, **12**, 1-19.

10. Block, H. W., Savits, T. H., and Singh, H. (2002). A criterion for burn-in which balances mean residual life and residual variance, *Operations Research*, **50**, 290–296.

11. Block, H. W., Savits, T. H., and Wondmagegnehu, E. (2003). Mixtures of distributions with linear failure rates, *Journal of Applied Probability*, **40**, 485–504.

12. Bretagnolle, J., and Huber-Carol, C. (1988). Effects of omitting covariates in Cox's model for survival data, *Scandinavian Journal of Statistics*, **15**, 125–138.

13. Chen, C. S., and Savits, T. H. (1992). Optimal age and block replacement for a general maintenance model, *Probability in Engineering and Information Sciences*, **6**, 81–98.

14. Glaser, R. E. (1980). Bathtub and related failure rate characterizations, *Journal of the American Statistical Association*, **75**, 667–672.

15. Gupta, R. C., and Warren, R. (2001). Determination of change points of non- monotonic failure rates, *Communications in Statistics—Theory and Methods*, **30**, 1903–1920.

16. Gurland, J., and Sethuraman, J. (1994). Reversal of increasing failure rates when pooling failure data, *Technometrics*, **36**, 416–418.

17. Gurland, J., and Sethuraman, J. (1995). How pooling failure data may reverse increasing failure rates, *Journal of the American Statistical Association*, **90**, 1416–1423.

18. Jensen, F., and Petersen, N.E. (1982). *Burn-in*, John Wiley & Sons, New York.

19. Jiang, R., and Murthy, D. N. P. (1998). Mixture of Weibull distributions: parametric characterization of failure rate functions, *Applied Stochastic Models in Data Analysis*, **14**, 47–65.

20. Klein, J. P., and Moeschberger, M. L. (1997). *Survival Analysis*, Springer-Verlag, New York.

21. Lynch, J. D. (1999). On conditions for mixtures of increasing failure rate distributions to have an increasing failure rate, *Probability in Engineering and Information Sciences*, **13**, 33–36.

22. Mi, J. (1996). Minimizing some cost functions related to both burn-in and field use, *Operations Research*, **44**, 497–500.

23. Navarro, J., and Hernandez, P. J. (2002). How to obtain bathtub-shaped failure rate models from normal mixtures *Probability in Engineering and Information Sciences*, **18**, 511–531.

24. Proschan, F. (1963). Theoretical explanation of observed decreasing failure rate, *Technometrics*, **5**, 373–383.

25. Puri, P. S., and Singh, H. (1986). Optimum replacement of a system subject to shocks: a mathematical lemma, *Operations Research*, **34**, 782–789.

26. Robertson, C. A., and Fryer, J. G. (1969). Some descriptive properties of normal mixtures, *Skandinavisk Aktuarietidskrift*, **52**, 137–146.

27. Samaniego, F. (1985). On the closure of the IFR class under the formation of coherent systems, *IEEE Transactions on Reliability*, **R-34**, 69–72.

28. Savits, T. H. (2003). Preservation of generalized bathtub functions, *Journal of Applied Probability*, **40**, 1–12.

29. Schilling, M. F., Watkins, A. E., and Watkins, W. (2002). Is human height bimodal? *The American Statistician*, **56**, 223–229.

30. Vaupel, T. W., and Yashin, A. I. (1985). Heterogeneity's ruse: some surprising effects of selection on population dynamics, *The American Statistician*, **39**, 176–185.

31. Wang, J.-L., Muller, H., and Capra, W. B. (1998). Analysis of oldest-old mortality, *The Annals of Statistics*, **26**, 126–133.

32. Wondmagegnehu, E. T. (2002). Mixture of distributions with increasing failure rates, Ph.D. Thesis, University of Pittsburgh, Pittsburgh, PA.

33. Wondmagegnehu, E. T., Navarro, J., and Hernandez, P. J. (2005). Bathtub shaped failure rates from mixtures: A practical point of view, *IEEE Transactions on Reliability*, **54**, 270–275.

18

Characterizations of the Relative Behavior of Two Systems via Properties of Their Signature Vectors

Henry Block,[1] **Michael R. Dugas,**[2] **and Francisco J. Samaniego**[2]

[1] *University of Pittsburgh, Pittsburgh, PA, USA*
[2] *University of California, Davis, CA, USA*

Abstract. The signature of a system of components with independent and identically distributed (iid) lifetimes is a probability vector whose ith element represents the probability that the ith component failure causes the system to fail. Samaniego (1985) introduced the concept and used it to characterize the class of systems that have increasing failure rates (IFR) when the components are iid IFR. Kochar *et al.* (1999) showed that when signatures are viewed as discrete probability distributions, the stochastic, hazard rate or likelihood ratio ordering of two signature vectors implies the same ordering of the lifetimes of the corresponding systems in iid components. In this paper, these latter results are extended in a variety of ways. For example, conditions on system signatures are identified that are not only sufficient for such orderings of lifetimes to hold, but are also necessary. More generally, given any two coherent systems whose iid components have survival functions $S_i(t)$ and failure rates $r_i(t)$, respectively, for $i = 1, 2$, the number and locations of crossings of the systems' survival functions or failure rates in $(0, \infty)$ can be fully specified in terms of the two system signatures. One is thus able to deduce how these systems compare to each other in real time, in contrast to the asymptotic comparisons one finds in the literature.

Keywords and phrases: Coherent system, mixed system, hazard rate ordering, stochastic ordering, likelihood ratio ordering, survival, k-out-of-n systems, reliability, crossing properties

18.1 Introduction

Modern reliability theory tends to restrict its attention to the study of "coherent systems," that is, systems (1) that are monotone (i.e., for which the replacement of a failed component by a functioning component cannot make the system worse) and (2) in which every component is relevant (i.e., its failure can, under specific circumstances, cause the system to fail). The structure function of a coherent system—the function that expresses the state (i.e., the success or failure) of a system in terms of the states of its components—is perhaps the most fundamental tool for studying the relationship between the design of a system of interest and that system's performance. Because it is a fairly complex algebraic object, however, the structure function has not proven particularly useful in the study of the comparative performance of competing systems. In this paper, we will focus on systems in iid components; for such systems, the notion of the "signature" $\mathbf{s}$ of a coherent system [see Samaniego (1985)] has proven to be much more manageable in comparative studies.

The signature of a system in iid components is an n-dimensional probability vector $\mathbf{s}$ whose ith element represents the probability that the failure of the system occurs upon the ith component failure. Generally, the computation of a system's signature involves combinatorial arguments identifying the number of permutations of the indexes of the n component failure times that result in system failure upon the ith component failure, where $i = 1, 2, \ldots, n$. The signatures of the five possible coherent systems of order 3 are easily found to be

$$(1, 0, 0), \quad (0, 1, 0), \quad (0, 0, 1), \quad (1/3, 2/3, 0), \quad \text{and} \quad (0, 2/3, 1/3).$$

Component performance is typically characterized by the *cumulative distribution function* F of the component's lifetime or by standard alternatives such as the *survival function* $\overline{F} = 1 - F$, the *density function* f, or the *failure rate* r, defined as the ratio $r(t) = f(t)/\overline{F}(t)$, the latter two functions being well defined when F is absolutely continuous. The performance of a system can similarly be described in terms of the system lifetime's cdf or survival function, its density, or its failure rate. For complex systems, these functions tend to be difficult to represent explicitly in terms of the behavior of the system's components; see, for example, Barlow and Proschan (1981) for further details.

For coherent systems in iid components, Samaniego (1985) obtained useful representations for the system's survival function $\overline{F}_T$, the system's density function f_T and the system failure rate r_T in terms the system's signature vector and the common distribution F of its components. In the next section, we will briefly review these results as well as those obtained by Kochar, Mukerjee, and Samaniego (1999) which provide sufficient conditions, in terms of the respective signatures, for two coherent systems in iid components to be ordered in some stochastic sense. The purpose of the present note is to extend these latter results. More specifically, we will extend the results of Kochar, Mukerjee, and Samaniego (1999) by providing, in each of the three scenarios considered in the latter paper, conditions that are both necessary and sufficient for the desired relationships (stochastic, hazard rate, and likelihood ratio ordering) between system lifetimes to hold.

In the sequel, we will utilize the notion of "mixed systems" as introduced in Boland and Samaniego (2004). A mixed system is a stochastic mixture of several coherent systems, and can be physically realized through a randomization process that selects a coherent system at random according to predetermined probabilities. The signature of a mixed system is clearly the corresponding mixture of the signatures of the systems involved. For example, for $n = 3$, the 50-50 mixture of a series and a parallel system results in a mixed system with signature (1/2, 0, 1/2). This agrees with our intuition that, because the series system is selected only 1/2 of the time, the chances that the first component failure results in the mixed system's failure is precisely 1/2.

Any probability vector $\mathbf{s}$ of length n can be interpreted as the signature of a mixed system. For example, the mixture of k-out-of-n:F systems (i.e., n-component systems that necessarily fail upon the kth component failure) according to the probabilities in $\mathbf{s}$ results in the probability that the system fails upon the ith component failure being equal to the ith element of $\mathbf{s}$. A given signature may correspond to more than one system. For instance, the signature (0,2/3,1/3) can be obtained as the signature of a single coherent system or as the mixture, with probabilities 2/3 and 1/3, of a 2-out-of-3:F system and a parallel system. Kochar, Mukerjee, and Samaniego (1999) provide an example of two coherent systems in four iid components that have the same signature vectors. The notion of mixed systems extends the finite class of coherent systems of order n to an infinite collection of systems indexed by the class of all n-dimensional probability vectors.

In Section 18.2, we provide the basic definitions and background results needed in our study. In Section 18.3 we establish our main results: necessary and sufficient conditions for one system to dominate another in a specified sense. Specifically, necessary and sufficient conditions, in terms of the two system signatures, are given for the lifetime distributions of two systems in iid components to be stochastically ordered, hazard-rate ordered, or likelihood ratio ordered. More generally, necessary and sufficient conditions for the two survival functions or failure rates to cross precisely k times, for any fixed k, are also given. In the final section, we explore the practical implications of these results.

18.2 Background Results for the Comparison of System Life

In this section, we give background results on signatures from Samaniego (1985) and on their use in obtaining results on comparison of system life from Kochar, Mukerjee, and Samaniego (1999). We first give a formal definition of the signature of a coherent system in iid components. Numerous examples of system signatures are given in the two papers cited above.

Definition 18.2.1 The signature of a coherent system with n iid component lifetimes is the probability vector $\mathbf{s} = (s_1, s_2, \ldots, s_n)$, where s_i is the probability the system fails upon the ith component failure.

Consider a coherent system with n iid components, with survival distribution $\overline{F}$, density function f and failure rate r. The following representations of $\overline{F}_T$, f_T, and r_T, the corresponding survival function, density and failure rate of the system lifetime T, in terms of the system's signature $\mathbf{s}$, are given in Samaniego (1985):

$$\overline{F}_T(t) = \sum_{i=1}^{n} s_i \sum_{j=0}^{i-1} \binom{n}{j} \{F(t)\}^j \{\overline{F}(t)\}^{n-j}, \tag{18.1}$$

or alternatively,

$$\overline{F}_T(t) = \sum_{j=0}^{n-1} \left(\sum_{i=j+1}^{n} s_i \right) \binom{n}{j} \{F(t)\}^j \{\overline{F}(t)\}^{n-j}, \tag{18.2}$$

$$f_T(t) = \sum_{i=0}^{n-1} (n-i) s_{i+1} \binom{n}{i} \{F(t)\}^i \{\overline{F}(t)\}^{n-i} r(t), \tag{18.3}$$

and

$$r_T(t) = \frac{\sum_{i=0}^{n-1} (n-i) s_{i+1} \binom{n}{i} \{F(t)\}^i \{\overline{F}(t)\}^{n-i}}{\sum_{i=0}^{n-1} (\sum_{j=i+1}^{n} s_j) \binom{n}{i} \{F(t)\}^i \{\overline{F}(t)\}^{n-i}} r(t). \tag{18.4}$$

It will be useful in the sequel to utilize the ratio

$$G(t) = \frac{F(t)}{\overline{F}(t)}.$$

Notice that as t goes from 0 to ∞, $G(t)$ increases from 0 to ∞. Utilizing $G(t)$, we may rewrite Eqs. (18.1)–(18.4) in a way that is more useful for our purposes. Specifically,

$$\overline{F}_T(t) = \{\overline{F}(t)\}^n \sum_{i=1}^{n} s_i \sum_{j=0}^{i-1} \binom{n}{j} \{G(t)\}^j, \tag{18.5}$$

or alternatively,

$$\overline{F}_T(t) = \{\overline{F}(t)\}^n \sum_{j=0}^{n-1} \left(\sum_{i=j+1}^{n} s_i \right) \binom{n}{j} \{G(t)\}^j, \tag{18.6}$$

$$f_T(t) = \{\overline{F}(t)\}^n \sum_{i=0}^{n-1} (n-i) s_{i+1} \binom{n}{i} \{G(t)\}^i r(t) \tag{18.7}$$

and

$$r_T(t) = \frac{\sum_{i=0}^{n-1} (n-i) s_{i+1} \binom{n}{i} \{G(t)\}^i}{\sum_{i=0}^{n-1} (\sum_{j=i+1}^{n} s_j) \binom{n}{i} \{G(t)\}^i} r(t). \tag{18.8}$$

We note that the eight representations above are equally applicable if $\mathbf{s}$ is the signature of a mixed system based on coherent systems with n iid components with cdf F, density f and failure rate r. For completeness, we include below the definitions of the three stochastic relationships on which we will concentrate.

The random variables X_1 and X_2, discrete or continuous, are stochastically ordered (i.e., $X_1 \leq_{st} X_2$) if the survival functions $\overline{F}_i(x) = P(X_i > x)$ are suitably ordered, that is, if $\overline{F}_1(x) \leq \overline{F}_2(x)$ for all x. We say that X_1 is smaller than X_2 in the hazard rate (or uniform stochastic) ordering if the ratio of survival functions $\overline{F}_2(x)/\overline{F}_1(x)$ is increasing in x. This ordering will be denoted by $X_1 \leq_{hr} X_2$. When the underlying distributions are absolutely continuous, the hr ordering is equivalent to the ordering of the failure rates, with X_2 having the smaller failure rate. Finally, X_1 is said to be smaller than X_2 in the likelihood ratio ordering ($X_1 \leq_{lr} X_2$) if the ratio $f_2(x)/f_1(x)$ is nondecreasing in x, where f_i represents the density or probability mass function of X_i. The implications $lr \Rightarrow hr \Rightarrow st$ are well-known.

The proposition below gives sufficient conditions on the signatures of two systems in iid components for specific stochastic relationships to hold between the lifetimes of these systems. These results are established in Kochar, Mukerjee, and Samaniego (1999) for coherent systems, but apply to mixed systems as well, assuming that all components involved have iid lifetimes distributed according to a common distribution F. The proposition given here covers this more general situation.

Proposition 18.2.1 *Let* $\mathbf{s_1}$, $\mathbf{s_2}$ *be signatures of two mixed systems based on coherent systems with* n *iid components with common distribution* F, *and let* T_1, T_2 *be the corresponding system lifetimes.*

(1.1) If $\mathbf{s_1} \leq_{st} \mathbf{s_2}$, *then* $T_1 \leq_{st} T_2$;

(1.2) If $\mathbf{s_1} \leq_{hr} \mathbf{s_2}$, *then* $T_1 \leq_{hr} T_2$;

(1.3) If $\mathbf{s_1} \leq_{lr} \mathbf{s_2}$, *then* $T_1 \leq_{lr} T_2$.

In the next section, necessary and sufficient conditions are presented for the ordering of system lifetimes as in the proposition above.

18.3 New Signature Conditions and Associated System Behavior

While the various ordering conditions on signatures of Section 18.2 (Proposition 18.2.1) are sufficient to imply corresponding orderings of the system lifetimes, they are not, in general, necessary. Counterexamples to their necessity are explicitly displayed in Block, Dugas, and Samaniego (2004). The question that remains is: Can necessary and sufficient conditions be identified in any problems of practical interest? An affirmative answer is provided below.

We first consider the stochastic ordering of system lifetimes. Suppose that two different systems with iid components with the same cdf F have stochastically ordered lifetimes. This condition can simply be written as

$$\overline{F}_{T_1}(t) \leq \overline{F}_{T_2}(t) \quad \text{for all} \quad t \geq 0 \tag{18.9}$$

which, in light of (18.6), becomes

$$\sum_{i=1}^{n} s_{1i} \sum_{j=0}^{i-1} \binom{n}{j} \{G(t)\}^j \leq \sum_{i=1}^{n} s_{2i} \sum_{j=0}^{i-1} \binom{n}{j} \{G(t)\}^j, \tag{18.10}$$

where $\mathbf{s_1}$ and $\mathbf{s_2}$ are the respective system signatures. The inequality (18.10) is, in turn, equivalent to

$$\sum_{i=1}^{n} (s_{2i} - s_{1i}) \sum_{j=0}^{i-1} \binom{n}{j} \{G(t)\}^j \geq 0 \tag{18.11}$$

or

$$\sum_{j=0}^{n-1} \binom{n}{j} \sum_{i=j+1}^{n} (s_{2i} - s_{1i}) \{G(t)\}^j \geq 0. \tag{18.12}$$

Now, define the function g to be the polynomial of degree $(n-1)$ given by

$$g(x) = \sum_{j=0}^{n-1} \binom{n}{j} \sum_{i=j+1}^{n} (s_{2i} - s_{1i}) x^j \quad \text{for} \quad x \geq 0. \tag{18.13}$$

From the equivalence of the conditions (18.9)–(18.12), the following result follows immediately.

Theorem 18.3.1 *Let $\mathbf{s_1}$ and $\mathbf{s_2}$ be the signatures of two arbitrary mixed systems based on coherent systems in n iid components and the same component* *distributions, and let T_1 and T_2 denote the system lifetimes. Then $T_1 \leq_{st} T_2$ if and only if*

$$g(x) \geq 0 \ \ \textit{for all} \ \ x \geq 0, \tag{18.14}$$

where $g(x)$ is the polynomial given in (18.13).

Some remarks on Theorem 18.3.1 are in order. First, it should be noted that, when signatures $\mathbf{s_1}$ and $\mathbf{s_2}$ are stochastically ordered, that is, $\mathbf{s_1} \leq_{st} \mathbf{s_2}$, then (18.14) clearly holds, because g is a polynomial with nonnegative coefficients. Thus, this new result contains Theorem 3 of Kochar, Mukerjee, and Samaniego (1999). It also extends it, as the condition given, namely (18.14), is also necessary for the stochastic ordering of T_1 and T_2 to hold. Just as the polynomial $x^2 - 2x + 2$ is positive for all real x even though not all its coefficients are positive, so too can the polynomial in (18.13) be positive for all positive x without the restrictive condition $\mathbf{s_1} \leq_{st} \mathbf{s_2}$.

The condition (18.14) is, admittedly, a complex statement concerning the relationship between the two system signatures involved. However, because the condition is a necessary and sufficient condition, no essential simplification is possible. Fortunately, it is a condition that is, in fact, numerically, if not algebraically, simple, because it is essentially equivalent to the problem of finding the minimum of a continuous function over a bounded interval. In practice, condition (18.14) can be checked without great strain.

Let us now consider a necessary and sufficient condition for the hazard rate ordering among system lifetimes. From (18.8), we know that the lifetimes T_1 and T_2 of two mixed systems based on coherent systems in n iid components, both having the same component distributions, satisfy $T_1 \leq_{hr} T_2$ if and only if for all $t \geq 0$,

$$\frac{\sum_{i=0}^{n-1}(n-i)s_{2,i+1}\binom{n}{i}\{G(t)\}^i}{\sum_{i=0}^{n-1}(\sum_{j=i+1}^{n} s_{2,j})\binom{n}{i}\{G(t)\}^i} \leq \frac{\sum_{i=0}^{n-1}(n-i)s_{1,i+1}\binom{n}{i}\{G(t)\}^i}{\sum_{i=0}^{n-1}(\sum_{j=i+1}^{n} s_{1,j})\binom{n}{i}\{G(t)\}^i}. \tag{18.15}$$

Now, let h be the rational function defined by

$$h(x) = \frac{\sum_{i=0}^{n-1}(n-i)s_{i+1}\binom{n}{i}x^i}{\sum_{i=0}^{n-1}(\sum_{j=i+1}^{n} s_j)\binom{n}{i}x^i}, \tag{18.16}$$

where $\mathbf{s}$ is an arbitrary n-dimensional probability vector (or signature in the setting of interest here). From the above, we see that a second necessary and sufficient condition has been identified, as it is clear that the ordering of the failure rates of two mixed systems in n iid components will occur precisely when the difference of the two corresponding h functions in (18.16) is non-negative (or nonpositive) for all $x \geq 0$. We record this result as follows.

Theorem 18.3.2 *Let* $\mathbf{s_1}$ *and* $\mathbf{s_2}$ *be the signatures of two arbitrary mixed systems based on coherent systems in* n *iid components and the same component distribution, and let* T_1 *and* T_2 *be the respective system lifetimes. Then* $T_1 \leq_{hr} T_2$ *if and only if*

$$h_1(x) - h_2(x) \geq 0 \quad \text{for all} \quad x \geq 0, \tag{18.17}$$

where h_j *represents the rational functions of* x *given in (18.16) with* $\mathbf{s} = \mathbf{s_j}$, $j = 1, 2$.

Condition (18.17) is again a complex but both necessary and sufficient condition for the hazard rate ordering of system lifetimes. It can be shown (though it is not immediately transparent) that $\mathbf{s_1} \leq_{hr} \mathbf{s_2}$ implies condition (18.17). As with the necessary and sufficient condition for stochastic ordering of lifetimes, condition (18.17) is mathematically complex, but, after cross-multiplying in the inequality $h_1(x) \geq h_2(x)$, reduces to checking that a certain polynomial of degree $2n - 3$ is non-negative for all $x \geq 0$. In a given problem of interest, this can be determined via standard numerical methods.

A similar result can be obtained for the likelihood ratio ordering. We omit the details, as they are similar to those in the developments above. A formal statement of the relevant result requires reference to the polynomial m defined as follows:

$$m(x) = \sum_{i=0}^{n-1} (n - i) s_{i+1} \binom{n}{i} x^i, \tag{18.18}$$

where $\mathbf{s}$ is an n-dimensional probability vector (or signature).

Theorem 18.3.3 *Let* $\mathbf{s_1}$ *and* $\mathbf{s_2}$ *be the signatures of two arbitrary mixed systems based on coherent systems in* n *iid components and the same component distribution, and let* T_1 *and* T_2 *be the respective system lifetimes. Then* $T_1 \leq_{lr} T_2$ *if and only if the rational function*

$$\frac{m_2(x)}{m_1(x)}, \tag{18.19}$$

is increasing in $x \geq 0$, *where* $m_i(x) = m(x)$ *in (18.18), with* $\mathbf{s} = \mathbf{s_i}$, $i = 1, 2$.

18.4 Practical Implications

Further inspection of the tools utilized in Section 18.3 leads to some interesting and, as yet, unexploited, insights. While our main interest in Theorems 18.3.1

and 18.3.2 was the development of necessary and sufficient condition for two curves (be they survival curves or failure rates) not crossing, one can see that the polynomial g in (18.13) and the rational function h in (18.16) completely determine the crossing properties of the survival function $\overline{F}$ and the failure rate r. Specifically, two survival functions will be equal at time t precisely when the identity $g(G(t)) = 0$ obtains, and two failure rates will be equal precisely when the functions $h_1(x)$ and $h_2(x)$ of Theorem 18.3.2 are equal. Thus by studying the functions g and h, which depend solely on the signatures of the two systems involved, one can characterize the crossing behavior of the associated survival functions or failure rates. This represents an important complement to existing literature that focuses on the behavior of survival functions and failure rates as $t \to \infty$; see, for example, Block and Joe (1997) and Block, Li, and Savits (2003).

We will present two examples of analyses that benefit from these type of considerations. Consider first the survival functions $\overline{F}_1(t)$ and $\overline{F}_2(t)$ of two mixed systems based on coherent systems in n iid components with common lifetime distribution F. If $g(x)$ is the polynomial in (18.13), then it is easy to see that the number of crossings (from $+$ to $-$ or from $-$ to $+$) of the survival functions $\overline{F}_1(t)$ and $\overline{F}_2(t)$ correspond exactly to the number of crossings of the polynomial $y = g(x)$ and the line $y = 0$ for $x \in (0, \infty)$. Theorem 18.3.1 says that $\overline{F}_1$ and $\overline{F}_2$ have no crossings (a condition equivalent to stochastic ordering) if and only if $g(x)$ has no crossings of the x-axis. The number of crossings of $\overline{F}_1$ and $\overline{F}_2$ can now be studied in terms of the function g. Indeed, if the polynomial $g(x)$ crosses the x-axis exactly k times in the interval $(0, \infty)$, then the survival functions $\overline{F}_1(t)$ and $\overline{F}_2(t)$ will cross exactly k times in that interval. Similar remarks apply to the crossing of failure rates. If $h_1(x)$ and $h_2(x)$ are the functions introduced in Theorem 18.3.2, then the number of crossings of the failure rates $r_1(t)$ and $r_2(t)$ are equal to the number of crossings of the functions $h_1(x)$ and $h_2(x)$ for $x \in (0, \infty)$.

We close with a simple example of two systems in iid components for which both the survival functions and the failure rates of two competing systems cross exactly once. The crossing time can be identified, in each case, as a specific quantile of the common component lifetime distribution F. In both of these illustrations, the two systems to be compared are the same, viz., the three-component systems having signatures $\mathbf{s_1} = (1/2, 0, 1/2)$ and $\mathbf{s_2} = (0, 1, 0)$, respectively. The first system results from selecting a series or a parallel system at random, each with probability 1/2, while the second system is simply a 2-out-of-3:F system (i.e., it fails upon the second component failure).

Example 18.4.1 To compare the survival functions of the two systems above, we identify the polynomial g in this problem as

$$g(x) = -1.5x^2 + 1.5x. \tag{18.20}$$

The function g has precisely one positive root, namely $x = 1$. From Theorem 18.3.1, it follows that the two survival functions will cross at the time t_0 for which $G(t_0) = 1$. This is equivalent to $t_0 = F^{-1}(1/2)$, which leads to the conclusion that the 2-out-of-3 system is as good as or better than the mixed system if and only if $t \leq t_0$. This would be a particularly important finding if the mission time for the chosen system happens to be smaller that t_0, as it would then serve to identify a system that is uniformly superior to the other in the time internal of interest. If the system's mission time is substantially longer than t_0, then one might well prefer the mixed system because it has superior performance if and when it has survived beyond time t_0.

Example 18.4.2 A comparison of the failure rates of these same two systems would proceed as follows. From (18.16) we see that the relevant functions h_1 and h_2 are given by

$$h_1(x) = \frac{3 + 3x^2}{2 + 3x + 3x^2} \tag{18.21}$$

and

$$h_2(x) = \frac{6x}{1 + 3x}. \tag{18.22}$$

The inequality $h_1(x) \geq h_2(x)$ is equivalent to

$$3x^3 + 5x^2 + x - 1 \leq 0. \tag{18.23}$$

For positive x, the inequality in (18.23) is valid if and only if x is sufficiently small, as the cubic involved has only one sign change and thus, by Descartes' Rule of Signs, can have at most one positive root. Because the polynomial is negative at 0 and positive at 1 it has exactly one root. That root is easily determined to be $x = 1/3$. We thus conclude that the functions h_1 and h_2 in (18.20) and (18.21) cross exactly once, as do the failure rates of the two systems involved. The crossing time of the two failure rates will of course depend upon the underlying component distribution F. In general, the two failure rates will cross at the time t_0 for which $G(t_0) = 1/3$ or, equivalently, when $F(t_0) = 1/4$. It follows that the 2-out-of-3 system has a smaller failure rate than the mixed system for $0 \leq t < F^{-1}(1/4)$ and has a larger failure rate than the mixed system if $t > F^{-1}(1/4)$.

Acknowledgement. The work of Henry Block was supported in part by NSF Grant DMS-0072207. The work of M. R. Dugas and F. J. Samaniego was supported in part by Army contracts AR019-02-1-0377 and WN11NF05-1-0118.

References

1. Barlow, R. E., and Proshan, F. (1981). *Statistical Theory of Reliability*, To Begin With Press, Silver Springs, MD.

2. Block, H., Dugas, M., and Samaniego F. J. (2004). Signature-related results on system failure rates and lifetimes, *Technical Report No. 405*, Department of Statistics, University of California, Davis.

3. Block, H., and Joe, H. (1997). Tail behavior of the failure rate functions of mixtures, *Lifetime Data Analysis*, **3**, 269–288.

4. Block, H., Li, Y., and Savits, T. (2003). Initial and final behavior of the failure rate functions for mixtures and systems, *Journal of Applied Probability*, **40**, 721–740.

5. Boland, P., and Samaniego, F. J. (2004). The signature of a coherent system and its applications in reliability, In *Mathematical Reliability: An Expository Perspective* (Eds., R. Soyer, T. A. Mazzuchi, and N. D. Singpurwalla), Kluwer Academic Publishers, Boston.

6. Kochar, S., Mukerjee, H., and Samaniego, F. J. (1999). The signature of a coherent system and its application to comparisons among systems, *Naval Research Logistics*, **46**, 507–523.

7. Samaniego, F. J. (1985). On closure of the IFR class under formation of coherent systems, *IEEE Transactions on Reliability*, **34**, 69–72.

19

Systems with Exchangeable Components and Gumbel Exponential Distribution

Jorge Navarro,[1] Jose M. Ruiz,[1] and Carlos J. Sandoval[2]

[1] *Universidad de Murcia, Murcia, Spain*
[2] *Universidad Catolica San Antonio de Murcia, Murcia, Spain*

Abstract: The life lengths of some possibly dependent components in a system can be modelled by a multivariate distribution. In this paper, we suppose that the joint distribution of the units is a symmetric multivariate Gumbel exponential distribution (GED). Hence, the components are exchangeable and have exponential (marginal) distributions. For this model, we obtain basic reliability properties for k-out-of-n systems (order statistics) and, in particular, for series and parallel systems. We pay special attention to systems with two components. Some results are extended to coherent systems with n exchangeable components.

Keywords and phrases: Reliability, failure rate, mean residual life, k-out-of-n systems, coherent systems, order statistics

19.1 Introduction

We consider a system with n possibly dependent components whose life lengths are represented by a random vector $(X_1, \ldots, X_n)$ with joint reliability (or survival) function $R(x_1, \ldots, x_n) = \Pr(X_1 \geq x_1, \ldots, X_n \geq x_n)$. Hence, the series system is represented by $X_{(1,n)} = \min(X_1, \ldots, X_n)$, the parallel system by $X_{(n,n)} = \max(X_1, \ldots, X_n)$ and, in general, the k-out-of-n system by $X_{(n-k+1,n)}$, where $X_{(i,n)}$ is the ith order statistic from $(X_1, \ldots, X_n)$.

If the components are similar and they are in the same environment, then we can suppose that the random vector is exchangeable, that is, the reliability function $R(x_1, \ldots, x_n)$ is the same for any permutation of $x_1, \ldots, x_n$. Hence, the units have the same distribution, that is, the marginal reliability functions $R_i(t) = \Pr(X_i \geq t)$, $i = 1, \ldots, n$, are the same. Note that the independent and identically distributed (i.i.d.) components case is also included.

Reliability properties for systems with i.i.d. components are given in Barlow and Proschan (1975), Cox and Oakes (1984), Kanjo and Abouammoh (1995), Belzunce *et al.* (2001), and Shaked and Suarez-Llorens (2003), and in the references given therein. Some of these properties were extended to nonidentically distributed components.

Recently, special attention has been paid to the study of systems with dependent components; see, for example, Baggs and Nagaraja (1996), Gupta (2001), Gupta and Gupta (2001), Roy (2001), Rychlik (2001a,b), and Mi and Shaked (2002).

In this chapter, we suppose that $(X_1, \ldots, X_n)$ is exchangeable and has a multivariate Gumbel exponential distribution GED [see Gumbel (1960) or Kotz *et al.* (2000, p. 350)] and then, we study reliability properties for coherent systems obtained from this model. The results can also be applied to order statistics from dependent samples.

Specifically, in Section 19.2, we give some preliminary general results. In Section 19.3, we study reliability and moments properties. In Section 19.4, we study aging measures such as the failure rate and mean residual life functions. In Section 19.5, we discuss ordering and classification properties and, in Section 19.6, parameter estimation for systems with bivariate symmetric Gumbel distribution. Some of these results are extended to coherent systems with n exchangeable components in Section 19.7.

19.2 General Properties

We denote the reliability function of $X_{(i,n)}$ by $R_{(i,n)}(t)$, the failure rate by $r_{(i,n)}(t)$, the mean residual life by $e_{(i,n)}(t)$, and so on. We use the definitions and properties for the stochastic ($\leq_{st}$), failure rate ($\leq_{fr}$), mean residual life ($\leq_{mrl}$) and likelihood ratio ($\leq_{lr}$) orders and IFR (DFR), DMRL (IMRL) and ILR (DLR) classes given in Shaked and Shanthikumar (1994). In particular, it is well-known that

$$X \leq_{lr} Y \Rightarrow X \leq_{fr} Y \Rightarrow X \leq_{mrl} Y \text{ and } X \leq_{st} Y \tag{19.1}$$

and

$$\text{ILR (DLR)} \Rightarrow \text{IFR (DFR)} \Rightarrow \text{DMRL (IMRL)}. \tag{19.2}$$

Moreover, we will use the following lemma.

Lemma 19.2.1 *If $f_\theta(t)$ is the density function of $X(\theta)$ for all $\theta \in (\alpha, \beta)$, then*

$$X(\theta) \leq_{lr} X(\theta') \ (\geq_{lr}) \text{ for } \alpha < \theta \leq \theta' < \beta$$

if and only if

$$\frac{\partial}{\partial\theta}\frac{\partial}{\partial t}\log f_\theta(t) \geq 0 \ (\leq) \ \textit{for } t \in \mathbb{R} \textit{ and } \alpha < \theta < \beta.$$

We will also use the fact that the component (marginal) distribution of a system with two identically distributed (i.d.) components is a 50% mixture between series and parallel systems, i.e.,

$$\frac{1}{2}R_{(1,2)}(t) + \frac{1}{2}R_{(2,2)}(t) = R_1(t). \tag{19.3}$$

Thus, from Block and Joe (1997), we have that, asymptotically, the behavior of the failure rate of the mixture is equal to that of the strongest component of the mixture. Moreover, the failure rate of the mixture is between the failure rates of the strongest and the weakest components in each point. Hence, in this case, if $X_{(1,2)} \leq_{fr} X_{(2,2)}$, then $X_{(1,2)} \leq_{fr} X_i \leq_{fr} X_{(2,2)}$ and $\lim_{t\to\infty} r_{(1,2)}(t) - r_i(t) = 0$ for $i = 1, 2$.

Expression (19.3) can be extended to the multivariate case supposing that $(X_1, \ldots, X_n)$ has an exchangeable distribution.

Lemma 19.2.2 *If $(X_1, \ldots, X_n)$ is an exchangeable random vector, then*

$$R_{(i,n)}(t) = \sum_{j=n-i+1}^{n}\sum_{s=0}^{n-j}(-1)^s\frac{n!}{j!s!(n-j-s)!}\Pr(X_1 \geq t, \ldots, X_{j+s} \geq t). \tag{19.4}$$

The proof is obtained by using the inclusion-exclusion formula in a similar way to that of formula 5.5.3 in David (1970) [see also Maurer and Margolin (1976) or Balakrishnan *et al.* (1992)]. This expression can be written as

$$R_{(i,n)}(t) - \sum_{s=n-i+1}^{n}\binom{n}{s}c_{n-i+1,s}R_{(1,s)}(t), \tag{19.5}$$

where $c_{i,s} = \sum_{j=i}^{s}(-1)^{s-j}\binom{s}{j}$. Hence, all the k-out-of-n systems are generalized mixtures (i.e., mixtures with possible negative weights) of series systems. In a similar way, we obtain

$$R_{(i,n)}(t) = \sum_{s=i}^{n}\binom{n}{s}c_{i,s}R_{(s,s)}(t), \tag{19.6}$$

that is, it is also a generalized mixture of parallel systems.

Samaniego (1985) [see also Kochar *et al.* (1999) or Shaked and Suarez-Llorens (2003)] defined the signature of a coherent system $T = \Phi(X_1, \ldots, X_n)$ as the vector $\mathbf{p} = (p_1, \ldots, p_n) \in \mathbb{R}^n$, where

$$p_i = \Pr(T = X_{(i,n)}), \tag{19.7}$$

showing that for systems with exchangeable components

$$p_i = \frac{\text{\# of orderings for which the } i\text{th failure causes system failure}}{n!}. \tag{19.8}$$

Samaniego (1985) also showed, in the absolutely continuous case, that the coherent system is a mixture of k-out-of-n systems with weights p_i, that is,

$$R_T(t) = \sum_{i=1}^{n} p_i R_{(i,n)}(t). \tag{19.9}$$

As a consequence, from (19.5), we have that any coherent system is a generalized mixture of series systems. We define the minimal signature as the vector $\mathbf{a} = (a_1, \ldots, a_n) \in \mathbb{R}^n$, such that

$$R_T(t) = \sum_{i=1}^{n} a_i R_{(1,i)}(t). \tag{19.10}$$

Analogously, from (19.6), we have that any coherent system is a generalized mixture of parallel systems and we define the maximal signature as the vector $\mathbf{b} = (b_1, \ldots, b_n) \in \mathbb{R}^n$, such that

$$R_T(t) = \sum_{i=1}^{n} b_i R_{(i,i)}(t). \tag{19.11}$$

19.3 Reliability and Moments

First, we study systems with two components (X_1, X_2) and exchangeable Gumbel exponential distribution $GED(a, b)$, with reliability function given by

$$R(x_1, x_2) = e^{-a(x_1+x_2)-ba^2x_1x_2}, \tag{19.12}$$

for $x_1, x_2 \geq 0$, $a > 0$ and $0 \leq b \leq 1$ [see Gumbel (1960) or Kotz *et al.* (2000, p. 350)]. Hence, X_i has an exponential distribution with mean $\mu = E(X_i)$ $= 1/a$ and

$$\rho = Corr(X_1, X_2) = -1 - \frac{1}{b} e^{1/b} \operatorname{Ei}(-1/b),$$

where $\operatorname{Ei}(x) = \int_{-\infty}^{x} t^{-1} e^t dt$; see Gumbel (1960). The correlation coefficient is zero for $b = 0$ and it decreases to -0.40365 as b increases to 1. Hence, the model can be reparametrized in terms of μ and ρ.

Gumbel defined other bivariate distributions with exponential marginal distributions. Baggs and Nagaraja (1996) studied some properties of series and parallel systems obtained from these models.

For simplicity, as $n = 2$, we use $X_{(1)}$ and $X_{(2)}$ for the series and parallel system life lengths, respectively. Hence,

$$aX_{(1)} = \min(X_1^*, X_2^*) = X_{(1)}^*$$

and

$$aX_{(2)} = \max(X_1^*, X_2^*) = X_{(2)}^*,$$

where $X_i^* = aX_i$, $i = 1, 2$, and (X_1^*, X_2^*) has a standard Gumbel exponential distribution $GED(1, b)$. Moreover, $R_{(i)}(t) = R_{(i)}^*(at)$, $i = 1, 2$. Clearly, a is a scale parameter and b is a shape parameter. Note that the reliability for $X_{(1)}^*$ only depends on b (ρ) and hence, it can be easily computed from

$$R_{(1)}^*(t) = \Pr(X_{(1)}^* \geq t) = \Pr(X_1^* \geq t, X_2^* \geq t) = e^{-2t-bt^2},\ t > 0. \tag{19.13}$$

Analogously, from (19.3),

$$R_{(2)}^*(t) = 2e^{-t} - R_{(1)}^*(t),\ t > 0, \tag{19.14}$$

that is, the parallel system is a negative mixture between series system and an exponential distribution.

From the preceding results, we have the following immediate properties.

Proposition 19.3.1 *If* $(X_1, X_2) \equiv GED(a, b)$, *then*

1. $E(X_{(i)}^k) = a^{-k}E((X_{(i)}^*)^k)$, $k = 1, 2, \ldots$ *and* $i = 1, 2$;

2. $E(X_{(1)}^k) + E(X_{(2)}^k) = 2a^{-k}k!$, $k = 1, 2, \ldots$;

3. $\sigma_{(i)} = \sigma_{(i)}^*/a$, $i = 1, 2$;

4. $\sigma_{(1)}^2 + \sigma_{(2)}^2 = 2\mu_{(1)}(2\mu - \mu_{(1)})$.

Moreover, we obtain all the moments for $X_{(1)}^*$ in the following proposition.

Proposition 19.3.2 *If* $(X_1, X_2) \equiv GED(a, b)$, *then*

1. $E(X_{(1)}^*) = \sqrt{\frac{\pi}{b}}e^{1/b}\Phi\left(-\sqrt{\frac{2}{b}}\right)$;

2. $E((X_{(1)}^*)^2) = \frac{1}{b} - \frac{2}{b}E(X_{(1)}^*)$;

3. $E((X_{(1)}^*)^k) = \frac{k}{2b}E((X_{(1)}^*)^{k-2}) - \frac{k}{b(k-1)}E((X_{(1)}^*)^{k-1})$, *for* $k = 3, 4, \ldots$

hold for all $0 < b \leq 1$, *where* Φ *is the standard normal distribution.*

PROOF. From (19.13), we have

$$E(X^*_{(1)}) = \int_0^\infty R^*_{(1)}(t)dt = e^{1/b}\int_0^\infty \exp\left(-\frac{(t+1/b)^2}{2/(2b)}\right)dt,$$

and hence, the first property holds. Analogously,

$$\begin{aligned} E((X^*_{(1)})^2) &= 2\int_0^\infty tR^*_{(1)}(t)dt \\ &= \int_0^\infty \frac{2+2bt-2}{b}e^{-2t-bt^2}dt \\ &= \frac{1}{b} - \frac{2}{b}\int_0^\infty R^*_{(1)}(t)dt \\ &= \frac{1}{b} - \frac{2}{b}E(X^*_{(1)}). \end{aligned}$$

Finally, if $k \geq 3$, then

$$\begin{aligned} E((X^*_{(1)})^k) &= k\int_0^\infty t^{k-1}R^*_{(1)}(t)dt \\ &= k\int_0^\infty t^{k-2}\frac{2+2bt-2}{2b}e^{-2t-bt^2}dt \\ &= -\frac{k}{b}\int_0^\infty t^{k-2}R^*_{(1)}(t)dt + k(k-2)\int_0^\infty t^{k-3}R^*_{(1)}(t)dt \\ &= -\frac{k}{b(k-1)}E((X^*_{(1)})^{k-1}) + \frac{k}{2b}E((X^*_{(1)})^{k-2}). \end{aligned}$$

■

Remark 19.3.1 The moments for $X^*_{(2)}$, $X_{(1)}$ and $X_{(2)}$, can be obtained from Propositions 19.3.1 and 19.3.2. In particular, we have

$$\mu_{(1)} = E(X_{(1)}) = \frac{1}{a}\sqrt{\frac{\pi}{b}}e^{1/b}\Phi\left(-\sqrt{\frac{2}{b}}\right), \tag{19.15}$$

$$\mu_{(2)} = E(X_{(2)}) = \frac{2}{a} - \frac{1}{a}\sqrt{\frac{\pi}{b}}e^{1/b}\Phi\left(-\sqrt{\frac{2}{b}}\right),$$

$$Var(X_{(1)}) = \frac{1}{a^2b} - \frac{2}{a^2b}\sqrt{\frac{\pi}{b}}e^{1/b}\Phi\left(-\sqrt{\frac{2}{b}}\right) - \frac{1}{a^2b}\pi e^{2/b}\Phi^2\left(-\sqrt{\frac{2}{b}}\right)$$

and

$$Var(X_{(2)}) = -\frac{1}{a^2b} - \frac{\pi}{a^2b}e^{2/b}\Phi^2\left(-\sqrt{\frac{2}{b}}\right) + \frac{2(1+b)}{a^2b}\sqrt{\frac{\pi}{b}}e^{1/b}\Phi\left(-\sqrt{\frac{2}{b}}\right),$$

where Φ is the standard normal distribution. The expression for $\mu_{(2)}$ was given in Kotz *et al.* (2003). It is easy to show that $\mu_{(1)}$ decreases in a and b, $\mu_{(2)}$ decreases in a and increases in b, and $Var(X_{(i)})$ decreases in a for $i = 1, 2$. Moreover, we can obtain bounds for the expected life lengths of series and parallel systems as

$$0.378936\mu \le \mu_{(1)} \le \frac{1}{2}\mu < \mu < \frac{3}{2}\mu \le \mu_{(2)} \le 1.62106\mu.$$

Tables 19.1 and 19.2 give the reliability function, the mean, the variance, and the skewness and kurtosis coefficients for a standard series system with GED joint distribution.

Table 19.1: t for some values of the standard series reliability function $R^*_{(1)}(t)$ $(0.99, 0.95, \ldots, 0.05, 0.01)$ and correlation coefficient ρ $(-0.4, \ldots, -0.03)$ for a system with two dependent components and $GED(1, \rho)$ joint distribution

$R^*_{(1)}(t)$	$\rho = -0.4$	-0.35	-0.30	-0.25	-0.20	-0.15	-0.10	-0.05	-0.03
0.99	0.005	0.005	0.005	0.005	0.005	0.005	0.005	0.005	0.005
0.95	0.025	0.025	0.025	0.025	0.025	0.025	0.026	0.026	0.026
0.90	0.051	0.051	0.051	0.052	0.052	0.052	0.052	0.053	0.053
0.85	0.078	0.078	0.079	0.079	0.080	0.080	0.081	0.081	0.081
0.80	0.106	0.107	0.108	0.109	0.109	0.110	0.111	0.111	0.111
0.75	0.134	0.136	0.138	0.139	0.140	0.141	0.142	0.143	0.143
0.70	0.164	0.167	0.170	0.172	0. 173	0.175	0.176	0.177	0.178
0.65	0.196	0.200	0.203	0.206	0.208	0.210	0.213	0.214	0.215
0.60	0.229	0.234	0.238	0.242	0.246	0.249	0.252	0.253	0.254
0.55	0.264	0.271	0.276	0.282	0. 286	0.290	0.293	0.296	0.297
0.50	0.301	0.310	0.317	0.324	0.329	0.335	0.339	0.343	0.345
0.45	0.341	0.352	0.361	0.370	0.377	0.384	0.389	0.394	0.397
0.40	0.385	0.398	0.409	0.420	0. 429	0.438	0.446	0.452	0.454
0.35	0.432	0.448	0.463	0.476	0.488	0.499	0.508	0.517	0.521
0.30	0.485	0.505	0.523	0.540	0.555	0.568	0.581	0.592	0.596
0.25	0.546	0.570	0.592	0.613	0.632	0.649	0.665	0.680	0.685
0.20	0.617	0.646	0.674	0.700	0.724	0.747	0.768	0.787	0.794
0.15	0.704	0.741	0.775	0.809	0.840	0.870	0.898	0.924	0.934
0.10	0.820	0.867	0.912	0.956	0.999	1.040	1.079	1.116	1.130
0.05	1.002	1.067	1.130	1.194	1.256	1.319	1.380	1.439	1.463
0.01	1.323	1.478	1.582	1.693	1.805	1.923	2.045	2.170	2.223

Table 19.2: Mean, variance, skewness, and kurtosis coefficients for a standard series system with two dependent components and $GED(1,\rho)$ joint distribution

ρ	−0.4	−0.35	−0.30	−0.25	−0.20	−0.15	−0.10	−0.05	−0.03
b	0.985	0.755	0.575	0.425	0.305	0.205	0.123	0.056	0.032
$\mu^*_{(1)}$	0.380	0.395	0.413	0.429	0.444	0.459	0.473	0.487	0.492
$\sigma^*_{(1)}$	0.315	0.335	0.359	0.382	0.404	0.427	0.451	0.475	0.485
γ^*_1	1.257	1.325	1.395	1.473	1.555	1.646	1.748	1.862	1.815
γ^*_2	1.787	2.056	2.357	2.714	3.123	3.619	4.222	4.975	14.052

19.4 Aging Measures

From (19.13), the failure rate for $X_{(1)}$ is

$$r_{(1)}(t) = ar^*_{(1)}(at) = 2a(1+abt). \tag{19.16}$$

Hence, it is IFR and has linear failure rate. This model can also be obtained as the minimum of two independent random variables having exponential and Rayleigh distributions [see, Sen and Bhattacharyya (1995) and the references given therein].

Analogously, from (19.14), we have

$$r^*_{(2)}(t) = \frac{1+bt-e^{t+bt^2}}{0.5-e^{t+bt^2}} \tag{19.17}$$

and $r_{(2)}(t) = ar^*_{(2)}(at)$. Moreover, the mean residual life functions are given by

$$e^*_{(1)}(t) = \sqrt{\frac{\pi}{b}}\exp(2t+bt^2+b^{-1})\Phi\left(-\frac{1+bt}{\sqrt{b/2}}\right)$$

and

$$e^*_{(2)}(t) = \frac{2e^{-t} - \sqrt{\pi/b}e^{1/b}\Phi\left(-\frac{1+bt}{\sqrt{b/2}}\right)}{2e^{-t} - e^{-2t-bt^2}}, \tag{19.18}$$

where Φ is the standard normal distribution and $e_{(i)}(t) = e^*_{(i)}(at)/a$.

From a practical point of view, it is very interesting to compute the regression-mean residual life function, defined by

$$m_{(2)}(t) = E(X_{(2)} - t \mid X_{(1)} = t)$$

which gives the expected life length for the two-component parallel system from the first failure. We have the following result.

Proposition 19.4.1 *If* $(X_1, X_2) \equiv GED(a, b)$, *then*

$$m_{(2)}(t) = \frac{1}{a + a^2bt} + \frac{a^2b}{(a + a^2bt)^3}. \tag{19.19}$$

PROOF. From (19.12), the joint pdf of $(X_{(1)}, X_{(2)})$ is

$$f(x_1, x_2) = 2(a^2(1-b) + ba^3(x_1 + x_2) + b^2a^4x_1x_2)\exp\{-a(x_1 + x_2) - ba^2x_1x_2\}$$

for $0 < x_1 < x_2$. Hence, from (19.13), the pdf of $(X_{(2)} \mid X_{(1)} = t)$ is

$$f_{(2)|(1)}(x_2 \mid t) = a\left(1 + abx_2 - \frac{b}{1 + bat}\right)\exp(-a(1 + abt)(x_2 - t)).$$

Integrating, we obtain (19.19). ∎

19.5 Stochastic Orders and Classes

For the series and parallel systems, we have the following classification results.

Proposition 19.5.1 *If* $(X_1, X_2) \equiv GED(a, b)$, *then*

1. $X_{(1)}$ *is ILR (IFR, DMRL);*
2. $X_{(1)}(a, b) \geq_{lr} X_{(1)}(a', b)$ *for* $a \leq a'$;
3. $X_{(1)}(a, b) \geq_{fr} X_{(1)}(a, b')$ *for* $b \leq b'$;
4. $X_{(2)}$ *is ILR (IFR, DMRL);*
5. $X_{(2)}(a, b) \geq_{fr} X_{(2)}(a', b)$ *for* $a \leq a'$;
6. $X_{(2)}(a, b) \leq_{st} X_{(2)}(a, b')$ *for* $b \leq b'$.

PROOF. From (19.13), we have

$$f_{(1)}(t) = (2a + 2a^2bt)\exp(-2at - a^2bt^2),$$

$$\frac{\partial}{\partial t}\log f_{(1)}(t) = \frac{ab}{1 + abt} - 2ab - 2a^2bt$$

and

$$\frac{\partial}{\partial t}\frac{\partial}{\partial t}\log f_{(1)}(t) = -\left(\frac{ab}{1 + abt}\right)^2 - 2a^2b < 0$$

which implies that $f_{(1)}(t)$ is log-concave and hence $X_{(1)}$ is ILR. Moreover,

$$\frac{\partial^2}{\partial a \partial t} \log f_{(1)}(t) = \frac{b}{(1+abt)^2} - 2b - 4abt < 0,$$

since $b \leq 1$ and hence, from Lemma 19.2.1, we obtain the second property. The third one is obtained from (19.16).

To prove that $X_{(2)}$ is ILR, we use (19.13) and (19.14) to obtain

$$f^*_{(2)}(t) = 2e^{-t} - 2(1+bt)e^{-2t-bt^2}.$$

Hence,

$$\frac{\partial}{\partial t} \log f^*_{(2)}(t) = \frac{2 - b + 4bt + 2b^2t^2 - e^{t+bt^2}}{e^{t+bt^2} - 1 - bt}$$

and

$$\frac{\partial}{\partial t}\frac{\partial}{\partial t} \log f^*_{(2)}(t) = \frac{m(t,b)e^{t+bt^2}}{(e^{t+bt^2} - 1 - bt)^2},$$

where

$$m(t,b) = -1 + b(4-5t) + b^2t(6-8t) - 4b^3t^3 - b(b+2+4bt+2b^2t^2)e^{-t-bt^2}.$$

Obviously, $m(t,b) \leq 0$, when $t \geq 1$. Let us suppose $0 \leq t < 1$, then $m(0,b) = -(b-1)^2 \leq 0$ and using that $0 \leq b \leq 1$ and $e^{-t-bt^2} \leq 1$, we have

$$\frac{\partial}{\partial t} m(t,b) = b\psi(t,b) + 2b^2t(bt-1) + 2b^3t^2(b-1) \leq b\psi(t,b),$$

where

$$\psi(t,b) = 6b - 5 - 6bt + (2 - 3b - 2b^2t)e^{-t-bt^2}.$$

To complete the proof, it is sufficient to prove that $\psi(t,b) \leq 0$ for $0 \leq t < 1$ and $0 < b \leq 1$. But, if

$$\psi_1(t,b) = 6b - 5 - 6bt + (2-3b)e^{-t-bt^2},$$

then $\psi_1(0,b) = 3(b-1) \leq 0$ and

$$\frac{\partial}{\partial t}\psi_1(t,b) = -6b + (3b-2)(1+2bt)e^{-t-bt^2} \leq 0,$$

since, if $0 < b \leq 2/3$, we have $3b - 2 \leq 0$, and if $2/3 < b \leq 1$, then

$$\frac{\partial}{\partial t}\psi(t,b) \leq -6\frac{2}{3} + 3be^{-t-bt^2} \leq -4 + 3b < 0.$$

The two last properties are obtained from $r_{(2)}(t) = ar^*_{(2)}(at)$, $R_{(2)}(t) = R^*_{(2)}(at)$ and (19.14). ■

Remark 19.5.1 In particular, from (19.1), we have

$$X_{(1)}(a,b) \geq_{fr,mrl,st} X_{(1)}(a',b')$$

for $a \leq a'$ and $b \leq b'$ and $X_{(2)}(a,b) \geq_{mrl,st} X_{(2)}(a',b)$ for $a \leq a'$. Note that the ordering properties can also be written using parameters μ and ρ. Also note that

$$\frac{\partial^2}{\partial b \partial t} \log f_{(1)}(t) = \frac{a}{(1+abt)^2} - 2a^2 t$$

has positive and negative values (e.g., $t=0$ and $t=1/a$) and hence $X_{(1)}$ is not ordered with respect to b in the lr-order. Moreover, from (19.18), $X_{(2)}$ is not mrl-ordered (hr, lr) in b. It is easy to show that

$$r_{(1)}(t) \geq 2a \geq r_i(t) = a \geq r_{(2)}(t)$$

and $\lim_{t\to\infty} r_{(2)}(t) = a$ (see Figure 19.1). Note that, from (19.3), we obtain a constant failure rate (exponential distribution) from a mixture of two strictly IFR distributions. We have also obtained that

$$X_{(1)} \leq_{lr} X_i \leq_{lr} X_{(2)}.$$

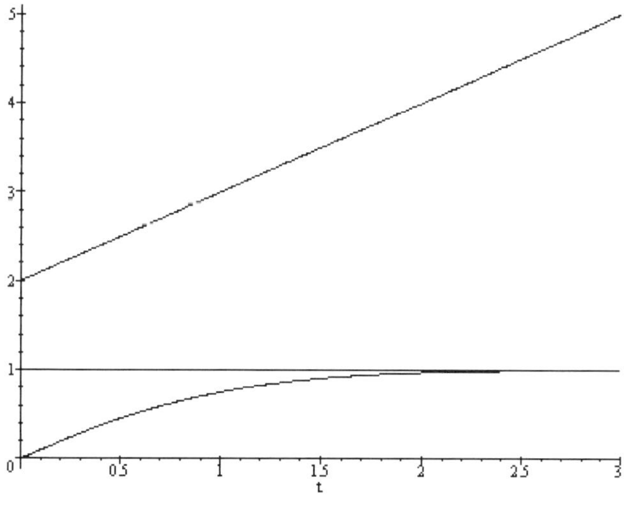

Figure 19.1: Failure rates for series, components, and parallel system from GED ($a=1$, $b=0.5$). Note that $r_{(2)}(t) \leq r_i(t) = 1 \leq r_{(1)}(t)$ and $r_{(2)}(t)$ tends to $r_i(t) = 1$ as $t \to \infty$

19.6 Parameters Estimation

Castillo *et al.* (1997) discussed an estimation method for $1/a$ and b by using a bivariate sample from (X_1, X_2). Here, we suppose that we have a sample from $X_{(1)}$ or $X_{(2)}$ (i.e., we only have information about system life length). The distributions of $X_{(1)}$ and $X_{(2)}$ only depend on a and b. So, we can estimate a and b from (19.15) by applying Pearson's estimation method using the sample mean and variance from $X_{(1)}$. The estimators are given in the following proposition.

Proposition 19.6.1 *If* $(X_1, X_2) \equiv GED(a, b)$ *and* $T_1, \ldots, T_n$ *is a sample of i.i.d. r.v.'s from* $X_{(1)}$*, then the Pearson's estimators for* a *and* b *are obtained from*

$$\frac{1 - 2\sqrt{\frac{\pi}{b}}e^{1/b}\Phi\left(-\sqrt{\frac{2}{b}}\right)}{\pi e^{2/b}\Phi^2\left(-\sqrt{\frac{2}{b}}\right)} = \frac{\overline{T^2}}{\overline{T}^2} \tag{19.20}$$

and

$$a = \frac{1}{\overline{T}}\sqrt{\frac{\pi}{b}}e^{1/b}\Phi\left(-\sqrt{\frac{2}{b}}\right),$$

where $\overline{T} = \frac{1}{n}\sum_{i=1}^n T_i$, $\overline{T^2} = \frac{1}{n}\sum_{i=1}^n T_i^2$ *and (19.20) has a unique solution* $b_0 \in [0, 1]$ *when* $1.6862 \leq \overline{T^2}/\overline{T}^2 \leq 2$.

The proof is easy. We note that $1.6862 \leq E(X_{(1)}^2)/E^2(X_{(1)}) \leq 2$. The estimators based on a sample from $X_{(2)}$ can be obtained in a similar way.

The parameter estimation from $X_{(1)}$ is equivalent to that of a distribution with linear failure rate $r(t) = \lambda_1 + 2\lambda_2 t$, which was studied by Bain (1974), Ashour and Youssef (1991), and Sen and Bhattacharyya (1995). They obtained maximum likelihood estimators for λ_1 and λ_2 for type II censored samples. If $\widehat{\lambda}_1$ and $\widehat{\lambda}_2$ are these estimators, then $\widehat{a} = \widehat{\lambda}_1/2$ and $\widehat{b} = 4\widehat{\lambda}_2/\widehat{\lambda}_1^2$ are maximum likelihood estimators for a and b for type II censored samples from $X_{(1)}$. They also gave exact confidence sets for $\eta = 2a$ and $\theta = ab$. The maximum likelihood estimators are given in the following proposition.

Proposition 19.6.2 *If* $(X_1, X_2) \equiv GED(a, b)$ *and* $T_1, \ldots, T_n$ *is a sample of i.i.d. r.v.'s from* $X_{(1)}$*, then the maximum likelihood estimators for* a *and* b *are determined by*

$$\frac{1}{n}\sum_{i=1}^n \frac{T_i}{a\overline{T^2} - (2a\overline{T} - 1)T_i} = 1 \tag{19.21}$$

and

$$b = \frac{1 - 2a\overline{T}}{a^2\overline{T^2}}, \tag{19.22}$$

where $\overline{T} = \frac{1}{n}\sum_{i=1}^{n} T_i$ *and* $\overline{T^2} = \frac{1}{n}\sum_{i=1}^{n} T_i^2$.

PROOF. To obtain maximum likelihood estimators, we must solve

$$\max_{a,b} \log l_{(1)} = n\log(2a) + \sum_{i=1}^{n}\log(1+abT_i) - 2an\overline{T} - ba^2 n\overline{T^2},$$

where $a > 0$ and $0 \leq b \leq 1$. Differentiating, we have

$$\frac{\partial}{\partial a}\log l_{(1)} = \frac{n}{a} + b\sum_{i=1}^{n}\frac{T_i}{1+abT_i} - 2n\overline{T} - 2abn\overline{T^2} = 0$$

and

$$\frac{\partial}{\partial b}\log l_{(1)} = a\sum_{i=1}^{n}\frac{T_i}{1+abT_i} - a^2 n\overline{T^2} = 0.$$

By solving this system, we obtain (19.21) and (19.22). ■

19.7 Systems with n Exchangeable Components

In this section we obtain some results for coherent systems with n exchangeable components and GED distribution. Let $(X_1, \ldots, X_n)$ be a random vector representing the life lengths of n components in a system with a multivariate GED defined by the reliability function

$$R(x_1, \ldots, x_n) = \exp\left\{-\sum_{s\in\xi_n}\lambda_s\left(\prod_{i=1}^{n} x_i^{s_i}\right)\right\} \tag{19.23}$$

[see, e.g., Kotz *et al.* (2000)], where $\xi_n = \{s = (s_1, \ldots, s_n) : s_i = 0 \text{ or } 1 \text{ for every } i \in \{1, \ldots, n\}\}$ and $\lambda_s \geq 0$ for all $s \in \xi_n$. Hence, the k-out-of-n system is represented by the order statistic $X_{(n-k+1,n)}$. If we suppose that $(X_1, \ldots, X_n)$ has an exchangeable distribution, then the GED can be reparametrized as

$$R(x_1, \ldots, x_n) = \exp\left(-\lambda_1\sum_{i=1}^{n} x_i - \lambda_2\sum_{i<j} x_i x_j - \cdots - \lambda_n x_1 x_2 \cdots x_n\right), \tag{19.24}$$

where $\lambda_1 > 0$ and $\lambda_i \geq 0$ for $i = 2, 3, \ldots, n$. We denote this model by $GED(\lambda_1, \ldots, \lambda_n)$. Obviously, $R_i(t) = \exp(-\lambda_1 t)$ for $i = 1, 2, \ldots, n$.

Thus, the reliability function for the series system is given by

$$R_{(1,n)}(t) = R(t, \ldots, t) = \exp\left(-\sum_{k=1}^{n}\binom{n}{k}\lambda_k t^k\right), \tag{19.25}$$

and the failure rate by

$$r_{(1,n)}(t) = n \sum_{k=1}^{n} \binom{n-1}{k-1} \lambda_k t^{k-1}.$$

Hence, $X_{(1,n)}$ is IFR, $X_{(1,n)} \leq_{fr} X_i$ and

$$X_{(1,n)}(\lambda_1, \ldots, \lambda_n) \geq_{fr,mrl,st} X_{(1,n)}(\lambda'_1, \ldots, \lambda'_n),$$

for $\lambda_i \leq \lambda'_i$, $i = 1, 2, \ldots, n$. Moreover, from (19.25), we have

$$X_{(1,n)}(\lambda_1, \ldots, \lambda_n) \geq_{lr} X_{(1,n)}(\lambda'_1, \lambda_2, \ldots, \lambda_n),$$

for $\lambda_1 \leq \lambda'_1$ (i.e., for $\mu_i \geq \mu'_i$ for $i = 1, 2, \ldots, n$). However, in general, this property does not hold for $\lambda_2, \ldots, \lambda_n$.

The reliability function for the k-out-of-n system can be computed from the following result.

Proposition 19.7.1 *If* $(X_1, \ldots, X_n) \equiv GED(\lambda_1, \ldots, \lambda_n)$, *then*

$$R_{(i,n)}(t) = \sum_{s=n-i+1}^{n} \binom{n}{s} c_{n-i+1,s} \exp\left(-\sum_{k=1}^{s} \binom{s}{k} \lambda_k t^k\right), \qquad (19.26)$$

for $t > 0$ *and* $i = 1, \ldots, n$, *where* $c_{i,s} = \sum_{j=i}^{s} (-1)^{s-j} \binom{s}{j}$.

The proof is obtained from (19.5), (19.24), and (19.25).

Remark 19.7.1 If $T = \Phi(X_1, \ldots, X_n)$ is a coherent system based on exchangeable components with GED, then the system reliability can be obtained from the representation (19.10) of a coherent system as a mixture of series systems and from (19.25). For example, the Samaniego's signature of the coherent system $T = \min(X_1, \max(X_2, X_3))$ is $(1/3, 2/3, 0)$ [see Samaniego (1985)], that is

$$R_T(t) = \frac{1}{3} R_{(1,3)}(t) + \frac{2}{3} R_{(2,3)}(t).$$

From (19.5), we have

$$R_{(2,3)}(t) = 3R_{(1,2)}(t) - 2R_{(1,3)}(t).$$

and hence, its minimal signature is $(0, 2, -1)$ and the system reliability can be computed as

$$\begin{aligned} R_T(t) &= 2R_{(1,2)}(t) - R_{(1,3)}(t) \\ &= 2\exp\left(-2\lambda_1 t - \lambda_2 t^2\right) - \exp\left(-3\lambda_1 t - 3\lambda_2 t^2 - \lambda_3 t^3\right). \end{aligned}$$

References

1. Ashour, S. K., and Youssef, A. (1991). Bayesian estimation of a linear failure rate, *Journal of the Indian Association for Production, Quality and Reliability*, **16**, 9–16.

2. Baggs, G. E., and Nagaraja, H. N. (1996). Reliability properties of order statistics from bivariate exponential distributions, *Communications in Statistics – Stochastic Models*, **12**, 611–631.

3. Bain, L. J. (1974). Analysis for the linear failure-rate life-testing distribution, *Technometrics*, **16**, 551–559.

4. Balakrishnan, N., Bendre, S. M., and Malik, H. J. (1992). General relations and identities for order statistics from non-independent non-identical variables, *Annals of the Institute of Statistical Mathematics*, **44**, 177–183.

5. Barlow, R. E., and Proschan, F. (1975). *Statistical Theory of Reliability and Life Testing*, Holt, Rinehart and Winston, New York.

6. Belzunce, F. Franco, M., Ruiz, J. M., and Ruiz, M. C. (2001). On partial orderings between coherent systems with different structures, *Probability in the Engineering and Informational Sciences*, **15**, 273–293.

7. Block, H. W., and Joe, H. (1997). Tail behavior of the failure rate functions of mixtures, *Lifetime Data Analysis*, **3**, 269–288.

8. Castillo, E., Sarabia, J. M., and Hadi, A. S. (1997). Fitting continuous bivariate distributions to data, *The Statistician*, **46**, 355–369.

9. Cox, D. R., and Oakes, D. (1984). *Analysis of Survival Data*, Chapman and Hall, London.

10. David, H. A. (1970). *Order Statistics*, John Wiley & Sons, New York.

11. Gumbel, E. J. (1960). Bivariate exponential distributions, *Journal of the American Statistical Association*, **55**, 698–707.

12. Gupta, R. C. (2001). Reliability studies of bivariate distributions with Pareto conditionals, *Journal of Multivariate Analysis*, **76**, 214–225.

13. Gupta, P. L., and Gupta, R. C. (2001). Failure rate of the minimum and maximum of a multivariate normal distribution, *Metrika*, **53**, 39–49.

14. Kanjo, A. I., and Abouammoh, A. M. (1995). Closure of mean remaining life classes under formation of parallel systems, *Pakistan Journal of Statistics*, **11** (2), 153–158.

15. Kochar, S., Mukerjee, H., and Samaniego, F. J. (1999). The "signature" of a coherent system and its application to comparison among systems, *Naval Research Logistics*, **46**, 507–523.

16. Kotz, S., Balakrishnan, N., and Johnson, N. L. (2000). *Continuous Multivariate Distributions*, John Wiley & Sons, New York.

17. Kotz, S., Lai, C. D. and Xie, M. (2003). On the effect of redundancy for systems with dependent components, *IIE Transactions*, **35**, 1103–1110.

18. Maurer, W., and Margolin, B. H. (1976). The multivariate inclusion-exclusion formula and order statistics from dependent variates, *Annals of Statistics*, **4**, 1190–1199.

19. Mi, J., and Shaked, M. (2002). Stochastic dominance of random variables implies the dominance of their order statistics, *Journal of the Indian Statistical Association*, **40**, 161–168.

20. Roy, D. (2001). Some properties of a classification system for multivariate life distributions, *IEEE Transactions on Reliability*, **50**, 214–220.

21. Rychlik, T. (2001a). Stability of order statistics under dependence, *Annals of the Institute of Statistical Mathematics*, **53**, 877–894.

22. Rychlik, T. (2001b). Mean-variance bounds for order statistics from dependent DFR, IFR, DFRA and IFRA samples, *Journal of Statistical Planning and Inference*, **92**, 21–38.

23. Samaniego, F. (1985). On closure of the IFR class under formation of coherent systems. *IEEE Transactions on Reliability*, **34**, 69–72.

24. Sen, A., and Bhattacharyya, G. K. (1995). Inference procedures for the linear failure rate model, *Journal of Statistical Planning and Inference*, **46**, 59–76.

25. Shaked, M., and Shanthikumar, J. G. (1994). *Stochastic Orders and Their Applications*, Academic Press, San Diego.

26. Shaked, M., and Suarez-Llorens, A. (2003). On the comparison of reliability experiments based on the convolution order. *Journal of the American Statistical Association*, **98**, 693–702.

20

Estimating the Mean of Exponential Distribution from Step-Stress Life Test Data

Zhenmin Chen, Jie Mi, and YanYan Zhou

Florida International University, Miami, FL, USA

Abstract: This paper considers the step-stress accelerated life tests (ALT) on an exponential population with mean θ. The MLEs of θ are studied for different data structures that include grouped data and censored data. Here, by grouped data we mean that instead of observing the exact failure times, only numbers of failures in some predetermined subintervals are available. Applying the tampered failure rate (TFR) model, we show the existence, uniqueness, strong consistency, and the asymptotic normality of the MLE of θ. An upper bound of the MLE of θ based on the grouped data is also derived.

Keywords and phrases: Exponential distribution, step-stress ALT, TFR model, type-I censored data, type-II censored data, grouped data

20.1 Introduction

The purpose of the accelerated life test (ALT) is to obtain information on the lifetime distribution of products under normal working stress with a short test time so as to save expense and manpower. Particularly, it is very useful when the products under test are highly reliable because it would take an extremely long time to complete life testing under a normal stress level. The step-stress accelerated life test (SSALT) is one of the ALT. The test is organized as follows: Choose $m \geq 2$ different stress levels $S_1 < S_2 < \cdots < S_m$, and $m-1$ termination points $t_0 \equiv 0 < t_1 < t_2 < \cdots < t_{m-1} < \infty \equiv t_m$. First, the test units are tested under stress level S_1 until time t_1, then the survived test units will be tested under stress level S_2 during the time interval $(t_1, t_2]$, and the test will be continued to the last stage at which the survived test units, if there are any, will be tested under stress level S_m on the time interval (t_{m-1}, ∞). If the lowest stress level S_1 is the same as the normal stress level S_0, then this SSALT is called a partial SSALT; otherwise, if $S_1 > S_0$, then it is called a complete SSALT. An ALT is called a simple ALT if the number of different stress levels $m = 2$.

A key for implementing ALT and obtaining information on the lifetime distribution of products under normal stress level is the relationship between the lifetime distribution at stress level S_0 and that at higher stress level S_i $(1 \leq i \leq m)$. In the literature, there are three different models. DeGroot and Goel (1979) proposed the tampered random variable (TRV) model. Nelson (1980) proposed the cumulative exposure (CE) model. Bhattacharyya and Zanzawi (1989) proposed tampered failure rate (TFR) model which assumes that the effect of changing stress level is to multiply the initial failure rate function $h_1(x)$ by a factor subsequent to the change point t. More specifically, let the failure rate function of the step-stress lifetime X^* by $h^*(x)$, then $h^*(x) = h_1(x)$ if $x \leq t$, and $h^*(x) = \alpha h_1(x)$ if $x > t$, where the accelerator factor α depends on stress levels S_1 and S_2 and possibly also on t, but not on x. Madi (1993) generalized the TFR model from the simple $(m = 2)$ step-stress setting to the multiple setting $(m \geq 2)$. Khamis and Higgins (1998) considered the same generalization.

Xiong (1998) discussed the MLE of the mean of exponential distribution based on a simple SSALT with type-II censored data. Xiong and Milliken (1999) studied the MLE of the mean based on general SSALT with type-I censored exponential data.

Throughout this chapter, it is assumed that all the accelerator factors α_i $(1 \leq i \leq m)$ are known. In addition to the type-I and type-II censored data in which the exact failure times in each interval $[t_{j-1}, t_j)$, $1 \leq j \leq m$, are known except those that are censored, we will also consider grouped data. By grouped data, we mean that instead of observing the exact failure times we can only observe the number of failed test items in each interval $[t_{j-1}, t_j)$. The paper is organized as follows. Section 20.2 studies the MLE $\hat{\theta}$ of θ with type-I censored data. The case of grouped data is discussed in Section 20.3. The case of type-II censored data will be investigated in Section 20.4. In all these cases, the existence and uniqueness of the MLE of θ are derived. Moreover, the strong consistency of the MLE is also obtained. In Section 20.5, we give our simulation findings, which indicates that the performance of the MLEs of θ based on grouped sample is almost the same as that based on complete sample. This justifies the application of the step-stress accelerated life tests.

20.2 Type I Censored Data

Let Y be the step-stress life. Then its failure rate function $h(t)$ under the tampered failure rate (TFR) model is determined as

$$h(t) = \alpha_i \lambda \equiv \lambda_i \ \forall\ t_{i-1} \leq t < t_i, \ 1 \leq i \leq m, \tag{20.1}$$

where $1 \leq \alpha_1 < \alpha_2 < \cdots < \alpha_m < \infty$ are known constants. Based on the TFR model (1), the survival function $\bar{F}(t)$ of Y is thus given by

$$\begin{aligned} \bar{F}(t) = e^{-\int_0^t h(u)du} &= \exp\left\{ -\sum_{i=1}^{\ell-1} \int_{t_{i-1}}^{t_i} h(u)du - \int_{t_{\ell-1}}^{t} h(u)du \right\} \\ &= \exp\left\{ -\sum_{i=1}^{\ell-1} \lambda_i (t_i - t_{i-1}) - \lambda_\ell (t - t_{\ell-1}) \right\} \ \forall\ t_{\ell-1} \leq t < t_\ell. \end{aligned} \tag{20.2}$$

Throughout this paper, $\sum_{i=1}^{0} a_i$ is defined as 0 for any a_i.

In this section, it is assumed that n units will be on the step-stress test, but only the failure times before t_{m-1} are available. That is, the data are censored at time t_{m-1}. Let n_i be the number of test units failed in the interval $[t_{i-1}, t_i)$, $1 \leq i \leq m$, $N_j = \sum_{i=1}^{j} n_i$, $1 \leq j \leq m$, and $N_0 = 0$ by convention. Note that $N_m = n$ and $N_{m-1} = 1 - n_m$. Denote the order statistics from $Y_1, \ldots, Y_n$ as $Y_{(i)}$, $1 \leq i \leq n$.

Lemma 20.2.1 *As $n \to \infty$, it holds that $Y_{(N_{j-1}+1)} \to t_{j-1}$ and $Y_{(N_j)} \to t_j$ with probability one.*

PROOF. First, let us show the following results: If $X, X_1, X_2, \ldots$ are iid random variables with support set that has infinimum $a > -\infty$, then $X_{1:n} \to a$ with probability one as $n \to \infty$. To see this, with $\epsilon > 0$ being arbitrary, observe that

$$\sum_{n=1}^{\infty} P(X_{1:n} > a + \epsilon) = \sum_{n=1}^{\infty} (P(X > a + \epsilon))^n < \infty$$

because $P(X \leq a + \epsilon) > 0$. Hence, $X_{1:n} \to a$ with probability one.

Similarly, it can be shown that $X_{n:n} \to b$ with probability one, where $b < \infty$ is the supremum of the support set of x.

Define $X_i = Y_i I_{[t_{j-1}, t_j)}(Y_i)$. It is easy to see that $Y_{(N_{j-1}+1)} = X_{1:n}$. Obviously t_{j-1} is the infinitum of the support set of X_i, and thus from the above results it follows that $Y_{(N_{j-1}+1)} \to t_{j-1}$. Similarly, notice that t_j is the supremum of the support set of X_i and $Y_{(N_j)} = X_{n:n}$, so $Y_{(N_j)} \to t_j$ a.s.. ■

Theorem 20.2.1 *Under the TFR model characterized by (20.1) if $n_m < n$, then the MLE of $\theta = 1/\lambda$ exists and is given by*

$$\hat{\theta} = \frac{\sum_{j=1}^{m} n_j \left(\sum_{\ell=1}^{j-1} \alpha_\ell (t_\ell - t_{\ell-1}) \right) + \sum_{j=1}^{m-1} \alpha_j \left(\sum_{i=N_{j-1}+1}^{N_j} (Y_{(i)} - t_{j-1}) \right)}{N_{m-1}} \tag{20.3}$$

based on the sample censored at time t_{m-1}.

PROOF. From the fact that the pdf of Y is $f(t) = h(t)\bar{F}(t)$ and (2), the likelihood function is obtained as

$$\begin{aligned} L &= c \prod_{j=1}^{m-1} \left\{ \prod_{i=N_{j-1}+1}^{N_j} \left[\alpha_j \lambda e^{-\sum_{\ell=1}^{j-1} \alpha_\ell (t_\ell - t_{\ell-1})\lambda - \alpha_j (y_{(i)} - t_{j-1})\lambda} \right] \right\} \\ &\quad \times \left(e^{-\sum_{\ell=1}^{m-1} \alpha_\ell (t_\ell - t_{\ell-1})\lambda} \right)^{n_m}, \end{aligned}$$

where $y_{(i)}$ is the observed value of $Y_{(i)}$. The log-likelihood function is thus given by

$$\begin{aligned} \ln L &= \ln c + N_{m-1} \ln \lambda - \lambda \sum_{j=1}^{m} n_j \left(\sum_{\ell=1}^{j-1} \alpha_\ell (t_\ell - t_{\ell-1}) \right) \\ &\quad - \lambda \sum_{j=1}^{m-1} \alpha_j \left(\sum_{i=N_{j-1}+1}^{N_j} (y_{(i)} - t_{j-1}) \right) \end{aligned} \tag{20.4}$$

It then follows that

$$\frac{d \ln L}{d\lambda} = \frac{N_{m-1}}{\lambda} - \sum_{j=1}^{m} n_j \left(\sum_{\ell=1}^{j-1} \alpha_\ell (t_\ell - t_{\ell-1}) \right) - \sum_{j=1}^{m-1} \alpha_j \left(\sum_{i=N_{j-1}+1}^{N_j} (y_{(i)} - t_{j-1}) \right).$$

Setting $d \ln L / d\lambda$ to be zero, we obtain $\hat{\theta} = 1/\hat{\lambda}$ given by (20.3). ■

Remark 20.2.1 If $n_m = n$, then the MLEs of λ and θ do not exist. The probability of the event $A_n \equiv \{n_m = n\}$ is

$$P(A_n) = (P(Y > t_m))^n = \exp\left\{ -n \sum_{\ell=1}^{m-1} \alpha_\ell (t_\ell - t_{\ell-1})\lambda \right\}.$$

From this, it follows that

$$P(\limsup_{n \to \infty} A_n) = \lim_{n \to \infty} P\left(\bigcup_{k=n}^{\infty} A_k \right)$$

$$
\begin{aligned}
&\leq \lim_{n\to\infty}\sum_{k=n}^{\infty} P(A_k) \\
&= \lim_{n\to\infty}\sum_{k=n}^{\infty}\left(e^{-\sum_{\ell=1}^{m-1}\alpha_\ell(t_\ell-t_{\ell-1})\lambda}\right)^k = 0.
\end{aligned}
$$

Therefore, in the following study on $\hat{\theta}$ obtained in Theorem 20.2.1 we can ignore the set $\limsup_{n\to\infty} A_n$ of probability zero.

Theorem 20.2.2 *For a given sample size n, denote the MLE of θ by $\hat{\theta}_n$. It holds that*

(i) $\dfrac{\frac{U_n}{n} - \theta F(t_{m-1})}{S_n/\sqrt{n}} \xrightarrow{L} N(0,1)$ *as* $n\to\infty$;

(ii) $\hat{\theta}_n \to \theta$ *with probability one, where U_n is the numerator in (20.3), and S_n^2 is the sample variance of $W_1,\ldots,W_n$ defined in (20.6) below.*

PROOF. We have

$$
\begin{aligned}
&U_n \\
&= \sum_{j=1}^{m} n_j\left(\sum_{\ell=1}^{j-1}\alpha_\ell(t_\ell-t_{\ell-1})\right)+\sum_{j=1}^{m-1}\alpha_j\sum_{i=N_{j-1}+1}^{N_j}Y_{(i)}-\sum_{j=1}^{m-1}\alpha_j n_j t_{j-1} \\
&= \sum_{j=1}^{m}\left(\sum_{i=1}^{n} I_{[t_{j-1},t_j)}(Y_i)\right)\left(\sum_{\ell=1}^{j-1}\alpha_\ell(t_\ell-t_{\ell-1})\right)+\sum_{j=1}^{m-1}\alpha_j\left(\sum_{i=1}^{n} Y_i I_{[t_{j-1},t_j)}(Y_i)\right) \\
&\quad -\sum_{j=1}^{m-1}\alpha_j t_{j-1}\left(\sum_{i=1}^{n} I_{[t_{j-1},t_j)}(Y_i)\right) \\
&= \sum_{i=1}^{n}\left(\sum_{j=1}^{m} I_{[t_{j-1},t_j)}(Y_i)\right)\left(\sum_{\ell=1}^{j-1}\alpha_\ell(t_\ell-t_{\ell-1})\right)+\sum_{i=1}^{n}\left(\sum_{j=1}^{m-1}\alpha_j Y_i I_{[t_{j-1},t_j)}(Y_i)\right) \\
&\quad -\sum_{i=1}^{n}\left(\sum_{j=1}^{m-1}\alpha_j t_{j-1} I_{[t_{j-1},t_j)}(Y_i)\right). \qquad (20.5)
\end{aligned}
$$

Define

$$
\begin{aligned}
W_i &= \sum_{j=1}^{m} I_{[t_{j-1},t_j)}(Y_i)\left(\sum_{\ell=1}^{j-1}\alpha_\ell(t_\ell-t_{\ell-1})\right)+\sum_{j=1}^{m-1}\alpha_j Y_i I_{[t_{j-1},t_j)}(Y_i) \\
&\quad - \sum_{j=1}^{m-1}\alpha_j t_{j-1} I_{[t_{j-1},t_j)}(Y_i), \quad 1\leq i\leq n. \qquad (20.6)
\end{aligned}
$$

Clearly, $W_1, \cdots, W_n$ are i.i.d. random variables. Also,

$$\begin{aligned}E(W_i) &= \sum_{j=1}^{m} P(t_{j-1} < Y_i \le t_j)\left(\sum_{\ell=1}^{j-1} \alpha_\ell (t_\ell - t_{\ell-1})\right) + \sum_{j=1}^{m-1} \alpha_j E(Y_i I_{[t_{j-1},t_j)}(Y_i)) \\ &\quad - \sum_{j=1}^{m-1} \alpha_j t_{j-1} P(t_{j-1} < Y_i \le t_j).\end{aligned}$$

Note that

$$E(Y_i I_{[t_{j-1},t_j)}(Y_i)) = -t_j \bar{F}(t_j) + t_{j-1}\bar{F}(t_{j-1}) + \int_{t_{j-1}}^{t_j} \bar{F}(y)\, dy.$$

Hence,

$$\begin{aligned}E(W_i) &= \sum_{j=1}^{m}(\bar{F}(t_{j-1}) - \bar{F}(t_j))(\sum_{\ell=1}^{j-1} \alpha_\ell (t_\ell - t_{\ell-1})) - \sum_{j=1}^{m-1} \alpha_j t_j \bar{F}(t_j) \\ &\quad + \sum_{j=1}^{m-1} \alpha_j t_{j-1} \bar{F}(t_{j-1}) + \sum_{j=1}^{m-1} \alpha_j \int_{t_{j-1}}^{t_j} \bar{F}(y)\, dy \\ &= \sum_{\ell=1}^{m-1} \sum_{j=\ell+1}^{m} (\bar{F}(t_{j-1}) - \bar{F}(t_j))(\alpha_\ell (t_\ell - t_{\ell-1})) - \sum_{j=1}^{m-1} \alpha_j t_j \bar{F}(t_j) \\ &\quad + \sum_{j=1}^{m-1} \alpha_j t_{j-1} \bar{F}(t_j) + \sum_{j=1}^{m-1} \alpha_j \int_{t_{j-1}}^{t_j} \bar{F}(y)\, dy \\ &= \sum_{\ell=1}^{m-1} (\alpha_\ell (t_\ell - t_{\ell-1}))\bar{F}(t_\ell) - \sum_{j=1}^{m-1} \alpha_j (t_j - t_{j-1}) \bar{F}(t_j) \\ &\quad + \sum_{j=1}^{m-1} \alpha_j \int_{t_{j-1}}^{t_j} \bar{F}(y)\, dy \\ &= \sum_{j=1}^{m-1} \alpha_j \int_{t_{j-1}}^{t_j} \bar{F}(y)\, dy. \end{aligned} \tag{20.7}$$

Note that

$$\begin{aligned}\alpha_j \int_{t_{j-1}}^{t_j} \bar{F}(t)dt &= \alpha_j \int_{t_{j-1}}^{t_j} e^{-\sum_{\ell=1}^{j-1} \alpha_\ell \lambda (t_\ell - t_{\ell-1}) - \alpha_j \lambda (t - t_{j-1})} dt \\ &= e^{-\sum_{\ell=1}^{j-1} \alpha_\ell \lambda (t_\ell - t_{\ell-1})} \int_{t_{j-1}}^{t_j} \alpha_j e^{-\alpha_j \lambda (t - t_{j-1})} dt\end{aligned}$$

$$= \theta e^{-\sum_{\ell=1}^{j-1} \alpha_\ell \lambda (t_\ell - t_{\ell-1})} \left(1 - e^{-\alpha_j \lambda (t - t_{j-1})}\right)$$
$$= \theta \left(\bar{F}(t_{j-1}) - \bar{F}(t_j)\right)$$

and thus

$$\sum_{j=1}^{m-1} \alpha_j \int_{t_{j-1}}^{t_j} \bar{F}(t)dt = \theta \sum_{j=1}^{m-1} \left(\bar{F}(t_{j-1}) - \bar{F}(t_j)\right) = \theta(1 - \bar{F}(t_{m-1})) = \theta F(t_{m-1}). \tag{20.8}$$

Therefore, we obtain $E(W_i) = \theta F(t_{m-1})$ by (20.7) and (20.8). Denote $Var(W_i) = \sigma^2$. From (20.5) and (20.6), we see that $U_n = \sum_{i=1}^n W_i$, and so by the central limit theorem it follows that

$$\frac{\frac{U_n}{n} - \theta F(t_{m-1})}{\frac{\sigma}{\sqrt{n}}} \xrightarrow{L} N(0,1).$$

If we denote the sample variance of sample $W_i, 1 \leq i \leq n$, by S_n^2, then by Slutsky theorem we obtain

$$\frac{\frac{U_n}{n} - \theta F(t_{m-1})}{\frac{S_n}{\sqrt{n}}} \xrightarrow{L} N(0,1).$$

The a.s. convergence in (ii) then readily follows from (i). ■

Corollary 20.2.1 *A $100(1-\alpha)\%$ confidence interval for θ can be obtained as*

$$(\psi^{-1}(U_n/n - (S_n/\sqrt{n})z_{\alpha/2}), \psi^{-1}(U_n/n + (S_n/\sqrt{n})z_{\alpha/2}))$$

provided

$$0 < U_n/n - (S_n/\sqrt{n})z_{\alpha/2} < U_n/n + (S_n/\sqrt{n})z_{\alpha/2}) < \sum_{\ell=1}^{m-1} \alpha_\ell (t_\ell - t_{\ell-1}),$$

where the function $\psi(\theta)$ is defined as

$$\psi(\theta) = \theta \left(1 - e^{-\sum_{\ell=1}^{m-1} \alpha_\ell (t_\ell - t_{\ell-1})/\theta}\right) \tag{20.9}$$

and $z_{\alpha/2}$ is the upper $\alpha/2$ percentile of the standard normal distribution.

PROOF. It is easy to verify that $\psi(0+) = 0$, $\psi(\infty) = \sum_{\ell=1}^{m-1} \alpha_\ell (t_\ell - t_{\ell-1})$, and $\psi(\theta)$ strictly increases in $\theta > 0$. Thus, from

$$P\left(\frac{U_n}{n} - z_{\alpha/2}\frac{S_n}{\sqrt{n}} < \theta F(t_{m-1}) < \frac{U_n}{n} + z_{\alpha/2}\frac{S_n}{\sqrt{n}}\right) \approx 1 - \alpha,$$

we obtain

$$P\left(\frac{U_n}{n} - z_{\alpha/2}\frac{S_n}{\sqrt{n}} < \theta(1 - e^{-\sum_{\ell=1}^{m-1}\alpha_\ell(t_\ell - t_{\ell-1})/\theta}) < \frac{U_n}{n} + z_{\alpha/2}\frac{S_n}{\sqrt{n}}\right) \approx 1 - \alpha$$

and

$$P\left(\frac{U_n}{n} - z_{\alpha/2}\frac{S_n}{\sqrt{n}} < \psi(\theta) < \frac{U_n}{n} + z_{\alpha/2}\frac{S_n}{\sqrt{n}}\right) \approx 1 - \alpha.$$

The desired confidence interval thus follows. ∎

Remark 20.2.2 Note that

$$\begin{aligned}
&\lim_{n\to\infty} \frac{U_n}{n} \pm z_{\alpha/2}\frac{S_n}{\sqrt{n}} \\
&= \theta F(t_{m-1}) = \theta\left(1 - \exp\left\{-\sum_{\ell=1}^{m-1}\alpha_\ell(t_\ell - t_{\ell-1})\right\}\right)\Big/\theta < \sum_{\ell=1}^{m-1}\alpha_\ell(t_\ell - t_{\ell-1})
\end{aligned}$$

with probability one. Hence, the conditions in Corollary 20.2.1 will be satisfied for large n.

20.3 Grouped Data

In this section we assume that only n_j, $1 \le j \le m$, are available.

Theorem 20.3.1 *The MLE of θ based on the grouped data $\{n_1, \ldots, n_m\}$ uniquely exists if and only if $n_m < n$. If $n_m < n$, then the MLE of θ is given as the unique solution of the equation*

$$\sum_{j=1}^{m-1} n_j \frac{\alpha_j(t_j - t_{j-1})}{1 - e^{-\alpha_j(t_j - t_{j-1})/\theta}} = \sum_{j=1}^{m-1} n_j\alpha_j(t_j - t_{j-1}) + \sum_{j=1}^{m} n_j\left(\sum_{\ell=1}^{j-1}\alpha_\ell(t_\ell - t_{\ell-1})\right). \tag{20.10}$$

PROOF. Based on the grouped data $\{n_1, \ldots, n_m\}$, the likelihood function is given by

$$\begin{aligned}
L &= c\left\{\prod_{j=1}^{m-1}\Big(P(t_{j-1} < Y \le t_j)\Big)^{n_j}\right\}\Big(P(Y > t_{m-1})\Big)^{n_m} \\
&= c\, e^{-\lambda\sum_{j=1}^{m-1} n_j\left(\sum_{\ell=1}^{j-1}\alpha_\ell(t_\ell - t_{\ell-1})\right)} \prod_{j=1}^{m-1}\left[1 - e^{-\lambda\alpha_j(t_j - t_{j-1})}\right]^{n_j}
\end{aligned}$$

$$\times\, e^{-\lambda n_m \sum_{\ell=1}^{m-1} \alpha_\ell(t_\ell - t_{\ell-1})}$$

$$= \;\; c\, e^{-\lambda \sum_{j=1}^{m} n_j \left(\sum_{\ell=1}^{j-1} \alpha_\ell(t_\ell - t_{\ell-1})\right)} \prod_{j=1}^{m-1} \left[1 - e^{-\lambda\alpha_j(t_j - t_{j-1})}\right]^{n_j}, \qquad (20.11)$$

where c is a constant free of $\lambda = 1/\theta$. From (20.11), the log-likelihood function is

$$\ln L = -\lambda \sum_{j=1}^{m} n_j \left(\sum_{\ell=1}^{j-1} \alpha_\ell(t_\ell - t_{\ell-1})\right) + \sum_{j=1}^{m-1} n_j \ln\left[1 - e^{-\lambda\alpha_j(t_j - t_{j-1})}\right]$$

and the derivative of $\ln L$ with respect to λ is

$$\frac{d \ln L}{d\lambda} = -\sum_{j=1}^{m} n_j \left(\sum_{\ell=1}^{j-1} \alpha_\ell(t_\ell - t_{\ell-1})\right) + \sum_{j=1}^{m-1} n_j \alpha_j (t_j - t_{j-1}) \frac{e^{-\lambda\alpha_j(t_j - t_{j-1})}}{1 - e^{-\lambda\alpha_j(t_j - t_{j-1})}}.$$

Setting $(\ln L)' = 0$, we obtain the likelihood equation

$$\sum_{j=1}^{m-1} n_j \alpha_j (t_j - t_{j-1}) \frac{e^{-\lambda\alpha_j(t_j - t_{j-1})}}{1 - e^{-\lambda\alpha_j(t_j - t_{j-1})}} = \sum_{j=1}^{m} n_j \left(\sum_{\ell=1}^{j-1} \alpha_\ell(t_\ell - t_{\ell-1})\right). \qquad (20.12)$$

Eq. (20.12) can be further simplified as

$$\sum_{j=1}^{m-1} n_j \frac{\alpha_j (t_j - t_{j-1})}{1 - e^{-\lambda\alpha_j(t_j - t_{j-1})}} = \sum_{j=1}^{m-1} n_j \alpha_j (t_j - t_{j-1}) + \sum_{j=1}^{m} n_j \left(\sum_{\ell=1}^{j-1} \alpha_\ell(t_\ell - t_{\ell-1})\right). \qquad (20.13)$$

Define $\varphi_j(\lambda) = \frac{\alpha_j(t_j - t_{j-1})}{1 - e^{-\lambda\alpha_j(t_j - t_{j-1})}}$. It is easy to see that $\varphi_j(\lambda)$ strictly decreases in $\lambda > 0$, $\varphi_j(\infty) = \alpha_j(t_j - t_{j-1})$, and $\varphi_j(0+) = \infty$. Denote the left-hand side of (20.13) by $\varphi(\lambda)$. Then we see that $\varphi(\infty) = \sum_{j=1}^{m-1} n_j \alpha_j (t_j - t_{j-1})$, and if $n_m < n$, then $\varphi(0+) = \infty$ and $\varphi(\lambda)$ strictly decreases in $\lambda > 0$. Note that the right-hand side of (20.13) is greater than $\varphi(\infty) = \sum_{j=1}^{m-1} n_j \alpha_j (t_j - t_{j-1})$, and so Eq. (20.13) has a unique solution and the result follows. ■

Remark 20.3.1 If $n_m = n$, then the log-likelihood function is

$$\ln L = -n \sum_{\ell=1}^{m-1} \alpha_\ell(t_\ell - t_{\ell-1})\lambda$$

and so the MLE for λ and consequently the MLE for θ do not exist.

The result below shows the asymptotic normality of the MLE of θ determined by (20.13).

Theorem 20.3.2 *Denote the MLE of θ determined from (20.13) by $\hat{\theta}_n$.*
(i) $\hat{\theta}_n$ is the BAN estimator of θ, that is,

$$\sqrt{n}(\hat{\theta}_n - \theta) \xrightarrow{L} N(0, I^{-1}(\theta)), \tag{20.14}$$

where

$$I(\theta) = \sum_{j=1}^{m} \left(\frac{\partial p_j(\theta)}{\partial \theta} \right)^2 \frac{1}{p_j(\theta)},$$

and $p_j(\theta) = P(t_{j-1} < Y \le t_j), j = 1, \ldots, n$.
(ii) $\hat{\theta}_n \to \theta$ with probability one as $n \to \infty$.

Proof. It is easy to verify that the following conditions are satisfied:
1. $0 < \theta_1 \neq \theta_2$ implies $\sum_{j=1}^{m} |p_j(\theta_1) - p_j(\theta_2)| > 0$;
2. The derivative $p_j'(\theta)$ is continuous in $\theta > 0$, $1 \le j \le m$;
3. $I(\theta) > 0$, $\forall \theta > 0$.

Thus from the general properties of MLE [see, e.g., Rao (1973)], we know that for any $\theta > 0$ the likelihood equation (20.10) has at least one solution that satisfies (20.14). However, it is shown in Theorem 20.3.1 that the likelihood equation (20.10) has a unique solution $\hat{\theta}_n$ and that is the MLE of θ. Hence, $\hat{\theta}_n$ must be the BAN estimator of θ and satisfies (20.14). ∎

The following result gives an upper bound to $\hat{\theta}$ based on grouped data, and thus greatly facilitate the numerical solution of the Eq. (20.13).

Theorem 20.3.3 *An upper bound to $\hat{\theta}$ determined by (20.13) is given by*

$$\hat{\theta} < \frac{\sum_{j=1}^{m-1} n_j \alpha_j (t_j - t_{j-1}) + \sum_{j=1}^{m} n_j \left(\sum_{\ell=1}^{j-1} \alpha_\ell (t_\ell - t_{\ell-1}) \right)}{\sum_{j=1}^{m-1} n_j} \tag{20.15}$$

if $n_m < n$.

Proof. Note that

$$\frac{\lambda x}{1 - e^{-\lambda x}} > 1 \quad \forall\, x > 0.$$

Hence,

$$\begin{aligned} \sum_{j=1}^{m-1} n_j \frac{\alpha_j (t_j - t_{j-1})}{1 - e^{-\hat{\lambda} \alpha_j (t_j - t_{j-1})}} &= \hat{\theta} \sum_{j=1}^{m-1} n_j \cdot \frac{\hat{\lambda} \alpha_j (t_j - t_{j-1})}{1 - e^{-\hat{\lambda} \alpha_j (t_j - t_{j-1})}} \\ &> \hat{\theta} \sum_{j=1}^{m-1} n_j \end{aligned} \tag{20.16}$$

because $\sum_{j=1}^{m-1} n_j > 0$. From the inequality (20.16) and Eq. (20.13), we immediately obtain

$$\hat{\theta} < \frac{\sum_{j=1}^{m-1} n_j \alpha_j (t_j - t_{j-1}) + \sum_{j=1}^{m} n_j \left(\sum_{\ell=1}^{j-1} \alpha_\ell (t_\ell - t_{\ell-1}) \right)}{\sum_{j=1}^{m-1} n_j}.$$

■

Remark 20.3.2 Recall expression (20.3) which gives the MLE of θ based on the sample censored at the time t_{m-1}. It is clear that

$$\sum_{i=N_{j-1}+1}^{N_j} (Y_{(i)} - t_{j-1}) < \sum_{i=N_{j-1}+1}^{N_j} (t_j - t_{j-1}) = n_j (t_j - t_{j-1})$$

and hence

$$\hat{\theta} < \frac{\sum\limits_{j=1}^{m} n_j \left(\sum\limits_{\ell=1}^{j-1} \alpha_\ell (t_\ell - t_{\ell-1}) \right) + \sum\limits_{j=1}^{m-1} \alpha_j n_j (t_j - t_{j-1})}{\sum\limits_{j=1}^{m-1} n_j}.$$

This indicates that (20.15) is a common upper bound to the MLE of θ obtained from grouped data and data censored at time t_{m-1}.

20.4 Type II Censored Data

In the present section, we study the MLE of θ based type-II censored data. Let $1 \leq r \leq n$ be a predetermined integer. The life test will be terminated upon observing the rth failure time. In addition to the notation introduced in the previous sections, we assume $N_{k-1} < r \leq N_k$, $1 \leq k \leq m$. Obviously, $k = k(n)$ is a random variable. We also express $r = N_{k-1} + q$, where $0 < q \leq N_k - N_{k-1} = n_k$ is also a random variable.

According to the design of the life test, the likelihood function is given by

$$\begin{aligned} L &= c \prod_{j=1}^{k-1} \left\{ \prod_{i=N_{j-1}+1}^{N_j} \left[\alpha_j \lambda e^{-\lambda \sum\limits_{\ell=1}^{j-1} \alpha_\ell (t_\ell - t_{\ell-1}) - \alpha_j \lambda (Y_{(i)} - t_{j-1})} \right] \right\} \\ &\quad \times \left\{ \prod_{i=N_{k-1}+1}^{r} \left(\alpha_k \lambda e^{-\lambda \sum\limits_{\ell=1}^{k-1} \alpha_\ell (t_\ell - t_{\ell-1}) - \alpha_k \lambda (Y_{(i)} - t_{k-1})} \right) \right\} \end{aligned}$$

$$
\begin{aligned}
&\times \left(e^{-\lambda \sum_{\ell=1}^{k-1} \alpha_\ell (t_\ell - t_{\ell-1}) - \alpha_k \lambda (Y_{(r)} - t_{k-1})} \right)^{n-r} \\
&= c\lambda^r \exp\Bigg\{ -\lambda \sum_{j=1}^{k-2} n_j \left(\sum_{\ell=1}^{j-1} \alpha_\ell (t_\ell - t_{\ell-1}) \right) - \lambda (n - N_{k-1}) \sum_{\ell=1}^{k-1} \alpha_\ell (t_\ell - t_{\ell-1}) \\
&\quad -\lambda \sum_{j=1}^{k-1} \left(\sum_{i=N_{j-1}+1}^{N_j} (Y_{(i)} - t_{j-1}) \right) - \lambda \alpha_k \sum_{i=N_{k-1}+1}^{r} (Y_{(i)} - t_{k-1}) \\
&\quad -\lambda \alpha_k (n-r)(Y_{(r)} - t_{k-1}) \Bigg\}
\end{aligned}
$$

and the log-likelihood function is

$$
\begin{aligned}
\ln L &= \ln c + r \ln \lambda - \lambda \Bigg\{ \sum_{\ell=1}^{k-1} n_j \left(\sum_{\ell=1}^{j-1} \alpha_\ell (t_\ell - t_{\ell-1}) \right) \Bigg) \\
&\quad + (n - N_{k-1}) \sum_{\ell=1}^{k-1} \alpha_\ell (t_\ell - t_{\ell-1}) \\
&\quad + \sum_{j=1}^{k-1} \alpha_j \sum_{i=N_{j-1}+1}^{N_j} (Y_{(i)} - t_{j-1}) + \alpha_k \sum_{i=N_{k-1}+1}^{r} (Y_{(i)} - t_{k-1}) \\
&\quad + \alpha_k (n-r)(Y_{(r)} - t_{k-1}) \Bigg\}.
\end{aligned}
$$

Setting $(\ln L(\lambda))'$ equal to 0, we obtain $\hat{\theta} = 1/\hat{\lambda} = U_n/r$, where

$$
\begin{aligned}
U_n &= \sum_{j=1}^{k-1} n_j \left(\sum_{\ell=1}^{j-1} \alpha_\ell (t_\ell - t_{\ell-1}) \right) + (n - N_{k-1}) \sum_{\ell=1}^{k-1} \alpha_\ell (t_\ell - t_{\ell-1}) \\
&\quad + \sum_{j=1}^{k-1} \alpha_j \sum_{i=N_{j-1}+1}^{N_j} (Y_{(i)} - t_{j-1}) \\
&\quad + \alpha_k \sum_{i=N_{k-1}+1}^{r} (Y_{(i)} - t_{k-1}) + \alpha_k (n-r)(Y_{(r)} - t_{k-1}). \qquad (20.17)
\end{aligned}
$$

Summarizing the above, we obtain the following result.

Theorem 20.4.1 *Under the type II censoring, if only the first r failure times are available, then the MLE of θ is given by*

$$
\hat{\theta} = \hat{\theta}_n = \frac{U_n}{r} \qquad (20.18)
$$

where U_n is defined by (20.17).

The strong consistency of $\hat{\theta}_n$ obtained in Theorem 20.4.1 is shown below.

Theorem 20.4.2 *If $r/n = \alpha + o(n^{-1/2})$ with $\alpha \in (0,1]$, then $\lim_{n\to\infty} \hat{\theta}_n = \theta_0$ with probability one as $n \to \infty$, where θ_0 is the true value of the parameter θ.*

PROOF. Two cases need to be considered separately. In one case $\alpha \neq P(0 < Y \leq t_j)$, $\forall\ 1 \leq j \leq m-1$, and in the other case there exists an index $1 \leq j \leq m-1$ such that $\alpha = P(0 < Y \leq t_j)$.

We first consider case 1, that is $\alpha \neq P(0 < Y \leq t_j)$, $\forall\ 1 \leq j \leq m-1$. Denote $t^* = F^{-1}(\alpha)$, i.e., $P(0 < Y \leq t^*) = \alpha$. Suppose that

$$P(0 < Y \leq t_{k^*-1}) < \alpha < P(0 < Y \leq t_{k^*})$$

for a k^* satisfying $1 \leq k^* \leq m$. This means that $t_{k^*-1} < t^* < t_{k^*}$. From $r/n = \alpha + o(n^{-1/2})$, we have $Y_{(r)} \to t^* = F^{-1}(\alpha)$ a.s. [see David and Nagaraja (2003)]. This implies that $Y_{(r)} \in (t_{k^*-1}, t_{k^*})$ if n is sufficiently large. Hence, the sequence $\{k = k(n)\}$ satisfying $N_{k-1} < r \leq N_k$ must have limit $\lim_{n\to\infty} k(n) = k^*$, and actually $k = k(n) = k^*$ for sufficiently large n since k is an integer. From this, we further have

$$\frac{N_{k-1}}{n} \to P(0 < Y \leq t_{k^*-1}) \tag{20.19}$$

and

$$\begin{aligned} \lim_{n\to\infty} \frac{q}{n} &= \lim_{n\to\infty} \frac{r - N_{k-1}}{n} \\ &= \alpha - P(0 < Y \leq t_{k^*-1}) = P(0 < Y \leq t^*) - P(0 < Y \leq t_{k^*-1}) \\ &= P(t_{k^*-1} < Y \leq t^*). \end{aligned} \tag{20.20}$$

To obtain $\lim_{n\to\infty} U_n/n$, we observe the following limits as $n \to \infty$.

$$\sum_{j=1}^{k-1} \frac{n_j}{n} \left(\sum_{\ell=1}^{j-1} \alpha_\ell (t_\ell - t_{\ell-1}) \right) \quad \to \quad \sum_{j=1}^{k^*-1} P(t_{j-1} < Y \leq t_j) \left(\sum_{\ell=1}^{j-1} \alpha_\ell (t_\ell - t_{\ell-1}) \right), \tag{20.21}$$

$$\frac{n - N_{k-1}}{n} \left(\sum_{\ell=1}^{k-1} \alpha_\ell (t_\ell - t_{\ell-1}) \right) \quad \to \quad P(Y > t_{k^*-1}) \left(\sum_{\ell=1}^{k^*-1} \alpha_\ell (t_\ell - t_{\ell-1}) \right), \tag{20.22}$$

$$\begin{aligned} &\sum_{j=1}^{k-1} \alpha_j \sum_{i=N_{j-1}+1}^{N_j} \frac{1}{n} Y_{(i)} \\ &= \sum_{j=1}^{k-1} \alpha_j \int_{[Y_{(N_{j-1}+1)}, Y_{(N_j)}]} t dF_n(t) \end{aligned}$$

$$
\begin{aligned}
&= \sum_{j=1}^{k-1} \alpha_j \{Y_{(N_{j-1}+1)} \bar{F}_n(Y_{(N_{j-1}+1)}) - Y_{(N_j)} \bar{F}_n(Y_{(N_j)})\} \\
&\quad + \sum_{j=1}^{k-1} \alpha_j \int_{Y_{(N_{j-1}+1)}}^{Y_{(N_j)}} \bar{F}_n(t)dt \\
&= \sum_{j=1}^{k-1} \alpha_j \left\{Y_{(N_{j-1}+1)} \left(1 - \frac{N_{j-1}+1}{n}\right) - Y_{(N_j)} \left(1 - \frac{N_j}{n}\right)\right\} \\
&\quad + \sum_{j=1}^{k-1} \alpha_j \int_{Y_{(N_{j-1}+1)}}^{Y_{(N_j)}} \bar{F}_n(t)dt \to \sum_{j=1}^{k^*-1} \alpha_j \{t_{j-1} P(Y > t_{j-1}) - t_j P(Y > t_j)\} \\
&\quad + \sum_{j=1}^{k^*-1} \alpha_j \int_{t_{j-1}}^{t_j} \bar{F}(t)dt, \qquad (20.23)
\end{aligned}
$$

$$
-\frac{1}{n} \sum_{j=1}^{k-1} \alpha_j \sum_{i=N_{j-1}+1}^{N_j} t_{j-1} = -\sum_{j=1}^{k-1} \alpha_j t_{j-1} \frac{n_j}{n} \to -\sum_{j=1}^{k^*-1} \alpha_j t_{j-1} P(t_{j-1} < Y \le t_j), \qquad (20.24)
$$

$$
\begin{aligned}
&\frac{1}{n} \alpha_k \sum_{i=N_{k-1}+1}^{r} Y_{(i)} \\
&= \alpha_k \int_{[Y_{(N_{k-1}+1)}, Y_{(r)}]} t dF_n(t) \\
&= \alpha_k \left\{-Y_{(r)} \bar{F}_n(Y_{(r)}) + Y_{(N_{k-1}+1)} \bar{F}_n(Y_{(N_{k-1}+1)})\right\} + \alpha_k \int_{Y_{(N_{k-1}+1)}}^{Y_{(r)}} \bar{F}_n(t)dt \\
&= \alpha_k \left\{-Y_{(r)} \left(1 - \frac{r}{n}\right) + Y_{(N_{k-1}+1)} \left(1 - \frac{N_{k-1}+1}{n}\right)\right\} \\
&\quad + \alpha_k \int_{Y_{(N_{k-1}+1)}}^{Y_{(r)}} \bar{F}_n(t)dt \\
&\to \alpha_{k^*} \{-t^*(1-\alpha) + t_{k^*-1} P(Y > t_{k^*-1})\} + \alpha_{k^*} \int_{t_{k^*-1}}^{t^*} \bar{F}(t)dt, \qquad (20.25)
\end{aligned}
$$

$$
-\frac{1}{n} \alpha_k \sum_{i=N_{k-1}+1}^{r} t_{k-1} = -\alpha_k t_{k-1} \frac{r - (N_{k-1}+1)}{n}
$$

$$
\begin{aligned}
&= -\alpha_k t_{k-1}\frac{q}{n} \\
&\to -\alpha_{k^*} t_{k^*-1} P(t_{k^*-1} < Y \leq t^*), \qquad (20.26)
\end{aligned}
$$

$$
\begin{aligned}
\frac{1}{n}\alpha_k(n-r)(Y_{(r)} - t_{k-1}) &= \alpha_k(1-\frac{r}{n})(Y_{(r)} - t_{k-1}) \\
&\to \alpha_{k^*}(1-\alpha)(t^* - t_{k^*-1}). \qquad (20.27)
\end{aligned}
$$

From (20.17) and (20.21)–(20.27) we have

$$
\begin{aligned}
\frac{U_n}{n} \to\ & \sum_{j=1}^{k^*-1} P(t_{j-1} < Y \leq t_j)\left(\sum_{\ell=1}^{j-1}\alpha_\ell(t_\ell - t_{\ell-1})\right) \\
& + P(Y > t_{k^*-1})\left(\sum_{\ell=1}^{k^*-1}\alpha_\ell(t_\ell - t_{\ell-1})\right) + \sum_{j=1}^{k^*-1}\alpha_j t_{j-1} P(Y > t_{j-1}) \\
& - \sum_{j=1}^{k^*-1}\alpha_j t_j P(Y > t_j) - \sum_{j=1}^{k^*-1}\alpha_j t_{j-1} P(t_{j-1} < Y \leq t_j) - \alpha_k^* t^*(1-\alpha) \\
& + \alpha_{k^*} t_{k^*-1} P(Y > t_{k^*-1}) \\
& - \alpha_{k^*} t_{k^*-1} P(t_{k^*-1} < Y \leq t^*) + \alpha_k^*(1-\alpha)t^* - \alpha_{k^*}(1-\alpha)t_{k^*-1} + I \\
= \ & \sum_{\ell=1}^{k^*-2}\alpha_\ell(t_\ell - t_{\ell-1}) \sum_{j=\ell+1}^{k^*-1} P(t_{j-1} < Y \leq t_j) \\
& + P(Y > t_{k^*-1})\left(\sum_{\ell=1}^{k^*-1}\alpha_\ell(t_\ell - t_{\ell-1})\right) - \sum_{j=1}^{k^*-1}\alpha_j(t_j - t_{j-1})P(Y > t_j) + I \\
= \ & \sum_{\ell=1}^{k^*-2}\alpha_\ell(t_\ell - t_{\ell-1})P(t_\ell < Y \leq t_{k^*-1}) \\
& + P(Y > t_{k^*-1})\left(\sum_{\ell=1}^{k^*-1}\alpha_\ell(t_\ell - t_{\ell-1})\right) - \sum_{j=1}^{k^*-1}\alpha_j(t_j - t_{j-1})P(Y > t_j) + I \\
= \ & \sum_{\ell=1}^{k^*-2}\alpha_\ell(t_\ell - t_{\ell-1})\left[P(t_\ell < Y \leq t_{k^*-1}) + P(Y > t_{k^*-1})\right] \\
& + P(Y > t_{k^*-1})\alpha_{k^*-1}(t_{k^*-1} - t_{k^*-2}) \\
& - \sum_{j=1}^{k^*-1}\alpha_j(t_j - t_{j-1})P(Y > t_j) + I \\
= \ & \sum_{\ell=1}^{k^*-2}\alpha_\ell(t_\ell - t_{\ell-1})P(Y > t_\ell) + P(Y > t_{k^*-1})\alpha_{k^*-1}(t_{k^*-1} - t_{k^*-2}) \\
& - \sum_{j=1}^{k^*-1}\alpha_j(t_j - t_{j-1})P(Y > t_j) + I = I, \qquad (20.28)
\end{aligned}
$$

where

$$I = \sum_{j=1}^{k^*-1} \alpha_j \int_{t_{j-1}}^{t_j} \bar{F}(t)dt + \alpha_{k^*} \int_{t_{k^*-1}}^{t^*} \bar{F}(t)dt.$$

Note that

$$\begin{aligned}
\alpha_j \int_{t_{j-1}}^{t_j} \bar{F}(t)dt &= e^{-\sum_{\ell=1}^{j-1} \alpha_\ell \lambda_0 (t_\ell - t_{\ell-1})} \int_{t_{j-1}}^{t_j} \alpha_j e^{-\alpha_j \lambda_0 (t - t_{j-1})} dt \\
&= e^{-\sum_{\ell=1}^{j-1} \alpha_\ell \lambda_0 (t_\ell - t_{\ell-1})} \theta_0 [1 - e^{-\alpha_j \lambda_0 (t_j - t_{j-1})}] \\
&= \theta_0 [\bar{F}(t_{j-1}) - \bar{F}(t_j)]; \qquad (20.29)
\end{aligned}$$

similarly,

$$\begin{aligned}
&\alpha_{k^*} \int_{t_{k^*-1}}^{t^*} \bar{F}(t)dt \\
&= -e^{-\sum_{\ell=1}^{k^*-1} \alpha_\ell \lambda_0 (t_\ell - t_{\ell-1})} \theta_0 [e^{-\alpha_k^* \lambda_0 (t^* - t_{k^*-1})} - 1] \\
&= \theta_0 [\bar{F}(t^*) - \bar{F}(t_{k^*-1})] = \theta_0 [-(1-\alpha) + \bar{F}(t_{k^*-1})]. \qquad (20.30)
\end{aligned}$$

Therefore,

$$\begin{aligned}
\lim_{n\to\infty} \frac{U_n}{n} &= \sum_{j=1}^{k^*-1} \theta_0 [\bar{F}(t_{j-1}) - \bar{F}(t_j)] + \theta_0 [-(1-\alpha) + \bar{F}(t_{k^*-1})] \\
&= \theta_0 [1 - \bar{F}(t_{k^*-1}) - (1-\alpha) + \bar{F}(t_{k^*-1})] = \alpha\theta_0
\end{aligned}$$

and finally

$$\lim_{n\to\infty} \hat{\theta}_n = \lim_{n\to\infty} \frac{U_n/n}{r/n} = \frac{\alpha\theta_0}{\alpha} = \theta_0.$$

This shows the desired result.

In the above we consider the case when $\alpha \neq P(0 < Y \leq t_j)$, for any $1 \leq j \leq m-1$. In the following we shall assume that there is an index k^* such that $\alpha = P(0 < Y \leq t_{k^*})$ i.e., $t_{k^*} = t^*$. This time the sequence $\{k = k(n)\}$ can have two limit points, namely k^* and $k^* + 1$. The discussion when a subsequence of $\{k(n)\}$ converges to k^* is the same as the above. Below we consider a subsequence of $\{k(n)\}$ that converges to $k^* + 1$. Without loss of generality, let us assume $k = k(n) \to k^* + 1$. This time, from $r/n = \alpha + o(n^{-1/2})$ we have $Y_{(r)} \to t_{k^*}$.

If $k \to k^* + 1$, then

$$\frac{N_{k-1}}{n} \to P(Y \le t_{k^*}) = \alpha$$

and

$$\frac{q}{n} = \frac{r - N_{k-1}}{n} \to 0.$$

Moreover, as $n \to \infty$

$$\sum_{j=1}^{k-1} \frac{n_j}{n} \left(\sum_{\ell=1}^{j-1} \alpha_\ell (t_\ell - t_{\ell-1}) \right) \to \sum_{j=1}^{k^*} P(t_{j-1} < Y \le t_j) \left(\sum_{\ell-1}^{j-1} \alpha_\ell (t_\ell - t_{\ell-1}) \right), \tag{20.31}$$

$$\frac{n - N_{k-1}}{n} \left(\sum_{\ell=1}^{k-1} \alpha_\ell (t_\ell - t_{\ell-1}) \right) \to P(Y > t_{k^*}) \left(\sum_{\ell=1}^{k^*} \alpha_\ell (t_\ell - t_{\ell-1}) \right), \tag{20.32}$$

$$\begin{aligned} \sum_{j=1}^{k-1} \alpha_j \sum_{i=N_{j-1}+1}^{N_j} \frac{1}{n} Y_{(i)} \to & \sum_{j=1}^{k^*} \alpha_j \{ t_{j-1} P(Y > t_{j-1}) - t_j P(Y > t_j) \} \\ & + \sum_{j=1}^{k^*} \alpha_j \int_{t_{j-1}}^{t_j} \bar{F}(t) dt, \end{aligned} \tag{20.33}$$

$$-\frac{1}{n} \sum_{j=1}^{k-1} \alpha_j \sum_{i=N_{j-1}+1}^{N_j} t_{j-1} \to -\sum_{j=1}^{k^*} \alpha_j t_{j-1} P(t_{j-1} < Y \le t_j), \tag{20.34}$$

and it can be shown that the following three limits equal to zero:

$$\begin{aligned} \lim_{n\to\infty} \frac{1}{n} \alpha_k \sum_{i=N_{k-1}+1}^{r} Y_{(i)} &= \lim_{n\to\infty} -\frac{1}{n} \alpha_k \sum_{i=N_{k-1}+1}^{r} t_{k-1} \\ &= \lim_{n\to\infty} \frac{1}{n} \alpha_k (n - r)(Y_{(r)} - t_{k-1}) = 0. \end{aligned} \tag{20.35}$$

Combining (20.17) with (20.31)–(20.34), we have

$$\begin{aligned} \lim_{n\to\infty} \frac{U_n}{n} = & \sum_{j=1}^{k^*} P(t_{j-1} < Y \le t_j) \left(\sum_{\ell=1}^{j-1} \alpha_\ell (t_\ell - t_{\ell-1}) \right) \\ & + P(Y > t_{k^*}) \left(\sum_{\ell=1}^{k^*} \alpha_\ell (t_\ell - t_{\ell-1}) \right) + \sum_{j=1}^{k^*} \alpha_j t_{j-1} P(Y > t_{j-1}) \\ & + \sum_{j=1}^{k^*} \alpha_j \int_{t_{j-1}}^{t_j} \bar{F}(t) dt - \sum_{j=1}^{k^*} \alpha_j t_j P(Y > t_j) \\ & - \sum_{j=1}^{k^*} \alpha_j t_{j-1} P(t_{j-1} < Y \le t_j). \end{aligned}$$

In the same way as before, we can further show that

$$\lim_{n\to\infty} \frac{U_n}{n} = \sum_{j=1}^{k^*} \alpha_j \int_{t_{j-1}}^{t_j} \bar{F}(t)dt = \theta_0 P(0 < Y \leq t_{k^*}) = \alpha\theta_0$$

and consequently

$$\lim_{n\to\infty} \hat{\theta}_n = \lim_{n\to\infty} \frac{U_n/n}{r/n} = \frac{\alpha\theta_0}{\alpha} = \theta_0. \tag{20.36}$$

We thus have shown that (20.36) holds for any converging subsequence of $\{k(n)\}$. Therefore, $\hat{\theta}_n$ converges to θ_0 with probability one.

20.5 Simulation Study

Selected sample sizes of $n = 40$, 70, and 100 were used in this simulation study. We generated 100,000 replicates for each sample size. The quantity of interest was the performance of the MLE of θ provided in Theorem 20.3.1 in this paper. In the simulation study, $m = 6$ stages were used with $\lambda_1 = 0.2$, $\lambda_2 = 0.4$, $\lambda_3 = 0.6$, $\lambda_4 = 0.8$, $\lambda_5 = 1.0$, and $\lambda_6 = 1.2$, respectively. Two sets of inspection times are selected:

Set I: 1, 2, 3, 4, 5
Set II: 1, 3, 5, 7, 9

The following table presents the results of the average values of the MLE's of θ and the mean square errors from the complete sample and grouped data with time sets I and II; see Table 20.1. As one can see that with a sample size of only 40, the MLE of θ from grouped data differ from those based on the complete sample only by 0.4%. The two estimators are almost identical as the sample size increases. This shows that with a very small sacrifice in MSE, we can use

Table 20.1: The results of the average values of the MLE's of θ and the mean square errors from the complete sample and grouped data with time sets I and II

	$n = 40$		$n = 70$		$n = 100$	
	Average	MSE	Average	MSE	Average	MSE
Complete Sample	5.021	0.673	5.011	0.381	5.008	0.265
Grouped Data with Time Set I	5.019	0.693	5.009	0.392	5.007	0.272
Grouped Data with Time Set II	4.999	0.676	4.999	0.388	4.999	0.271

SSALT to save much valuable experiment time. This large gain justifies the use of SSALT.

References

1. Bhattacharyya, G. K., and Zanzawi, S. (1989). A tempered failure rate model for step-stress accelerated life test, *Communications in Statistics–Theory and Methods*, **18**, 1627–1643.

2. David, H. A., and Nagaraja, H. N. (2003). *Order Statistics*, 3rd ed., John Wiley & Sons, New York.

3. DeGroot, M. H., and Goel, P. K. (1979). Bayesian estimation and optimal design in partially accelerated life testing, *Naval Research Logistics Quarterly*, **26**, 223–235.

4. Khamis, I. H., and Higgins, J. J. (1998). A new model for step-stress testing, *IEEE Transactions in Reliability*, **47**, 131–134.

5. Madi, M. T. (1993). Multiple step-stress accelerated life test: The tampered failure rate model, *Communications in Statistics–Theory and Methods*, **22**, 2631–2639.

6. Nelson, W. (1980). Accelerated life testing: Step-stress models and data analysis, *IEEE Transactions on Reliability*, **29**, 103–108.

7. Nelson, W. (1990). *Accelerated Testing: Statistical Models, Test Plans, and Data Analysis*, John Wiley & Sons, New York.

8. Rao, C. R. (1973). *Linear Statistical Inference and Its Applications*, John Wiley & Sons, New York.

9. Xiong, C. (1998). Inferences on a simple step-stress model with type-II censored exponential data, *IEEE Transactions on Reliability*, **47**, 142–146.

10. Xiong, C., and Milliken, G. A. (1999). Step-stress life testing with random stress-change times for exponential data, *IEEE Transactions in Reliability*, **48**, 141–148.

21

Random Stress-Dependent Strength Models Through Bivariate Exponential Conditionals Distributions

Ashis SenGupta

Indian Statistical Institute, Kolkata, India
University of California, Riverside, CA, USA

Abstract: The bivariate exponential conditionals (BEC) distribution here is proposed as a probability model for accelerated life testing. For the conditional experiments, the exponentiality of its conditionals, nonpositivity of its correlation, and nonlinearity of its regressions along with its amenability to development of elegant statistical inference procedures, provide sufficient motivation. It is also shown that this model enhances derivation and statistical inference for unconditional reliability when random stress is also envisaged in the experiments, as in many real-life scenarios.

Keywords and phrases: Accelerated life testing, bivariate exponential conditionals distribution, conditional and unconditional reliability, negatively likelihood ratio dependent density

21.1 Introduction

In life testing problems in general and in accelerated life testing problems in particular, one often encounters situations where the strength, Y, of the system is influenced by the stress, X, it has to undergo. Usually, in such stress-dependent strength (SDS) experiments, Y will be negatively dependent on X. In the case of controlled experiments, this phenomenon may be studied through the regression of Y for fixed levels of X. Here a natural choice for the associated conditional distribution of Y for each given $X = x$ is the exponential distribution. In the case where both Y and X are subject to observational errors, as in errors-in-variables set-up, the study of the reliability measure will be instructive. The bivariate exponential conditionals (BEC) distribution is proposed for this case, which subsumes the former case. The essential (and rare) feature of only nonpositive correlation along with a simple exponential family structure for the BEC distribution enhances it as a reasonable choice for modelling the SDS data. Some properties of the BEC distribution are presented. Inference procedures related to accelerated life testing problems are also developed.

21.2 Bivariate Exponential Conditionals Distribution

Among bivariate distributions defined on the positive quadrant we restrict our search to those with only nonpositive correlation. There is a paucity of even such distributions. Further, for modelling SDS data, specification of conditionals as exponentials is quite natural. The construction of such a family of distributions is achieved by appealing to the following theorem, which yields the BEC distribution.

Theorem 21.2.1 (Arnold, Castillo, and Sarabia (1999, Theorem 4.1)) *Let $f_1(x;\eta)$ and $f_2(y;\tau)$ denote members of two l_1- and l_2- parameter exponential families. Let $f(x,y)$ be a bivariate density whose conditional densities satisfy*

$$f(x|y) = f_1(x;\underline{\eta}(y))$$

and

$$f(x|y) = f_2(y;\underline{\tau}(x))$$

for some function $\underline{\eta}(y)$ and $\underline{\tau}(x)$. Then,

$$\begin{aligned} f(x,y) &= r_1(x)r_2(y)\exp\{\underline{q}^{(1)}(x)'M\underline{q}^{(2)}(y)\}, \\ \underline{q}^{(1)}(x) &= (q_{10}(x), q_{11}(x), q_{12}(x), \ldots, q_{1l_1}(x)), \\ \underline{q}^{(2)}(y) &= (q_{20}(y), q_{21}(y), q_{22}(y), \ldots, q_{2l_2}(y)), \end{aligned}$$

where $q_{10}(x) = q_{20}(y) \equiv 1$ and M is a matrix of constant parameters of appropriate dimensions (i.e., $(l_1+1)\times(l_2+1)$) subject to the requirement that

$$\int_{D_1}\int_{D_2} f(x,y)d\mu_1(x)d\mu_2(y) = 1.$$

For convenience we can partition the matrix M as follows:

$$M = \begin{pmatrix} m_{00} & | & m_{01} & \cdots & m_{0l_2} \\ -- & + & -- & -- & -- \\ m_{10} & | & & & \\ \cdots & | & & \tilde{M} & \\ m_{l_10} & & & & \end{pmatrix}$$

Observe that independence results when $\tilde{M} \equiv 0$.

To get exponential conditionals, in Theorem 21.2.1 we put $l_1 = l_2 = 1, r_1(t) = r_2(t) = I(t > 0)$, and $q_{12}(t) = q_{21}(t) = -t$. The densities then take the form

$$f(x,y) = \exp(m_{00} - m_{10}x - m_{01}y + m_{11}xy), x > 0, y > 0.$$

For convergence, we must have $m_{10} > 0, m_{01} > 0$ and $m_{11} \le 0$. For simplicity we will often also use the notation: $\lambda_1 = m_{10} > 0, \lambda_2 = m_{01} > 0, \lambda_{12} = -m_{11} \ge 0$. The resulting form of the density is then,

$$f(x,y) = \exp[-(\lambda_0 + \lambda_1 x + \lambda_2 y + \lambda_{12} xy)], \lambda_1, \lambda_2 > 0, \lambda_{12} \ge 0.$$

Another simple parametrization is

$$f(x,y) = \frac{k(c)}{\sigma_1 \sigma_2} \exp[-x/\sigma_1 - y/\sigma_2 - cxy/(\sigma_1 \sigma_2)] \tag{21.1}$$

where $k(c)$ is obtained in terms of the classical exponential integral function (appearing in its denominator) as

$$k(c) = \frac{ce^{-\frac{1}{c}}}{\int_{\frac{1}{c}}^{\infty} \frac{e^{-w}}{w} dw}.$$

Note that $\sigma_i^{-1} = \lambda_i, i = 1, 2, c = \lambda_{12}/\lambda_1\lambda_2$. Then, the normalizing constant $\exp(m_{00})$ is $\lambda_1\lambda_2 k(c)$.

This density has been discussed extensively in Arnold and Strauss (1988, 1991). We will see that this joint density yields nonpositive correlation and non-linear regressions between X and Y.

Both the conditional densities are exponentials:

$$X|(Y = y) \sim \exp[(1 + cy/\sigma_2)/\sigma_1]. \tag{21.2}$$

$$Y|(X = x) \sim \exp[(1 + cx/\sigma_1)/\sigma_2]. \tag{21.3}$$

The regression of Y on X is given by

$$E(Y|(X = x)) = \theta_x^{-1}, \theta_x = [(\sigma_1 + cx)/\sigma_1\sigma_2]^{-1} \equiv (\alpha + \beta x),$$

$$\alpha = 1/\sigma_2 = \lambda_2, \quad \beta = c/\sigma_1\sigma_2 = \lambda_{12}.$$

The marginal densities have simple, though not any popularly known, forms:

$$f_X(x) = \frac{k(c)}{\sigma_1}(1 + cx/\sigma_1)^{-1} e^{-x/\sigma_1}, \ x > 0,$$

$$f_Y(y) = \frac{k(c)}{\sigma_2}(1 + cy/\sigma_2)^{-1} e^{-y/\sigma_2}, \ y > 0,$$

Remark 21.2.1 Another bivariate distribution defined on the positive quadrant and possessing only nonpositive correlation is the Gumbel Type I distribution. But, its marginals, and not conditionals, are exponentials and there is no nontrivial sufficient statistic for the parameter vector. Further, its conditionals are not members of the regular exponential family and are somewhat complicated; see, for example, Arnold *et al.* (1999, p. 259). Also, it yields linear while BEC yields nonlinear regression.

Remark 21.2.2 Here we are modelling pdf's, unlike Arnold *et al.* (1999) who attempted to model hazard/survival functions, through conditional specifications. The latter approach, however, led to severe unsolved difficulties.

21.3 Properties of BEC

We first recall below some basic dependency properties inherited by a bivariate distribution and extend and establish similar properties for a BEC distribution. Some statistical properties of the BEC are also presented.

Definition 21.3.1 A distribution is said to be *positively likelihood ratio dependent* (PLRD) if the density $f(x, y)$ satisfies

$$f(x_1, y_1)f(x_2, y_2) \geq f(x_1, y_2)f(x_2, y_1)$$

for all $x_1 > x_2, y_1 > y_2$ [see Tong (1980, pp. 78–82)].

The PLRD has several implications:

$$\begin{aligned}\text{PLRD} \quad \Rightarrow \quad & P\{X \leq x | Y = y\} \text{ is } \downarrow = \text{in } y \text{ for all } x, \text{ and similarly} \\ & P\{Y \leq y | X = x\} \text{ is } \downarrow = \text{in } x \text{ for all } y.\end{aligned}$$

This property is called *positively regression dependent* (PRD).

$$\begin{aligned}\text{PRD} \quad \Rightarrow \quad & P\{Y > y | X > x\} \text{ is } \uparrow = \text{in } x \text{ for all } y, \text{ and} \\ & P\{Y \leq y | X \leq x\} \text{ is } \downarrow = \text{in } x \text{ for all } y.\end{aligned}$$

Further details are available from Lai (2004) and Lai and Xie (2003).

The following definition is now introduced.

Definition 21.3.2 A distribution is said to be negatively likelihood ratio dependent (NLRD) if the density $f(x, y)$ satisfies

$$f(x_1, y_1)f(x_2, y_2) \leq f(x_1, y_2)f(x_2, y_1) \tag{21.4}$$

for all $x_1 \geq x_2, y_1 \geq y_2$.

21.3.1 Dependency properties of BEC

Result 21.3.1 The BEC distribution possesses the following dependency properties:

(i) The correlation coefficient ρ is always nonpositive;
(ii) X and Y are independent if and only if $\rho = 0$;
(iii) It is a negative likelihood ratio dependent (NLRD) family
(iv) It is a negative regression dependent (NRD) or equivalently stochastic decreasing (SD) family.

PROOF. (i) The coefficient of correlation is

$$\rho(X,Y) = \frac{c + k(c) - k^2(c)}{k(c)[1 + c - k(c)]}, \tag{21.5}$$

and it can be seen [Arnold *et al.* (1999, p. 82)] that $-0.32 < \rho \leq 0$.

(ii) Consider (21.5). Note that $k(0) = 1$. It follows that $\rho = 0$ iff $c = 0$, that is, iff $\lambda_{12} = 0$. The necessity part is then obvious.

Consider the sufficiency part. Because of independence,

$$f(x,y) = g(x)h(y), \tag{21.6}$$

where $g(.)$ and $h(.)$ are marginal densities of X and Y, respectively. Then, (21.6) implies

$$\exp(-\lambda_{12}xy) = k(c)[1 + (\lambda_{12}/\lambda_2)x]^{-1}[1 + (\lambda_{12}/\lambda_1)y]^{-1}$$

or,

$$[1 - \lambda_{12}xy + \cdots] = k(c)[1 - (\lambda_{12}/\lambda_2)x + \cdots][1 - (\lambda_{12}/\lambda_1)y + \cdots]. \tag{21.7}$$

Equating coefficients of x, y or xy on both sides of (21.7), and because $k(c) > 0, \lambda_1 > 0$ and $\lambda_2 > 0$, we must have $\lambda_{12} = 0$.

(iii) For the BEC distribution, (21.4) gives,

$$\begin{aligned} & e^{-[\hat{\lambda}_1(x_1+x_2)+\hat{\lambda}_2(y_1+y_2)+\hat{\lambda}_{12}(x_1y_1+x_2y_2)]} \\ & \quad \leq e^{-[\hat{\lambda}_1(x_1+x_2)+\hat{\lambda}_2(y_1+y_2)+\hat{\lambda}_{12}(x_1y_1+x_2y_2)]} \\ & \quad \Rightarrow x_1y_1 + x_2y_2 \geq x_1y_2 + x_2y_1 \\ & \quad \Rightarrow (x_1 - x_2)(y_1 - y_2) \geq 0, \text{ for all } \lambda_1, \lambda_2 > 0, \lambda_{12} \geq 0 \end{aligned}$$

which holds for all $x_1 \geq x_2, y_1 \geq y_2$.

(iv) $P(Y > y | X = x) = \exp(-\theta_x y)$, which, by definition of θ_x is decreasing in x. ■

Result 21.3.2 (a) Let (X_i, Y_i), $i = 1, ..., n$, be a random sample from the BEC distribution. Then the BEC distribution is a member of the regular exponential family with sufficient statistic $(\Sigma x_i, \Sigma y_i, \Sigma x_i y_i)$ for the parameter vector (σ_1, σ_2, c), or equivalently $(\lambda_1, \lambda_2, \lambda_{12})$.

(b) The failure rate at inception is higher at a higher stress level.

PROOF. (a) Result follows trivially.

(b) Let $S_x(t)$ and $\lambda_x(t)$ denote the survival function and hazard rate for given stress level $X = x$ at time t.

$$\begin{aligned} S_x(t) &= P(Y > t|X = x) = S_0\left(\frac{(\alpha + \beta x)t}{\alpha}\right) \\ &\Rightarrow \lambda_x(t) = \lambda_0\left(\frac{(\alpha + \beta x)t)}{\alpha}\right)\left(\frac{\alpha + \beta x}{\alpha}\right) \qquad (21.8) \\ &\Rightarrow \lambda_x(0) = \lambda_0(0)((\alpha + \beta x)/\alpha), \qquad (21.9) \end{aligned}$$

which implies that the failure rate at inception, that is, at $t = 0$, increases with the stress level x. ∎

Corollary 21.3.1 *Equality of the failure rates at inception or at initial time-point for different stress levels implies that they should be the same at all time points.*

PROOF. From (21.8) note that $\lambda_x(0) = \lambda_0(0)\forall x$ iff $\beta = 0$. Then, from (21.9) we must have $\lambda_x(t) = \lambda_0(t)\forall t$ and for every x, and hence the corollary. ∎

Remark 21.3.1 Result 21.3.2(b) may be viewed as a common-sense requirement for most studies in reliability engineering. However, in survival analysis involving response times to drug effects, a deviation from the consequence given in the corollary, may be envisaged; see, for example, Chen and Wang (2000).

21.4 Model Representations in ALT

We expose now the implications of the BEC model in representing the strength (life-time) Y at a given stress level $X = x$ and study its relationships with some currently used representations in ALT. In the following, we denote the random variable $Y|(X = x)$ by Y_x.

One popular representation, known as *accelerated failure time model* (AFTM), gives

$$\log Y_x = x \log \beta + \epsilon,$$

where $\beta < 1$ and ϵ is a random error component, with

$$\begin{aligned} E(Y_x) &\approx \beta^x, \quad 0 < \beta < 1, \\ &\equiv \eta_1(1-\delta_1)^x, \eta_1 = 1, \delta_1 = 1-\beta, 0 < \delta_1 < 1. \end{aligned}$$

So,

$$E(Y_x) \approx \eta_1(1 - \delta_1 x + 0(x^2)). \tag{21.10}$$

Then, $S_x(t) = S_0(t/\beta^x)$ and $\lambda_x(t) = \lambda_0(t/\beta^x)/\beta^x$.

A more familiar representation, which can be shown to encompass the Arrhenius and Power models related to acceleration factor (AF), is given as

$$\begin{aligned} \log Y_x &= \alpha + \beta x + \epsilon, \quad \beta < 0, \\ &= \alpha - \delta_2 x + \epsilon, \quad \delta_2 = -\beta > 0. \end{aligned}$$

So,

$$E(Y_x) \approx \eta_2(1 - \delta_2 x + 0(x^2)), \quad \eta_2 = e^{\alpha}. \tag{21.11}$$

For the BEC distribution, we have

$$E(Y_x) = (\alpha + \beta x)^{-1} \approx \eta_3(1 - \delta_3 x + 0(x^2)), \quad \eta_3 = 1/\alpha, \delta_3 = \beta/\alpha > 0. \tag{21.12}$$

Thus, by modelling through the BEC distribution, we can encompass the current representations for ALT models at least upto the first order.

21.5 Statistical Inference Under Normal Stress

The main aim in ALT is to derive inference procedures at the normal stress level from data collected more easily (early) at the accelerated stress level. This may be approached in several ways. First the estimates of the parameters α and β are to be obtained from the given data. It may be felt that the statistical model, that is, the probability distribution, used at the accelerated stress levels will continue to hold at the normal stress level. Then, inference can be carried out by working with the relevant parametric functions with x_s, the accelerated stress level, replaced simply by x_0, the normal stress level. This approach has been suggested by Xiong (2003), for example. The paucity of plausible distributions amenable to statistical inference seems to have been a major obstacle here. However, we show below that use of BEC distribution can be quite useful for this approach. When the model cannot be relied upon to be valid at x_0, appeal to models from physics such as the Power law, Arrhenius law, etc. has been suggested. Here also, the parameters of such (deterministic) laws need to be estimated and this may be done again through the estimated α and β in the same lines as in Sections 22.3.1–22.3.4 of Elsayed (2003). However, related inference procedures do not seem to be well known. We follow the first approach here.

21.5.1 Estimation of α and β

We can estimate α and β by several methods. One method may be by the method of maximum likelihood (ML) through conditional likelihood. A second approach, much simpler, may be to exploit the method of least squares. These two methods are briefly discussed below.

Let (x_i, y_{ij}), $j = 1, \ldots, n_i$, $i = 1, \ldots, m$, be independent random observations on the strength Y at given values of stress $X = x_i$.

Method of conditional maximum likelihood

The log-conditional likelihood for given stress $X = x_i$, $i = 1, \ldots, m$, is

$$\begin{aligned} \ln L &= \sum_{i=1}^{m} \sum_{j=1}^{n_i} [\ln(\alpha + \beta x_i) - y_{ij}(\alpha + \beta x_i)] \\ &= \sum n_i \ln(\alpha + \beta x_i) - \alpha n \bar{y} - \beta \sum n_i x_i \bar{y}_i, \end{aligned}$$

where $\bar{y} = \sum_{i=1}^{m} \sum_{j=1}^{n_i} y_{ij}/n = \sum_{i=1}^{n} n_i y_i/n$, $n = \sum_{i=1}^{m} n_i$. Then the ML estimators of α and β are given by the maximizer of the likelihood function among the roots of the likelihood equations,

$$\sum n_i (\alpha + \beta x_i)^{-1} = n\bar{y} \tag{21.13}$$

$$\sum n_i x_i (\alpha + \beta x_i)^{-1} = \sum n_i x_i \bar{y}_i. \tag{21.14}$$

Method of moment based least squares

Recalling that $E(Y_x) = \theta_x$, method of least squares may be implemented with $(\bar{y}_i, (\alpha + \beta x_i)^{-1})$, $i = 1, \ldots, m$. The normal equations are given by

$$\sum (\alpha + \beta x_i)^{-2} [\bar{y}_i - (\alpha + \beta x_i)^{-1}] = 0, \tag{21.15}$$

$$\sum x_i (\alpha + \beta x_i)^{-2} [\bar{y}_i - (\alpha + \beta x_i)^{-1}] = 0. \tag{21.16}$$

The systems of nonlinear equations for both the methods above can be solved, leading to possibly multiple solutions, for α and β iteratively, through bivariate Newton-Raphson method, for example. Among the roots (pairs) lying in the admissible support, the one (if unique) may be obtained as that which yields the maximum value of the (conditional) likelihood function. Contrary to the usual, here the likelihood based method is simpler than the moment based one. However, remembering that the admissible support is simply defined by the inequality constraints $\alpha > 0, \beta \geq 0$, the problem is really that of linear programming and standard subroutines in software packages such as *Mathematica* and MATLAB, etc. may be invoked.

21.5.2 Asymptotic inference for θ_0

For the BEC model, denote the expected mean lifetime at normal condition (stress) x_0 by $\theta_{x_0} \equiv \theta_0 = (\alpha + \beta x_0)^{-1}$. Also denote the information matrix by $I = E(-\frac{\partial^2 lnL}{\partial\alpha\partial\beta})$ and the observed/empirical information matrix by $\hat{I}$. Then, writing $\theta_{x_i} \equiv \theta_i$ for convenience, we have

$$\hat{I} = \sum_{i=1}^{m} n_i \hat{\theta}_i^2 \begin{pmatrix} 1 & x_i \\ x_i & x_i^2 \end{pmatrix}, \hat{\theta}_i = (\hat{\alpha} + \hat{\beta} x_i)^{-1}, \qquad i = 1, \ldots, m,$$

where $\hat{\alpha}$ and $\hat{\beta}$ are the conditional MLEs. Let $\Sigma = I^{-1}$ and $\hat{\Sigma} = \hat{I}^{-1}$. Then, asymptotically $(\hat{\alpha}, \hat{\beta})' \sim N((\alpha, \beta)', \Sigma)$.

For the BEC model, denote the expected mean lifetime at normal condition (stress) x_0 by $\theta_{x_0} \equiv \theta_0 = (\alpha + \beta x_0)^{-1}$. By the invariance principle for ML estimation, the conditional MLE $\hat{\theta}^0$ of θ_0 is given by $\hat{\theta}_0 = (\hat{\alpha} + \hat{\beta} x_0)^{-1}$. By the delta method, it follows that asymptotically, $\hat{\theta}_0 \xrightarrow{L} N(\theta_0, \theta_0^2.(1, x_0)'\Sigma(1, x_0))$. With large samples, statistical inference for θ_0 may then be based on this result. For example, a $100(1-\gamma)\%$ confidence interval for θ_0 is given by

$$\hat{\theta}_0[1 \pm z_{\gamma/2}\{(1, x_0)'\hat{\Sigma}(1, x_0)\}^{1/2}],$$

where $z_{\gamma/2}$ is the upper $\gamma/2\%$ point of the standard normal distribution. Similarly, asymptotic tests of significance for θ_0 may be derived.

It is worth noting the similarity, the exact representations, in the forms of the asymptotic distributions of our $\hat{\theta}_0$ and the currently used $\hat{\theta}_0 = \exp(\hat{\alpha} + \hat{\beta} x_0)$ by Xiong (2001, p. 462). (However, the expression therein for the confidence interval has to be corrected to the form given here.) Further, the conditional reliability $R_{x_0}(t)$ at the given normal stress level $X = x_0$ can be obtained from the above result by noting that $R_{x_0}(t) = 1 - F(t/\theta_0)$, where $F(y)$ is the c.d.f. of an exponential random variable with unit expectation.

It may be of interest to have a preliminary test for independence in BEC. This is available from SenGupta (1995) as also referred to in Arnold *et al.* (1999, pp. 358–359).

21.6 Unconditional Reliability Function and Measure

Some definitions are presented below.

Definitions.

(a) The *conditional reliability function,* $R_x(t)$, is defined as the probability of survival beyond time t at a given stress level $X = x$;
(b) The *unconditional reliability function,* or simply the *reliability function,* $R(t)$, is defined as the probability of survival beyond time t in the face of any possible stress level;
(c) The *reliability measure,* R, is defined as the probability of withstanding, or equivalently, of surviving any possible stress level.

When Y and X are measured in commensurable units, we may take $R = P(Y > X)$. However, if this is not the case as in experiments where stress can be temperature or pressure while strength can be lifetime (in hours), it may be more meaningful to take $R = P(Y > \theta_X^{-1})$.

For an application of R, see Ferdous *et al.* (1995), for example.

Using the BEC distribution as a random stress-strength model, we now obtain the above reliability functions and measure:

$$R_x(t) \equiv P(Y > t|X = x) = \exp[-(\lambda_2 + \lambda_{12}x)t]. \tag{21.17}$$

$$R(t) \equiv P(Y > t) = \int_0^\infty R_x(t) f_X(x) dx = \frac{\lambda_1 k(c)}{\lambda_1^* k(c^*)} \exp(-\lambda_2 t), \tag{21.18}$$

where $\lambda_1^* = \lambda_1 + \lambda_{12}t$ and c^* is obtained by replacing λ_1 by λ_1^* in c. Note that $R(t)$ is $\downarrow t$. Finally,

$$R_x = P(Y > ((\alpha + \beta x)^{-1}) = \exp(-1)\forall x,$$

and hence, $R = \exp(-1)$ also.

When observations (x_i, y_i), $i = 1, \ldots, n$, are available from the joint BEC density, $(\lambda_1, \lambda_2, \lambda_{12})$ may be estimated through generalized Stein's identity as suggested in Section 5.1 of Arnold *et al.* (2001). An estimator $\hat{R}(t)$ of $R(t)$ may then be obtained by substituting these estimators in (21.18) – the classical exponential integral therein can be computed numerically.

Variations of $R(t)$ with λ_{12} are exhibited in Figure 21.1 for $t = 1, 3$, and 7, where we have taken $\lambda_1 = 1$, $\lambda_2 = 2$. It is seen that $R(t)$ increases with λ_{12} in a similar form, though its magnitude decreases greatly (from about 4 x 10^{-1} to 3

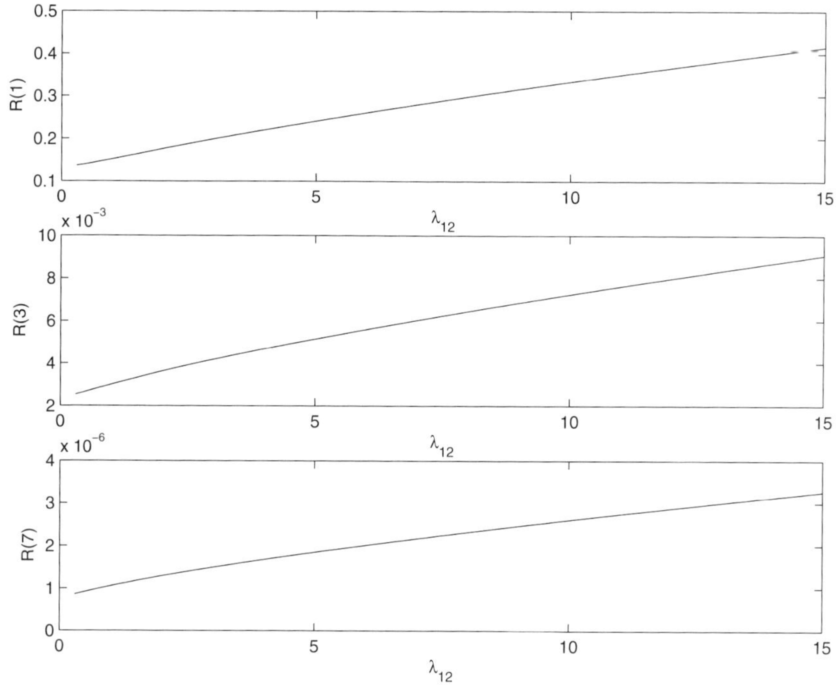

Figure 21.1: Plots of unconditional reliability against λ_{12}

x 10^{-6}), as t is increased (in a moderate range from 1 through 7). Because the correlation coefficient, ρ, increases in absolute magnitude with c, which again is just a constant multiplier of λ_{12} for fixed λ_1 and λ_2, the same can be inferred about the effect on $R(t)$ for variations in ρ.

21.7 Conclusions

We have noted here that the BEC model induces a regression that encompasses, at least up to the first order, the relationships currently being used in the literature. The advantage for the approach presented here is, of course, the enhancement of this relationship through a parametric formulation that uses an elegant probability distribution model thus facilitating the development of objective statistical inference procedures. It has also been shown that this model further yields an elegant form for the unconditional reliability in stress dependent strength models where both strength and stress are subject to random variations as in errors-in-variables models, for example.

Acknowledgements. The author would like to thank Prof. Barry C. Arnold for discussions and encouragements to work on conditionally specified models.

References

1. Arnold, B. C., Castillo, E., and Sarabia, J. M. (1999). *Conditional Specification of Statistical Models,* Springer-Verlag, New York.

2. Arnold, B. C., Castillo, E., and Sarabia, J. M. (2001). A multivariate version of Stein's identity with applications to moment calculations and estimation of conditionally specified distributions, *Communications in Statistics—Theory and Methods,* **12,** 2517–2542.

3. Arnold, B. C., and Strauss, D. (1988). Bivariate distributions with exponential conditionals, *Journal of the American Statistical Association,* **83,** 522–527.

4. Arnold, B. C., and Strauss, D. (1991). Bivariate distributions with conditionals in prescribed exponential families, *Journal of the Royal Statistical Society, Series B,* **53,** 365–375.

5. Basu, A. P. (1995). Accelerated life testing with applications, In *The Exponential Distributions: Theory, Methods, and Applications* (Eds., N. Balakrishnan and A. P. Basu), pp. 377–383, Gordon and Breach, Newark, NJ.

6. Chen, Y. Q., and Wang, M-C. (2000). Analysis of accelerated hazard models, *Journal of the American Statistical Association,* **95,** 608–618.

7. Elsayed, E. A. (2003). Accelerated life testing, In *Handbook of Reliability Engineering* (Ed., H. Pham), pp. 415–428, Springer-Verlag, New York.

8. Ferdous, J., Uddin, M. B., and Pandey, M. (1995). Estimation of reliability of a component under multiple stresses, *Microelectronics and Reliability,* **35,** 279–283.

9. Lai, C. D. (2004). Constructions of continuous bivariate distributions. *Journal of the Indian Society for Probability and Statistics,* **8,** 21–44.

10. Lai, C. D., and Xie, M. (2003). Concepts of stochastic dependence in reliability analysis, In *Handbook of Reliability Engineering* (Ed., H. Pham), Springer-Verlag, New York.

11. SenGupta, A. (1995). Optimal tests in multivariate exponential distributions, In *The Exponential Distributions: Theory, Methods and Applications* (Eds., N. Balakrishnan and A. P. Basu), pp. 351–376, Gordon and Breach, Newark, NJ.

12. Tong, Y. L. (1980). *Probability Inequalities in Multivariate Distributions*, Academic Press, New York.

13. Xiong, C. (2003). Step-stress accelerated life testing, In *Handbook of Reliability Engineering* (Ed., H. Pham), pp. 457–469, Springer-Verlag, New York.

PART V
Inference

22

Some New Methods for Local Sensitivity Analysis in Statistics

Enrique Castillo,[1] Carmen Castillo,[2] Ali S. Hadi,[3] and J. M. Sarabia[1]

[1] *University of Cantabria, Santander, Spain*
[2] *University of Granada, Santander, Spain*
[3] *The American University in Cairo, Cairo, Egypt*

Abstract: This chapter deals with the problem of local sensitivity analysis, that is, how sensitive the results of a statistical analysis are to a change in the data. A closed formula for the calculation of local sensitivities in optimization problems is applied to some optimization problems in statistics, including regression, maximum likelihood, and other situations involving ordered and data constrained parameters. In addition, a general method for evaluating the sensitivities for the method of moments is obtained. The methods are illustrated with several examples.

Keywords and phrases: Data constrained parameters, exponential families, local sensitivity, mathematical programming, duality, maximum likelihood, method of moments, ordered parameters

22.1 Introduction and Motivation

Statisticians work with mathematical models to analyze data and describe the reality being observed. They frequently use parametric models and estimate their parameters based on different items of information, so that the finally selected model becomes fairly dependent on the available data. Because conclusions drawn from an analysis are sensitive to changes in a model, one needs to know the influence of each data item on the final results so as to make the adequate corrections when necessary. It is therefore essential for data analysts to be able to assess the sensitivity of their conclusions to various perturbations in the inputs; see Saltelli, Chan, and Scott (2000). This is known as *sensitivity analysis.* Today we cannot satisfy people only with solutions to their problems and we need to specify how sensitive these solutions are to data. Thus, sensitivity analysis adds quality to statistical studies and is becoming more and more frequently demanded.

There is a large literature on sensitivity analysis and outlier detection; see, for example, the books by Hawkins (1980), Belsley, Kuh, and Welsch (1980), Cook and Weisberg (1982), Atkinson (1985), Chatterjee and Hadi (1988), and Barnett and Lewis (1994), and the works by Welsch and Kuh (1977), Pregibon (1981), Gray and Ling, (1984), Gray (1986), Cook (1977, 1986), Jones and Ling (1988), Weissfeld and Schneider (1990a, 1990b), Schwarzmann (1991), Paul and Fung (1991), Simonoff (1991), Escobar and Meeker (1992), Nyquist (1992), Hadi (1992a,b), Hadi and Simonoff (1993), Atkinson (1984), Hadi (1994), Peña and Yohai (1995), Barrett and Gray (1997), Mayo and Gray (1997), Billor, Hadi and Velleman (2000), Chatterjee, Hadi, and Price (2000), Billor, Chatterjee and Hadi (2001), and Winsnowski, Montgomery, and James (2001).

However, sensitivity analysis has been almost exclusively applied to regression and is rare in other statistical applications. In this paper, we deal with the problem of local sensitivity analysis in general. The paper is structured as follows. Section 22.2 discusses local sensitivities when the estimation problem can be expressed as a nonlinear programming problem. Section 22.3 presents some existing results of sensitivity analysis in regression and introduces a new regression method that combines least squares and minimax methods. Section 22.4 discusses the maximum likelihood sensitivity with respect to data. Section 22.5 describes an example of ordered linear model parameters and data constrained parameters. Section 22.6 is devoted to the problem of local sensitivity of the method of moments estimates. Some examples are used to illustrate the methods. Finally, Section 22.7 offers some concluding remarks.

22.2 Sensitivities of the Objective Function

Consider the following general nonlinear programming *primal problem* (P):

$$\underset{\mathbf{x}}{\text{Minimize}}\ Q_P = f(\mathbf{x};\mathbf{a}) \tag{22.1}$$

subject to

$$\mathbf{h}(\mathbf{x};\mathbf{a}) = \mathbf{0} : \boldsymbol{\lambda}, \tag{22.2}$$

$$\mathbf{g}(\mathbf{x};\mathbf{a}) \leq \mathbf{0} : \boldsymbol{\mu}, \tag{22.3}$$

where boldface letters refer to vectors, $\mathbf{x} \in \mathbb{R}^m$, $\mathbf{h}(\mathbf{x};\mathbf{a}) \in \mathbb{R}^t$, $\mathbf{g}(\mathbf{x};\mathbf{a}) \in \mathbb{R}^q$, and $\boldsymbol{\lambda}$ and $\boldsymbol{\mu}$ are the dual variables associated with the equality and inequality constraints, respectively.

Every primal nonlinear programming problem P, which is stated as in (22.1)–(22.3), has an associated dual problem D, defined as:

$$\underset{\boldsymbol{\lambda},\boldsymbol{\mu}}{\text{Maximize}}\ Q_D = \ \text{Inf}_{\mathbf{x}}\ \{\mathcal{L}(\mathbf{x}, \boldsymbol{\lambda}, \boldsymbol{\mu}; \mathbf{a})\} \tag{22.4}$$

subject to

$$\boldsymbol{\mu} \geq \mathbf{0}, \tag{22.5}$$

where

$$\mathcal{L}(\mathbf{x}, \boldsymbol{\lambda}, \boldsymbol{\mu}; \mathbf{a}) = f(\mathbf{x}; \mathbf{a}) + \boldsymbol{\lambda}^T \mathbf{h}(\mathbf{x}; \mathbf{a}) + \boldsymbol{\mu}^T \mathbf{g}(\mathbf{x}; \mathbf{a}) \tag{22.6}$$

is the Lagrangian function associated with the primal problem (22.1)–(22.3), and $\boldsymbol{\lambda}$ and $\boldsymbol{\mu}$ are the dual variables associated with the equality and inequality constraints, respectively. They are vectors of dimensions p and q, the number of equality and inequality constraints, respectively.

Given some regularity conditions [see Luenberger (1989), Bazaraa, Sherlai, and Shetty (1993) and Castillo *et al.* (2001)], if the primal problem (22.1)–(22.3) has a local optimal solution $\mathbf{x}^*$, the dual problem (22.4)–(22.5) also has a local optimal solution $(\boldsymbol{\lambda}^*, \boldsymbol{\mu}^*)$, and the optimal values of the objective functions of both problems coincide.

The following theorem provides a general method for obtaining closed-form formulas for local sensitivity analysis; see Castillo *et al.* (2004b) and Conejo *et al.* (2005).

Theorem 22.2.1 (Objective function sensitivities to a) *The sensitivity of the objective function of the primal problem (22.1)–(22.3) with respect to the parameter* $\mathbf{a}$ *is given by*

$$\frac{\partial Q_P^*}{\partial \mathbf{a}} = \nabla_{\mathbf{a}} \mathcal{L}(\mathbf{x}^*, \boldsymbol{\lambda}^*, \boldsymbol{\mu}^*; \mathbf{a}), \tag{22.7}$$

which is the partial derivative of its Lagrangian function in (22.6) with respect to $\mathbf{a}$ *evaluated at the optimal solution* $\mathbf{x}^*, \boldsymbol{\lambda}^*$, *and* $\boldsymbol{\mu}^*$.

Because these sensitivities may be difficult to interpret because they may depend on unknown parameters and/or on the unit of measurement and may also have different variances, we approximate them by replacing the parameters by their estimated values, and we also convert them into a dimensional measures.

To standardize the sensitivities, we subtract the mean and divide by the standard deviation:

$$S(a_i) = \frac{(\partial Q_P^*/\partial a_i) - E\left[(\partial Q_P^*/\partial a_i)\right]}{\sqrt{\text{Var}\left[(\partial Q_P^*/\partial a_i)\right]}}. \tag{22.8}$$

If the mean and/or the standard deviation are unknown, we replace them by the sample mean and the sample standard deviation of $\partial Q_P^*/\partial a_i$, $i = 1, 2, \ldots, t$. So, instead of $\lambda_i = \partial Q_P/\partial a_i$, we use the standardized version. The standardized sensitivities, $S(a_i)$, are easier to interpret because they are unitless.

22.3 Applications to Regression

We start with two existing applications (the least squares and minimax regressions); then we present a new application (mixed least-squares and minimax regression).

22.3.1 Least-squares regression

Castillo *et al.* (2004b) consider the standard linear regression problem

$$\underset{\boldsymbol{\beta}}{\text{Minimize}}\; Q_{LS} = \sum_{i=1}^{n} (y_i - \mathbf{x}_i^T \boldsymbol{\beta})^2. \tag{22.9}$$

where $\mathbf{y} = (y_1, \ldots, y_n)^T$ is an $n \times 1$ vector of response variables, $\mathbf{X}^T$ is an $n \times k$ matrix of rank k of predictor variables, $\mathbf{x}_i^T$ is the ith row in $\mathbf{X}$, $\boldsymbol{\beta}$ is a $k \times 1$ vector of regression parameters, and $\boldsymbol{\varepsilon} = (\varepsilon_1, \ldots, \varepsilon_n)^T$ is an $n \times 1$ vector of independent random errors $N(0, \sigma^2)$.

Using the proposed method, that is, Theorem 22.2.1, they obtain the sensitivity:

$$\frac{\partial Q_{LS}^*}{\partial y_i} = 2(y_i - \mathbf{x}_i^T \boldsymbol{\beta}) = 2e_i, \tag{22.10}$$

which is twice the residual value e_i and leads to the standardized sensitivity for y_i,

$$S_{LS}(y_i) = \frac{e_i}{\hat{\sigma}\sqrt{1 - p_{ii}}} \tag{22.11}$$

where p_{ii} is the ith leverage value (the ith diagonal element of $\mathbf{X}(\mathbf{X}^T\mathbf{X})^{-1}\mathbf{X}^T$) and

$$\hat{\sigma}^2 = \frac{\mathbf{e}^T\mathbf{e}}{n-k}. \tag{22.12}$$

Similarly, the sensitivity of Q_{LS}^* with respect to the prediction variables x_{ij} becomes

$$\frac{\partial Q_{LS}^*}{\partial x_{ij}} \;=\; 2e_i\hat{\beta}_j, \tag{22.13}$$

that leads to the standardized sensitivities of the least squares objective function with respect to x_{ij}:

$$S_{LS}(x_{ij}) = \frac{e_i \hat{\beta}_j}{\hat{\sigma}\sqrt{(1-p_{ii})\left[\hat{\sigma}^2 c_{jj} + \hat{\beta}_j^2\right]}}, \quad j = 1, 2, \ldots, k, \tag{22.14}$$

where c_{jj} is the jth diagonal element of $(\mathbf{X}^T\mathbf{X})^{-1}$.

22.3.2 Minimax regression

Castillo *et al.* (2004b) also consider the minimax (MM) regression problem:

$$\underset{\boldsymbol{\beta}}{\text{Minimize}}\ Q_{MM} = \max_i |y_i - \mathbf{x}_i^T \boldsymbol{\beta}|, \tag{22.15}$$

that is equivalent to the linear programming problem:

$$\underset{\boldsymbol{\beta},\varepsilon}{\text{Minimize}}\ Q_{MM} = \varepsilon \tag{22.16}$$

subject to

$$\begin{aligned} y_i - \mathbf{x}_i^T \boldsymbol{\beta} &\le \varepsilon : \mu_i^{(1)}, \quad i = 1, \ldots, n, & (22.17)\\ \mathbf{x}_i^T \boldsymbol{\beta} - y_i &\le \varepsilon : \mu_i^{(2)}, \quad i = 1, \ldots, n, & (22.18)\\ \varepsilon &\ge 0, & (22.19) \end{aligned}$$

where $\mu_i^{(1)}$ and $\mu_i^{(2)}$ are the corresponding dual variables. Then, the sensitivities of Q_{MM} with respect to the response values are

$$\frac{\partial Q_{MM}^*}{\partial y_i} = \mu_i^{(1)} - \mu_i^{(2)} = \begin{cases} \mu_i^{(1)}, & \text{if } \ y_i - \mathbf{x}_i^T \boldsymbol{\beta} = \varepsilon, \\ -\mu_i^{(2)}, & \text{if } \ \mathbf{x}_i^T \boldsymbol{\beta} - y_i = \varepsilon, \\ 0, & \text{otherwise,} \end{cases} \tag{22.20}$$

and the sensitivities with respect to the predictor values $x_{ij}, j = 1, 2, \ldots, k$,

$$\frac{\partial Q_{MM}^*}{\partial x_{ij}} = -\mu_i^{(1)} \beta_j + \mu_i^{(2)} \beta_j = \begin{cases} -\mu_i^{(1)} \beta_j, & \text{if } \ y_i - \mathbf{x}_i^T \boldsymbol{\beta} = \varepsilon, \\ \mu_i^{(2)} \beta_j, & \text{if } \ \mathbf{x}_i^T \boldsymbol{\beta} - y_i = \varepsilon, \\ 0, & \text{otherwise,} \end{cases} \tag{22.21}$$

where $\mu_i^{(1)}$ and $\mu_i^{(1)}$ are the dual variables in (22.17) and (22.18), respectively.

22.3.3 Mixed least-squares and minimax regression

To illustrate the use of Theorem 22.2.1, we apply it to a regression problem that combines least squares and minimax constraints. We solve the following optimization problem:

$$\underset{\boldsymbol{\beta}, \varepsilon}{\text{Minimize}}\ Q_1 = \varepsilon \tag{22.22}$$

subject to

$$\begin{aligned} y_i - \mathbf{x}_i^T \boldsymbol{\beta} &\leq \varepsilon : \mu_i^{(1)}, \quad i = 1, \ldots, n, &(22.23)\\ \mathbf{x}_i^T \boldsymbol{\beta} - y_i &\leq \varepsilon : \mu_i^{(2)}, \quad i = 1, \ldots, n, &(22.24)\\ \sum_{i=1}^{n} (y_i - \mathbf{x}_i^T \boldsymbol{\beta})^2 &\leq B_{LS} : \mu^{(3)} &(22.25)\\ \varepsilon &\geq 0 : \mu^{(4)}. &(22.26) \end{aligned}$$

where B_{LS} is the sum of squares bound. The Lagrangian function becomes

$$\begin{aligned} L(\boldsymbol{\beta}, \varepsilon; \boldsymbol{\mu}^{(1)}, \boldsymbol{\mu}^{(2)}, \mu^{(3)}, \mu^{(4)}) &= \varepsilon + \sum_{i=1}^{n} \mu_i^{(1)} \left(y_i - \mathbf{x}_i^T \boldsymbol{\beta} - \varepsilon \right) \\ &\quad + \sum_{i=1}^{n} \mu_i^{(2)} \left(\mathbf{x}_i^T \boldsymbol{\beta} - y_i - \varepsilon \right) \\ &\quad + \mu^{(3)} \left(\sum_{i=1}^{n} (y_i - \mathbf{x}_i^T \boldsymbol{\beta})^2 - B_{LS} \right) - \mu^{(4)} \varepsilon \end{aligned} \tag{22.27}$$

and then, the sensitivities are:

$$\frac{\partial Q_1^*}{\partial y_s} = \left(\mu_s^{(1)} - \mu_s^{(2)} \right) + 2\mu^{(3)} \left(y_s - \mathbf{x}_s^T \boldsymbol{\beta} \right) = \left(\mu_s^{(1)} - \mu_s^{(2)} \right) + 2\mu^{(3)} e_s, \tag{22.28}$$

$$\frac{\partial Q_1^*}{\partial x_{rs}} = \left[\left(\mu_s^{(2)} - \mu_s^{(1)} \right) - 2\mu^{(3)} \left(y_s - \mathbf{x}_s^T \boldsymbol{\beta} \right) \right] \beta_r = \left[\left(\mu_s^{(2)} - \mu_s^{(1)} \right) - 2\mu^{(3)} e_s \right] \beta_r. \tag{22.29}$$

22.3.4 Example: Simulated data

We consider the simulated example in Castillo *et al.* (2004). The data has been simulated using the model

$$y_i = \beta_0 + \beta_1 X_1 - \beta_2 X_2 + \epsilon_i, \tag{22.30}$$

where $\beta_0 = 1,\ \beta_1 = 2,\ \beta_2 = -3,\ X_1 \sim U(0,1),\ X_2 \sim U(0,1),\ \epsilon_i \sim N(0, 0.05)$ and observations $18, 19$ and 20 have been replaced by outliers (three standard deviations from the mean) by

$$\begin{aligned} y_{18} &= 1 + 2x_1 - 3x_2 - 0.05 \times 3, \\ y_{19} &= 1 + 2x_1 - 3x_2 + 0.05 \times 3, \\ y_{20} &= 1 + 2x_1 - 3x_2 - 0.05 \times 3. \end{aligned} \tag{22.31}$$

The data appear in Table 22.1.

Table 22.1: Simulated data generated using the models in (22.30) and (22.31)

i	x_{i1}	x_{i2}	y_i
1	0.8433	0.5504	1.0439
2	0.2922	0.2241	0.9947
3	0.8563	0.0671	2.5746
4	0.9981	0.5787	1.3503
5	0.7623	0.1307	2.0700
6	0.1595	0.2501	0.4891
7	0.4354	0.3597	0.7510
8	0.1315	0.1501	0.7726
9	0.8309	0.2308	1.9422
10	0.7759	0.3037	1.7049
11	0.5024	0.1602	1.5052
12	0.2651	0.2858	0.6371
13	0.7227	0.6282	0.5642
14	0.4133	0.1177	1.4674
15	0.0466	0.3386	−0.0106
16	0.6457	0.5607	0.6092
17	0.2978	0.6611	0.4035
18	0.6274	0.2839	1.2533
19	0.1025	0.6413	−0.5687
20	0.0315	0.7924	−1.4640

Using the GAMS (general algebraic modelling system) software, we obtained the following estimates of the regression coefficients according to the least squares, the minimax, and the mixed methods:

$$\begin{array}{llllll} \text{LS:} & \hat{\beta}_0 = 0.94, & \hat{\beta}_1 = 2.07, & \hat{\beta}_2 = -2.97 & Q_{LS} = 0.1009, \\ \text{MM:} & \hat{\beta}_0 = 1.07, & \hat{\beta}_1 = 1.93, & \hat{\beta}_2 = -3.09 & Q_{MM} = 0.1461, \\ \text{MX:} & \hat{\beta}_0 = 1.01, & \hat{\beta}_1 = 1.98, & \hat{\beta}_2 = -3.02 & Q_{MX} = 0.1492. \end{array}$$

Note that the MX method gives parameter estimates that are closest to the true values of the parameters. The sensitivities of the minimax in (22.28) and

Table 22.2: The dual variables $\mu_i^{(1)}$ and $\mu_i^{(2)}$ associated with the problem (22.22)–(22.26) and the sensitivities of Q_1^* with respect to y_i, x_{1i}, and x_{2i}

i	$-\mu_i^{(1)}$	$-\mu_i^{(2)}$	$\partial Q_1^*/\partial y_i$	$\partial Q_1^*/\partial x_{1i}$	$\partial Q_1^*/\partial x_{2i}$
1	0.00	0.00	0.005	–0.010	0.015
2	0.00	0.00	0.021	–0.042	0.064
3	0.00	0.00	0.017	–0.035	0.053
4	0.00	0.00	0.028	–0.056	0.085
5	0.00	0.00	–0.017	0.034	–0.051
6	0.00	0.00	–0.024	0.047	–0.072
7	0.00	0.00	–0.011	0.022	–0.034
8	0.00	0.00	–0.014	0.027	–0.041
9	0.00	0.00	–0.007	0.013	–0.020
10	0.00	0.00	0.019	–0.037	0.056
11	0.00	0.00	–0.006	0.012	–0.019
12	0.00	0.00	–0.011	0.022	–0.033
13	0.00	0.00	0.003	–0.007	0.010
14	0.00	0.00	–0.003	0.006	–0.010
15	0.00	0.00	–0.026	0.052	–0.079
16	0.00	0.00	0.002	–0.003	0.005
17	0.00	0.00	–0.004	0.007	–0.011
18	0.00	–0.08	**–0.124**	**0.247**	**–0.375**
19	–0.53	0.00	**0.575**	**–1.141**	**1.736**
20	0.00	–0.38	**–0.423**	**0.840**	**–1.277**

(22.29) are shown in Table 22.2, together with the values of the dual variables $\mu_i^{(1)}$ and $\mu_i^{(2)}$. The two other dual variables are

$$\mu^{(3)} = \frac{\partial Q^*}{\partial B} = 0.137 \quad \text{and} \quad \mu^{(4)} = 0.$$

It is clear from Table 22.2 that the mixed method estimates are most sensitive to the three planted outliers (observations 18, 19, and 20).

22.4 The Maximum Likelihood Function

It is of interest to know the sensitivities of the likelihood, that is, how much the optimal likelihood changes when a data point is marginally modified, that is, what is the value of the partial derivative of the optimal likelihood with respect to each data point. The proposed method can also be used to derive these sensitivities. We derive the sensitivities for the exponential family and illustrate it using the gamma and the beta families.

22.4.1 Local sensitivities

For the exponential family, the likelihood becomes

$$L(\boldsymbol{\theta}|\mathbf{x}) = \prod_{i=1}^{n} h(x_i)[c(\boldsymbol{\theta})]^n \exp\left(\sum_{j=1}^{k} w_j(\boldsymbol{\theta}) \sum_{i=1}^{n} t_j(x_i)\right) \tag{22.32}$$

and the corresponding log-likelihood

$$\ell(\boldsymbol{\theta}|\mathbf{x}) = \log L(\boldsymbol{\theta}|\mathbf{x}) = \sum_{i=1}^{n} \log h(x_i) + n \log c(\boldsymbol{\theta}) + \sum_{j=1}^{k} w_j(\boldsymbol{\theta}) \sum_{i=1}^{n} t_j(x_i), \tag{22.33}$$

and the sensitivity of the log-likelihood optimal value to the data point x_s becomes

$$\frac{\partial \ell(\boldsymbol{\theta}|\mathbf{x})}{\partial x_s} = \frac{h'(x_s)}{h(x_s)} + \sum_{j=1}^{k} w_j(\boldsymbol{\theta}) t'_j(x_s). \tag{22.34}$$

22.4.2 Examples: The gamma and beta families

For the gamma family the log-likelihood sensitivities in (22.34) become

$$\frac{\partial \ell(\boldsymbol{\theta}|\mathbf{x})}{\partial x_s} = -\frac{1}{x_s} + \frac{\hat{k}}{x_s} - \hat{\lambda}, \tag{22.35}$$

and for the beta family the log-likelihood sensitivities are

$$\frac{\partial \ell(\boldsymbol{\theta}|\mathbf{x})}{\partial x_s} = -\frac{2x_s - 1}{x_s(1 - x_s)} + \frac{\hat{r}}{x_s} - \frac{\hat{t}}{1 - x_s}. \tag{22.36}$$

22.5 Ordered and Data Constrained Parameters

Suppose that $Y_{ij}; j = 1, 2, \ldots, k,\ i = 1, 2, \ldots, n_j$ are conditionally independent observations from location and scale exponential models, so that Y_{ij} has a density

$$f(Y_{ij}|\theta_j, \sigma_j) = \frac{1}{\sigma_j} \exp\left[-(Y_{ij} - \theta_j)/\sigma_j\right], \quad Y_{ij} \geq \theta_j > 0,\ \sigma_j > 0. \tag{22.37}$$

The constrained maximum likelihood method leads to the optimization problem

$$\underset{\boldsymbol{\theta},\boldsymbol{\sigma}}{\text{Maximize}}\ Q = \sum_{j=1}^{k}\sum_{i=1}^{n_j}[(\theta_j - Y_{ij})/\sigma_j - \log\sigma_j] \tag{22.38}$$

subject to

$$\begin{aligned} \theta_j - Y_{ij} &\le 0 : \mu_{ij}^{(1)};\ j = 1,2,\ldots,k;\ i = 1,2,\ldots,n_j, && (22.39)\\ -\theta_j &\le 0; : \mu_j^{(2)};\ j = 1,2,\ldots,k, && (22.40)\\ -\sigma_j &\le 0; : \mu_j^{(3)};\ j = 1,2,\ldots,k, && (22.41) \end{aligned}$$

where the μ's are the corresponding dual variables. Then, the Lagrangian function becomes

$$\begin{aligned} L(\boldsymbol{\theta},\boldsymbol{\sigma};\boldsymbol{\mu}^{(1)},\boldsymbol{\mu}^{(2)},\boldsymbol{\mu}^{(3)}) &= \sum_{j=1}^{k}\sum_{i=1}^{n_j}[(\theta_j - Y_{ij})/\sigma_j - \log\sigma_j]\\ &+\sum_{j=1}^{k}\sum_{i=1}^{n_j}\mu_{ij}^{(1)}(\theta_j - Y_{ij}) - \sum_{j=1}^{k}\mu_j^{(2)}\theta_j - \sum_{j=1}^{k}\mu_j^{(3)}\sigma_j \end{aligned} \tag{22.42}$$

and the sensitivities of the likelihood function with respect to the data Y_{ij} becomes

$$\frac{\partial Q^*}{\partial Y_{ij}} = -\frac{1}{\sigma_j} - \mu_{ij}^{(1)}, \quad j = 1,2,\ldots,k;\ \ i = 1,2,\ldots,n_j. \tag{22.43}$$

To illustrate, we have simulated 20 data points from 5 different exponential populations $\exp(\theta_j,\sigma_j)$, $j = 1,2,\ldots,5$, with the parameter values shown in Table 22.3, columns 2 and 4. The resulting data values are shown in Table 22.4. We have estimated the parameters solving the problem (22.38)–(22.41) and obtained the parameter estimates $\{\theta_j,\sigma_j\}$ in Table 22.3, columns 3 and 5.

Finally, we have calculated the sensitivities $\partial Q^*/\partial Y_{ij}$ using (22.43), which are shown in Table 22.5. The nonzero sensitivities are given in boldface. Note that they are the data points coincident with the corresponding θ_j estimates. The sensitivities must be interpreted as follows. If the data point $Y_{1,5}$ is increased in a small amount ϵ, the optimal maximum likelihood value will increase in 25.501ϵ units. Similar conclusions can be obtained for data points $Y_{2,2}, Y_{5,6}, Y_{3,8}$, and $Y_{4,18}$ using their corresponding sensitivities, also given in boldface in Table 22.5. Small changes in the remaining data values will produce no change in the likelihood. According to this, the most influential data point is $Y_{2,2}$ because its associated sensitivity 30.897 is the largest.

Table 22.3: Population and estimated parameters

j	θ_j	$\hat{\theta}_j$	σ_j	$\hat{\sigma}_j$
1	1.000	1.069	1.00	0.784
2	2.000	2.044	1.00	0.647
3	3.000	3.000	1.00	0.874
4	4.000	4.030	1.00	0.933
5	5.000	5.215	1.00	1.262

Table 22.4: Data values Y_{ij} simulated from $\exp(\theta_j, \sigma_j)$, where (θ_j, σ_j), $j = 1, 2, \ldots, 5$ are given in Table 22.3

	j				
i	1	2	3	4	5
1	1.20	2.53	3.23	5.53	6.31
2	1.36	2.04	3.18	4.94	6.59
3	1.33	2.40	3.39	4.46	5.68
4	2.93	2.83	6.31	4.72	5.78
5	1.07	3.46	8.04	4.20	5.60
6	1.87	3.42	4.46	4.14	5.22
7	5.73	2.99	3.51	8.31	7.99
8	2.02	2.10	3.00	5.51	5.31
9	1.16	3.03	3.52	6.69	6.01
10	1.58	3.57	3.28	4.21	5.50
11	1.44	2.06	3.50	4.29	6.65
12	1.15	3.39	5.75	4.10	5.79
13	1.90	2.19	3.54	4.16	7.71
14	2.09	2.97	3.47	5.17	5.41
15	2.49	2.49	3.31	5.43	9.12
16	1.69	2.27	3.51	5.19	10.26
17	1.18	2.15	3.10	5.87	5.86
18	1.33	3.53	3.20	4.03	5.43
19	1.92	2.35	3.15	4.20	7.42
20	1.63	2.08	3.05	4.14	5.91

Table 22.5: The sensitivities $\partial Q^*/\partial Y_{rs}$ for the data Y_{ij} in Table 22.4

i	j				
	1	2	3	4	5
1	0.00	0.00	0.00	0.00	0.00
2	0.00	**30.90**	0.00	0.00	0.00
3	0.00	0.00	0.00	0.00	0.00
4	0.00	0.00	0.00	0.00	0.00
5	**25.50**	0.00	0.00	0.00	0.00
6	0.00	0.00	0.00	0.00	**15.84**
7	0.00	0.00	0.00	0.00	0.00
8	0.00	0.00	**22.90**	0.00	0.00
9	0.00	0.00	0.00	0.00	0.00
10	0.00	0.00	0.00	0.00	0.00
11	0.00	0.00	0.00	0.00	0.00
12	0.00	0.00	0.00	0.00	0.00
13	0.00	0.00	0.00	0.00	0.00
14	0.00	0.00	0.00	0.00	0.00
15	0.00	0.00	0.00	0.00	0.00
16	0.00	0.00	0.00	0.00	0.00
17	0.00	0.00	0.00	0.00	0.00
18	0.00	0.00	0.00	**21.43**	0.00
19	0.00	0.00	0.00	0.00	0.00
20	0.00	0.00	0.00	0.00	0.00

22.6 The Method of Moments Estimates

One common theme in the above problems is that they can be formulated as optimization problems. In some estimation problems, the estimates are obtained as a result of solving a system of equations rather than optimizing a given function. An example of such a situation is the method of moments estimators. In this section, we first show how the local sensitivities of the moment estimates can be obtained; then we give two illustrative examples (the gamma and beta families).

22.6.1 Local sensitivities

Consider an iid sample $\mathbf{x}$ coming from a k-parameter family of univariate pdf $f(x;\boldsymbol{\theta})$, where $\boldsymbol{\theta} = (\theta_1, \theta_2, \ldots, \theta_k)$ is the associated parameter vector, and let $\{\mu^r_{a_r} : r = 1, 2, \ldots, k\}$ be a selected set of moments, where $\mu^r_{a_r}$ is the moment of order b_r taken with respect to the point a_r. Let $g_r(\boldsymbol{\theta})$ and $h_r(\mathbf{x}) = n^{-1}\sum_{i=1}^n (x_i - a_r(\mathbf{x}))^{b_r}$ be the corresponding population and sample versions, respectively. Then, the moment estimates are given by the system of equations:

$$g_r(\ddot{\boldsymbol{\theta}}) = h_r(\mathbf{x}), \qquad r = 1, 2, \ldots, k. \tag{22.44}$$

We look for the local sensitivities λ_{si} of $\hat{\theta}_s$ with respect to each single data value x_i, that is, the partial derivative $\frac{\partial \hat{\theta}_s}{\partial x_i}$. From (22.44) we have

$$\sum_{s=1}^{k} \frac{\partial g_r(\hat{\boldsymbol{\theta}})}{\partial \hat{\theta}_s} \frac{\partial \hat{\theta}_s}{\partial x_i} = \frac{\partial h_r(\mathbf{x})}{\partial x_i}, \quad i = 1, 2, \ldots, n, \quad r = 1, 2, \ldots, k, \tag{22.45}$$

which can be written in matrix form as

$$\sum_{s=1}^{k} g_{rs}\lambda_{si} = h_{ri}, \quad i = 1, 2, \ldots, n, \quad r = 1, 2, \ldots, k, \tag{22.46}$$

where $g_{rs} = \frac{\partial g_r(\hat{\boldsymbol{\theta}})}{\partial \hat{\theta}_s}$, $\lambda_{si} = \frac{\partial \hat{\theta}_s}{\partial x_i}$ and $h_{ri} = \frac{\partial h_r(\mathbf{x})}{\partial x_i}$. Then, we have

$$\lambda_{si} = g^*_{sr} h_{ri}, \tag{22.47}$$

where $g_{rs}g^*_{sj} = \delta_{rj}$, that is, the Kronecker's delta.

22.6.2 Example 1: The gamma family

Consider a Gamma $G(k, \lambda)$ random variable with density proportional to $e^{-\lambda x}x^{k-1}$ if $x > 0$ with $k, \lambda > 0$. Then, selecting the moments μ_0^1 (mean) and $\mu_{\bar{x}}^2$ (variance), we have

$$g_1(k, \lambda) = \frac{k}{\lambda}, \qquad g_2(k, \lambda) = \frac{k}{\lambda^2}, \tag{22.48}$$

$$h_1(\mathbf{x}) = \frac{1}{n}\sum_{i=1}^{n} x_i, \qquad h_2(\mathbf{x}) = \frac{1}{n}\sum_{i=1}^{n}(x_i - \bar{x})^2 \tag{22.49}$$

and the method of moments leads to the system of equations

$$\frac{\hat{k}}{\hat{\lambda}} = \frac{1}{n}\sum_{i=1}^{n} x_i = \bar{x}, \quad \frac{\hat{k}}{\hat{\lambda}^2} = \frac{1}{n}\sum_{i=1}^{n}(x_i - \bar{x})^2 = \sigma^2 \tag{22.50}$$

with solution

$$\hat{k} = \frac{\left(\sum\limits_{i=1}^{n} x_i\right)^2}{n\sum\limits_{i=1}^{n}(x_i - \bar{x})^2}; \quad \hat{\lambda} = \frac{\sum\limits_{i=1}^{n} x_i}{\sum\limits_{i=1}^{n}(x_i - \bar{x})^2} \tag{22.51}$$

Since

$$G = \begin{pmatrix} \frac{1}{\lambda} & -\frac{k}{\lambda^2} \\ \lambda^{-2} & \frac{-2k}{\lambda^3} \end{pmatrix} \quad \Leftrightarrow \quad G^{-1} = \begin{pmatrix} 2\lambda & -\lambda^2 \\ \frac{\lambda^2}{k} & -\frac{\lambda^3}{k} \end{pmatrix},$$

where G is the matrix with elements $g_{rs} = \frac{\partial g_r(\hat{\boldsymbol{\theta}})}{\partial \hat{\theta}_s}$ and

$$h_{1i} = \frac{\partial h_1(\mathbf{x})}{\partial x_i} = \frac{1}{n}, \quad h_{2i} = \frac{\partial h_2(\mathbf{x})}{\partial x_i} = \frac{2(x_i - \bar{x})}{n}, \tag{22.52}$$

we obtain the local sensitivities of the moment estimates with respect to x_i:

$$\begin{pmatrix} \lambda_{1i} \\ \\ \lambda_{2i} \end{pmatrix} = \begin{pmatrix} \frac{\partial \hat{k}}{\partial x_i} \\ \frac{\partial \hat{\lambda}}{\partial x_i} \end{pmatrix} = \begin{pmatrix} 2\hat{\lambda} & -\hat{\lambda}^2 \\ \\ \frac{\hat{\lambda}^2}{k} & -\frac{\hat{\lambda}^3}{k} \end{pmatrix} \begin{pmatrix} 1/n \\ \\ 2(x_i - \hat{k}/\hat{\lambda})/n \end{pmatrix}$$
$$= \frac{1}{n\hat{k}} \begin{pmatrix} 2\hat{k}(\hat{\lambda} - \hat{\lambda}^2(x_i - \bar{x})) \\ \\ \hat{\lambda}^2 - 2\hat{\lambda}^3(x_i - \bar{x}) \end{pmatrix}. \tag{22.53}$$

Because the k and λ sensitivities in (22.53) are affine transformations of order statistics, their plots apart from scale and location changes coincide, and for the sake of measuring the degree of outlyingness of the data points, based on the fact that the ith order statistic of a sample of size n from a standard uniform population has a beta $\mathrm{B}(i, n-i+1)$ distribution, we can use the following indices, which range on $(0,1)$:

$$\lambda_i = F_{B(i,n-i+1)}\left(F_{gamma(\hat{k},\hat{\lambda})}(x_{(i)})\right), \qquad i = 1, 2, \ldots, n, \tag{22.54}$$

where a value close to 0 or 1 means a high degree of outlyingness. For the sake of illustration, we have simulated a sample of size $n = 100$ from a gamma(2,3) and the sensitivities have been calculated. In order to have a reference for these sensitivities, we have also calculated the local sensitivities associated with the 0.5-quantiles of the order statistic $x_{(i)}$. Figure 22.1 shows the outlyingness indices λ_i based on the sensitivities. Note that the smallest order statistics can be considered as outliers (too small values).

22.6.3 Example 2: The beta family

Consider a beta $\mathrm{B}(r,t)$ random variable with parameters $r, t > 0$. Then, selecting the moments μ_0^1 (mean) and $\mu_{\bar{x}}^2$ (variance), we have

$$g_1(k, \lambda) = \frac{r}{t+r}, \qquad g_2(k, \lambda) = \frac{rt}{(t+r)^2(t+r+1)} \tag{22.55}$$

$$h_1(\mathbf{x}) = \frac{1}{n}\sum_{i=1}^{n} x_i, \qquad h_2(\mathbf{x}) = \frac{1}{n}\sum_{i=1}^{n}(x_i - \bar{x})^2 \tag{22.56}$$

and the method of moments leads to the system of equations

$$\frac{r}{t+r} = \frac{1}{n}\sum_{i=1}^{n} x_i, \quad \frac{rt}{(t+r)^2(t+r+1)} = \frac{1}{n}\sum_{i=1}^{n}(x_i - \bar{x})^2 \tag{22.57}$$

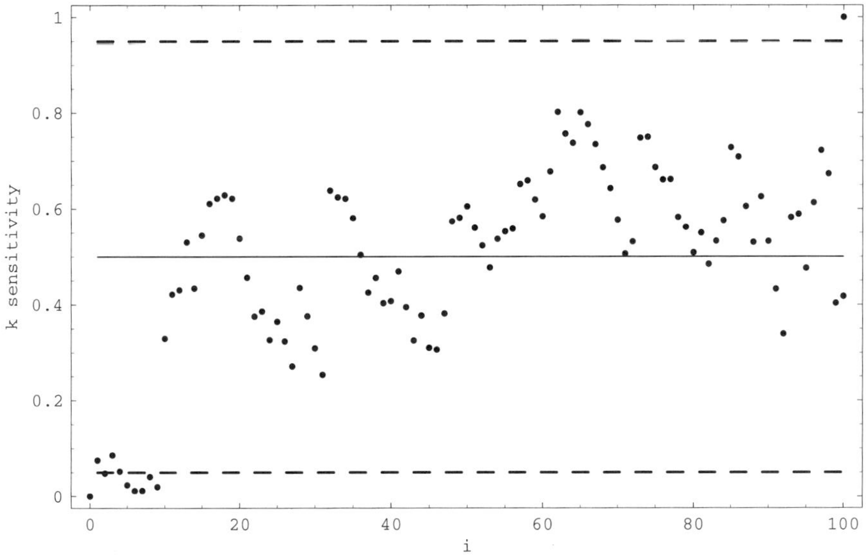

Figure 22.1: Outlyingness indices of sample points

with solution

$$\hat{r} = \frac{\bar{x}(\bar{x} - \bar{x}^2 - \sigma^2)}{\sigma^2}, \quad \hat{t} = \frac{(1 - \bar{x})(\bar{x} - \bar{x}^2 - \sigma^2)}{\sigma^2}. \tag{22.58}$$

Since

$$G = \begin{pmatrix} a & b \\ c & d \end{pmatrix},$$

where

$$\begin{aligned}
a &= \frac{\hat{t}}{\left(\hat{r} + \hat{t}\right)^2}, \\
b &= -\left(\frac{\hat{r}}{\left(\hat{r} + \hat{t}\right)^2}\right), \\
c &= \frac{\hat{t}\left(-2\,\hat{r}^2 - \hat{r}\left(1 + \hat{t}\right) + \hat{t}\left(1 + \hat{t}\right)\right)}{\left(\hat{r} + \hat{t}\right)^3\left(1 + \hat{r} + \hat{t}\right)^2}, \\
d &= \frac{\hat{r}\left(\hat{r} + \hat{r}^2 - \hat{r}\,\hat{t} - \hat{t}\left(1 + 2\,\hat{t}\right)\right)}{\left(\hat{r} + \hat{t}\right)^3\left(1 + \hat{r} + \hat{t}\right)^2}
\end{aligned}$$

and

$$G^{-1} = \begin{pmatrix} 1+\hat{r}-\dfrac{\hat{r}}{\hat{t}}-\dfrac{\hat{r}^2}{\hat{t}}+2\,\hat{t} & -\left(\dfrac{\left(\hat{r}+\hat{t}\right)\,\left(1+\hat{r}+\hat{t}\right)^2}{\hat{t}}\right) \\ \dfrac{-2\,\hat{r}^2-\hat{r}\,\left(1+\hat{t}\right)+\hat{t}\,\left(1+\hat{t}\right)}{\hat{r}} & -\left(\dfrac{\left(\hat{r}+\hat{t}\right)\,\left(1+\hat{r}+\hat{t}\right)^2}{\hat{r}}\right) \end{pmatrix},$$

where G is the matrix with elements $g_{rs} = \frac{\partial g_r(\hat{\boldsymbol{\theta}})}{\partial \hat{\theta}_s}$ and

$$\frac{\partial h_1(\mathbf{x})}{\partial x_i} = \frac{1}{n}, \quad \frac{\partial h_2(\mathbf{x})}{\partial x_i} = \frac{2(x_i-\bar{x})}{n}, \tag{22.59}$$

we obtain the local sensitivities of the moment estimates with respect to x_i:

$$\begin{pmatrix} \lambda_{1i} \\ \lambda_{2i} \end{pmatrix} = \begin{pmatrix} \frac{\partial \hat{r}}{\partial x_i} \\ \frac{\partial \hat{t}}{\partial x_i} \end{pmatrix}$$

$$= -\frac{\hat{r}+\hat{t}}{n} \begin{pmatrix} \frac{-1+2\,\hat{r}^2\,(-1+x_i)+2\,x_i+2\,\hat{t}^2\,x_i+\hat{t}\,(-2+4\,x_i)+\hat{r}\,\left(-3+4\,x_i+\hat{t}\,(-2+4\,x_i)\right)}{\hat{t}} \\ \frac{2\,\hat{r}^2\,(-1+x_i)+2\,\hat{r}\,\left(1+\hat{t}\right)\,(-1+2\,x_i)+\left(1+\hat{t}\right)\,\left(-1+2\,\left(1+\hat{t}\right)\,x_i\right)}{\hat{r}} \end{pmatrix}.$$

Note that they are also affine transformations of order statistics and then the previous treatment applies.

22.7 Conclusions

In this chapter a general method for sensitivity analysis, which is applicable to any model that can be formulated as an optimization problem or as a system of equations, is given. Theorem 22.2.1 provides a powerful tool to derive the analytical expression for the sensitivities when the problem can be stated as an optimization problem and the method described in Section 22.6.1 gives the sensitivities for parameters resulting from a system of equations. The power of the method has been proved and illustrated by its application to several examples in regression, maximum likelihood, ordered and data-constrained parameters, and the method of moments. Several numerical examples are used to illustrate the sensitivities.

Acknowledgments. The authors are indebted to the Spanish Ministry of Science and Technology (Projects DPI2002-04172-C04-02 and DPI2003-01362) for partial support.

References

1. Atkinson, A. C. (1984). Fast very robust methods for the detection of multiple outliers, *Journal of the American Statistical Association*, **89**, 1329–1339.

2. Atkinson, A. C. (1985). *Plots, Transformations, and Regression: An Introduction to Graphical Methods of Diagnostic Regression Analysis*, Clarendon Press, Oxford, England.

3. Barnett, V., and Lewis, T. (1994). *Outliers in Statistical Data*, 3rd ed., John Wiley & Sons, New York.

4. Barrett, B. E., and Gray, J. B. (1997). On the use of robust diagnostics in least squares regression analysis, *Proceedings of the Statistical Computing Section of the American Statistical Association*, 130–135.

5. Bazaraa M. S., Sherali, H. D., and Shetty C. M. (1993). *Nonlinear Programming, Theory and Algorithms*, 2nd ed., John Wiley & Sons, New York.

6. Belsley, D. A., Kuh, E., and Welsch, R. E. (1980). *Regression Diagnostics: Identifying Influential Data and Sources of Multicollinearity*, John Wiley & Sons, New York.

7. Billor, N., Chatterjee, S., and Hadi, A. S. (2001). Iteratively re-weighted least squares method for outlier detection in linear regression, *Bulletin of the International Statistical Institute*, **1**, 470–472.

8. Billor, N., Hadi, A. S., and Velleman, P. F. (2000). BACON: Blocked adaptive computationally-efficient outlier nominators, *Computational Statistics and Data Analysis*, **34**, 279–298.

9. Castillo, E., Conejo, A. J., Mínguez, R., and Castillo, C. (2006). A closed formula for local sensitivity analysis in mathematical programming, *Engineering Optimization*.

10. Castillo, E., Conejo, A. J., Pedregal, P., García, R., and Alguacil, N. (2001). *Building and Solving Mathematical Programming Models in Engineering and Science*, John Wiley & Sons, New York.

11. Castillo, E., Hadi, A. S., Conejo, A., and Fernández-Canteli, A. (2004b). A general method for local sensitivity analysis with application to regression models and other optimization problems, *Technometrics*, **46**, 433–444.

12. Conejo, A. J., Castillo, E., Mínguez, R., and García-Bertrand, R. (2005). *Decomposition Techniques in Mathematical Programming: Engineering and Science Applications*, Springer-Verlag, New York.

13. Chatterjee, S., and Hadi, A. S. (1988). *Sensitivity Analysis in Linear Regression*, John Wiley & Sons, New York.

14. Chatterjee, S., Hadi, A. S., and Price B. (2000). *Regression Analysis by Example*, Third edition, John Wiley & Sons, New York.

15. Cook, R. D. (1977). Detection of influential observations in linear regression, *Technometrics*, **19**, 15–18.

16. Cook, R. D. (1986). Assessment of local influence (with discussion), *Journal of the Royal Statistical Society, Series B*, **48**, 133–169.

17. Cook, R. D., and Weisberg, S. (1982). *Residuals and Influence in Regression*, Chapman and Hall, London.

18. Escobar, L. A., and Meeker, W. Q. (1992). Assessing influence in regression analysis with censored data, *Biometrics*, **48**, 507–528.

19. Gray, J. B. (1986). A simple graphic for assessing influence in regression, *Journal of Statistical Computation and Simulation*, **24**, 121–134.

20. Gray, J. B., and Ling, R. F. (1984). K-clustering as a detection tool for influential subsets in regression (with discussion), *Technometrics*, **26**, 305–330.

21. Hadi, A. S. (1992a). Identifying multiple outliers in multivariate data, *Journal of the Royal Statistical Society, Series B*, **54**, 761–771.

22. Hadi, A. S. (1992b). A new measure of overall potential influence in linear regression, *Computational Statistics & Data Analysis*, **14**, 1–27.

23. Hadi, A. S. (1994). A modification of a method for the detection of outliers in multivariate samples, *Journal of the Royal Statistical Society, Series B*, **56**, 393–396.

24. Hadi, A. S., and Simonoff, J. S. (1993). Procedures for the identification of multiple outliers in linear models, *Journal of the American Statistical Association*, **88**, 1264–1272.

25. Hawkins, D. M. (1980). *Identification of Outliers*, Chapman and Hall, London.

26. Jones, W. D., and Ling, R. F. (1988). A new unifying class of influence measures for regression diagnostics, *Proceedings of the Statistical Computing Section of the American Statistical Association*, 305–310.

27. Luenberger, D. G. (1989). *Linear and Nonlinear Programming*, 2nd ed., Addison-Wesley, Reading, MA.

28. Mayo, M. S., and Gray, J. B. (1997). Elemental subsets: The building blocks of regression, *Journal of the American Statistical Association*, **51**, 122–129.

29. Nyquist, H. (1992). Sensitivity analysis in empirical studies, *Journal of Official Statistics*, **8**, 167–182.

30. Paul, S. R. and Fung, K. Y. (1991). A generalized extreme studentized residual multiple-outlier-detection procedure in linear regression, *Technometrics*, **33**, 339–348.

31. Peña, D., and Yohai, V. (1995). The detection of influential subsets in linear regression by using an influence matrix, *Journal of the Royal Statistical Society, Series B*, **57**, 145–156.

32. Pregibon, D. (1981). Logistic regression diagnostics, *Annals of Statistics*, **9**, 705–724.

33. Saltelli, A., Chan, K. and Scott, E. M. (2000). *Sensitivity Analysis*, John Wiley & Sons, New York.

34. Schwarzmann, B. (1991). A connection between local-influence analysis and residual diagnostics, *Technometrics*, **33**, 103–104.

35. Simonoff, J. S. (1991). General approaches to stepwise identification of unusual values in data analysis, In *Directions in Robust Statistics and Diagnostics: Part II* (Eds., W. Stahel and S. Weisberg), pp. 223–242, Springer-Verlag, New York.

36. Weissfeld, I., and Schneider, H. (1990a). Influence diagnostics for the normal linear model with censored data, *Australian Journal of Statistics*, **32**, 11–20.

37. Weissfeld, I., and Schneider, H. (1990b). Influence diagnostics for the Weibull model fit to censored data, *Statistics & Probability Letters*, **9**, 67–73.

38. Welsch, R. E., and Kuh, E. (1977). Linear regression diagnostics, *Technical Report 923-77*, Sloan School of Management, Massachusetts Institute of Technology, Boston, MA.

39. Winsnowski, W. J., Montgomery, D. C., and James, R. S. (2001), A comparative analysis of multiple outlier detection procedures in the linear regression model, *Computational Statistics & Data Analysis*, **36**, 351–382.

23

t-Tests with Models Close to the Normal Distribution

Alfonso García-Pérez

Universidad Nacional de Educación a Distancia, Madrid, Spain

Abstract: The t-distribution is a very usual distribution for several test statistics because a normal distribution is frequently assumed as underlying model. Even in some tests based on robust statistics, such as the test based on the sample trimmed mean, a t-distribution is used as distribution for the standardized sample trimmed mean if the underlying model is normal. Nevertheless, it is necessary to have a deeper understanding of the behaviour of these kind of tests and the computations of their key elements, such as the p-value and the critical value, with small samples, when the underlying model is close but different from the normal distribution. In this paper, we obtain good analytic approximations, with small samples, for the p-value and the critical value of a t-test (i.e., a test with a t-distribution for the test statistic under a normal model), studying its behaviour when the underlying distribution is close but different from the normal model. We conclude the paper with a discussion on some robustness properties of t-tests.

Keywords and phrases: Robustness in hypotheses testing, von Mises expansion, tail area influence function, saddlepoint approximation, robustness of t-tests

23.1 Introduction

Many classical parametric tests were obtained assuming a normal distribution as underlying model. This is the reason why the χ^2, Student's t-, and F-distributions play a prominent role in statistics as distributions for test statistics.

Also in some tests based on robust statistics, such as the test based on the α-trimmed mean, a t-distribution is used as the distribution for the standardized trimmed mean if the underlying model is normal, even with a small sample size; see, for instance, Tukey and McLaughlin (1963), Patel *et al.* (1988), Staudte and Sheather (1990), and Wilcox (1997).

Nevertheless, it is necessary to have a deeper understanding of the behaviour of these kind of tests and the computations of their key elements, such as the p-value and the critical value, with small samples, when the underlying distribution is not normal but a slight deviation from it. Previous studies are, for instance, by Benjamini (1983), Cressie (1980), Chen and Loh (1990), and Sawilowsky and Blair (1992). Really, the distribution of the Student's statistic under other models is usually obtained through simulations, except in the paper by Lee and Gurland (1977) where this distribution is obtained only under contaminated normal models.

Here, we obtain good closed-form approximations of some key elements of a t-test, such as the critical value and the p-value, in a close to normal situation, developing a method proposed in García-Pérez (2003), which is based on considering all these elements as functionals of the model distribution, and that makes use of the von Mises expansion of a functional plus, in some cases, saddlepoint approximations.

This method is especially useful in robustness studies where the model distribution is, frequently, a slight deviation from the normal distribution (e.g., a contaminated normal) but complicated enough to render an exact calculation of these elements impossible.

With these aims, in Section 23.2 we briefly explain the method that we will use in the following sections. We obtain, in Section 23.3, von Mises approximations for t-tests (i.e., tests in which the test statistic follows a t-distribution under a normal model), but now when the model distribution is close to the normal distribution. In Section 23.4 we obtain saddlepoint approximations of the von Mises approximations and some interesting results; for instance, a complementary result of the one obtained by Benjamini (1983) or the conclusions drawn by Cressie (1980), that "a light-tailed parent distribution causes a heavy-tailed t-distribution." We also study the robustness of t-tests, obtaining some results that confirm the idea that, also with small samples sizes, the t-test has robustness of validity, at least in the tails, with slight departures from the normality.

23.2 Preliminaries

Although the method that we explain in this chapter can be extended to a more general setting, we will consider it in a one-dimensional test based on a test statistic $T_n = T_n(X_1, \ldots, X_n)$ that rejects the null hypothesis H_0 when

T_n is larger than the critical value k_n^F, where F is the distribution that the X_i's follow under H_0. If $T_n = t$, the p-value will be then the tail probability $p_n^F = P_F\{T_n > t\}$.

In particular, we will consider t-tests here, that is, tests in which T_n follows a t-distribution under a normal model, studying its behaviour under a model F, close but different, from the normal.

In these tests we will just consider two elements, the critical value k_n^F and the p-value p_n^F, although the method can be used to approximate other elements like the power. One of the key points is to consider these elements as functionals of the model distribution F.

We will suppose that T_n is real valued although the sample $X_1, \ldots, X_n$ can be one- or multidimensional. The only restriction is that, under the null hypothesis, both the critical value k_n^F and the p-value p_n^F must be functionals of only one distribution function F that we will assume to be univariate.

In a one-dimensional parametric test of the null hypothesis $H_0 : \theta = \theta_0$, if $X_1, \ldots, X_n$ is a sample from a random variable X with distribution function F_θ and $F_{n;\theta}$ is the cumulative distribution function of the test statistic T_n, the critical value of the level-α test

$$k_n^F = F_{n;\theta_0}^{-1}(1-\alpha)$$

and the p-value

$$p_n^F = P_{F_{\theta_0}}\{T_n > t\}$$

will be considered as functionals of F_{θ_0}. (Throughout the paper, the inverse of any distribution function G is defined, as usual, by $G^{-1}(s) = \inf\{y | G(y) \geq s\}$, $0 < s < 1$.)

For instance, if $T_n = M$ is the sample median, then

$$k_n^F = F_{\theta_0}^{-1}(B^{-1}(1-\alpha))$$

and

$$p_n^F = 1 - B(F_{\theta_0}(t)),$$

where B is the cumulative distribution function of a beta $\beta((n+1)/2, (n+1)/2)$.

If $T_n = \overline{x}$ is the sample mean and $F_{\theta_0} \equiv \Phi_{\theta_0,\sigma}$, the normal distribution $N(\theta_0, \sigma)$, we have

$$k_n^F = \frac{1}{\sqrt{n}} \left(\Phi_{\theta_0,\sigma}^{-1}(1-\alpha) + \theta_0 \left(\sqrt{n} - 1\right) \right)$$

and the p-value as

$$p_n^F = 1 - \Phi_{\theta_0,\sigma}\left(t\sqrt{n} - \theta_0(\sqrt{n} - 1)\right).$$

The most common t-test is the one based on the usual t-statistic

$$T_n = \frac{\sqrt{n}(\overline{x} - \theta_0)}{S}$$

with a t_{n-1} distribution under a $N(\theta_0, \sigma)$ model.

Besides, if $k = [n\alpha]$ is the integer portion of $n\alpha$, another t-test is the standardized sample α-trimmed mean (removing the k largest and k smallest observations)

$$T_n = \frac{(1-2\alpha)\sqrt{n}(\overline{x}_\alpha - \mu_{\alpha,0})}{S_w}$$

with an approximate t_{n-2k-1} distribution under a normal model, where S_w^2 is the sample Winsorized variance, if the null hypothesis is about the parameter α-trimmed mean, $H_0 : \mu_\alpha = \mu_{\alpha,0}$; see Tukey and McLaughlin (1963), Wilcox (1997, p. 75), Staudte and Sheather (1990, pp. 105, 156, 186), and Patel *et al.* (1988).

Finally, let us observe that it does not matter that the functionals k_n^F and p_n^F depend on n because we are not interested in the asymptotic (in n) distributional properties of these functionals. Actually, n is what Reeds (1976, p. 39) calls *an auxiliary parameter*.

23.2.1 Influence functions of p_n^F and k_n^F

To obtain the von Mises expansions of the functionals p_n^F and k_n^F, we will need their influence functions with respect to a model G (that later we will assume it to be the normal distribution).

We will represent these influence functions, respectively, by $\overset{\bullet}{p}{}_n^G$ and $\overset{\bullet}{k}{}_n^G$; they will be based on the *tail area influence function* (TAIF) defined by Field and Ronchetti (1985). This one is just the influence function of the tail probability of a statistic T_n at a distribution G and is defined as

$$\text{TAIF}(x; t; T_n, G) = \frac{\partial}{\partial \epsilon} P_{G^\epsilon}\{T_n > t\}\bigg|_{\epsilon=0}$$

for all $x \in I\!R$ where the right-hand side exists, being $G^\epsilon := (1-\epsilon)G + \epsilon\delta_x$ the contaminated model, and δ_x the point mass distribution at $x \in I\!R$.

The TAIF is really the influence function of the p-value,

$$\overset{\bullet}{p}{}_n^G = \text{TAIF}(x; t; T_n, G)$$

and after some computations [see García-Pérez (2003) for details], it is

$$\overset{\bullet}{k}{}_n^G = \frac{\text{TAIF}(x; k_n^G; T_n, G)}{g_n(k_n^G)}$$

assuming that the distribution function G_n of the test statistic, under the model G, has a density g_n with respect to the Lebesgue measure and that $g_n(k_n^G) \neq 0$.

Because, in this discussion, the distribution G (called *pivotal distribution* in the sequel) will be the normal distribution, we will have no problem about this with k_n^G.

23.2.2 Von Mises expansions of p_n^F and k_n^F

Let T be a functional defined on a convex set $\mathcal{F}$ of distribution functions and with range the set of the real numbers.

If F and G are two members of $\mathcal{F}$ and $s \in [0,1]$ is a real number, let us define the function A of the real variable s by

$$A(s) = T\left((1-s)G + sF\right) = T\left(G + s(F-G)\right).$$

Considering the viewpoint adopted by Filippova (1961) and Reeds (1976), the (*low-brow way* of the) von Mises expansion of the functional T is just the ordinary Taylor expansion of the real function $A(s)$, assuming that A satisfies the usual conditions for a Taylor expansion to be valid if $s \in [0,1]$; see, for instance, Serfling (1980, p. 43, Theorem 1.12.1A).

Then, expanding $A(s)$ about $s = 0$ and evaluating the resultant expansion at $s = 1$, we obtain the *von Mises expansion of the functional T at the distribution* $F \in \mathcal{F}$ as

$$T(F) = T(G) + \sum_{k=1}^{m} \frac{A^{k)}(0)}{k!} + Rem., \tag{23.1}$$

where $A^{k)}(0)$ is the ordinary kth derivative of A at the point 0,

$$A^{k)}(0) = \left.\frac{d^k}{dt^k} A(t)\right|_{t=0}, \qquad k = 1, \ldots, m,$$

and where the remainder term *Rem* depends on F and G, and on the $(m+1)$th derivative of A (i.e., on the influence function of T, if it exists).

Considering the sum in (23.1) up to the first or second term, we have, respectively, the *first-order von Mises expansion*

$$T(F) = T(G) + A^{1)}(0) + Rem_1$$

and the *second-order von Mises expansion*

$$T(F) = T(G) + A^{1)}(0) + \frac{1}{2} A^{2)}(0) + Rem_2$$

of T, with the second one having a higher degree of accuracy than the first one. Because we will obtain very accurate approximations just considering the first-order expansion, we consider this one in the rest of the discussion, omitting in the sequel the subscript of the remainder term.

Moreover, if the influence function of the functional T exists, usually represented by $\overset{\bullet}{T}(x)$ or just by $IF(x;T,G)$, it is

$$T(F) = T(G) + \int IF(x;T,G)\, dF(x) + Rem,$$

with the remainder term being

$$Rem = \frac{1}{2} \int\int T_H(x,y)\, dF(x) dF(y),$$

where

$$T_H(x,y) = \frac{\partial}{\partial \epsilon} IF(x;T,H_{\epsilon,y})\bigg|_{\epsilon=0} + IF(y;T,H)$$

and $H(x) = G(x) + \lambda(F(x) - G(x))$ with λ some constant in $[0,1]$ depending on F, G, T, and $H_{\epsilon,y} = (1-\epsilon)H + \epsilon\, \delta_y$ the H-contaminated distribution; see García-Pérez (2003) for more details.

Then, if there exists $\overset{\bullet}{p_n^G}$, the (first-order) *von Mises expansion of the p-value* will be

$$p_n^F = p_n^G + \int \overset{\bullet}{p_n^G}(x)\, dF(x) + Rem;$$

and, if there exists $\overset{\bullet}{k_n^G}$, the (first-order) *von Mises expansion of the critical value* will be

$$k_n^F = k_n^G + \int \overset{\bullet}{k_n^G}(x)\, dF(x) + Rem,$$

where the remainder terms, usually different in both expansions, will be smaller as F and G are closer. This can be formalized with the usual sup-norm or with a tail ordering on distributions like the $<_t$-ordering defined by Loh (1984).

23.2.3 Von Mises approximations of p_n^F and k_n^F with a model F close to the normal distribution

From the previous von Mises expansions, we define the approximations we were looking for, using the normal distribution $\Phi_{\mu,\sigma}$, as distribution G.

So, we define the (first-order) *von Mises (VOM) approximation of* p_n^F *by* p_n^Φ as

$$p_n^F \simeq p_n^\Phi + \int \mathrm{TAIF}(x;t;T_n,\Phi_{\mu,\sigma})\, dF(x) \tag{23.2}$$

and the (first-order) *von Mises (VOM) approximation of* k_n^F *by* k_n^Φ as

$$k_n^F \simeq k_n^\Phi + \frac{1}{\phi_n(k_n^\Phi)} \int \mathrm{TAIF}(x; k_n^\Phi; T_n, \Phi_{\mu,\sigma})\, dF(x) \tag{23.3}$$

where ϕ_n is the density of T_n under the normal model $\Phi_{\mu,\sigma}$.

In these equations we see explicitly the extra term that we add to the usual asymptotic normal approximations p_n^Φ and k_n^Φ, that improve them.

To simplify the notation we will omit the parameters of the normal distribution when it appears as subscript or superscript. We will represent the distribution and density functions of the standard normal $N(0,1)$, respectively, by Φ_s and ϕ_s.

23.3 Von Mises Approximations for *t*-Tests

In this section, we consider *t*-tests, that is, tests such that the test statistic T_n follows a *t*-distribution under the null hypothesis, when the underlying model is the normal distribution $N(\mu, \sigma)$. Here we will determine the VOM approximations (23.2) and (23.3), for their *p*-value and critical value, when the underlying model distribution F is not normal but a slight deviation from it.

The key element in the VOM approximations (23.2) and (23.3) is the TAIF under the normal model. To obtain this, we express first the tail probability of a *t*-test as a functional of the cumulative distribution function $\Phi_{\mu,\sigma}$ of the normal distribution $N(\mu, \sigma)$.

If the test statistic T_n follows a *t*-distribution with n degrees of freedom, t_n, we can express the tail probability of T_n as

$$P_\Phi\{T_n > t\} = \frac{1}{2} \int_{-\infty}^{\infty} P_\Phi \left\{ \chi_n^2 \le \frac{n(y-\mu)^2}{t^2\sigma^2} \right\} d\Phi_{\mu,\sigma}(y),$$

where χ_n^2 is a random variable with a chi-square distribution with n degrees of freedom.

So, under the contaminated model $\Phi^\epsilon = (1-\epsilon)\,\Phi_{\mu,\sigma} + \epsilon\,\delta_x$, we have

$$\begin{aligned} P_{\Phi^\epsilon}\{T_n > t\} \;&=\; \frac{1}{2}\left((1-\epsilon) \int_{-\infty}^{\infty} \left[1 - P_{\Phi^\epsilon}\left\{ \chi_n^2 > \frac{n(y-\mu)^2}{t^2\sigma^2} \right\} \right] d\Phi_{\mu,\sigma}(y) \right. \\ &\left. +\epsilon \left[1 - P_{\Phi^\epsilon}\left\{ \chi_n^2 > \frac{n(x-\mu)^2}{t^2\sigma^2} \right\} \right] \right). \end{aligned}$$

Now, we express the $\mathrm{TAIF}(x; t; t_n, \Phi_{\mu,\sigma})$ in terms of the TAIF of the χ^2, $\mathrm{TAIF}(x; t; \chi_n^2, \Phi_{\mu,\sigma})$, as

$$
\begin{aligned}
\mathrm{TAIF}(x;t;t_n,\Phi_{\mu,\sigma}) &= \frac{\partial}{\partial\epsilon}P_{\Phi^\epsilon}\{T_n>t\}\Big|_{\epsilon=0}\\
&= \frac{1}{2}\left(-\int_{-\infty}^{\infty}P\left\{\chi_n^2\le\frac{n(y-\mu)^2}{t^2\sigma^2}\right\}\,d\Phi_{\mu,\sigma}(y)\right.\\
&\quad-\int_{-\infty}^{\infty}\mathrm{TAIF}(x;\frac{n(y-\mu)^2}{t^2\sigma^2};\chi_n^2,\Phi_{\mu,\sigma})\,d\Phi_{\mu,\sigma}(y)\\
&\quad\left.+P\left\{\chi_n^2\le\frac{n(x-\mu)^2}{t^2\sigma^2}\right\}\right).
\end{aligned}
$$

García-Pérez (2004) obtained that the TAIF, under a normal model, of the functional χ_n^2 test is, if $n>1$,

$$
\mathrm{TAIF}(x;t;\chi_n^2,\Phi_{\mu,\sigma}) = n\,P\left\{\chi_{n-1}^2>t-\left(\frac{x-\mu}{\sigma}\right)^2\right\}-n\,P\{\chi_n^2>t\}.
$$

Then, if $n>1$, the TAIF under a normal model, of the functional t-test considered is

$$
\begin{aligned}
\mathrm{TAIF}(x;t;t_n,\Phi_{\mu,\sigma}) &= \frac{n}{2}-(n+1)P\{t_n>t\}+\frac{1}{2}P\left\{\chi_n^2\le\frac{n(x-\mu)^2}{t^2\sigma^2}\right\}\\
&\quad-\frac{n}{2}\int_{-\infty}^{\infty}P\left\{\chi_{n-1}^2>\frac{n(y-\mu)^2}{t^2\sigma^2}-\frac{(x-\mu)^2}{\sigma^2}\right\}\,d\Phi_{\mu,\sigma}(y).
\end{aligned}
$$

Because García-Pérez (2004) showed that

$$
\int_{-\infty}^{\infty}P\left\{\chi_{n-1}^2>t-\frac{(x-\mu)^2}{\sigma^2}\right\}\,d\Phi_{\mu,\sigma}(x)=P\{\chi_n^2>t\}
$$

it is easy to check that

$$
\int_{-\infty}^{\infty}\mathrm{TAIF}(x;t;t_n,\Phi_{\mu,\sigma})\,d\Phi_{\mu,\sigma}(x)=0.
$$

Finally, to obtain the VOM approximation of the p-value p_n^F and the critical value k_n^F when the model distribution for the observable random variable is F, we have to integrate the last TAIF with respect to F, as shown in Eqs. (23.2) and (23.3), obtaining

$$
\begin{aligned}
p_n^F \;&\simeq\; \frac{n}{2} - n\,P\{t_n > t\} + \frac{1}{2}\int_{-\infty}^{\infty} P\left\{\chi_n^2 \le \frac{n(x-\mu)^2}{t^2\sigma^2}\right\} dF(x) \\
&\quad -\frac{n}{2}\int_{-\infty}^{\infty}\int_{-\infty}^{\infty} P\left\{\chi_{n-1}^2 > \frac{n(y-\mu)^2}{t^2\sigma^2} - \frac{(x-\mu)^2}{\sigma^2}\right\} d\Phi_{\mu,\sigma}(y)\, dF(x)
\end{aligned}
\tag{23.4}
$$

and

$$
\begin{aligned}
k_n^F \;&\simeq\; t_{n;\alpha} + \frac{1}{g_{t_n}(t_{n;\alpha})}\left[\frac{n}{2} - (n+1)\,\alpha + \frac{1}{2}\int_{-\infty}^{\infty} P\left\{\chi_n^2 \le \frac{n(x-\mu)^2}{t_{n;\alpha}^2\,\sigma^2}\right\} dF(x)\right. \\
&\quad \left. -\frac{n}{2}\int_{-\infty}^{\infty}\int_{-\infty}^{\infty} P\left\{\chi_{n-1}^2 > \frac{n(y-\mu)^2}{t_{n;\alpha}^2\,\sigma^2} - \frac{(x-\mu)^2}{\sigma^2}\right\} d\Phi_{\mu,\sigma}(y)\, dF(x)\right]
\end{aligned}
\tag{23.5}
$$

where $t_{n;\alpha}$ is the $(1-\alpha)$-quantile of a t_n distribution and g_{t_n} the density function of this distribution.

Example 23.3.1 (t-tests under a scale contaminated normal model) If we assume a sample from a scale contaminated normal model $F = (1-\epsilon)N(\mu,\sigma) + \epsilon N(\mu, k\,\sigma)$, the VOM p-value (23.4) and the VOM critical value (23.5) of a t_n test are, respectively,

$$
\begin{aligned}
p_n^F \;&\simeq\; P\{t_n > t\} + \epsilon\left[\frac{n}{2} - (1+n)\,P\{t_n > t\} + P\{t_n > t/k\}\right. \\
&\quad \left. -\frac{n}{2}\int_{-\infty}^{\infty}\int_{-\infty}^{\infty} P\left\{\chi_{n-1}^2 > \frac{n(y-\mu)^2}{t^2\sigma^2} - \frac{(x-\mu)^2}{\sigma^2}\right\} d\Phi_{\mu,\sigma}(y)\, d\Phi_{\mu,k\sigma}(x)\right]
\end{aligned}
$$

and

$$
\begin{aligned}
k_n^F \;&\simeq\; t_{n;\alpha} + \frac{\epsilon}{g_{t_n}(t_{n;\alpha})}\left[\frac{n}{2} - (1+n)\,\alpha + P\{t_n > t_{n;\alpha}/k\}\right. \\
&\quad \left. -\frac{n}{2}\int_{-\infty}^{\infty}\int_{-\infty}^{\infty} P\left\{\chi_{n-1}^2 > \frac{n(y-\mu)^2}{t_{n;\alpha}^2\sigma^2} - \frac{(x-\mu)^2}{\sigma^2}\right\} d\Phi_{\mu,\sigma}(y)\, d\Phi_{\mu,k\sigma}(x)\right].
\end{aligned}
$$

We see in these two expressions the extra term we add to the usual p-value and critical value of a t-test under a normal model, and the influence of each element $(\epsilon, n, k, \ldots)$ in these extra terms.

To finish the example with numerical values, let us consider, for instance, a sample of size 4 from a distribution $0.95\,N(0,1) + 0.05\,N(0,\sqrt{4})$ instead of a

$N(0,1)$, and the usual Student's test statistic

$$\frac{\sqrt{4}\,\overline{x}}{S}$$

that follows a t_3 distribution, to test at level α, $H_0 : \mu = 0$ against $H_1 : \mu > 0$. The approximated p-value and critical value are, respectively,

$$\begin{aligned} p_n^F &\simeq P\{t_3 > t\} + 0'05 \left[\frac{3}{2} - 4\,P\{t_3 > t\} + P\{t_3 > t/2\} \right. \\ &\left. - \frac{3}{2} \int_{-\infty}^{\infty} \int_{-\infty}^{\infty} P\left\{ \chi_2^2 > \frac{3\,y^2}{t^2} - x^2 \right\} d\Phi_{0,1}(y)\, d\Phi_{0,2}(x) \right] \end{aligned}$$

and

$$\begin{aligned} k_n^F &\simeq t_{3;\alpha} + \frac{0'05}{g_{t_3}(t_{3;\alpha})} \left[\frac{3}{2} - 4\,\alpha + P\{t_3 > t_{3;\alpha}/2\} \right. \\ &\left. - \frac{3}{2} \int_{-\infty}^{\infty} \int_{-\infty}^{\infty} P\left\{ \chi_2^2 > \frac{3\,y^2}{t_{3;\alpha}^2} - x^2 \right\} d\Phi_{0,1}(y)\, d\Phi_{0,2}(x) \right]. \end{aligned}$$

Table 23.1: *Exact* and approximate p-values under a contaminated normal model and $n = 4$

t	"exact"	VOM
1	0.1979	0.1979
2	0.0688	0.0716
4	0.0138	0.0158
5	0.0071	0.0089

Table 23.2: *Exact* and approximate critical values under a contaminated normal model and $n = 4$

α	"exact"	VOM
0.01	4.445	4.718
0.05	2.330	2.378
0.1	1.629	1.631

Tables 23.1 and 23.2 present the *exact* values (obtained through simulation of a 30,000 samples and using the package 'stepfun' of the software R in different t's) and the (first-order) VOM approximations in this situation (using the package 'adapt' of R for the numerical integration).

However, to obtain these results we had to compute the approximations using numerical integration. Normally we would prefer to have analytic expressions for them that can be used as elements in other more complex problems; for instance, to study the robustness of the t-tests. For this reason, we will obtain saddlepoint approximations of these (first-order) VOM approximations in the next section.

23.4 Saddlepoint Approximations for t-Tests

Although it is possible to use known saddlepoint approximations for the tails of the χ^2 and t-distributions that appear in Eqs. (23.4) and (23.5)[see, e.g., Jensen (1995, pp. 49, 86)], these would be numerical again, or would depend on integrals of the normal cumulative distribution function with respect to the underlying model F, not obtaining, in this way, manageable analytic expressions of them. For this reason, we will approximate the TAIF, using the Lugannani and Rice formula, before integration in (23.2) and (23.3).

If T_n follows a t_n distribution, and Y_1, Y_2 are two independent gamma distributions, $\gamma(1/2, 1/2)$ and $\gamma(n/2, n/2)$, respectively, we can write

$$P\{T_n > t\} = P\{Y_1 - t^2 Y_2 > 0\},$$

where the random variable $Y = Y_1 - t^2 Y_2$ has cumulant generating function

$$K(\theta) = \log M(\theta) = \log M_\gamma(\theta) + n \log M_\gamma(-\theta t^2/n)$$

with

$$M_\gamma(\theta) = \int_{-\infty}^{\infty} e^{\theta\,(u-\mu)^2/\sigma^2}\, d\Phi_{\mu,\sigma}(u)$$

being the moment generating function of a gamma $\gamma(1/2, 1/2)$, a functional that depends on the distribution model $\Phi_{\mu,\sigma}$.

Now, we can use the Lugannani and Rice formula [see Lugannani and Rice (1980) or Daniels (1983] for the tail, in a sample of size one, of $Y = Y_1 - t^2 Y_2$, obtaining

$$P_\Phi\{Y > 0\} = 1 - \Phi_s(w) + \phi_s(w) \left\{ \frac{1}{r} - \frac{1}{w} + O(1) \right\}, \qquad (23.6)$$

where the functionals r and w are

$$w = sign(z_0)\, \sqrt{-2\, K(z_0)},$$

$$r = z_0\, \sqrt{K''(z_0)}$$

that depend on the saddlepoint z_0, which is the solution of the equation

$$K'(z_0) = 0$$

from where we obtain the saddlepoint

$$z_0 = \frac{t^2 - 1}{2\,t^2}\,\frac{n}{1+n}.$$

Now, from (23.6), we obtain

$$\begin{aligned}
\text{TAIF}\,(x;t;t_n,\Phi_{\mu,\sigma}) &= \frac{\partial}{\partial\epsilon} P_{\Phi^\epsilon}\left\{Y>0\right\}\Big|_{\epsilon=0} \\
&\simeq \frac{\phi_s(w)}{r}\left\{-w\,\overset{\bullet}{w} - \frac{\overset{\bullet}{r}}{r} + \frac{\overset{\bullet}{w}\,r}{w^2}\right\} \\
&= \frac{e^K}{\sqrt{2\pi}\,z_0\,\sqrt{K''}}\left\{\overset{\bullet}{K}\left[1 - \frac{z_0\sqrt{K''}}{(-2K)^{3/2}}\right] - \frac{\overset{\bullet}{z}_0}{z_0} - \frac{\overset{\bullet}{K''}}{2K''}\right\}.
\end{aligned}$$

After some algebraic computations and approximations, if

$$A_1 = \frac{t}{\sqrt{\pi}\,(t^2-1)}\,e^{-(t^2-1)/2}$$

and

$$A_2 = 1 - \frac{t^2-1}{\sqrt{2}(t^2-1-2\log t)^{3/2}},$$

we obtain $\forall x$ and $t > 1$, for the functional t-test considered,

$$\begin{aligned}
\text{TAIF}\,(x;t;t_n,\Phi_{\mu,\sigma}) = A_1 &\left\{\left(A_2 - \frac{3t^2+1}{4(t^2-1)}\right) t^{-1}\,e^{(t^2-1)(x-\mu)^2/(2t^2\sigma^2)}\right. \\
&+\frac{3t^2-1}{2(t^2-1)}\,t^{-3}\left(\frac{x-\mu}{\sigma}\right)^2 e^{(t^2-1)(x-\mu)^2/(2t^2\sigma^2)} \\
&-\frac{t^{-5}}{4}\left(\frac{x-\mu}{\sigma}\right)^4 e^{(t^2-1)(x-\mu)^2/(2t^2\sigma^2)} \\
&\left.-\frac{t^2}{t^2-1}\left(\frac{x-\mu}{\sigma}\right)^2 + \frac{t^2}{t^2-1} - A_2\right\}.
\end{aligned}$$

(It is easy to verify that $\int_{-\infty}^{\infty} \text{TAIF}\,(x;t;t_n,\Phi_{\mu,\sigma})\;d\Phi_{\mu,\sigma}(x) = 0$).

Now, integrating this TAIF with respect to a model F, from (23.2) we obtain the *VOM+SAD approximate p-value* of the test, under a model F, as

$$p_n^F \simeq P\{t_n > t\} + \int \text{TAIF}(x;t;t_n,\Phi_{\mu,\sigma})\,dF(x). \qquad (23.7)$$

From (23.3), we obtain the *VOM+SAD approximate critical value* of the test, under a model F, as

$$k_n^F \simeq t_{n;\alpha} + \frac{1}{g_{t_n}(t_{n;\alpha})} \int \mathrm{TAIF}(x; t_{n;\alpha}; t_n, \Phi_{\mu,\sigma})\, dF(x), \tag{23.8}$$

where $t_{n;\alpha}$ is the $(1-\alpha)$-quantile of a t_n distribution with density g_{t_n}.

Example 23.4.1 (t-tests under scale contaminated normal models) Let us consider a t-test, that is, a test in which the test statistic follows a t_n distribution under a normal model $N(0,1)$. Now, let us consider as model for this test, a scale contaminated normal model $F = (1-\epsilon)N(0,1) + \epsilon N(0,k) \equiv (1-\epsilon)\,\Phi_s + \epsilon\,\Phi_{0,k}$.

Because, for $t > 1$ and $a = 0, 2,$ or 4, we have

$$\int_{-\infty}^{\infty} x^a\, e^{(t^2-1)x^2/(2t^2)}\, d\Phi_{0,k}(x) = \frac{2^{a/2}\,\Gamma((a+1)/2)\, t^{a+1}\, k^a}{\sqrt{\pi}\,[t^2 - k^2(t^2-1)]^{(a+1)/2}}$$

if $k < \sqrt{1/(1-t^{-2})}$, the VOM+SAD p-value in (23.7) is

$$\begin{aligned} p_n^F \;\simeq\;& P\{t_n > t\} + \epsilon\, A_1 \left\{ \left(A_2 - \frac{3t^2+1}{4(t^2-1)} \right) [t^2 - k^2(t^2-1)]^{-1/2} \right. \\ &+ \frac{3t^2-1}{t^2-1}\,\frac{k^2}{2}\,[t^2 - k^2(t^2-1)]^{-3/2} - \frac{3\,k^4}{4}\,[t^2 - k^2(t^2-1)]^{-5/2} \\ &\left. + \frac{t^2(1-k^2)}{t^2-1} - A_2 \right\} \end{aligned}$$

and the VOM+SAD critical value in (23.8) is

$$\begin{aligned} k_n^F \;\simeq\;& t_{n;\alpha} + \frac{\epsilon\, A_1}{g_{t_n}(t_{n;\alpha})} \left\{ \left(A_2 - \frac{3t_{n;\alpha}^2+1}{4(t_{n;\alpha}^2-1)} \right) [t_{n;\alpha}^2 - k^2(t_{n;\alpha}^2-1)]^{-1/2} \right. \\ &+ \frac{3t_{n;\alpha}^2-1}{t_{n;\alpha}^2-1}\,\frac{k^2}{2}\,[t_{n;\alpha}^2 - k^2(t_{n;\alpha}^2-1)]^{-3/2} - \frac{3\,k^4}{4}\,[t_{n;\alpha}^2 - k^2(t_{n;\alpha}^2-1)]^{-5/2} \\ &\left. + \frac{t_{n;\alpha}^2(1-k^2)}{t_{n;\alpha}^2-1} - A_2 \right\}, \end{aligned}$$

where, as before, $t_{n;\alpha}$ is the $(1-\alpha)$-quantile of a t_n distribution with density g_{t_n}.

Now, if we consider a scale contaminated normal model $0.95\,N(0,1) + 0.05\,N(0,0.6)$ (a situation with inliers), a sample of size $n = 10$ and the usual Student's test statistic

$$\frac{\sqrt{n}\,\overline{x}}{S}$$

Table 23.3: *Exact* and approximate p-values under a scale contaminated normal model, and standard p-values when $n = 10$

t	"exact"	VOM+SAD	$P\{t_9 > t\}$
1.5	0.08533	0.09908	0.08393
2	0.03787	0.03978	0.03828
3	0.00760	0.00749	0.00748

Table 23.4: *Exact* and approximate critical values under a scale contaminated normal model, and standard critical values when $n = 10$

α	"exact"	VOM+SAD	$t_{9;\alpha}$
0.01	2.82284	2.82323	2.82144
0.05	1.84097	1.87263	1.83311
0.1	1.39572	1.58127	1.38303

that follows a t_9 distribution under a normal model, to test at level α, $H_0 : \mu = 0$ against $H_1 : \mu > 0$, the VOM+SAD p-values and critical values are presented in Tables 23.3 and 23.4 together with the *exact* ones (obtained with a simulation of 30,000 samples and using the package 'stepfun' of R), and the usual values obtained under a standard normal model $N(0, 1)$.

From these tables, we observe that the VOM+SAD approximations are quite good and, comparing these with the last column (values under a normal model) we see that, for most of the values, using a light-tailed model we obtain a long-tailed distribution for the test statistic.

Remark 23.4.1 One of the questions related with the behaviour of t-tests is if they are conservative or liberal with long-tailed and short-tailed distributions, that is, that, if it is $F >_t G$, with $>_t$ a (partial) ordering of distribution functions then, is $P_G\{t_n > t\} \geq P_F\{t_n > t\}$?

A complete answer to this question depends on the integrals of the TAIF with respect to the distribution model through Eq. (23.7). Because of the last example, we have a solution inside the class of scale contaminated normal models, complementary of the conclusion drawn by Benjamini (1983), which is that "a light-tailed parent distribution causes a heavy-tailed t-distribution." In other words, if we consider two distributions $F_{k_1} = (1 - \epsilon)\, \Phi_s + \epsilon\, \Phi_{0,k_1}$ and $F_{k_2} = (1 - \epsilon)\, \Phi_s + \epsilon\, \Phi_{0,k_2}$, where $0 < k_1 < k_2 < 1$, [i.e., $F_{k_2} >_t F_{k_1}$ with respect, for instance, to the tail ordering defined by Loh (1984)], we have

$$P_{F_{k_1}}\{t_n > t\} \geq P_{F_{k_2}}\{t_n > t\}$$

Table 23.5: Actual sizes of the one sample t-test when sampling from two scale contaminated normal models when $n = 3$

Nominal level of significance (α)	$0.98N(0,1) + 0.02N(0,0'6)$	$0.95N(0,1) + 0.05N(0,0.6)$
0.01	0.00999	0.00999
0.05	0.05013	0.05031
0.1	0.10306	0.10764

Table 23.6: Actual sizes of the one-sample t-test when sampling from two scale contaminated normal models when $n = 5$

Nominal level of significance (α)	$0.98N(0,1) + 0.02N(0,0'6)$	$0.95N(0,1) + 0.05N(0,0.6)$
0.01	0.00999	0.00999
0.05	0.05056	0.05141
0.1	0.10689	0.11723

at least if the critical value t is $1 < t \leq 1.747$.

Remark 23.4.2 From Table 23.3, we see that the size of the test does not change very much with a distribution $0.95\,N(0,1) + 0.05\,N(0,0.6)$ considering a t_9-distribution. In Tables 23.5 and 23.6, we obtain the same conclusions with a t_3- and a t_5 distributions, respectively.

From these computations we can state that, with small samples, the t-test has robustness of validity for small departures from a Gaussian population, at least in the tails. This is not, however, the most common situation we meet in real life. Sawilowsky and Blair (1992) agree with both these conclusions.

Example 23.4.2 (t-tests under location contaminated normal models) If we assume a sample from a location contaminated normal model $F = (1 - \epsilon)N(0,1) + \epsilon N(\mu, 1)$, the VOM+SAD p-value of a t_n test is

$$p_n^F \simeq P\{t_n > t\} + \epsilon\, A_1 \left\{ \left(e^{\mu^2(t^2-1)/2} - 1 \right) \left(A_2 + \frac{t^2\mu^2}{t^2 - 1} \right) - e^{\mu^2(t^2-1)/2}\, \frac{\mu^4 t^4}{4} \right\}.$$

(There is no problem to compute the critical value or to consider other more general location contaminated normal models, or even location-scale contaminated normal models).

Table 23.7: Actual sizes of the one-sample t-test when sampling from location contaminated normal models. $n = 3$

Nominal level of significance (α)	$0'98N(0,1)+$ $0'02N(\mp 0'5,1)$	$0'95N(0,1)+$ $0'05N(\mp 0'5,1)$	$0'9N(0,1)+$ $0'1N(\mp 0'5,1)$
$0'01$	$0'00999$	$0'00997$	$0'00995$
$0'05$	$0'04978$	$0'04945$	$0'04890$
$0'1$	$0'09862$	$0'09655$	$0'09310$

Remark 23.4.3 From Table 23.7, we obtain for location contaminated normal models the same conclusions as before in the sense that, with small samples, the t-test has robustness of validity, at least in the tails, for small departures from a Gaussian population.

References

1. Benjamini, Y. (1983). Is the t test really conservative when the parent distribution is long-tailed?, *Journal of the American Statistical Association*, **78**, 645–654.

2. Chen, H., and Loh, W. Y. (1990). Uniform robustness against nonnormality of the t and F tests, *Communications in Statistics—Theory and Methods*, **19**, 3707–3723.

3. Cressie, N. (1980). Relaxing assumptions in the one sample t-test, *Australian Journal of Statistics*, **22**, 143–153.

4. Daniels, H. E. (1983). Saddlepoint approximations for estimating equations, *Biometrika*, **70**, 89–96.

5. Field, C. A., and Ronchetti, E. (1985). A tail area influence function and its application to testing, *Communications in Statistics—Theory and Methods*, **14**, 19–41.

6. Filippova, A. A. (1961). Mises' theorem on the asymptotic behaviour of functionals of empirical distribution functions and its statistical applications, *Theory of Probability and Its Applications*, **7**, 24–57.

7. García-Pérez, A. (2003). Von Mises' approximation of the critical value of a test, *Test*, **12**, 385–411.

8. García-Pérez, A. (2004). Chi-square tests under models close to the normal distribution, *Technical Report*, Departamento de Estadística, Investigación Operativa y Cálculo Numérico, UNED, Madrid.

9. Jensen, J. L. (1995). *Saddlepoint Approximations*, Clarendon Press, New York.

10. Lee, A. F. S., and Gurland, J. (1977). One-sample *t*-test when sampling from a mixture of normal distributions, *The Annals of Statistics*, **5**, 803–807.

11. Loh, W. Y. (1984). Bounds on AREs for restricted classes of distributions defined via tail-orderings, *The Annals of Statistics*, **12**, 685–701.

12. Lugannani, R., and Rice, S. (1980). Saddle point approximation for the distribution of the sum of independent random variables, *Advances in Applied Probability*, **12**, 475–490.

13. Patel, K. R., Mudholkar, G. S., and Fernando, J. L. I. (1988). Student's *t* approximations for three simple robust estimators, *Journal of the American Statistical Association*, **83**, 1203–1210.

14. Reeds, J. A. (1976). *On the Definitions of Von Mises' Functionals*, Ph.D. Thesis, Department of Statistics, Harvard University, Cambridge, MA.

15. Sawilowsky, S. S., and Blair, R. C. (1992). A more realistic look at the robustness and type II error properties of the *t* test to departures from population normality, *Psychological Bulletin*, **111**, 352–360.

16. Serfling, R. J. (1980). *Approximation Theorems of Mathematical Statistics*, John Wiley & Sons, New York.

17. Staudte, R. G., and Sheather, S. J. (1990). *Robust Estimation and Testing*, John Wiley & Sons, New York.

18. Tukey, J. W., and McLaughlin, D. H. (1963). Less vulnerable confidence and significance procedures for location based on a single sample: Trimming/Winsorization 1, *Sankhyā, Series A*, **25**, 331–352.

19. Wilcox, R. R. (1997). *Introduction to Robust Estimation and Hypothesis Testing*, Academic Press, San Diego.

24

Computational Aspect of the Chi-Square Goodness-of-Fit Test Application

Michael Divinsky

Israel Public Works Department, Tel Aviv, Israel

Abstract: The purpose of the paper is to attract attention to the chi-square goodness-of-fit test computation employing the SAS System possibilities. The results of the analysis based on the chi-square goodness-of-fit test application prove that the limit value for the theoretical expectations should be taken into consideration while computing and interpreting the chi-square test results. A computational procedure should be analyzed. Typical examples including analysis of the actual data and modeled sample of the generated values have been considered, and comparative analyses of the output results have been carried out. Suggested additional options in regard to possibilities concerning the chi-square goodness-of-fit test application serve for increasing the reliability of the interpretation of the output results.

Keywords and phrases: Probability distribution, statistical hypothesis, goodness-of-fit test, chi-square test, computational method

24.1 Introduction

The chi-square goodness-of-fit test (χ^2-test) for the specified theoretical probability density function is one of the key features in the SAS System Capability Procedure [SAS/QC (1989)] that helps to identify a statistical model of the process data values. The histogram statement in the procedure provides an approximation of the distribution of the process data [King (1995)] and can be used in order to suggest an appropriate probability model for the process under investigation that plays essential role in quality improvement. The chi-square goodness-of-fit test provides a quantitative test of the discrepancies between the observed, f_i, and expected, F_i, frequencies [Snedecor and Cochran (1980)] under the null hypothesis that observations are randomly drawn from the specified theoretical distribution. The test can be applied to both continuous and discrete distributions.

In its current form, the χ^2-statistic is computed using the upper tail of the χ^2-distribution as critical region for the test statistic value [Stuart and Ord (1991)] even though early practice was to use both small and large χ^2-values for the critical region determination. In addition, the χ^2-statistic is asymptotically equivalent to the maximum likelihood statistic [Stuart and Ord (1991)] when the null hypothesis, H_0, holds.

The purpose of this paper is to attract attention to the additional options regarding the SAS System possibilities with respect to the chi-square goodness-of-fit test application.

24.2 On the Chi-Square Test Application

Sometimes expected hypothetical frequencies, F_i, may be too small at the tails of the distribution for unimodal distribution and equal-length classes. At the same time, the chi-square test results are based on the large sample theory, and therefore the expectations must not be too small for any class. The working rule of no expectations in the class less than 1 has been stressed. These expectations can be close to 1 provided that most of the remaining expectations exceed 5.

Usually, combination of the classes containing small hypothetical frequencies are recommended. If combination of the classes is needed, the number of classes k (after all the combinations) should be used for the computation of degrees of freedom. The requirement that each expectation should be at least 5 is noted in Afifi and Azen (1979). It has been stressed that the approximation is reasonable when this limit for the theoretical frequency $F_i \geq 2$ while the remaining F_i should be at least 5. When F_i increases, the accuracy of the χ^2-approximation improves. The preference of the limit of 5 for F_i is also given in Barnes (1994), but when $k \geq 3$ and no $F_i < 1$, as many as 20% of the classes can be presented by $F_i < 5$.

It can be taken that, in practice, the theoretical frequencies in each class should exceed 1 or 2 under above-mentioned conditions and restrictions without significant influence over the χ^2-test results.

Various computational procedures concerning the chi-square goodness-of-fit test application have been examined. In the χ^2-test computations, the Capability Procedure uses practically all expected frequencies greater than zero. Hence, this application needs to be compared with the test computation using the above-referenced restrictions for the theoretical expectations in classes at the tails of the specified distribution.

In order to demonstrate the differences in the test computations using various values of the limited theoretical expectations in the class, we have presented some examples that reflect the typical generalization of the chi-square goodness-of-fit test application.

24.3 An Actual Data Set

Laboratory tested tensile strength parameter values, *TS*, have been analyzed as a typical example. The histogram of the *TS* parameter values (in kg/cm^2) superimposed with the theoretical curves of the normal, lognormal, and gamma distributions, similar to Divinsky and Livneh (1999), is shown in Figure 24.1. The figure also presents summary statistics information, including sample size, N, mean, standard deviation, variance, the smallest (minimum) and the largest (maximum) values in the sample under study, skewness, kurtosis, and coefficient of variation.

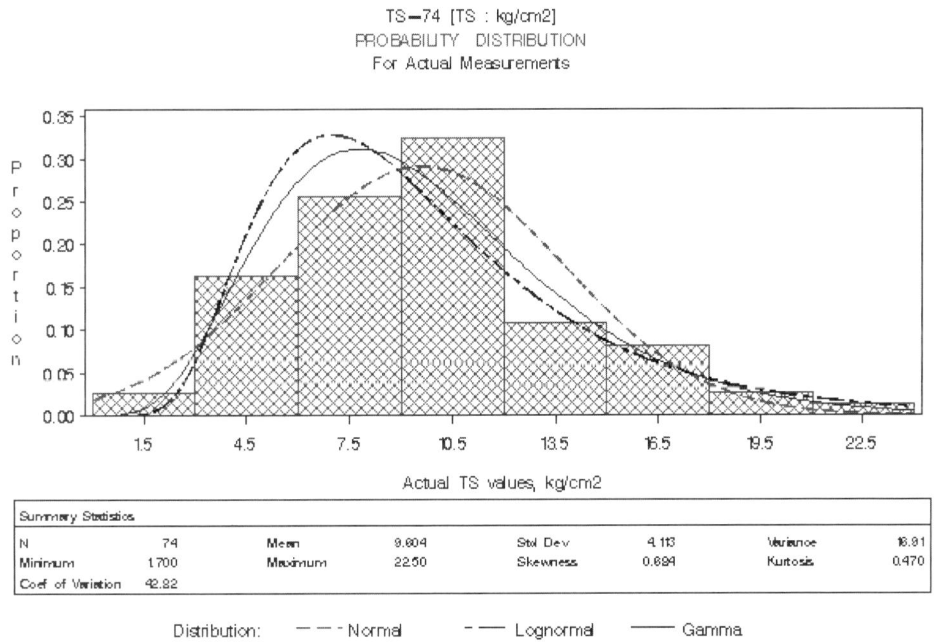

Figure 24.1: The histogram of the actual *TS* parameter values (in kg/cm^2) superimposed on the theoretical curves of the normal, lognormal, and gamma distributions

Observed and expected frequencies of the *TS* parameter values for the above theoretical distributions are presented in Table 24.1. Corresponding results of

Table 24.1: Observed and expected frequencies of actual *TS* parameter values for the normal, lognormal, and gamma theoretical distributions

Class limits	Class midpoints	Observed frequencies f_i	Expected frequencies, F_i, for theoretical distribution		
			Normal	Lognormal	Gamma
0–3.0	1.5	2	3.2855	0.8404	1.3861
3.0–6.0	4.5	12	10.0837	14.9199	13.2536
6.0–9.0	7.5	19	18.5877	23.3293	22.3414
9.0–12.0	10.5	24	20.5934	16.6793	18.4985
12.0–15.0	13.5	8	13.7147	9.1739	10.6293
15.0–18.0	16.5	6	5.4876	4.6011	4.9020
18.0–21.0	19.5	2	1.3178	2.2450	1.9528
21.0–24.0	22.5	1	0.1897	1.0955	0.7009

Table 24.2: The results of the goodness-of-fit test application corresponding to data presented in Table 24.1

F_m	Theoretical distribution								
	Normal			Lognormal			Gamma		
	DF	χ_c^2	$P(\chi^2)$	DF	χ_c^2	$P(\chi^2)$	DF	χ_c^2	$P(\chi^2)$
0	5	7.6841	0.1745	5	6.7988	0.2360	5	3.5515	0.6156
1	4	5.3465	0.2536	4	4.8237	0.3059	4	3.4679	0.4827
2	3	4.3956	0.2218	3	4.8234	0.1852	3	3.1054	0.3757
3	3	4.3956	0.2218	3	4.8234	0.1852	2	3.0904	0.2133

the goodness-of-fit test application for actual *TS* parameter values are shown in Table 24.2. The latter table includes limit values, F_m, for theoretical expectations, F_i, that satisfy the condition $F_i \geq F_m$. In addition, the table presents degrees of freedom, *DF*, computed χ^2-statistic value, χ_c^2 , and values of the probability for the test, $P(\chi^2)$, for the normal, lognormal and gamma distributions separately, computed using several limit values, F_m, for theoretical expectations.

Table 24.2 shows differences in the χ^2-test results about the computed χ_c^2 and the probability $P(\chi^2)$values with changing F_m values. The above differences are noticeable enough, but not crucial for presented theoretical distributions.

24.4 Modeled Sample of the Generated Values

In addition, the influence of the F_m value on the χ^2-test results has been analyzed using generated samples of the *TS* parameter values.

The histogram of 148 generated gamma distributed values superimposed on the theoretical curves of the normal, lognormal, and gamma distributions are shown in Figure 24.2. The parameters estimated for the actual (original) measurements of the TS values have been used for generation of the values. The figure also presents summary statistics information. Observed and expected frequencies of *TS* parameter values concerning Figure 24.2 are shown in Table 24.3. The corresponding results of the χ^2goodness-of-fit test application are presented in Table 24.4.

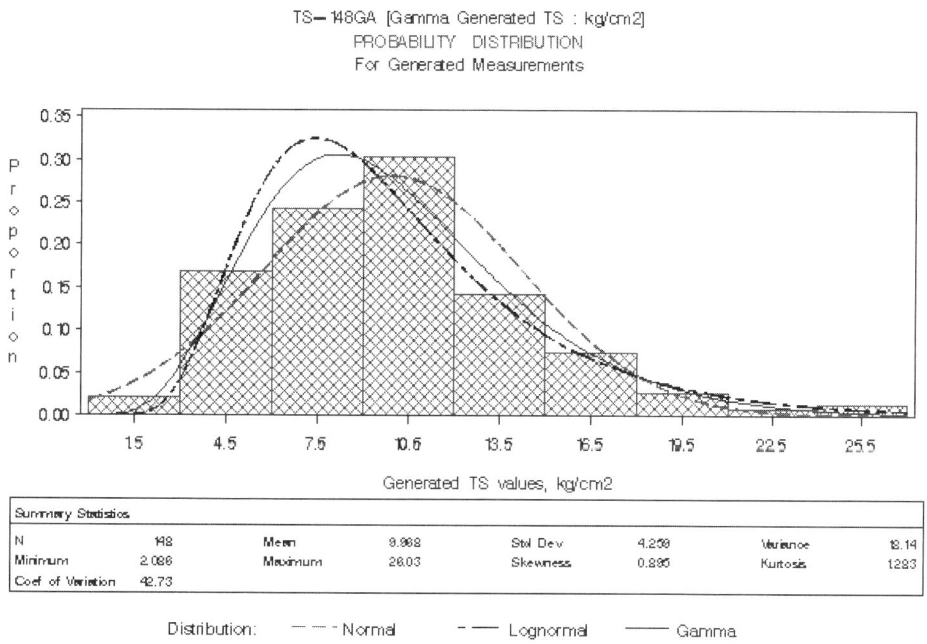

Figure 24.2: The histogram of the generated *TS* parameter values (in kg/cm^2) superimposed on the theoretical curves of the normal, lognormal, and gamma distributions

Again, the χ^2-test results show noticeable dependence on the limit F_m value for all theoretical distributions under study as approximation for the empirical distribution of generated *TS* parameter values. It should be pointed to the fact that for $F_m = 0$, the normal hypothesis is rejected at significance level

Table 24.3: Observed and expected frequencies regarding generated *TS* parameter values for the normal, lognormal, and gamma theoretical distributions

Class limits	Class midpoints	Observed frequencies f_i	Expected frequencies, F_i, for theoretical distribution		
			Normal	Lognormal	Gamma
0–3.0	1.5	3	6.1097	1.0003	2.0639
3.0–6.0	4.5	25	18.4768	25.3135	23.2088
6.0–9.0	7.5	36	34.6854	46.5398	43.2967
9.0–12.0	10.5	45	40.4432	35.6469	38.3393
12.0–15.0	13.5	21	29.2946	20.0577	23.0957
15.00–18.0	16.5	11	13.1775	10.0548	11.0276
18.0–21.0	19.5	4	3.6784	4.8421	4.5098
21.0–24.0	22.5	1	0.6365	2.3150	1.6515
24.0–27.0	25.5	2	0.0682	1.1159	0.5568

Table 24.4: The results of the goodness-of-fit test application corresponding to data presented in Table 24.3

F_m	Theoretical distributions								
	Normal			Lognormal			Gamma		
	DF	χ^2_c	$P(\chi^2)$	DF	χ^2_c	$P(\chi^2)$	DF	χ^2_c	$P(\chi^2)$
0	6	62.1386	0.0000	6	10.5698	0.1026	6	7.1955	0.3031
1	4	8.7199	0.0685	6	10.5698	0.1026	5	3.4814	0.6262
2	4	8.7199	0.0685	4	5.2828	0.2595	5	3.4814	0.6262
3	4	8.7199	0.0685	4	5.2828	0.2595	3	2.8832	0.4100

substantially less than 0.01. However, for the remaining F_m values, the normal model can be accepted at significance level greater than 0.05.

24.5 Conclusions

The results of the analysis based on the chi-square goodness-of-fit test application reveal that the limit value for the theoretical expectations should be taken into consideration while computing and interpreting the chi-square test results. Additional options for the chi-square goodness-of-fit test estimation employing the SAS System, depending on the limit value for the expected hypothetical frequencies, have also been examined.

Analysis of an actual data set as well as modeled sample of the generated values have been considered as illustrative examples. Significant differences in the output results with respect to the computational method have been demonstrated. The limit values F_m from 0 to 3 for theoretical expectations can be recommended as default, for the computational procedure, but optionally, the F_m value may be specified by user in order to extend or restrict the F_m value range. The above consideration concerning additional options regarding the SAS System possibilities for the chi-square goodness-of-fit test application will help in increasing the reliability of the interpretation of the output results of the procedure.

References

1. Afifi, A. A., and Azen, S. P. (1979). *Statistical Analysis. A Computer Oriented Approach*, Academic Press, New York.

2. Barnes, J. W. (1994), *Statistical Analysis for Engineers and Scientists: A Computer-Based Approach*, McGraw-Hill, New York.

3. Divinsky, M., and Livneh, M. (1999). Probabilistic model for the analysis of dynamic cone penetrometer test values in pavement structure evaluation, *Journal of Testing and Evaluation*, **27**, 7–14.

4. King, D. W. (1995). *Statistical Quality Control Using the SAS System*, SAS Institute, Cary, NC.

5. SAS/QC (1989). *SAS/QC Software: Reference*, Version 6, Second edition, SAS Institute, Cary, NC.

6. Snedecor, G. W., and Cochran, W. G. (1980). *Statistical Methods*, 7th ed., The Iowa State University Press, Ames, IA.

7. Stuart, A., and Ord, J. K. (1991). *Kendall's Advanced Theory of Statistics, Vol. 2*, 5th ed., Edward Arnold, London.

25

An Objective Bayesian Procedure for Variable Selection in Regression

F. Javier Girón,[1] Elías Moreno,[2] and M. Lina Martínez[1]

[1] *Universidad de Málaga, Málaga, Spain*
[2] *Universidad de Granada, Granada, Spain*

Abstract: The Bayesian analysis of the variable selection problem in linear regression when using objective priors needs some form of encompassing the class of all submodels of the full linear model as they are nonnested models. After we provide a nested setting, objective intrinsic priors suitable for computing model posterior probabilities, on which the selection is based, can be derived.

The way of encompassing the models is not unique and there is no clear indications for the optimal way. Typically, the class of linear models are encompassed into the full model.

In this paper, we explore a new way of encompassing the class of linear models that consequently produces a new method for variable selection. This method seems to have some advantages with respect to the usual one.

Specific intrinsic priors and model posterior probabilities are provided along with some of their main properties. Comparisons are made with R^2 and adjusted R^2, along with other frequentist methods for variable selection as *lasso*. Some illustrations on simulated and real data are provided.

Keywords and phrases: Calibration curve, determination coefficient, g-priors, intrinsic priors, lasso criterion, model selection, normal linear model, reference priors

25.1 Introduction

Suppose that Y represents an observable random variable and $X_1, X_2, \ldots, X_k$ a set of k potential explanatory covariates related through the normal linear model

$$Y = \alpha_1 X_1 + \alpha_2 X_2 + \cdots + \alpha_k X_k + \varepsilon, \quad \varepsilon \sim N(\cdot|0, \sigma^2). \tag{25.1}$$

The variable selection problem consists in reducing the complexity of model (25.1) by identifying a subset of the α_i coefficients that have a zero value based on the available dataset $(\mathbf{y}, \mathbf{X})$, where $\mathbf{y}$ is a vector of dimension n and $\mathbf{X}$ is a $n \times k$ design matrix of full rank. In variable selection it is customary to set $X_1 = 1$ and $\alpha_1 \neq 0$ to include the intercept in any model. Thus, the number of possible submodels of the above full model is 2^{k-1}.

Subjective prior information on the regression coefficients and the variance errors of the models can occasionally be considered, but the complexity of the required inputs suggests that objective priors are a more appropriate choice. We will use intrinsic priors that provide an automatic Bayesian set-up, and have been proved to behave extremely well in a wide variety of problems; see for example, Berger and Pericchi (1996b), Casella and Moreno (2005, 2006), Girón, Martínez, and Moreno (2003), Girón *et al.* (2003), Moreno, Bertolino and Racugno (2000, 2003), Moreno, Girón and Torres (2003), Moreno and Liseo (2003) and Moreno, Torres, and Casella (2005).

A fully objective analysis of model comparison in linear regression was given in Berger and Pericchi (1996a). They utilize an encompassing approach and an empirical measure—the intrinsic Bayes factor—that do not depend on any subjective prior information. For large sample sizes, this empirical measure closely approximates an *actual* Bayes factor for intrinsic priors.

A recently developed Bayesian procedure [see Casella and Moreno (2005)], consists of considering the pairwise model comparison between the full linear model M_F given in (25.1) and a generic submodel M_i having k_i ($< k$) nonzero regression coefficients. The posterior probability of M_i in the space of models $\{M_i, M_F\}$ is computed, and an ordering of the whole set of submodels in accordance to their posterior probabilities $P(M_i|\mathbf{y}, \mathbf{X})$, $i = 1, \ldots, 2^{k-1}$, is obtained.

The interpretation of this probability is the following: the submodel having the highest posterior probability is the most plausible reduction in complexity from the full model, the second highest the second most plausible reduction, and so on. Notice that any model M_i is nested in the full model M_F, a property that makes possible the derivation of intrinsic priors.

Although this procedure behaves extremely well, it is based on multiple pairwise comparisons. This implies that for ranking the models we compare probabilities coming from different probability spaces, that is, $(M_i, M_F, P(\cdot|\mathbf{y}, \mathbf{X}))$, $i = 1, \ldots, 2^{k-1}$. One can argue that this might not be coherent, even when all models are compared with the full.

A natural alternative procedure, which has not yet been considered, is the one based on the pairwise model comparison between a generic submodel M_i and the model

$$Y = \alpha_1 + \varepsilon, \quad \varepsilon \sim N(\cdot|0, \sigma^2),$$

the one containing the intercept only, which is denoted as M_1. In the space of models $\{M_1, M_i\}$ the posterior probability of M_i is computed, and a new ordering of the models M_i according to these posterior probabilities $\{P^*(M_i|\mathbf{y},\mathbf{X}), i = 1,\ldots,2^{k-1}\}$ is obtained. Notice that M_1 is nested in M_i, so that intrinsic priors can also be derived.

This procedure is formally based on multiple pairwise comparisons but, as we will show, it is equivalent to ordering the models according to model posterior probabilities computed in the space of all models by using intrinsic priors. Therefore, it is a fully Bayesian coherent procedure.

In this chapter, we mainly focus on this objective Bayesian approach to the variable selection problem. We analyze some of its theoretical properties, and the ordering of the models it provides. It will be seen that it is related to the frequentist criteria for variable selection based on R^2, or its adjusted version R^2_{adj}, and the C_p criterion [Mallows (1973)]. These frequentist criteria for model choice provide reasonable answers when comparing models that have the same dimension but usually fail when comparing models with different dimension. An explanation for this drawback is given and a calibration of R^2 is provided. The more recent frequentist criterion, the *lasso* [Tibshirani (1996)], which is intended to avoid this problem, is also briefly considered here for comparison purposes.

The chapter is organized as follows. Section 25.2 gives the intrinsic prior distributions involved in the analysis, and a summary of their main properties. Section 25.3 develops formulae for computing model posterior probabilities, some asymptotic properties, and the relationship with the R^2 statistic. This yields a calibration of this statistic through the model posterior probability. Section 25.4 summarizes some comparisons done through simulated and real data examples. Finally, Section 25.5 gives some concluding remarks and recommendations.

25.2 Intrinsic Priors for Variable Selection

The variable selection problem consists of testing whether the set of k explanatory variables can be reduced to a tentative subset with k_i regressors, for some $k_i \leq k$. From the point of view explained in the introduction, the variable selection problem consists in choosing the submodel M_i, which, compared with the intercept only model, provides the maximum posterior probability.

A Bayesian formulation of this problem is to choose between the two nested models

$$M_1 : N_n(\mathbf{y}|\alpha_1\mathbf{1}_n, \sigma_1^2\mathbf{I}_n),\ \pi^N(\alpha_1,\sigma_1) = \frac{c_1}{\sigma_1}, \tag{25.2}$$

and

$$M_i : N_n(\mathbf{y}|\mathbf{X}_i\gamma_i, \sigma_i^2\mathbf{I}_n),\ \pi^N(\gamma_i,\sigma_i) = \frac{c_i}{\sigma_i}, \tag{25.3}$$

where $\boldsymbol{\gamma}_i$ and $\mathbf{X}_i$ represent the vector of the regression coefficients and the corresponding $n \times k_i$ submatrix of $\mathbf{X}$ of submodel M_i, respectively, π^N is the usual improper reference prior distribution for estimating parameters, and c_1 and c_i are arbitrary positive constants.

The direct use of improper priors for computing model posterior probabilities is not possible, but they can be converted into suitable intrinsic priors [Berger and Pericchi (1996a) and Moreno *et al.* (1998)]. Intrinsic priors for the parameters of the above nested linear models provide *actual* Bayes factors [Moreno *et al.* (1998)], and, more importantly, posterior probabilities of the model M_1 and M_i, assuming that priors probabilities are assigned to them. An objective assessment of this latter prior is $P(M_1) = P(M_i) = 1/2$.

Applying the standard method [Moreno *et al.* (1998)], the intrinsic prior for the parameters $\boldsymbol{\gamma}_i, \sigma_i$, conditional on $\boldsymbol{\alpha}_1$, σ_1, turns out to be

$$\pi^I(\gamma_i, \sigma_i|\boldsymbol{\alpha}_1, \sigma_1) = \frac{2}{\pi\sigma_1(1+\sigma_i^2/\sigma_1^2)} N_{k_i}(\gamma_i|\tilde{\boldsymbol{\alpha}}_1, (\sigma_1^2+\sigma_i^2)\mathbf{W}_i^{-1}), \tag{25.4}$$

where $\tilde{\boldsymbol{\alpha}}_1 = (\alpha_1, 0, \ldots, 0)^t$.

For estimating the matrix $\mathbf{W}_i^{-1}$ two close related forms have been proposed [Casella and Moreno (2005) and Girón *et al.* (2006)]. Although both essentially give the same posterior answer, the computational simpler form is that given in Girón *et al.* (2006) as

$$\mathbf{W}_i^{-1} = \frac{n}{k_i+1}(\mathbf{X}_i^t\mathbf{X}_i)^{-1},$$

which resembles the covariance matrix of Zellner's g-prior [Zellner (1986)].

The conditional intrinsic prior for the parameter $\boldsymbol{\gamma}_i$ is a normal distribution "centered" at the intercept α_1, which plays the role of the null hypothesis for the multiple hypothesis testing we are considering. This is fixed across all models M_i. However, the matrix $\mathbf{W}_i^{-1}$ changes when the alternative model M_i changes.

The unconditional intrinsic prior for the parameters $(\boldsymbol{\gamma}_i, \sigma_i)$ is obtained from

$$\pi^I(\gamma_i, \sigma_i) = \int \pi^I(\gamma_i, \sigma_i|\alpha_1, \sigma_0)\pi^N(\alpha_1, \sigma_1)d\alpha_1, d\sigma_1.$$

Therefore, the intrinsic priors for comparing model (25.2) and (25.3) are $\{\pi^N(\alpha_1, \sigma_1), \pi^I(\gamma_i, \sigma_i)\}$. Note that both priors are improper, but they depend on the same arbitrary constant so that the Bayes factor for intrinsic priors is well defined. The tails of the intrinsic priors are very heavy, in fact they do not have moments, which seems to be a reasonable property for a nonsubjective prior distribution. We remark that intrinsic priors have been obtained in a completely automatic way, so that they do not need to adjust any hyperparameters.

25.3 Bayes Factors and Model Posterior Probabilities

Adapting the proof in Moreno *et al.* (2003), the Bayes factor to compare models M_1 and M_i using the intrinsic priors $\{\pi^N(\alpha_1, \sigma_1), \pi^I(\gamma_i, \sigma_i)\}$ turns out to be

$$B_{1i}(n, \mathcal{B}_i) = \frac{2(k_i+1)^{(k_i-1)/2}}{\pi} \int_0^{\pi/2} \frac{(\sin\varphi)^{k_i-1}(n+(k_i+1)\sin^2\varphi)^{(n-k_i)/2}}{(n\mathcal{B}_i+(k_i+1)\sin^2\varphi)^{(n-1)/2}} d\varphi, \tag{25.5}$$

where the statistic $\mathcal{B}_i$ is given by

$$\mathcal{B}_i = \frac{\mathbf{y}^t(\mathbf{I}_n - \mathbf{H}_i)\mathbf{y}}{ns_y^2} = \frac{\mathbf{y}^t(\mathbf{I}_n - \mathbf{H}_i)\mathbf{y}}{\mathbf{y}^t(\mathbf{I}_n - 1/n\mathbf{1}_n\mathbf{1}_n^t)\mathbf{y}},$$

with $\mathbf{H}_i = \mathbf{X}_i(\mathbf{X}_i^t\mathbf{X}_i)^{-1}\mathbf{X}_i^t$, $s_y^2 = \sum_{i=1}^n (y_i - \bar{y})^2/n$, and $\bar{y} = \sum_{i=1}^n y_i/n$.

The statistic $\mathcal{B}_i$ can be expressed as

$$\mathcal{B}_i = 1 - R_i^2, \tag{25.6}$$

where R_i^2 represents the popular coefficient of determination, the ratio of the sum of squares due to regression of model M_i to the total sum of squares.

Thus, the intrinsic posterior probability of M_i is given by

$$P^*(M_i|n, R_i^2) = P^*(M_i|n, \mathcal{B}_i) = \frac{B_{i1}(n, \mathcal{B}_i)}{1 + B_{i1}(n, \mathcal{B}_i)}, \tag{25.7}$$

where $B_{i1}(n, \mathcal{B}_i) = B_{1i}(n, \mathcal{B}_i)^{-1}$.

The set of all models is now ordered according to the intrinsic posterior probabilities $\{P^*(M_i|n, \mathcal{B}_i),\ i = 1, \ldots, 2^{k-1}\}$, where $P^*(M_1|n, \mathcal{B}_1) = 1/2$ as $\mathcal{B}_1 = 1$. These posterior probabilities measure the strength of adding the variables of each submodel M_i to the intercept-only model.

However, a serious interpretive difficulty of the above probabilities may arise when the number of covariates is large or even moderate. In fact, as we add new covariates the statistic R_i^2 increases rendering high values sometimes very close to 1, so that the Bayes factor increases to infinity. In this case, and also for moderate or large sample sizes, many of the posterior probabilities $P^*(M_i|n, \mathcal{B}_i)$ tend to 1, making it difficult to interpret them as measures of discrimination for variable selection. This becomes apparent when looking at the behavior of the *calibration curves* (see Figure 25.1), and not only happens with simulated data but even with real data as will be seen in the examples in Section 25.4.

A way to remedy this difficulty is by computing the posterior probabilities of M_i in the set of all models $\mathcal{M} = \{M_i, i = 1, \ldots, 2^{k-1}\}$, instead of the dichotomous set $\{M_1, M_i\}$. Assuming the objective uniform prior in the set of all models, that is, $P(M_i) = 1/2^{k-1}$, $i = 1, \ldots, 2^{k-1}$, the posterior probability of M_i in the space $\mathcal{M}$ is given by

$$\Pr(M_i|\mathbf{y}, \mathbf{X})) = \frac{m_i(\mathbf{y}|\mathbf{X})}{m_1(\mathbf{y}|\mathbf{X}) + \sum_{i=2}^{2^{k-1}} m_i(\mathbf{y}|\mathbf{X})}, \tag{25.8}$$

where $m_i(\mathbf{y}|\mathbf{X})$ represents the marginal of the data when using the intrinsic prior, that is,

$$m_i(\mathbf{y}|\mathbf{X}) = \int N_n(\mathbf{y}|\mathbf{X}_i\gamma_i, \sigma_i^2\mathbf{I}_n)\ \pi^I(\gamma_i, \sigma_i) d\gamma_i d\sigma_i,$$

and

$$m_1(\mathbf{y}|\mathbf{X}) = \int N_n(\mathbf{y}|\tilde{\alpha}_1\mathbf{1}_n, \sigma_1^2\mathbf{I}_n)\ \pi^N(\alpha_1, \sigma_1) d\alpha_1 d\sigma_1.$$

Dividing (25.6) by $m_1(\mathbf{y}|\mathbf{X})$, the posterior probability can be written as

$$\Pr(M_i|n, \mathcal{B}_i) = \frac{B_{i1}(n, \mathcal{B}_i)}{1 + \sum_{i=2}^{2^{k-1}-1} B_{i1}(n, \mathcal{B}_i)}. \tag{25.9}$$

Observe now that the posterior probability given in (25.8), obtained from the pairwise comparison, and that given in (25.9), which represents a probability in the space of all models, are both increasing functions of the Bayes factor $B_{i1}(n, \mathcal{B}_i)$. This implies that the ordering of the models given by the set $\{P^*(M_i|n, \mathcal{B}_i),\ i = 1 \geq 1\}$ is exactly the same as that obtained from $\{\Pr(M_i|n, \mathcal{B}_i),\ i \geq 1\}$. This last set of probabilities is a coherent set in the space of all models $\mathcal{M}$. Thus, the use of these *diluted* probabilities for ordering the set of all models—which may be denoted as the diluted intrinsic Bayes (DIB) criterion—justifies the procedure based on pairwise comparisons. Further, these diluted posterior probabilities, unlike the pairwise probabilities, can be used for model averaging.

The use of diluted probabilities is useful per se, as they provide a coherent criterion for model selection and for model averaging, and allows for recognizing possibly masked differences in the paired-wise probabilities when these are too close to unity.

25.4 Relation with the R^2 and Other Classical Criterion for Model Selection

A recent reference of frequentists, and some Bayesian, criteria is the book by Miller (2002). Here, due to the relation of the Bayes factor with the coefficient of determination, we mainly focus on those frequentist criteria most related to the behavior of the R^2 statistic.

Let $\mathcal{M}_i$ denote the set of models of $\mathcal{M}$ that have k_i regressors including the intercept. Thus, $\mathcal{M} = \bigcup_{i=1}^{k} \mathcal{M}_i$. The *best subset regression* chooses in each set $\mathcal{M}_i$ the model that minimizes the residual sum of squares, or equivalently the one that maximizes R^2 in that class. In general, this produces a set of $k-1$ *maximal* or *admissible* models each one with $1, \ldots, k-1$ covariates plus the intercept. The question of how to choose among the maximals usually involves considering the trade-off between bias and variance; therefore, one has to use some additional criterion. One such criterion, which takes into account the dimension of the model, is the adjusted R^2_{adj} defined as

$$R^2_{adj} = 1 - (1 - R^2)\,\frac{n-1}{n-k_i}.$$

Another related criterion is based on minimizing the C_{k_i} statistic [Mallows (1979)],

$$C_{k_i} = \frac{\mathbf{y}^t(\mathbf{I}_n - \mathbf{H}_i)\mathbf{y}}{\mathbf{y}^t(\mathbf{I}_n - \mathbf{H})\mathbf{y}}(n-k) + (2k_i - n), \quad 1 \leq k_i \leq k,$$

where $\mathbf{H} = \mathbf{X}(\mathbf{X}^t\mathbf{X})^{-1}\mathbf{X}^t$ is the *hat* matrix of the full model.

These two criteria applied to the whole set of models $\mathcal{M}$ always select *admissible* models because, for fixed n and k_i, they are monotonic increasing and decreasing of the R^2 statistic, respectively.

The Bayes factor $B_{i1}(n, \mathcal{B}_i)$ can be expressed in terms of the R_i^2 as

$$\begin{aligned} &B_{1i}(n, 1-R_i^2) \\ &= \frac{2(k_i+1)^{(k_i-1)/2}}{\pi} \int_0^{\pi/2} \frac{(\sin\varphi)^{k_i-1}(n+(k_i+1)\sin^2\varphi)^{(n-k_i)/2}}{(n(1-R_i^2)+(k_i+1)\sin^2\varphi)^{(n-1)/2}} d\varphi, \end{aligned} \tag{25.10}$$

and both the intrinsic posterior $P^*(M_i|n, R_i^2)$, and the diluted posterior probability $\Pr(M_i|n, R_i^2)$ of M_i can be written as

$$\begin{aligned} P^*(M_i|n, R_i^2) &= \frac{B_{i1}(n, 1-R_i^2)}{1+B_{i1}(n, 1-R_i^2)}; \\ \Pr(M_i|n, R_i^2) &= \frac{B_{i1}(n, 1-R_i^2)}{1+\sum_{i=2}^{2^{k-1}-1} B_{i1}(n, 1-R_i^2)}, \end{aligned} \tag{25.11}$$

respectively.

Thus, the posterior probabilities $P^*(M_i|n, R_i^2)$ and $\Pr(M_i|n, \mathcal{B}_i)$ are increasing functions of R_i^2 for any fixed n and k_i. This implies that the DIB criterion always select *admissible models*.

From (25.10) and (25.11) it follows that for the class of models $\mathcal{M}_i$ having a fixed number k_i of regressors, the ordering provided by the coefficient of determination, the adjusted version, C_{k_i} and the intrinsic posterior probabilities (25.7) or (25.9) coincide. However, this is no longer true when k_i varies, as in variable selection where $1 \leq k_i \leq k$. In this case, neither R^2_{adj} nor Mallows C_{k_i} generally produce the same ordering than that given by the DIB criterion. Thus, we are confronted with the problem of calibrating the R^2 scale when jumping among models of different dimensions.

For a fixed sample size n, and model dimension k_i, it follows from (25.11) that there is a one-to-one correspondence between R_i^2 and the posterior probability of M_i. This curve is termed the *calibration curve* of R^2. Figure 25.1 shows how this curve is affected by the sample size n for $n = 10, 20$ and 30 for a fixed value of $k_i = 5$ (left curve), and by the dimension of the model k_i for $k_i = 2, 4$, and 6 for a fixed sample size $n = 40$ (right curve), respectively. Other properties of this curve are discussed in Girón and Moreno (2005).

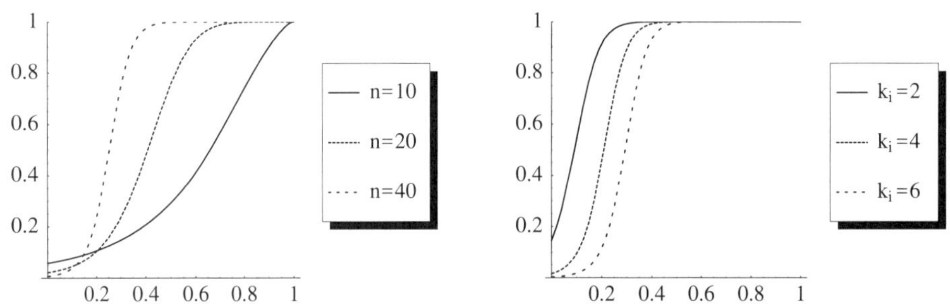

Figure 25.1: Calibration curves

When k is very large the exhaustive search over $\mathcal{M}$, *best subset regression*, becomes impractical. For moderate values of k, say about $k = 40$, the *leaps and bounds* algorithm of Furnival and Wilson (1974) provides an efficient way to obtain the class of *admissible models*. Other optimization algorithms for selecting the best model using the DIB criterion might be *simulated annealing* and *genetic* algorithms. If we want instead to compute those models with highest diluted probability, we need to resort to MCMC methods such as Gibbs sampling, the Metropolis-Hastings algorithm or for very high-dimensional models, to the reversible jump algorithm. All these algorithms adapted to the variable selection problem can be found in Chapter 2 of Denison *et al.* (2002). For a nice implementation of the Metropolis-Hastings algorithm to the Bayesian objective criterion for model selection briefly explained in the introduction, see Casella and Moreno (2005). The advantage of using MCMC algorithms over other search methods is that they allow for model averaging.

Finally, we want to recall that nonexhaustive subset selection methods such as *forward stepwise selection*, *backward stepwise selection*, and shrinkage methods such as *ridge regression* and the *lasso* do not usually render admissible models. For this reason, we do not make comparisons of our criterion with these two stepwise criteria.

25.5 Examples

To test the performance of the DIB criterion, and to compare it with other frequentist criteria, we first consider how they behave against simulated data by measuring their respective discriminatory power in some specific way to be explained below. Later in this section, we make comparisons based on two real data sets. Comparisons with the lasso procedure are only given for the two well-known real data sets described below.

25.5.1 Simulation study

For the simulation example that follows, we have considered a medium-sized linear regression problem with sample size $n = 40$ and $k = 6$, that is, five covariates. This is part of a more comprehensive simulation study, but it illustrates the performance of the new criterion, DIB, against some well established frequentist ones.

The model considered for simulation is

$$\mathbf{y} = \mathbf{X}\boldsymbol{\alpha} + \boldsymbol{\varepsilon},$$

where $\mathbf{y}$ is a vector of length 40, $\mathbf{X}$ is a 40×6 matrix whose entries were obtained by simulation from a standard normal distribution $N(0, 1)$, except the entries in the first column which were set equal to 1 to include the intercept, and $\boldsymbol{\alpha}= (\alpha_1, \alpha_2, \ldots, \alpha_6)^t$ is a vector of length 6. The error term coordinates of $\boldsymbol{\varepsilon}= (\varepsilon_1, \ldots, \varepsilon_n)^t$ are i.i.d. $\varepsilon_i \sim N(0, 1)$.

After fixing $\mathbf{X}$, samples of size 5,000 were simulated from the model for five different settings of the vector of regression coefficients $\boldsymbol{\alpha}$ including 1, 2, 3, 4, and 5 nonzero coefficients. In particular, we set

$$\begin{aligned}
\boldsymbol{\alpha}_1 &= (-1, -2, 0, 0, 0, 0)\\
\boldsymbol{\alpha}_2 &= (-1, -2, 2, 0, 0, 0)\\
\boldsymbol{\alpha}_3 &= (-1, -2, 2, -3, 0, 0)\\
\boldsymbol{\alpha}_4 &= (-1, -2, 3/2, -2, 2, 0)\\
\boldsymbol{\alpha}_5 &= (-1, -2, 3/2, -2, 2, -1).
\end{aligned}$$

Table 25.1: Comparison of different criteria for simulated data

	No. of covariates				
Criterion	**1**	**2**	**3**	**4**	**5**
DIB	0.991	0.910	0.962	0.977	1.000
Mallows C_p	0.500	0.563	0.692	0.850	1.000
Adjusted R^2	0.234	0.292	0.452	0.716	1.000

The entries in Table 25.1 represent the proportion of times that the true model is selected in the first place in the 5,000 simulations according to the three criteria and to the number of nonzero regression coefficients in the model.

Note how the DIB criterion outperforms the Mallows and the adjusted R^2 criteria for all choices of the number of covariates considered. As expected, the adjusted R^2 criterion performs very poorly except when the best model is the full model (the model with five covariates in our simulation study). In fact, the DIB criterion uniformly dominates the other criteria. This dominance is even more pronounced when the number of variables is small as we should expect from a Bayesian criterion.

Finally, we remark that all criteria perform best when the full model is the true one, a case that is usually of no interest in model selection.

25.5.2 Hald's data

Hald's data are a widely known data set used as a benchmark for comparing variable selection criteria. They are interesting because the sample size $n = 13$ is small and the number of regressors $k = 5$ is moderate, that is, there are 4 covariates. For this set, the admissible models are those containing the following covariates $\{x_4\}$, $\{x_1, x_2\}$, $\{x_1, x_2, x_4\}$, and the full model $\{x_1, x_2, x_3, x_4\}$.

Table 25.2, which only includes the seven more probable models, presents some comparative results, showing strong discrepancies in the ordering of the models.

The DIB criterion chooses model $\{x_1, x_2\}$, and so does the Mallows criterion, while the adjusted R^2 chooses model $\{x_1, x_2, x_4\}$. However, the most striking feature of the table is that model $\{x_1, x_4\}$ which is the second best according to DIB is the seventh according to R^2 and R^2_{adj}, and the sixth according to Mallow's.

Figure 25.2 shows the profiles of the lasso coefficients for the whole data. Selection of the shrinking parameter s presents difficulties for the usual ten-fold cross-validation due to the small sample size. Nevertheless, from this figure it is clear that variable x_4—the first entering variable—is always selected by the lasso for most values of the tuning parameter s except for values greater than

Table 25.2: Comparison of different criteria for Hald's data

Models	**Diluted Probs.**	R^2	**Adjusted** R^2	**Mallows** C_p
$\{x_1, x_2\}$	0.546571	0.978678	0.974414	2.678240
$\{x_1, x_4\}$	0.176554	0.972471	0.966965	5.495850
$\{x_1, x_2, x_4\}$	0.088912	0.982335	0.976447	3.018230
$\{x_1, x_2, x_3\}$	0.087916	0.982285	0.976380	3.041280
$\{x_1, x_3, x_4\}$	0.070828	0.981281	0.975041	3.496820
$\{x_2, x_3, x_4\}$	0.016538	0.972820	0.963760	7.337470
$\{x_1, x_2, x_3, x_4\}$	0.008199	0.982376	0.973563	5.000000

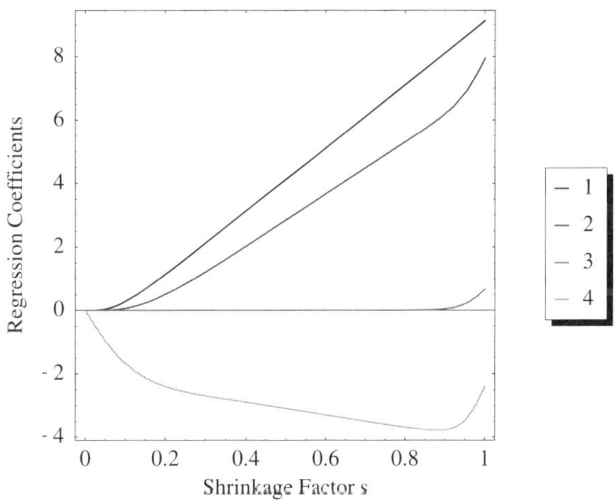

Figure 25.2: Profiles of the lasso coefficients for Hald's data

$s = 0.9$, so that the selected model is $\{x_1, x_2, x_4\}$.

As expected, the Bayesian criterion DIB, by choosing one model with two covariates with high diluted probability 0.547, obeys Occam's Razor principle.

25.5.3 Prostate cancer data

The data for this example come from a study by Stamey *et al.* (1989) and can be found in `http://www-stat.stanford.edu/ElemStatLearn`. These data are analysed, using different variable selection procedures, in Hastie *et al.* (2001). The response variable is the level of prostate-specific antigen (`lpsa`), and the eight covariates are the log cancer volume (`lcavol`), log prostate weight (`lweight`), `age`, log of the amount of benign prostatic hyperplasia (`lbph`), seminal vesicle invasion (`svi`), log of capsular penetration (`lcp`), Gleason score (`gleason`), and percent of Gleason scores 4 or 5 (`pgg45`). The sample size is $n = 97$ and the number of regressors, including the intercept, is $k = 9$.

The set of admissible models turns out to be:

$$\begin{gathered}\{x_1\}\\ \{x_1, x_2\}\\ \{x_1, x_2, x_5\}\\ \{x_1, x_2, x_4, x_5\}\\ \{x_1, x_2, x_3, x_4, x_5\}\\ \{x_1, x_2, x_3, x_4, x_5, x_8\}\\ \{x_1, x_2, x_3, x_4, x_5, x_6, x_8\}\\ \{x_1, x_2, x_3, x_4, x_5, x_6, x_7, x_8\}\end{gathered}$$

Using the whole data set, the DIB criterion selects model $\{x_1, x_2, x_5\}$, while the adjusted R^2 selects model $\{x_1, x_2, x_3, x_4, x_5, x_6, x_8\}$ and Mallows criterion selects model $\{x_1, x_2, x_4, x_5\}$.

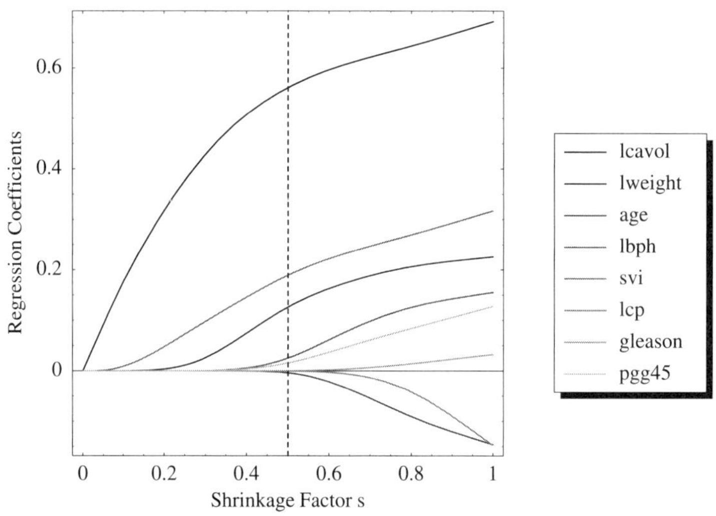

Figure 25.3: Profiles of the lasso coefficients for the prostate cancer data

We want to remark that our results are not comparable to those of Hastie *et al.* (2001) because they analyse a training subset on size 67 leaving as a test set the remaining 30. Further, they apply ten-fold cross-validation to the training subset and find that the best subset criterion selects model $\{x_1, x_2\}$, while the lasso selects model $\{x_1, x_2, x_4, x_5, x_8\}$. Note that the lasso selects an inadmissible model of five variables if the shrinkage parameter is set at about $s \approx 0.5$. Our Figure 25.3, computed using all the data, resembles Figure 3.8

of Hastie *et al.* (2001), obtained using the training subset, but it is slightly different.

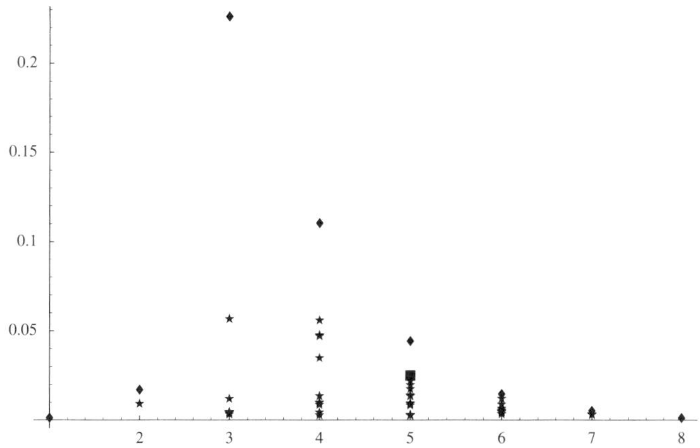

Figure 25.4: Plot of the diluted probabilities of the most probable models

For these data all criteria choose models of different dimension. Except for the best subset criterion—which was chosen using a subset of size 67—which selects the model with two covariates, viz. `lacavol` and `lweight`, the Bayesian criterion chooses the most parsimonious model including the three covariates `lacavol`, `lweight`, and `svi`.

Figure 25.4 displays the plot of number of covariates versus diluted probabilities for the admissible models (diamonds), the rest of models having highest diluted probabilities (stars), and the model selected by the lasso (a square). A more detailed discussion of this example can be found in Girón and Moreno (2005).

25.6 Conclusions

The objective *diluted intrinsic Bayesian* criterion we have developed in this chapter does not use subjective prior information on the model parameters, which otherwise is not generally available in variable selection problems. It is completely automatic and there is no need to adjust any hyperparameters.

We have seen from the simulation study that even for moderate sample sizes the DIB criterion selects the correct model with very high probability.

The R^2 criterion provides an ordering of the models without any subjective input, which only coincides with the Bayesian ordering in the class of models with the same fixed dimension, but when models of different dimension are to be compared the R^2 criterion obviously fails. Further, the R^2 scale has serious interpretative problems, thus the need to calibrate its scale. The alternative R^2_{adj} criterion does not necessarily adjust the jumps in dimensionality as seen, for instance, in the analysis of Hald's data. The simulation study summarized in Table 25.1 clearly shows its bad performance, too.

Turning to nonexhaustive variable selection criteria, one disadvantage of the *lasso* criterion, also shared by other frequentist methods, is that a threshold has to be adjusted to select one model usually via cross-validation. As a sequential or stepwise continuous selection criterion the lasso does not necessarily choose the optimal model in the class of models with a fixed dimension, as was seen in the prostate cancer example.

It seems that there is neither one reasonable frequentist criteria for variable selection nor the frequentist criteria obey Occams Razor principle. The existing procedures have serious difficulties in recognizing the jumps in dimensionality of the models. Apparently, the DIB criterion does not share these difficulties as exemplified by the simulated and real data sets examined in this chapter.

References

1. Berger, J. O., and Pericchi, L. R. (1996a). The intrinsic Bayes factor for model selection and prediction, *Journal of American Statistical Association*, **91**, 109–122.

2. Berger, J. O., and Pericchi, L. R. (1996b). On the justification of default and intrinsic Bayes factor, In *Modelling and Prediction* (Eds., J. C. Lee, W. Johnson, and A. Zellner), pp. 276–293, Springer-Verlag, New York.

3. Casella, G., and Moreno, E. (2005). Intrinsic meta analysis of contingency tables, *Statistics in Medicine*, **24**, 583–604.

4. Casella, G., and Moreno, E. (2006). Objective Bayesian variable selection, *Journal of the American Statistical Association* (to appear).

5. Denison, D. G. T., Holmes, C. C., Mallick, B. K., and Smith, A. F. M. (2002). *Bayesian Methods for Nonlinear Classification and Regression*, John Wiley & Sons, Chichester.

6. Furnival, G. M., and Wilson, R. W. (1974). Regression by leaps and bounds, *Technometrics*, **16**, 499–511.

7. Girón, F. J., Martínez, M. L., and Moreno, E. (2003). Bayesian analysis of matched pairs, *Journal of Statistical Planning and Inference*, **113**, 49–66.

8. Girón, F. J., Martínez, M. L., Moreno, E., and Torres, F. (2003). Bayesian analysis of matched pairs in the presence of covariates, In *Bayesian Statistics 7* (Eds., J. M. Bernardo, M. J. Bayarri, A. P. Dawid, D. Heckerman, A. F. M. Smith, and M. West), pp. 553–563, Oxford University Press, Oxford.

9. Girón, F. J., Martínez, M. L., Moreno, E., and Torres, F. (2006). Objective testing procedures in linear models: Calibration of the p-values, *Scandinavian Journal of Statistics* (to appear).

10. Girón, F. J. and Moreno, E. (2005). Comparative analysis of objective Bayesian procedures for variable selection in regression, (submitted).

11. Hastie, T., Tibshirani, R., and Friedman, J. (2001). *The Elements of Statistical Learning: Data Mining, Inference, and Prediction*, Springer-Verlag, New York.

12. Mallows, C. L. (1973). Some comments on C_p, *Technometrics*, **15**, 661–675.

13. Miller, A. (2002). *Subset Selection in Regression*, 2nd ed., Chapman and Hall/CRC, Boca Raton, FL.

14. Moreno, E., Bertolino, F., and Racugno, W. (1998). An intrinsic limiting procedure for model selection and hypothesis testing, *Journal of the American Statistical Association*, **93**, 1451–1460.

15. Moreno, E., Bertolino, F., and Racugno, W. (2003). Bayesian inference under partial prior information, *Scandinavian Journal of Statistics*, **30**, 565–580.

16. Moreno, E., Bertolino, F., and Racugno, W. (2000). Default Bayesian analysis of the Behrens-Fisher problem, *Journal of Statistical Planning and Inference*, **81**, 323–333.

17. Moreno, E., Girón, F. J., and Torres, F. (2003). Intrinsic priors for hypothesis testing in normal regression models, *Rev. R. Acad. Cien. Serie A, Mat.*, 53–61.

18. Moreno, E., and Liseo, B. (2003). A default Bayesian test for the number of components in a mixture, *Journal of Statistical Planning and Inference*, **111**, 129–142.

19. Moreno, E., Torres, F., and Casella, G. (2005). Testing equality of regression coefficients in heteroscedastic normal regression models, *Journal of Statistical Planning and Inference*, **131**, 117–134.

20. Stamey, T., Kabakin, J., McNeal, J., Johnstone, I. Freiha, F., Redwine, E., and Yang, N. (1989). Prostate-specific antigen in the diagnosis and treatment of adenocarcinoma of the prostate II: radical prostatectomy treated patients, *Journal of Urology*, **16**, 1076–1083.

21. Tibshirani, R. (1996). Regression shrinkage and the selection via lasso, *Journal of the Royal Statistical Society, Series B*, **58**, 267–288.

22. Zellner, A. (1986). On assessing prior distributions and Bayesian regression analysis with g-prior distributions, In *Bayesian Inference and Decision Techniques: Essays in Honor of Bruno deFinetti* (Eds., P. K. Goel and A. Zellner), North-Holland, Amsterdam.

26

On Bayesian and Decision-Theoretic Approaches to Statistical Prediction

Tapan K. Nayak and Abeer El-Baz

George Washington University, Washington, DC, USA

Abstract: Let Y and Z be two random vectors with joint density $f(y, z|\theta)$, where $\theta \in \Theta$ is an unknown parameter vector, and consider predicting Z based on y, the observed value of Y. We investigate Bayesian and decision-theoretic approaches to this problem, taking into account the loss function and the prior distribution of θ. Exploring connections between statistical prediction and decision theory, we find that a prediction problem can be reduced to a standard decision theory problem if the induced loss function is allowed to depend on the observed data y in addition to the unknown parameter θ and the decision d. In general, the predictive posterior density $f(z|y)$ may not contain all information necessary for obtaining optimum predictions, but the posterior density $f(\theta|y)$ is adequate for that purpose. Some admissibility results are also discussed.

Keywords and phrases: Admissibility, Bayes risk, loss function, predictive posterior distribution

26.1 Introduction

Let Y and Z be two random vectors with joint density $f(y, z|\theta)$, where $\theta \in \Theta$ is an unknown parameter vector. Suppose Z is unobserved and the data consist of y, the observed value of Y. Consider predicting z, the realized but unobserved value of Z, based on the data y. Our interest in this problem stems primarily from its generality, as briefly discussed below, and our main objective is to discuss suitable extensions of some standard results in Bayesian and decision-theoretic estimation to cover the prediction problem.

Parametric estimation theory regards the data y as the observed value of a random vector Y which is assumed to follow a probability distribution $f(y|\theta)$, where $\theta \in \Theta$ is an unknown parameter vector, and then deals with estimation of a parametric function $h(\theta)$ based on y. In some applications, however, the quantity of inferential interest is not a function only of θ. It may be the unobserved but realized value of a random vector W as in a prediction problem, or a function of θ and y (as in loss estimation, and estimation after selection), or more generally a function of θ, y, and an unobservable random vector W [as in the "species problem"; see, Robbins (1968) and Nayak (1996)]. Prediction under mixed linear models, and estimation under superpopulation models also fall outside the standard estimation framework as the quantities of inferential interest depend on θ or y, and some unobserved random quantities. Such non-standard problems motivated some researchers such as Hill (1990), Yatracos (1992), Bjornstad (1996), and Nayak (2000) to define more general structures of statistical inference problems and extend the theory of estimation for them. Nayak (2000) suggested that a statistical inference problem can be regarded as a prediction problem. In a parametric setup, the problem is to predict the realized but unobserved value, z, of a random vector Z based on the observed value, y, of a random vector Y, assuming that the joint density of Y and Z is $f(y, z|\theta)$, $\theta \in \Theta$. An estimation problem is a special case, where $Z = h(\theta)$, i.e., the distribution of Z given y and θ is degenerate and independent of y.

While statistical prediction is an old problem, most prediction methods are based on regression and time series models. In regression models, Y and Z are assumed to be independent given θ. Time series models incorporate dependence between Y and Z using stochastic processes that evolve over a "time" dimension. Many problems, such as, prediction of the order statistic $X_{(s)}$ based on the first r order statistics (where $r < s$) discussed, for example, by Takada (1979, 1981a), and prediction of the number of future failures discussed by many authors including Bain and Patel (1993), Escobar and Meeker (1999), Nelson (2000), and Nordman and Meeker (2002) have neither independence nor a "time" dimension. Thus, we believe further investigation of the general prediction problem will be helpful.

This chapter examines Bayesian and decision-theoretic aspects of statistical prediction and their connections to known results in estimation. We do not assume conditional independence of Y and Z given θ, which is fairly common in the general Bayesian prediction literature; see, for example, Aitchison and Dunsmore (1975) and Geisser (1993). In Section 26.2, we review the Bayesian approach and explore the possibility of reducing a prediction problem to a standard decision problem. We find that can be done if the loss function in the reduced problem is allowed to depend on the observed data y in addition to θ and the decision d. Then, Bayes predictions can be obtained from $\pi(\theta|y)$, the posterior distribution of θ, and the given loss function. Specific methods for deriving the Bayes predictors, from the posterior distribution, are presented for certain loss functions. In Section 26.3, we discuss some admissibility results, analogous to those in estimation theory.

26.2 Bayesian Prediction

Let $\pi(\theta)$ be the prior distribution of θ and $m(y) = \int f(y|\theta)\pi(\theta)d\theta$ be the marginal distribution of Y. Then, the predictive posterior distribution is usually calculated by [see Geisser (1993)]

$$\pi(z|y) = \int f(z|y,\theta)\pi(\theta|y)d\theta, \tag{26.1}$$

where $\pi(\theta|y)$ is the posterior distribution of θ given the data y, that is, $\pi(\theta|y) = [f(y|\theta)\pi(\theta)]/m(y)$. When $Y = (Y_1,\ldots,Y_n)$ and Z are iid random variables, Besag (1989) showed that

$$\pi(z|y) = f(z|\theta)\frac{\pi(\theta|y)}{\pi(\theta|y,z)}.$$

It can be seen easily that this result is more generally true, and $\pi(z|y)$ in (26.1) can also be derived using

$$\pi(z|y) = f(z|y,\theta)\frac{\pi(\theta|y)}{\pi(\theta|y,z)}. \tag{26.2}$$

It may be noted that as the left side of (26.2) is independent of θ, $\pi(z|y)$ may be obtained by evaluating the right-hand side of (26.2) for only one (any one) value of θ. Bayesian predictions, both point and interval, are often obtained from $\pi(z|y)$; see Geisser (1993). This implicitly assumes that the loss function depends on z and the prediction d but not on θ. However, the loss function may also depend on θ. For example, if θ is a scale parameter, $L(d,z,\theta) = [(d-z)/\theta]^2$ may be a reasonable loss function. If the loss depends also on θ, one would need $\pi(z,\theta|y)$, the joint posterior distribution of Z and θ given the data y, to calculate posterior risk and then minimize it to find a Bayes prediction.

For a more formal discussion of Bayesian and decision-theoretic prediction, let $L(d,z,\theta)$ denote the loss for using d as the predicted value of Z when the realized value is z and the true parameter value is θ. This allows the loss to depend on all of the three quantities d, z and θ. The risk function $R(\delta,\theta)$ of a predictor $\delta(Y)$ is then

$$R(\delta,\theta) = \int\int L(\delta(y),z,\theta))f(y,z|\theta)dzdy. \tag{26.3}$$

The Bayes risk of $\delta(Y)$ with respect to $\pi(\theta)$, to be denoted by $r(\pi,\delta)$, is

$$r(\pi,\delta) = \int R(\delta,\theta)\pi(\theta)d\theta.$$

From a decision-theoretic viewpoint, a Bayes predictor is obtained by minimizing the Bayes risk $r(\pi,\delta)$ with respect to δ.

Definition 26.2.1 A predictor $\delta_\pi(Y)$ is a Bayes predictor with respect to a prior distribution $\pi(\theta)$ if $r(\pi, \delta_\pi)$ is finite and

$$r(\pi, \delta_\pi) \leq r(\pi, \delta)$$

for all other predictors δ.

From a Bayesian point of view, a Bayes prediction is obtained by minimizing the posterior risk

$$E[L(\delta(y), Z, \theta)|y] = \int\int L(\delta(y), z, \theta)\pi(z, \theta|y)dzd\theta$$

with respect to $\delta(y)$ for given y. Although the Bayesian and decision-theoretic approaches are conceptually different, they result in the same predictor, because by minimizing the posterior risk (for each y) one also minimizes the Bayes risk.

We now proceed to give another construction of Bayes predictors. Because $\pi(z, \theta|y) = \pi(z|\theta, y)\pi(\theta|y)$, by interchanging the order of integrations, the posterior risk can be expressed as

$$E[L(\delta(y), Z, \theta)|y] = \int L_*(\delta(y), y, \theta)\pi(\theta|y)d\theta, \tag{26.4}$$

where $L_*(\delta(y), y, \theta) = \int L(\delta(y), z, \theta))f(z|y, \theta)dz$. Thus, a Bayes predictor can be obtained by minimizing $E[L_*(\delta(y), y, \theta)|y]$ with respect to $\delta(y)$ for each y. This shows that Bayes predictors can be obtained from the posterior distribution of θ given y without deriving the posterior predictive distribution $\pi(z|y)$. As we noted earlier, if the loss depends on θ, the posterior risk cannot be calculated only from the $\pi(z|y)$. However, $\pi(z|y)$ is easy to interpret and is useful for deriving general prediction intervals.

We shall now explore connections between statistical prediction and standard decision theory. A standard decision problem is specified by a sample space $\mathcal{Y}$, a parameter space Θ, a decision space $\mathcal{D}$, a family of distributions $\{f(y|\theta), \theta \in \Theta\}$ on $\mathcal{Y}$, a prior distribution $\pi(\theta)$ on Θ, and a loss function $L(d, \theta)$ defined on $\mathcal{D} \times \Theta$. A prediction problem involves the additional random element Z. Also a general prediction loss function $L(d, z, \theta)$ is defined on $\mathcal{D} \times \mathcal{Z} \times \Theta$, where $\mathcal{Z}$ is the sample space of Z. To reduce a prediction problem to a standard decision theory problem, one would need to eliminate the extra element Z. We may attempt to do that by averaging the loss with respect to $f(z|y, \theta)$ and considering the induced loss function

$$L_*(d, y, \theta) = E[L(d, Z, \theta)|y, \theta] = \int L(d, z, \theta)f(z|y, \theta)dz. \tag{26.5}$$

Then, the risk function in (26.3) can be expressed as

$$R(\delta, \theta) = \int L_*(\delta, y, \theta)f(y|\theta)dy,$$

which resembles the risk function in a standard decision problem; it is the average loss (induced) with respect to the distribution of Y. Thus, a prediction problem can be stated as a decision problem in terms of $\pi(\theta)$, $f(y|\theta)$ and $L_*(d, y, \theta)$. By Eq. (26.4) the posterior risk can also be calculated from these three elements. However, this formulation of a prediction problem differs from a standard decision theory problem in the structure of the (induced) loss function L_*, as it may depend on not only on d and θ but also on y. Moreover, L_* may be strictly positive, as can be seen for squared error loss discussed below. In the special case where Y and Z are independent given θ, L_* is independent of y. It can be seen that if $L(d, z, \theta)$ is strictly convex in d, then $L_*(d, y, \theta)$ is also so, and a Bayes predictor is unique if it exists.

We now discuss some specific loss functions and the corresponding Bayes predictors. For squared error loss, $L(d, z) = (d - z)^2$, the induced loss function L_* is

$$L_*(d, y, \theta) = E[(d - Z)^2|y, \theta] = Var(Z|y, \theta) + [d - \gamma(y, \theta)]^2, \tag{26.6}$$

where $\gamma(y, \theta) = E(Z|y, \theta)$ is the regression function. Note that for given y and θ, L_* may be positive for all d. The first term on the right side of (26.6) is due to inherent variation of Z and it is independent of d. The second term is the squared error loss from estimating the regression $\gamma(y, \theta)$ by d. It can be seen easily that the Bayes predictor under squared error loss is

$$\delta_B(y) = E[\gamma(y, \theta)|y] = \int \gamma(y, \theta)\pi(\theta|y)d\theta.$$

For weighted squared error loss $L(d, z, \theta) = w(\theta)(d - z)^2$, with $w(\theta) > 0$, the induced loss is

$$L_*(d, y, \theta) = w(\theta)\{Var(Z|y, \theta) + [d - \gamma(y, \theta)]^2\},$$

and the corresponding Bayes predictor is

$$\delta_B(y) = \frac{\int \gamma(y, \theta)w(\theta)\pi(\theta|y)d\theta}{\int w(\theta)\pi(\theta|y)d\theta}.$$

Now, suppose Z is a vector and the loss function is $L(d, z, \theta) = (d - z)'w(\theta)(d - z)$, where $w(\theta)$ is a positive definite matrix. Then, the induced loss function is

$$L_*(d, y, \theta) = E[\{Z - \gamma(y, \theta)\}'w(\theta)\{Z - \gamma(y, \theta)\}|y, \theta] + [d - \gamma(y, \theta)]'w(\theta)[d - \gamma(y, \theta)], \tag{26.7}$$

where $\gamma(y, \theta) = E[Z|y, \theta]$. As the first term on the right side of (26.7) is independent of d, the Bayes predictor is

$$\delta_B(y) = \left\{\int w(\theta)\pi(\theta|y)d\theta\right\}^{-1}\left\{\int w(\theta)\gamma(y, \theta)\pi(\theta|y)d\theta\right\}. \tag{26.8}$$

Note that if $w(\theta)$ is a fixed matrix, independent of θ, (26.8) reduces to $E[\gamma(y,\theta)|y]$, the posterior mean of the regression function.

Now, let Z be scalar and consider the following LINEX loss function:

$$L(d,z,\theta) = \exp\{\alpha(\theta)(d-z)\} - \alpha(\theta)(d-z) - 1, \tag{26.9}$$

where $\alpha(\theta)$ is a positive-valued function of θ. LINEX loss was introduced by Varian (1975) for modeling asymmetric losses, and it was used by Zellner (1986) for estimation and prediction. In Zellner (1986), $\alpha(\theta)$ is a constant, independent of θ, but we allow it to depend on θ for generality. The induced loss function for (26.9) is

$$L_*(d,y,\theta) = \gamma_1(y,\theta)\exp\{\alpha(\theta)d\} - \alpha(\theta)[d-\gamma(y,\theta)] - 1, \tag{26.10}$$

where $\gamma(y,\theta)$ is the regression function, and

$$\gamma_1(y,\theta) = \int e^{-\alpha(\theta)z} f(z|y,\theta)dz.$$

It can be seen that for any given y, the minimizer (i.e., the Bayes prediction) $d_B(y)$ of the posterior mean of (26.10) is the solution (for d) of the equation

$$E[e^{\alpha(\theta)d}\gamma_1(y,\theta)\alpha(\theta)|y] = E[\alpha(\theta)|y] \tag{26.11}$$

provided that the relevant expectations exist. If $\alpha(\theta)$ is a constant independent of θ, i.e., $\alpha(\theta)=\alpha$, the solution of (26.11) is given by

$$d_B(y) = -(1/\alpha)\log E[\gamma_1(y,\theta)|y] = -(1/\alpha)\log\left\{\int \gamma_1(y,\theta)\pi(\theta|y)d\theta\right\}.$$

26.3 Admissible Predictors

In standard decision theory, it is known that admissible decision rules are either Bayes rules or limits of Bayes rules. Similar results follow in prediction.

Definition 26.3.1 A predictor $\delta_1(Y)$ of Z is said to be better than (dominate) another predictor $\delta_2(Y)$ if $R(\delta_1,\theta) \le R(\delta_2,\theta)$ for all $\theta \in \Theta$ with '$<$' for some θ.

Definition 26.3.2 A predictor $\delta(Y)$ is said to be inadmissible if it is dominated by some other predictor $\delta_*(Y)$; otherwise, $\delta(Y)$ is said to be admissible.

Definition 26.3.3 A class C of predictors is said to be complete if for any $\delta(Y)$ not in C there exists a predictor $\delta'(Y)$ in C that dominates $\delta(Y)$.

The following three theorems can be proved using arguments similar to those in the proofs of analogous results in decision theory.

Theorem 26.3.1 *Any unique Bayes predictor is admissible.*

Theorem 26.3.2 *Suppose $\delta_\pi(Y)$ is a Bayes predictor having finite Bayes risk with respect to a prior density π which is positive for all $\theta \in \Theta$, and that the predictors with continuous risk functions form a complete class. Then, $\delta_\pi(Y)$ is an admissible predictor.*

Theorem 26.3.3 *(Blyth's Theorem) Suppose that the parameter space $\Theta \subset R^r$ is open, and all predictors with continuous risk functions form a complete class. Let $\delta(Y)$ be a predictor with a continuous risk function and let $\{\pi_m\}$ be a sequence of (possibly improper) prior densities such that*

(a) $r(\pi_m, \delta) < \infty$ *for all* m;

(b) for any nonempty open set $\Theta_\circ \subset \Theta$, *there exist constants* $B > 0$ *and* M *such that*

$$\int_{\Theta_\circ} \pi_m(\theta) d\theta \geq B$$

for all $m \geq M$;

(c) $r(\pi_m, \delta) - r(\pi_m, \delta_{\pi_m}) \to 0$ *as* $m \to \infty$.

Then, $\delta(Y)$ is an admissible predictor.

As we noted earlier, if $L(d, z, \theta))$ is strictly convex in d, the Bayes predictor is unique. So, for strictly convex loss functions, all Bayes predictors are admissible by Theorem 26.3.1. Regarding the continuity assumption of Theorems 26.3.2 and 26.3.3, we note that if (i) the loss is independent of θ, and (ii) $f(y, z|\theta)$, $\theta \in \Theta$ forms an exponential family, then any predictor $\delta(Y)$ with finite risk has a continuous risk function. This follows from a standard result for exponential families; see, for example, Lehmann (1986, p. 59).

Theorems 26.3.1 and 26.3.2 are not applicable if the predictor under consideration cannot be a Bayes predictor under any prior. We now give a simple example of a predictor that is best according to some criteria, but cannot be Bayes under any prior. Let

$$Y = Z + \epsilon,$$

where $Z \sim N(\theta, \sigma^2), \epsilon \sim N(0, \tau^2)$ are independent, and σ and τ are known. Then, $Y \sim N(\theta, \sigma^2 + \tau^2)$, and $\gamma(y, \theta) = E(Z|y, \theta) = ky + (1 - k)\theta$, where $k = \sigma^2/(\sigma^2 + \tau^2)$. Consider predicting Z under the squared error loss $L(d, z) = (d - z)^2$. Then, the Bayes predictor is

$$\delta_B(y) = ky + (1 - k)E[\theta|y]. \tag{26.12}$$

Consider the predictor $\delta(y) = y$. Then $E[\delta(Y)|z, \theta] = z$ for all z, θ and hence $E[\delta(Y)|\theta] = \theta$ for all θ. In fact, $\delta(Y)$ is the uniformly minimum variance unbiased estimator of θ. Because the distribution of Y is complete, from Theorem 3.2 of Nayak (2000), it follows that $\delta(Y)$ is the best unbiased predictor of Z. Now, by (26.12), $\delta(Y)$ can be a Bayes predictor if and only if there exists a prior distribution $\pi(\theta)$ for which

$$E[\theta|y] = \int \theta\pi(\theta|y)d\theta = y. \tag{26.13}$$

If (26.13) holds, $\delta(Y)$ would also be the Bayes estimator of θ under squared error loss. However, that cannot happen as $\delta(Y)$ is also an unbiased estimator of θ; see, Lehmann and Casella (1998, p. 234). Thus, there does not exist any prior distribution $\pi(\theta)$ for (26.13) to hold true. Here, Theorems 26.3.1 and 26.3.2 are not useful for proving admissibility of $\delta(Y)$. However, $\delta(Y)$ is admissible and that can be proved using Theorem 26.3.3 and a sequence of normal priors $N(0, c_m^2)$ for θ, where $c_m \to \infty$ as $m \to \infty$.

Blyth's theorem was generalized by Rukhin (1988), Johnstone (1988) and Lele (1993) for investigating admissibility of estimators of realized loss in an estimation problem. They considered the problem of estimating the realized loss $z = L(\delta(y), \theta)$ when $\delta(Y)$ is used to estimate θ (or $g(\theta)$) under the loss function $L(d, \theta)$. They investigated admissibility of estimators $T(Y)$ of z when the loss from estimating z by l is $\tilde{L}(l, z)$. Blyth's original theorem is not applicable to loss estimators as the quantity to be estimated (z) is the realized value of a random variable.

Rukhin (1988) introduced a loss function $Q(d, l, \theta)$ that accounts for the combined loss of using d and l as estimates of θ and z, respectively. He proposed the combined loss function

$$Q(d, l, \theta) = L(d, \theta)l^{-1/2} + l^{1/2}. \tag{26.14}$$

Let $\delta(Y)$ be an estimator of θ and $T(Y)$ be an estimator of the realized loss. Then, Rukhin (1988) presented necessary and sufficient conditions for admissibility of the pair (δ, T) under the combined loss (26.14) assuming certain regularity conditions and that $L(d, \theta)$ is convex in d. His proof relied on Farrell's (1968) work.

We believe that according to Farrell's theorem, the admissibility of the pair (δ, T) requires Q to be strictly convex in (d, l), which requires Q to be convex in l for a given d. For this to be true, we must have

$$\frac{\partial^2 Q}{\partial l^2} = \frac{l^{-3/2}}{4}\left\{\frac{3L(d, \theta)}{l} - 1\right\} > 0, \tag{26.15}$$

or equivalently $l < 3L(d, \theta)$. In most applications, however, $L(d, \theta) = 0$ when $d = \theta$, and (26.15) cannot hold for all l, d, and θ. Thus, we believe additional conditions on the combined loss function are needed for Rukhin's result to hold.

We now discuss a simple relationship between admissibility of estimators and admissibility of predictors in a specific setting. Suppose the loss is squared error and

$$\gamma(y,\theta) = T(y) + h(\theta), \tag{26.16}$$

where T and h are two functions. Then, the risk function of δ is

$$R(\delta,\theta) = E[(\delta(Y) - T(Y)) - h(\theta)]^2 + E[Var(Z|y,\theta)].$$

As the first term is the squared error risk of $\delta(Y) - T(Y)$ for estimating $h(\theta)$, and the second term is independent of δ, for two predictors δ_1 and δ_2, $R(\delta_1,\theta) \leq R(\delta_2,\theta)$ if and only if

$$E[(\delta_1(Y) - T(Y)) - h(\theta)]^2 \leq E[(\delta_2(Y) - T(Y)) - h(\theta)]^2.$$

Thus, δ_1 is a better predictor of Z than δ_2 if and only if $\delta_1 - T$ is a better estimator of $h(\theta)$ than $\delta_2 - T$, under squared error loss. So, the class of all admissible predictors of Z can be obtained easily from the class of all admissible estimators of $h(\theta)$. An application of this result is given next.

Let $X_{(1)} < X_{(2)} < \cdots < X_{(n)}$ be the order statistics from a sample of size n from an exponential distribution with density

$$f(x,\theta) = (1/\theta)\exp(-x/\theta), \quad x > 0,\ \theta > 0. \tag{26.17}$$

Consider predicting $Z = X_{(k)}$ based on the first r order statistics $\{X_{(1)},\ldots,X_{(r)}\}$, where $r < k \leq n$. Here, it can be seen easily that (i) $S = \sum_{i=1}^{r} X_{(i)} + (n-r)X_{(r)}$ is a complete sufficient statistic, (ii) $W_i = (n-i+1)(X_{(i)} - X_{(i-1)})$, $i = 1,\ldots,n$, are independent and identically distributed with density (26.17), and (iii) $X_{(i)} = \sum_{j=1}^{i} W_j/(n-j+1)$, $i = 1,\ldots,n$.

By these and the Markovian property of the order statistics, $\gamma(y,\theta)$ can be obtained as follows:

$$\begin{aligned}
\gamma(y,\theta) &= E(X_{(k)}|X_{(1)},\ldots,X_{(r)}) \\
&= E(X_{(k)}|X_{(r)}) \\
&= E\left(\sum_{i=1}^{k}\frac{W_i}{n-i+1}\middle|\sum_{i=1}^{r}\frac{W_i}{n-i+1}\right) \\
&= E\left(\sum_{i=1}^{r}\frac{W_i}{n-i+1} + \sum_{i=r+1}^{k}\frac{W_i}{n-i+1}\middle|\sum_{i=1}^{r}\frac{W_i}{n-i+1}\right) \\
&= X_{(r)} + a\theta,
\end{aligned}$$

where $a = \sum_{i=r+1}^{k}(n-i+1)^{-1}$. Hence $\gamma(y,\theta)$ is of the form (26.16) where $T = X_{(r)}$ and $h(\theta) = a\theta$, and so an admissible predictor δ of Z must be of the form

$$X_{(r)} + a\eta,$$

where $\eta(X_{(1)}, \ldots, X_{(r)})$ is an admissible estimator of θ. For this prediction problem, many predictors found in literature [see Takada (1981b, 1991) and Ebrahimi (1992)] are of the form $\delta_b = X_{(r)} + bS$, where b is a constant. For example, $b = a/r$ yields the uniformly minimum mean squared error unbiased predictor in Takada (1981b).

Note that $S = \sum_{i=1}^{r} W_i$ has gamma distribution with parameters (r, θ), where r is known. In this case, Karlin (1957) proved that (i) within the class of the estimators of θ of the form bS, the estimator $S/(r+1)$ has minimum mean squared error, and (ii) $S/(r+1)$ is an admissible estimator of θ under squared error loss. Those results now imply that the predictor

$$\delta(Y) = X_{(r)} + \frac{aS}{r+1} \tag{26.18}$$

is admissible for predicting $X_{(k)}$, and that among all predictors $\delta_b = X_{(r)} + bS$, $\delta(Y)$ in (26.18) minimizes the mean squared error for all θ.

References

1. Aitchison, J., and Dunsmore, I. R. (1975). *Statistical Prediction Analysis*, Cambridge University Press, New York.

2. Bain, L. J., and Patel, J. K. (1993). Prediction intervals based on partial observations for some discrete distributions, *IEEE Transactions on Reliability*, **42**, 459–463.

3. Besag, J. (1989). A candidate's formula: A curious result in Bayesian prediction, *Biometrika*, **76**, 183–183.

4. Bjornstad, J. F. (1996). On the generalization of the likelihood function and the likelihood principle, *Journal of the American Statistical Association*, **91**, 791-806.

5. Ebrahimi, N. (1992). Prediction intervals for future failures in the exponential distribution under Hybrid censoring, *IEEE Transactions on Reliability*, **41(1)**, 127–132.

6. Escobar, L. A., and Meeker, W. Q. (1999). Statistical prediction based on censored life data, *Technometrics*, **41**, 113–124.

7. Farrell, R. H. (1968). On a necessary and sufficient condition for admissibility of estimators when strictly convex loss is used, *Annals of Statistics*, **38**, 23–28.

8. Geisser, S. (1993). *Predictive Inference: An Introduction*, Chapman & Hall, New York.

9. Hill, J. R. (1990). A general framework for model-based statistics, *Biometrika*, **77**, 115–126.

10. Johnstone, I. M. (1988). On admissibility of unbiased estimates of loss, In *Statistical Decision Theory and Related Topics IV* (Eds., S. S. Gupta and J. O. Berger), pp. 361–379, Springer-Verlag, New York.

11. Karlin, S. (1957). Admissibility for estimation with quadratic loss, *Annals of Mathematical Statistics*, 29, 406–436.

12. Lehmann, E. L. (1986). *Testing Statistical Hypotheses*, 2nd ed., John Wiley & Sons, New York.

13. Lehmann, E. L., and Casella, G. (1998). *Theory of Point Estimation*, 2nd ed., Springer-Verlag, New York.

14. Lele, C. (1993). Admissibility results in loss estimation, *Annals of Statistics*, **21**, 378–390.

15. Nayak, T. K. (1996). On estimating the conditional probability of discovering a new species, *Communications in Statistics—Theory and Methods*, **25**, 2039–2056.

16. Nayak, T. K. (2000). On best unbiased prediction and its relationships to unbiased estimation, *Journal of Statisticae Planning and Inference*, **84**, 171–189.

17. Nelson, W. (2000). Weibull prediction of a future number of failures, *Quality and Reliability Engineering International*, **16**, 23–26.

18. Nordman, D., and Meeker, W. Q. (2002). Weibull prediction intervals for a future number of failures, *Technometrics*, **44 (1)**, 15–23.

19. Robbins, H. (1968). Estimating the total probability of the unobserved outcomes of an experiment, *Annals of Mathematical Statistics*, **39**, 256-257.

20. Rukhin, A. L. (1988). Estimated loss and admissible loss estimators, In *Statistical Decision Theory and Related Topics IV* (Eds., S. S. Gupta and J. O. Berger), pp. 409-418, Springer-Verlag, New York.

21. Takada, Y. (1979). The shortest interval for the largest observation from the exponential distribution, *Journal of the Japan Statistical Society*, **9**, 87–91.

22. Takada, Y. (1981a). Invariant prediction rules and an adequate statistic, *Annals of the Institute of Statistical Mathematics*, **33**, 91–100.

23. Takada, Y. (1981b). Relation of the best invariant predictor and the best unbiased predictor in location and scale families, *Annals of Statistics*, **9**, 917–921.

24. Takada, Y. (1991). Median unbiasedness in an invariant prediction problem, *Statistics & Probability Letters*, **12**, 281–283.

25. Varian, H. R. (1975). A Bayesian approach to real estate assessment, In *Studies in Bayesian Econometrics and Statistics in Honor of Leonard J. Savage* (Eds., S. E. Fienberg and A. Zellner), pp. 195–208, North-Holland, Amsterdam.

26. Yatracos, Y. J. (1992). On prediction and mean squared error, *Canadian Journal of Statistics*, **20**, 187–200.

27. Zellner A. (1986). Bayesian estimation and prediction using asymmetric loss functions, *Journal of the American Statistical Association*, **81**, 446–451.

27

Phi-Divergence-Type Test for Positive Dependence Alternatives in 2 × k Contingency Tables

L. Pardo and M.L. Menéndez

Complutense University of Madrid, Madrid, Spain
Politechnical University of Madrid, Madrid, Spain

Abstract: In this chapter, we consider $2 \times k$ contingency tables and derive a new family of test statistics for detecting positive dependence in them. The family of test statistics introduced here is based on the ϕ-divergence measures of which the likelihood ratio test is a special case.

Keywords and phrases: Asymptotic distributions, likelihood ratio test, ϕ-divergence test statistics, $2 \times k$ contingency tables

27.1 Introduction

Let X_1 and Y_1 denote two categorical response variables having 2 and k levels, respectively. The responses (X_1, Y_1) of a subject randomly chosen from some population have a probability distribution. Let $p_{ij} = \Pr(X_1 = i, Y_1 = j)$ with $p_{ij} > 0$, $i = 1, 2$, $j = 1, \ldots, k$. In the following, we denote this probability distribution by $\boldsymbol{P} = (p_{ij}), i = 1, 2,\ j = 1, \ldots, k$. We display this distribution in a rectangular table having 2 rows for the categories of X_1 and k columns for the categories of Y_1. Consider a random sample of size n on (X_1, Y_1) and we denote by n_{ij} the observed frequency in the (i, j)th cell for $(i, j) \in 2 \times k$ with $n = \sum_{i=1}^{2} \sum_{j=1}^{k} n_{ij}$ and the totals for the ith row and jth column by $n_{i*} = \sum_{j=1}^{k} n_{ij}$ and $n_{*j} = \sum_{i=1}^{2} n_{ij}$, $i = 1, 2$, $j = 1, \ldots, k$, respectively. Let $\widehat{\boldsymbol{p}} = (\widehat{p}_{11}, \widehat{p}_{21}, \ldots, \widehat{p}_{1k}, \widehat{p}_{2k})^T$ be the nonparametric estimator of the unknown probability vector $\boldsymbol{p} = (p_{11}, p_{21}, \ldots, p_{1k}, p_{2k})^T$, where $\widehat{p}_{ij} = n_{ij}/n$, $i = 1, 2$, $j = 1, \ldots, k$. In the following, we assume that n_{ij} is the observed value corresponding to a random variable N_{ij}, $i = 1, 2$, $j = 1, \ldots, k$, in such a way that the random vector $(N_{11}, N_{21}, \ldots, N_{1k}, N_{2k})$ is multinomially distributed with parameters n

and $(p_{11}, p_{21}, \ldots, p_{1k}, p_{2k})$. The hypothesis of independence is given by

$$H_0 : p_{ij} = p_{i*}p_{*j}, \; i = 1, 2, \; j = 1, \ldots, k, \tag{27.1}$$

where $p_{i*} = \sum_{j=1}^{k} p_{ij}$, $i = 1, 2$ and $p_{*j} = p_{1j} + p_{2j}$, $j = 1, \ldots, k$.

If we assume that X_1 and Y_1 are ordered categorical response variables, dependence between these variables is often viewed in terms of stochastic ordering among the conditional variables of X_1 given $Y_1 = y$ (denoted by X_{1y}). If the variable $X_{1y'}$ is stochastically larger (smaller) than X_{1y} for any $y < y'$, there is positive (negative) dependence between X_1 and Y_1. In $2 \times k$ contingency tables, positive dependence is equivalent to the inequalities

$$\frac{p_{2j}}{p_{*j}} \leq \frac{p_{2\,j+1}}{p_{*\,j+1}}, \quad j = 1, \ldots, k-1 \tag{27.2}$$

while negative dependence is equivalent to the inequalities

$$\frac{p_{2j}}{p_{*j}} \geq \frac{p_{2\,j+1}}{p_{*\,j+1}}, \quad j = 1, \ldots, k-1.$$

If we denote

$$\theta_{1j} = p_{2j}/(p_{1j} + p_{2j}), \; \theta_{2j} = p_{1j} + p_{2j}, \quad j = 1, \ldots, k, \tag{27.3}$$

and by $\theta_1 = (\theta_{11}, \ldots, \theta_{1k})^T$, $\theta_2 = (\theta_{21}, \ldots, \theta_{2k})^T$ the corresponding vectors, the hypothesis of independence given in (27.1) can be rewritten as

$$H_0 : \theta_{11} = \cdots = \theta_{1k} \tag{27.4}$$

and the hypothesis of positive dependence given in (27.2) as

$$H_1 : \theta_{11} \leq \cdots \leq \theta_{1k}. \tag{27.5}$$

The problems of testing positive (negative) dependence for a $2 \times k$ table, that is,

$$H_{Null} : H_0 \quad \text{versus} \quad H_{Alt} : H_1 - H_0 \tag{27.6}$$

have been considered by many authors; see, for example, Armitage (1955), Grove (1980), Patefield (1982), Lee (1989), Robertson *et al.* (1983, 1988), and the references therein, on the basis of the maximum likelihood estimator. If we denote by H_2 the hypothesis that imposes no restriction, sometimes it is interesting to test

$$H_{Null} : H_1 \quad \text{versus} \quad H_{Alt} : H_2 - H_1. \tag{27.7}$$

Recently, Park (2002) has assumed that ordered categorical variables (X_1, Y_1) and (X_2, Y_2) are expressed as two $2 \times k$ contingency tables given by $\boldsymbol{P} =$

(p_{ij}) and $\boldsymbol{Q} = (q_{ij})$, $i = 1, 2$, $j = 1, \ldots, k$, respectively. We denote by m_{ij} the observed frequency in the (i, j)th cell corresponding to the second $2 \times k$ contingency table, with $m = \sum_{i=1}^{2} \sum_{j=1}^{k} m_{ij}$, the totals for the ith row and jth column by $m_{i*} = \sum_{j=1}^{k} m_{ij}$ and $m_{1j} + m_{2j} = m_{*j}$, $i = 1, 2$, $j = 1, \ldots, k$, respectively, and $\tau_{1j} = q_{1j}/(q_{1j}+q_{2j})$, $j = 1, \ldots, k$. We are interested in studying which of these two contingency tables shows more positive dependence than the other. We define $\theta - \tau = (\theta_{11} - \tau_{11}, \ldots, \theta_{1k} - \tau_{1k})^T$ and we consider the hypotheses

$$H_0^* : \theta_{11} - \tau_{11} = \cdots = \theta_{1k} - \tau_{1k},$$

$$H_1^* : \theta_{11} - \tau_{11} \leq \cdots \leq \theta_{1k} - \tau_{1k}$$

and H_2^* the hypothesis that imposes no restriction. Then, we can test

$$H_{Null} : H_0^* \quad \text{versus} \quad H_{Alt} : H_1^* - H_0^* \tag{27.8}$$

and

$$H_{Null} : H_1^* \quad \text{versus} \quad H_{Alt} : H_2^* - H_1^*. \tag{27.9}$$

Some examples of the importance of these hypotheses can be seen in Park (2002) as well as the likelihood ratio tests for testing (27.8) and (27.9).

In this paper, we present a new family of test statistics, based on ϕ-divergence measures, which is a generalization of the likelihood ratio test for testing (27.6)–(27.9). The new families of test statistics, introduced here, are called *ϕ-divergence test statistics* and include as special case the likelihood ratio test. Some applications of the ϕ-divergence test statistics in different problems of testing with ordered alternatives can be seen in Menéndez *et al.* (2002, 2003a,b). In Section 27.2 we present the families of ϕ-divergence test statistics for the hypotheses test problems considered in (27.6)–(27.9), and in Section 27.3 we derive their asymptotic distribution.

27.2 Phi-Divergence Test Statistics

Given two probability vectors $\boldsymbol{p} = (p_{11}, p_{21}, \ldots, p_{1k}, p_{2k})^T$ and $\boldsymbol{q} = (q_{11}, q_{21}, \ldots, q_{1k}, q_{2k})^T$ the ϕ-divergence between them is defined as

$$D_{\phi}(\boldsymbol{p}, \boldsymbol{q}) = \sum_{i=1}^{2} \sum_{j=1}^{k} q_{ij} \phi\left(\frac{p_{ij}}{q_{ij}}\right), \ \phi \in \Phi^*, \tag{27.10}$$

where Φ^* is the class of all convex functions $\phi(x)$, $x > 0$, such that at $x = 1, \phi(1) = 0, \phi''(1) > 0$, and at $x = 0$, $0\phi(0/0) = 0$ and $0\phi(p/0) = \lim_{u \to \infty} \phi(u)/u$. For every $\phi \in \Phi^*$ that is differentiable at $x = 1$, the function $\psi(x) \equiv \phi(x) -$

$\phi'(1)(x-1)$, also belongs to Φ^*. Then we have $D_\psi(\boldsymbol{p},\boldsymbol{q}) = D_\phi(\boldsymbol{p},\boldsymbol{q})$, and ψ has the additional property that $\psi'(1) = 0$. Because the two divergence measures are equivalent, we can consider the set Φ^* to be equivalent to the set

$$\Phi \equiv \Phi^* \cap \left\{\phi : \phi'(1) = 0\right\}.$$

In what follows, we give the theoretical results for $\phi \in \Phi$ but we often apply them to choices of functions in Φ^*.

An important example of a family of ϕ-divergence measures in statistical problems is the power divergence family given by

$$\begin{array}{l} \phi_{(\lambda)}(x) = \left\{\lambda(\lambda+1)\right\}^{-1}\left\{x^{\lambda+1} - x + \lambda(1-x)\right\};\ \lambda \neq 0, \lambda \neq -1, \\ \phi_{(0)}(x) = \lim_{\lambda\to 0}\phi_{(\lambda)}(x),\ \phi_{(-1)}(x) = \lim_{\lambda\to -1}\phi_{(\lambda)}(x), \end{array} \tag{27.11}$$

which was introduced and studied by Cressie and Read (1984). Notice that $\phi_{(\lambda)} \in \Phi$. The divergence measure obtained with $\phi_{(0)}(x) = x\log x - x + 1$ corresponds to the well-known Kullback-Leibler divergence measure

$$D_{Kull}(\boldsymbol{p},\boldsymbol{q}) = \sum_{i=1}^{2}\sum_{j=1}^{k} p_{ij}\log\frac{p_{ij}}{q_{ij}}. \tag{27.12}$$

In the case of a $2\times k$ contingency table, using the reparametrization given in (27.3), the likelihood function is given by

$$L(\theta) = \prod_{j=1}^{k}\theta_{1j}^{n_{2j}}(1-\theta_{1j})^{n_{1j}}\prod_{j=1}^{k}\theta_{2j}^{n_{*j}}$$

and it is a simple exercise to check that the unconstrained maximum likelihood estimators of θ_{1j} and θ_{2j} are given by

$$\widehat{\theta}_{1j} = \frac{n_{2j}}{n_{1j}+n_{2j}},\quad \widehat{\theta}_{2j} = \frac{n_{*j}}{n}\quad j = 1,\ldots,k.$$

We can observe that the hypotheses H_0 and H_1 given in (27.5) and (27.6) do not impose any restriction on θ_{2j}, their corresponding MLE's under these hypotheses can be obtained by maximizing $\prod_{j=1}^{k}\theta_{2j}^{n_{*j}}$. For this reason, if we denote by $\overline{\theta}_2 = \left(\overline{\theta}_{21},\ldots,\overline{\theta}_{2k}\right)$ and $\theta_2^* = (\theta_{21}^*,\ldots,\theta_{2k}^*)$ the MLE's of θ_2 under H_0 and H_1 respectively, we have

$$\widehat{\theta}_{2j} = \overline{\theta}_{2j} = \theta_{2k}^*,\quad j = 1,\ldots,k.$$

It is easy to establish that, under H_0, the MLE of θ_1 is given by

$$\overline{\theta}_{11} = \cdots = \overline{\theta}_{1k} = \frac{n_{2*}}{n}$$

and the MLE of θ_1 under the hypothesis H_1 is the isotonic regression $\theta_1^* = (\theta_{11}^*, \ldots, \theta_{1k}^*)$ of $\widehat{\theta}_1$ with weights $w_j = n_{1j} + n_{2j}$. We denote $\boldsymbol{w} = (w_1, \ldots, w_k)$. This isotonic regression θ_1^* is the least-squares projection of $\widehat{\theta}_1$ onto the set $\mathbb{S} = \left\{x \in \mathbb{R}^k : x_1 \leq x_2 \leq \cdots \leq x_k\right\}$. There are several available algorithms in the literature for computing θ_1^*. The easiest to implement is the "pool adjacent violators algorithm" (PAVA), first published by Ayer *et al.* (1955). Focusing on our problem the PAVA can be described as follows. We wish to minimize

$$\sum_{j=1}^{k} \left(x_j - \frac{n_{2j}}{n_{1j} + n_{2j}} \right)^2 (n_{1j} + n_{2j})$$

over $\mathbb{S}$. The minimizing θ_{1j}^*, $j = 1, \ldots, k$, is termed the isotonic regression of $\frac{n_{21}}{n_{11}+n_{21}}, \ldots, \frac{n_{2k}}{n_{1k}+n_{2k}}$ with weights $n_{11} + n_{21}, \ldots, n_{1k} + n_{2k}$. PAVA provides the solution as follows. If

$$\frac{n_{21}}{n_{11} + n_{21}} \leq \cdots \leq \frac{n_{2k}}{n_{1k} + n_{2k}},$$

let $\theta_{1j}^* = \frac{n_{2j}}{n_{1j}+n_{2j}}$, $j = 1, \ldots, k$, and we are finished. Otherwise, find the first j so that

$$\frac{n_{2j}}{n_{1j} + n_{2j}} < \frac{n_{2\,j-1}}{n_{1\,j-1} + n_{2\,j-1}}$$

and define the new isotonic regression problem on $k - 1$ points as

$$\frac{n_{21}}{n_{11} + n_{21}}, \ldots, \frac{n_{2\,j-2}}{n_{1\,j-2} + n_{2\,j-2}}, \frac{n_{2\,j-1} + n_{2j}}{n_{1\,j-1} + n_{2\,j-1} + n_{1j} + n_{2j}}, \frac{n_{2\,j+1}}{n_{1\,j+1} + n_{2\,j+1}}, \ldots, \frac{n_{2k}}{n_{1k} + n_{2k}},$$

with weights $n_{11} + n_{21}, \ldots, n_{1\,j-2} + n_{2\,j-2}, n_{1\,j-1} + n_{2\,j-1} + n_{1j} + n_{2j}$, $n_{1\,j+1} + n_{2\,j+1}, \ldots, n_{1k} + n_{2k}$. If

$$\begin{aligned} \frac{n_{21}}{n_{11}+n_{21}} \leq \cdots &\leq \frac{n_{2\,j-2}}{n_{1\,j-2}+n_{2\,j-2}} \leq \frac{n_{2\,j-1}+n_{2j}}{n_{1\,j-1}+n_{2\,j-1}+n_{1j}+n_{2j}} \\ &\leq \frac{n_{2\,j+1}}{n_{1\,j+1}+n_{2\,j+1}} \leq \cdots \leq \frac{n_{2k}}{n_{1k}+n_{2k}}, \end{aligned}$$

let $\theta_{1l}^* = \frac{n_{2l}}{n_{1l}+n_{2l}}$, $l \neq j - 1$ and $\theta_{1j-1}^* = \theta_{1j}^* = \frac{n_{2\,j-1}+n_{2j}}{n_{1\,j-1}+n_{2\,j-1}+n_{1j}+n_{2j}}$. Otherwise, return to the previous step to find the first new violators and repeat until no violators are left. For the final solution, we term the adjacent integers with common value θ_{1i}^* as the collapsed levels. In the following, we shall denote $\theta_1^* = E_{\mathbf{w}}\left[\widehat{\theta}_1/\mathbb{S}\right]$.

For testing (27.6), the likelihood ratio test is given by

$$T_{01} = 2\sum_{j=1}^{k} \left\{ n_{2j} \log \frac{\theta_{1j}^*}{\bar{\theta}_{1j}} + n_{1j} \log \frac{1 - \theta_{1j}^*}{1 - \bar{\theta}_{1j}} \right\} \tag{27.13}$$

and for the problem considered in (27.7) by

$$T_{12} = 2\sum_{j=1}^{k}\left\{n_{2j}\log\frac{\widehat{\theta}_{1j}}{\theta^*_{1j}} + n_{1j}\log\frac{1-\widehat{\theta}_{1j}}{1-\theta^*_{1j}}\right\}. \tag{27.14}$$

The asymptotic distribution of the test statistic T_{01} under the null hypothesis given in (27.5) is chi-bar-squared. This distribution refers to a mixture of independent chi-squared random variables of the form $\sum_{j=1}^{k}\rho_j\chi^2_{j-1}$ where χ^2_j is a chi-squared variate with j degrees of freedom (with $\chi^2_0 \equiv 0$) and $\{\rho_j\}$ are a set of probabilities. If we denote by

$$\widehat{\boldsymbol{p}} = \left(\frac{n_{11}}{n}, \frac{n_{21}}{n}, \frac{n_{12}}{n}, \frac{n_{22}}{n}, \ldots, \frac{n_{1k}}{n}, \frac{n_{2k}}{n}\right)^T, \tag{27.15}$$

$$p(\overline{\theta}_1) = \left(\frac{(1-\overline{\theta}_{11})\, n_{*1}}{n}, \frac{\overline{\theta}_{11}n_{*1}}{n}, \frac{(1-\overline{\theta}_{12})\, n_{*2}}{n}, \frac{\overline{\theta}_{12}n_{*2}}{n}, \ldots, \frac{(1-\overline{\theta}_{1k})\, n_{*k}}{n}, \frac{\overline{\theta}_{1k}n_{*k}}{n}\right)^T$$

and

$$p(\theta^*_1) = \left(\frac{(1-\theta^*_{11})\, n_{*1}}{n}, \frac{\theta^*_{11}n_{*1}}{n}, \frac{(1-\theta^*_{12})\, n_{*2}}{n}, \frac{\theta^*_{12}n_{*2}}{n}, \ldots, \frac{(1-\theta^*_{1k})\, n_{*k}}{n}, \frac{\theta^*_{1k}n_{*k}}{n}\right)^T,$$

it is easy to check that T_{01} given in (27.13) can be written as

$$T_{01} = 2n\left(D_{Kull}\left(\widehat{\boldsymbol{p}}, p(\overline{\theta}_1)\right) - D_{Kull}\left(\widehat{\boldsymbol{p}}, p(\theta^*_1)\right)\right) \tag{27.16}$$

and T_{12} given in (27.14) as

$$T_{12} = 2nD_{Kull}\left(\widehat{\boldsymbol{p}}, p(\theta^*_1)\right), \tag{27.17}$$

where $D_{Kull}(\boldsymbol{p}, \boldsymbol{q})$ is the Kullback divergence measure defined in (27.12) between the probability vectors $\boldsymbol{p}$ and $\boldsymbol{q}$. Based on expressions (27.16) and (27.17), we can consider the following family of test statistics based on the ϕ-divergence measures for the hypothesis testing problems in (27.6) and (27.7):

$$T^{\phi}_{01} = \frac{2n}{\phi''(1)}\left\{D_{\phi}\left(\widehat{\boldsymbol{p}}, p(\overline{\theta}_1)\right) - D_{\phi}\left(\widehat{\boldsymbol{p}}, p(\theta^*_1)\right)\right\}, \tag{27.18}$$

$$T^{\phi}_{12} = \frac{2n}{\phi''(1)}D_{\phi}\left(\widehat{\boldsymbol{p}}, p(\theta^*_1)\right). \tag{27.19}$$

If we take $\phi(x) = \phi_{(0)}(x) = x\log x - x + 1$, we obtain the likelihood ratio tests given in (27.13) and (27.14).

For the case of two $2 \times k$ contingency tables, the likelihood function can be expressed by

$$L(\theta, \tau) = \prod_{j=1}^{k} \theta_{1j}^{n_{2j}} (1 - \theta_{1j})^{n_{1j}} \tau_{1j}^{m_{2j}} (1 - \tau_{1j})^{m_{1j}} \prod_{j=1}^{k} \theta_{2j}^{n_{*j}} \tau_{2j}^{m_{*j}},$$

where $\tau_{2j} = q_{1j} + q_{2j}$, $j = 1, \ldots, k$. It is easy to get the MLE's as

$$\widehat{\theta}_{1j} = \frac{n_{2j}}{n_{*j}}, \quad \widehat{\theta}_{2j} = \frac{n_{*j}}{n}, \quad \widehat{\tau}_{1j} = \frac{m_{2j}}{m_{*j}}, \quad \widehat{\tau}_{2j} = \frac{m_{*j}}{m} \quad j = 1, \ldots, k.$$

If we denote by $\overline{\overline{\theta}}_2 = \left(\overline{\overline{\theta}}_{21}, \ldots, \overline{\overline{\theta}}_{2k}\right)$ and $\overline{\overline{\tau}}_2 = (\overline{\overline{\tau}}_{21}, \ldots, \overline{\overline{\tau}}_{2k})$ the MLE's of θ_2 and τ_2 under H_0^*, respectively, and by $\theta_2^{**} = (\theta_{21}^{**}, \ldots, \theta_{2k}^{**})$ and $\tau_2^{**} = (\tau_{21}^{**}, \ldots, \tau_{21}^{**})$ the MLE's of θ_2 and τ_2 under H_1^*, respectively, we have

$$\overline{\overline{\theta}}_{2j} = \theta_{2j}^{**} = \widehat{\theta}_{2j}, \quad \overline{\overline{\tau}}_{2j} = \tau_{2j}^{**} = \widehat{\tau}_{2j}, \quad j = 1, \ldots, k.$$

Let $\left(\overline{\overline{\theta}}_1, \overline{\overline{\tau}}_1\right)$, $(\theta_1^{**}, \tau_1^{**})$ and $\left(\widehat{\theta}_1, \widehat{\tau}_1\right)$ the be MLE's of (θ_1, τ_1) under H_0^*, H_1^* and H_2^* respectively. Park (2002) introduced an algorithm to get the estimators $\left(\overline{\overline{\theta}}_1, \overline{\overline{\tau}}_1\right)$ and $(\theta_1^{**}, \tau_1^{**})$, which is based on a previous result of Park (1998). Park (2002) obtained, for testing

$$H_{Null} : H_0^* \quad \text{versus} \quad H_{Alt} : H_1^* - H_0^*, \tag{27.20}$$

the likelihood ratio test, which has the expression:

$$S_{01} = 2 \left\{ \log L_1 (\theta_1^{**}, \tau_1^{**}) - \log L_1 \left(\overline{\overline{\theta}}_1, \overline{\overline{\tau}}_1\right) \right\},$$

where

$$L_1 (\theta_1, \tau_1) = \prod_{j=1}^{k} \theta_{1j}^{n_{2j}} (1 - \theta_{1j})^{n_{1j}} \tau_{1j}^{m_{2j}} (1 - \tau_{1j})^{m_{1j}}$$

and

$$S_{12} = 2 \left\{ \log L_1 \left(\widehat{\theta}_1, \widehat{\tau}_1\right) - \log L_1 (\theta_1^{**}, \tau_1^{**}) \right\}$$

for testing

$$H_{Null} : H_1^* \quad \text{versus} \quad H_{Alt} : H_2^* - H_1^*. \tag{27.21}$$

We consider the following probability vectors,

$$\widehat{\widehat{\boldsymbol{p}}} = \left(\left(\frac{n}{n+m} \widehat{\boldsymbol{p}} \right)^T, \left(\frac{m}{n+m} \widehat{\boldsymbol{q}} \right)^T \right)^T,$$

where $\widehat{\boldsymbol{p}}$ is given in (27.15),

$$\widehat{\boldsymbol{q}} = \left(\frac{m_{11}}{m}, \frac{m_{21}}{m}, \ldots, \frac{m_{1k}}{m}, \frac{m_{2k}}{m}\right)^T$$

and

$$p(\theta_1, \tau_1) = \left(\left(\frac{n}{n+m}p(\theta_1)\right)^T, \left(\frac{m}{n+m}p(\tau_1)\right)^T\right)^T$$

being

$$p(\theta_1) = \left(\frac{(1-\theta_{11})\, n_{*1}}{n}, \frac{\theta_{11} n_{*1}}{n}, \frac{(1-\theta_{12})\, n_{*2}}{n}, \frac{\theta_{12} n_{*2}}{n}, \ldots, \frac{(1-\theta_{1k})\, n_{*k}}{n}, \frac{\theta_{1k} n_{*k}}{n}\right)^T,$$

and

$$p(\tau_1) = \left(\frac{(1-\tau_{11})\, m_{*1}}{m}, \frac{\tau_{11} m_{*1}}{m}, \frac{(1-\tau_{12})\, m_{*2}}{m}, \frac{\tau_{12} m_{*2}}{m}, \ldots, \frac{(1-\tau_{1k})\, m_{*k}}{m}, \frac{\tau_{1k} m_{*k}}{m}\right).$$

Using this notation, we have the likelihood ratio test, S_{01}, for testing (27.20), to be

$$S_{01} = 2\,(n+m)\left\{D_{Kull}\left(\widehat{\widehat{\boldsymbol{p}}}, p(\overline{\overline{\theta}}_1, \overline{\overline{\tau}}_1)\right) - D_{Kull}\left(\widehat{\widehat{\boldsymbol{p}}}, p(\theta_1^{**}, \tau_1^{**})\right)\right\} \tag{27.22}$$

and the likelihood ratio test, S_{12}, for testing (27.21), to be

$$S_{12} = 2\,(n+m)\, D_{Kull}\left(\widehat{\widehat{\boldsymbol{p}}}, p(\theta_1^{**}, \tau_1^{**})\right). \tag{27.23}$$

It is not difficult to establish that

$$(n+m) D_{Kull}\left(\widehat{\widehat{\boldsymbol{p}}}, p(\theta_1, \tau_1)\right) = n D_{Kull}\left(\widehat{\boldsymbol{p}}, p(\theta_1)\right) + m D_{Kull}\left(\widehat{\boldsymbol{q}}, p(\tau_1)\right). \tag{27.24}$$

Based on expressions (27.22) and (27.23), we introduce here the following families of test statistics for testing (27.8) and (27.9):

$$S_{01}^{\phi} = 2\frac{n+m}{\phi''(1)}\left[D_{\phi}\left(\widehat{\widehat{\boldsymbol{p}}}, p(\overline{\overline{\theta}}_1, \overline{\overline{\tau}}_1)\right) - D_{\phi}\left(\widehat{\widehat{\boldsymbol{p}}}, p(\theta_1^{**}, \tau_1^{**})\right)\right]$$

and

$$S_{12}^{\phi} = 2\frac{n+m}{\phi''(1)} D_{\phi}\left(\widehat{\widehat{\boldsymbol{p}}}, p(\theta_1^{**}, \tau_1^{**})\right).$$

It is clear that they represent an extension of the likelihood ratio tests, because we obtain them for $\phi(x) = x \log x - x + 1$.

27.3 Asymptotic Distribution of the ϕ-Divergence Test Statistics

In this section, we obtain the asymptotic distribution of the ϕ-divergence test statistics T_{01}^{ϕ}, T_{12}^{ϕ}, S_{01}^{ϕ} and S_{12}^{ϕ} introduced in the last section. In the first theorem, we obtain the asymptotic distribution of the ϕ-divergence test statistics T_{01}^{ϕ} and T_{12}^{ϕ}.

Theorem 27.3.1 *Let T_{01}^{ϕ} and T_{12}^{ϕ} be the ϕ-divergence test statistics for testing H_0 against $H_1 - H_0$ and H_1 against $H_2 - H_1$, respectively. Under H_0, we have for any real number c,*

$$\lim_{n\to\infty} \Pr\left(T_{01}^{\phi} > c\right) = \sum_{j=1}^{k} p(j,k,\boldsymbol{w}') \Pr\left(\chi_{j-1}^2 > c\right)$$

and

$$\lim_{n\to\infty} \Pr\left(T_{12}^{\phi} > c\right) = \sum_{j=1}^{k} p(j,k,\boldsymbol{w}') \Pr\left(\chi_{k-j}^2 > c\right),$$

where $\boldsymbol{w}' = \lim_{n\to\infty} \boldsymbol{w}/n$, $p(j,k,\boldsymbol{w}')$ is the probability that $E_{\boldsymbol{w}'}(\boldsymbol{U}/\mathbb{S})$ has exactly j distinct values and $\boldsymbol{U} = (U_1, \ldots, U_k)$ is a k-dimensional normal distribution with mean vector zero and variance-covariance matrix given by $\boldsymbol{D} = diag\left(\theta_{21}^{-1}, \ldots, \theta_{2k}^{-1}\right)$.

The asymptotic least favorable distribution associated with T_{01}^{ϕ} is given by

$$\sup_{H_0} \lim_{n\to\infty} \Pr\left(T_{01}^{\phi} > c\right) = \sum_{j=1}^{k} \binom{k-1}{j-1} 2^{-k+1} \Pr\left(\chi_{j-1}^2 > c\right). \tag{27.25}$$

PROOF. A second-order Taylor expansion of the function

$$g(\theta_1) = D_{\phi}\left(\widehat{\boldsymbol{p}}, p(\theta_1)\right)$$

around $\theta_1 = \widehat{\theta}$, and taking $\theta = \theta_1^*$ and $\theta = \overline{\theta}_1$, yields

$$D_{\phi}\left(\widehat{\boldsymbol{p}}, p(\theta_1^*)\right) = \frac{\phi''(1)}{2} \sum_{j=1}^{k} \left(\theta_{1j}^* - \widehat{\theta}_{1j}\right)^2 \frac{\widehat{\theta}_{2j}}{\widehat{\theta}_{1j}\left(1 - \widehat{\theta}_{1j}\right)} + o_p(n^{-1}) \tag{27.26}$$

and

$$D_{\phi}\left(\widehat{\boldsymbol{p}}, p(\overline{\theta}_1)\right) = \frac{\phi''(1)}{2} \sum_{j=1}^{k} \left(\overline{\theta}_{1j} - \widehat{\theta}_{1j}\right)^2 \frac{\widehat{\theta}_{2j}}{\widehat{\theta}_{1j}\left(1 - \widehat{\theta}_{1j}\right)} + o_p(n^{-1}).$$

Then, we obtain

$$\begin{aligned} T_{01}^{\phi} &= \frac{2n}{\phi''(1)}\left(D_{\phi}\left(\widehat{\boldsymbol{p}}, p(\overline{\theta}_1)\right) - D_{\phi}\left(\widehat{\boldsymbol{p}}, p(\theta_1^*)\right)\right) \\ &= n\sum_{j=1}^{k}\left\{\left(\overline{\theta}_{1j}-\widehat{\theta}_{1j}\right)^2 - \left(\theta_{1j}^*-\widehat{\theta}_{1j}\right)^2\right\}\frac{\widehat{\theta}_{2j}}{\widehat{\theta}_{1j}\left(1-\widehat{\theta}_{1j}\right)} + o_p(1). \end{aligned}$$

Therefore, the asymptotic distribution of the ϕ-divergence test statistic T_{01}^{ϕ} coincides with the asymptotic distribution of the random variable

$$\sum_{j=1}^{k}\left\{\left(\sqrt{n}\left(\overline{\theta}_{1j}-\widehat{\theta}_{1j}\right)\right)^2 - \left(\sqrt{n}\left(\theta_{1j}^*-\widehat{\theta}_{1j}\right)\right)^2\right\}\frac{\theta_{2j}}{\theta_{1j}(1-\theta_{1j})},$$

but this expression can be rewritten as

$$\sum_{j=1}^{k}\left(\frac{\sqrt{n}\left(\overline{\theta}_{1j}-\widehat{\theta}_{1j}\right)}{\sqrt{\theta_{1j}\left(1-\theta_{1j}\right)}}\right)^2\theta_{2j} - \sum_{j=1}^{k}\left(\frac{\sqrt{n}\left(\theta_{1j}^*-\widehat{\theta}_{1j}\right)}{\sqrt{\theta_{1j}\left(1-\theta_{1j}\right)}}\right)^2\theta_{2j}.$$

But

$$\begin{aligned} &\sum_{j=1}^{k}\left(\frac{\sqrt{n}\left(\theta_{1j}^*-\widehat{\theta}_{1j}\right)}{\sqrt{\theta_{1j}\left(1-\theta_{1j}\right)}}\right)^2 \\ &\quad= \sum_{j=1}^{k}\left(\frac{\sqrt{n}\left(\theta_{1j}^*-\theta_{1j}\right)}{\sqrt{\theta_{1j}\left(1-\theta_{1j}\right)}} - \frac{\sqrt{n}\left(\widehat{\theta}_{1j}-\theta_{1j}\right)}{\sqrt{\theta_{1j}\left(1-\theta_{1j}\right)}}\right)^2 \\ &\quad= \sum_{j=1}^{k}\left(E_{\frac{\boldsymbol{w}}{n}}\left(\frac{\sqrt{n}\left(\theta_{1j}^*-\theta_{1j}\right)}{\sqrt{\theta_{1j}\left(1-\theta_{1j}\right)}}/\mathbb{S}\right)_j - \frac{\sqrt{n}\left(\widehat{\theta}_{1j}-\theta_{1j}\right)}{\sqrt{\theta_{1j}\left(1-\theta_{1j}\right)}}\right)^2. \end{aligned}$$

It is easy to establish that

$$\begin{aligned} E\left[\frac{\partial^2}{\partial\theta_{1j}^2}\log L\left(\theta_1\right)\right] &= \frac{-n\theta_{2j}}{\theta_{1j}\left(1-\theta_{1j}\right)}, \quad j=1,\ldots,k, \\ E\left[\frac{\partial^2}{\partial\theta_{1j}\partial\theta_{1i}}\log L\left(\theta_1\right)\right] &= 0, \quad j\neq i, \end{aligned}$$

and therefore the Fisher information matrix is given by

$$I_F\left(\theta_1\right) = diag\left(\frac{\theta_{21}}{\theta_{11}\left(1-\theta_{11}\right)},\ldots,\frac{\theta_{2k}}{\theta_{1k}\left(1-\theta_{1k}\right)}\right)$$

and

$$\sqrt{n}\left(\frac{\widehat{\theta}_1-\theta_1}{\sqrt{\theta_1(1-\theta_1)}}\right)\underset{n\to\infty}{\overset{L}{\to}} N(\mathbf{0},\boldsymbol{D}),$$

where $\boldsymbol{D}=diag(\theta_{21}^{-1},\ldots,\theta_{2k}^{-1})$.

Hence,

$$\sum_{j=1}^{k}\left(\frac{\sqrt{n}\left(\theta_{1j}^{*}-\widehat{\theta}_{1j}\right)}{\sqrt{\theta_{1j}\left(1-\theta_{1j}\right)}}\right)^{2}\theta_{2j}\underset{n\to\infty}{\overset{L}{\to}}\sum_{j=1}^{k}\left(E_{\theta_2}\left(\boldsymbol{U}/\mathbb{S}\right)_j-U_j\right)^2\theta_{2j},$$

where $\boldsymbol{U}=(U_1,\ldots,U_k)$ is a k-dimensional normal random variable with mean vector zero and variance-covariance matrix $\boldsymbol{D}$.

Similarly, taking into account that we are obtaining the asymptotic distribution under H_0, we have

$$\sqrt{n}\left(\widehat{\theta}_1-\overline{\theta}_1\right)\underset{n\to\infty}{\overset{L}{\to}}\boldsymbol{U}-\overline{\boldsymbol{U}},$$

where $\overline{\boldsymbol{U}}=\sum_{j=1}^{k}\theta_{2j}U_j$. Therefore,

$$\sum_{j=1}^{k}\left(\frac{\sqrt{n}\left(\overline{\theta}_{1j}-\widehat{\theta}_{1j}\right)}{\sqrt{\theta_{1j}\left(1-\theta_{1j}\right)}}\right)^{2}\theta_{2j}\underset{n\to\infty}{\overset{L}{\to}}\sum_{j=1}^{k}\theta_{2j}\left(U_j-\overline{\boldsymbol{U}}\right)^2.$$

Using Theorem 7.8 of Barlow *et al.* (1972), we have

$$\begin{aligned}\sum_{j=1}^{k}\theta_{2j}\left(U_j-\overline{\boldsymbol{U}}\right)^2-\sum_{j=1}^{k}\left(E_{\theta_2}\left(\boldsymbol{U}/\mathbb{S}\right)_j-U_j\right)^2\theta_{2j}\\=\sum_{j=1}^{k}\left(E_{\theta_2}\left(\boldsymbol{U}/\mathbb{S}\right)_j-\overline{\boldsymbol{U}}\right)^2\theta_{2j}.\end{aligned}$$

So, the asymptotic distribution of the ϕ-divergence test statistic T_{01}^{ϕ} coincides with the distribution of the random variable

$$\sum_{j=1}^{k}w'_j\left(E_{\boldsymbol{w}'}\left(\boldsymbol{U}/\mathbb{S}\right)_j-\overline{\boldsymbol{U}}\right)^2,$$

where $\boldsymbol{w}'_j=\lim_{n\to\infty}\boldsymbol{w}_j/n$.

This distribution can be obtained by using the corollary on p. 70 of Robertson *et al.* (1988). The probability that $E_{\boldsymbol{w}'}\left(\boldsymbol{U}/\mathbb{S}\right)$, weighted least-squares projection of $\boldsymbol{U}$ onto $\mathbb{S}$ with weights $\boldsymbol{w}'_j$, has exactly l distinct values is denoted by $p(l, k, \boldsymbol{w}')$, and is called *level probability*. Using Theorem 3.6.1 of Robertson *et al.* (1988), the least favorable distribution given in (27.25) is obtained.

It is an easy exercise to see that under H_0 the asymptotic distribution of the ϕ-divergence test statistic T_{12}^{ϕ} coincides with the distribution of the random variablc

$$\sum_{j=1}^{k} \boldsymbol{w}'_j \left(E_{\boldsymbol{w}'} \left(\boldsymbol{U}/\mathbb{S} \right)_j - U_j \right)^2$$

and this is given in corollary on p. 70 of Robertson *et al.* (1988). ■

Remark 27.3.1 The result presented in Theorem 27.3.1 can be used for testing the existence of a monotonic dose-response relationship in clinical and epidemiological studies. The purpose is to test the null hypotheses H_0 of no relationship between a binary response Y (e.g., disease occurrence) and an ordered categorical exposure X, with values $x_1 \leq x_2 \leq \cdots \leq x_k$, against the alternative hypothesis H_1 of a positive dose-response relationship between response Y and exposure X. We denote by $\pi(x_j) = \Pr(Y = 1|X = x_j)$ and we assume that $\pi(x_j) = \exp(\alpha + \beta x_j) \{1 + \exp(\alpha + \beta x_j)\}^{-1}$. We consider the probability distribution $\boldsymbol{P} = (p_{ij})$, $i = 1, 2$, $j = 1, \ldots, k$, with

$$p_{ij} = \begin{cases} (1 - \pi(x_j)) \frac{n_{*j}}{n} & \text{if } \ i = 1 \\ \pi(x_j) \frac{n_{*j}}{n} & \quad\;\; i = 2 \end{cases}.$$

It is immediate to prove that testing

$$H_{Null} : H_0 \quad \text{versus} \quad H_{Alt} : H_1 - H_0$$

is equivalent to testing

$$H_{Null} : \beta = 0 \quad \text{versus} \quad H_{Alt} : \beta > 0.$$

Remark 27.3.2 The asymptotic distribution of T_{01}^{ϕ} depends on $\boldsymbol{w}'$ through the level probabilities $\mathrm{p}(j, k, \boldsymbol{w}')$. When the level probabilities are equal (it is customary to omit the weights when they are equal), then they can be calculated recursively through the formula given in Corollary B on p. 145 of Barlow *et al.* (1972)

$$p(l, k) = \frac{1}{k} p(l - 1, k - 1) + \frac{k - 1}{k} p(l, k - 1)$$

for $l = 2, 3, \ldots, k - 1$ with $p(1, k) = \frac{1}{k}$ and $p(k, k) = \frac{1}{k}$. However, if there is no serious deviation among the w'_js, for example, if $R = \max_j w'_j / \min w'_j \leq 1.4$, the equal weights case provides a good approximation, and if $R = \max_j w'_j / \min w'_j \leq 3.4$ the equal weights case provides an adequate approximation. For arbitrary weights and $k \leq 4$, the calculations for the level probabilities are given in Robertson et al. (1988), but for $k \leq 5$, no closed-form expressions for the level probabilities exist.

Now we present the asymptotic distribution of S_{01}^{ϕ} and S_{12}^{ϕ}.

Theorem 27.3.2 *When H_0^* is true for any real number c and assuming that*

$$\lim_{m,n\to\infty} \frac{m}{n} = \widehat{\gamma} > 0,$$

we have

$$\lim_{n,m\to\infty} \Pr\left(S_{01}^{\phi} > c\right) = \sum_{j=1}^{k} p(j,k,\boldsymbol{u}) \Pr(\chi_{j-1}^2 > c)$$

and

$$\lim_{n,m\to\infty} \Pr\left(S_{12}^{\phi} > c\right) = \sum_{j=1}^{k} p(j,k,\boldsymbol{u}) \Pr(\chi_{k-j}^2 > c)$$

where $\boldsymbol{u} = (u_1,\ldots,u_k)$, $u_j = \gamma\alpha_j\beta_j/(\alpha_j+\gamma\beta_j)$, $j = 1,\ldots,k$, $\gamma = m/n$, $\alpha_j = \theta_{2j}/\left\{\theta_{1j}(1-\theta_{1j})\right\}$ and $\beta_j = \tau_{2j}/\left\{\tau_{1j}(1-\tau_{1j})\right\}$, $j = 1,\ldots,k$.

Proof. As in (27.24), we have

$$D_{\phi}\left(\widehat{\widehat{\boldsymbol{p}}}, p(\theta_1,\tau_1)\right) = \frac{n}{n+m} D_{\phi}\left(\widehat{\boldsymbol{p}}, p(\theta_1)\right) + \frac{m}{n+m} D_{\phi}\left(\widehat{\boldsymbol{q}}, p(\tau_1)\right),$$

and using the second-order Taylor expansion given in (27.26) we get

$$\begin{aligned} D_{\phi}\left(\widehat{\widehat{\boldsymbol{p}}}, p(\theta_1,\tau_1)\right) &= \frac{\phi''(1)}{2}\left[\frac{n}{n+m}\left\{\sum_{j=1}^{k}\left(\widehat{\theta}_{1j}-\theta_{1j}\right)^2\widehat{\alpha}_j + o_p(n^{-1})\right\}\right. \\ &\left.+ \frac{m}{n+m}\left(\sum_{j=1}^{k}\left\{\widehat{\tau}_{1j}-\tau_{1j}\right)^2\widehat{\beta}_j + o_p(m^{-1})\right\}\right], \end{aligned}$$

where

$$\widehat{\alpha}_j = \frac{\widehat{\theta}_{2j}}{\widehat{\theta}_{1j}(1-\widehat{\theta}_{1j})}, \quad \widehat{\beta}_j = \frac{\widehat{\tau}_{2j}}{\widehat{\tau}_{1j}(1-\widehat{\tau}_{1j})}.$$

Therefore,

$$\begin{aligned} & 2\frac{n+m}{\phi''(1)} D_{\phi}\left(\widehat{\widehat{\boldsymbol{p}}}, p(\theta_1,\tau_1)\right) \\ &= n\sum_{j=1}^{k}\left(\widehat{\theta}_{1j}-\theta_{1j}\right)^2\widehat{\alpha}_j + m\sum_{j=1}^{k}\left(\widehat{\tau}_{1j}-\tau_{1j}\right)^2\widehat{\beta}_j + o_p(1) \\ &= n\left\{\sum_{j=1}^{k}\left(\widehat{\theta}_{1j}-\theta_{1j}\right)^2\widehat{\alpha}_j + \frac{m}{n}\sum_{j=1}^{k}\left(\widehat{\tau}_{1j}-\tau_{1j}\right)^2\widehat{\beta}_j\right\} + o_p(1). \end{aligned}$$

Hence, the asymptotic distribution of $2\frac{n+m}{\phi''(1)}D_\phi\left(\widehat{\boldsymbol{p}}, p(\theta_1,\tau_1)\right)$ coincides with the asymptotic distribution of the quadratic form

$$Q(\theta_1,\tau_1)=n\left\{\sum_{j=1}^{k}\left(\widehat{\theta}_{1j}-\theta_{1j}\right)^2\widehat{\alpha}_j+\widehat{\gamma}\sum_{j=1}^{k}\left(\widehat{\tau}_{1j}-\tau_{1j}\right)^2\widehat{\beta}_j\right\}];$$

then, we have

$$S_{01}^{\phi}=2\left(Q(\overline{\overline{\theta}}_1,\overline{\overline{\tau}}_1)-Q(\theta_1^{**},\tau_1^{**})\right)$$

and

$$S_{12}^{\phi}=2Q(\theta_1^{**},\tau_1^{**}).$$

Now the result follows from Theorem 1 of Park (2002). ■

Remark 27.3.3 As in Theorem 27.3.1, Theorem 27.3.2 can be used for testing the existence of two monotonic dose-response relationships. If we denote the models by $\pi_1(x_j)=\exp(\alpha_1+\beta_1x_j)\{1+\exp(\alpha_1+\beta_1x_j)\}^{-1}$ and $\pi_2(x_j)=\exp(\alpha_2+\beta_2x_j)\{1+\exp(\alpha_2+\beta_2x_j)\}^{-1}$, we can use the previous theorem for testing

$$H_0:\beta_1=\beta_2 \text{ versus } H_1:\beta_1-\beta_2>0.$$

Acknowledgement. This work was supported partially by Grant BFM 2003-0892.

References

1. Armitage, P. (1955). Tests for linear trends in proportions and frequencies, *Biometrics* **11**, 375–386.

2. Ayer, M., Brunk, H. O., Ewing, G. M., Reid, W. I., and Silverman, E. (1955). An empirical distribution function for sampling with incomplete information, *Annals of Mathematical Statistics*, **26**, 641–647.

3. Barlow, R. E., Bartholomew, D. J., Bremmer, J. M., and Brunk, H. D. (1972). *Statistical Inference Under Order Restrictions*, John Wiley & Sons, New York.

4. Cressie, N., and Read, T. R. C. (1984). Multinomial goodness-of-fit tests, *Journal of the Royal Statistical Society, Series B*, **46**, 440–464.

5. Grove, D. M. (1980). A test of independence against a class of ordered alternatives in a $2\times C$ contingency table, *Journal of the American Statistical Association*, **75**, 454–459.

6. Lee, M. T. (1989). Some cross-product difference statistics and a test for trends in ordered contingency tables, *Statistics & Probability Letters*, **7**, 41–46.

7. Menéndez, M. L., Pardo, L., and K. Zografos (2002). Tests of hypotheses for and against order restrictions on multinomial parameters based on ϕ-divergences, *Utilitas Mathematica*, **61**, 209–223.

8. Menéndez, M. L., Pardo, J. A., and Pardo, L. (2003a). Tests for bivariate symmetry against ordered alternatives in square contingency tables. *Australian and New Zealand Journal of Statistics,* **45, 1,** 115-124.

9. Menéndez, M. L., Morales, D., and Pardo, L. (2003b). Tests based on divergences for and against ordered alternatives in cubic contingency tables, *Applied Mathematics and Computation*, **134**, 207–216.

10. Park, C. G. (1998). Testing for unimodal dependence in an ordered contingency table with restricted marginal probabilities, *Statistics & Probability Letters*, **37**, 121–129.

11. Park, C. G. (2002). Testing for ordered trends of binary responses between contingency tables, *Journal of Multivariate Analysis*, **81**, 229–241.

12. Patefield, W. M. (1982). Exact tests for trends in ordered contingency tables, *Applied Statistics*, **31**, 32-43.

13. Robertson, T., and Wright, F. T. (1983). On approximation of the level probabilities and associated distributions in order restricted inference, *Biometrika*, **70**, 597–606.

14. Robertson, T., Wright, F. T., and Dykstra, R. L. (1988). *Order-Restricted Statistical Inference*, John Wiley & Sons, New York.

28

Dimension Reduction in Multivariate Time Series

Daniel Peña and Pilar Poncela

Universidad Carlos III de Madrid, Madrid, Spain
Universidad Autónoma de Madrid, Madrid, Spain

Abstract: This chapter compares models for dimension reduction in time series and tests of the dimension of the dynamic structure. We consider both stationary and nonstationary time series and discuss principal components, canonical analysis, scalar component models, reduced rank models, and factor models. The unifying view of canonical correlation analysis between the present and past values of the series is emphasized. Then, we review some of the tests based on canonical correlation analysis to find the dimension of the dynamic relationship among the time series. Finally, the procedures are compared through a real data example.

Keywords and phrases: Canonical correlation analysis, dimension reduction, vector time series

28.1 Introduction

Dimension reduction is very important in vector time series because the number of parameters in a model grows very fast with the dimension m of the vector of time series $\boldsymbol{y}_t$. Linear models usually have a number of parameters that grows with m^2 and, for instance, a VARMA(p, q) model contains $m^2(p+q)$ parameters. This problem can be even more important in a nonlinear vector time series and, for instance, in a bilinear vector model or a threshold AR vector the number of parameters can easily be very large. The same problem appears in models with changing conditional variance as multivariate ARCH or GARCH models. Finding simplifying structures or factors in these models is important to reduce the number of parameters required to apply them to real data. In this article, we will consider linear time series models and we will concentrate in the time domain approach. See Brillinger (1981) and Shumway and Stoffer (2000) for analysis in the frequency domain. The first approach for reducing the dimension of a dynamic linear system is, by analogy with standard multivariate statistical analysis, finding linear combinations of the time series variables with simple properties. In a stationary time series, we would be interested in finding linear combinations that are white noise because then the dynamics of the vector time series can be expressed by a number of components smaller than its dimension, m. In a nonstationary series, we also would be interested in finding linear combinations that are stationary, reducing the dimension of the nonstationary space. This has been an important topic of research in the econometric literature under the name of cointegration; see, for instance, Engle and Granger (1987), Banarjee *et al.* (1993), and Johansen (1995). For VARMA models, dimension reduction was already analyzed in the pioneering work of Quenouille (1968). Some seminal contributions to this problem are the canonical analysis [Box and Tiao (1977)], the scalar component models, SCM, [Tiao and Tsay (1989)] and the reduced-rank models [Velu *et al.* (1986), Ahn and Reinsel (1990), Ahn (1997) and Reinsel and Velu (1998)]. A second approach for dimension reduction is by using dynamic factor models; see Anderson (1963), Priestly *et al.* (1974), Geweke and Singleton (1981), Brillinger (1981), Peña and Box (1987), Stock and Watson (1988), Molenaar *et al.* (1992), Forni *et al.* (2000) and Peña and Poncela (2004, 2006), among others. Factor models are very related to cointegration as it can be shown that the number of cointegration relations among the components of a vector of a time series is the dimension of the vector minus the number of nonstationary common factors [Escribano and Peña (1994)].

In the state space approach [see Durbin and Koopman (2001)] dimension reduction appears in a natural way in defining the dimension of the state. Akaike (1974) in a seminal work introduced canonical correlation between the present and the future to determine the dimension of the state variables. Aoki (1987) made also important contributions. The dynamic factor model in state space form has been considered by Harvey (1989). State space models for multivariate time series have two advantages over the VARMA representation. First, the number of parameters in the model depends on the dimension of the state vector, and when the series can be represented by a low-dimension state vector the number of parameters is automatically reduced. Second, the state space representation provides a direct interpretation of the time series vector in components such as trend, cycle, seasonal, and disturbance terms. In this way, we have the additional flexibility of searching for dimension reduction in the components, instead of trying a simplifying structure of the whole vector of time series.

One of the main tools for building tests for the dimension of a linear system is canonical correlation analysis. It can be shown that both linear combinations that are white noise and linear combinations which are stationary or nonstationary can be obtained from this approach. Also, it provides dimension tests that are invariant to affine transformations of the time series variables. The test proposed by Tiao and Tsay (1989) for SCM, the test used by Ahn and Reinsel (1988) and Reinsel and Ahn (1992) for the reduced rank autoregressive model, the cointegration test by Johansen (1988, 1991), and the tests proposed by Hu and Chou (2004) and Peña and Poncela (2006) for dynamic factor models are all based on canonical correlation analysis between the vector of time series or some of its differences and its lags. Related tests are the principal component test of Stock and Watson (1988) and Harris (1997).

This article is organized as follows. Section 28.2 presents different approaches for finding simplifying linear combinations in a time series. Section 28.3 discusses tests for finding the dimension of the system based on canonical correlation analysis. Section 28.4 applies the procedures to an example and Section 28.5 includes some final remarks.

28.2 Models for Dimension Reduction

Suppose a $m \times 1$ vector $\boldsymbol{y}_t$ follows a linear time series process. We are interested in finding linear combinations $x_{1t} = \boldsymbol{m}'\boldsymbol{y}_t$ of the vector of time series with useful properties for model simplification and dimension reduction. Also, we will consider dynamic factor models in which the factors are not necessarily linear combinations of the observed time series.

28.2.1 Principal components

Let $\boldsymbol{y}_t$ be a stationary process with mean $\boldsymbol{\mu}$. Define the covariance matrices by

$$\boldsymbol{\Gamma}_y(k) = E\left\{(\boldsymbol{y}_{t-k} - \boldsymbol{\mu})(\boldsymbol{y}_t - \boldsymbol{\mu})'\right\},$$

and suppose that we are interested in linear combinations, $x_{1t} = \boldsymbol{m}'\boldsymbol{y}_t$, with maximum variance. Let $x_{it} = \psi(B)u_t$ be the model for the linear combination x_{1t}; then, as $\mathrm{Var}(x_{it}) = \sigma_u^2 \sum \psi_i^2$, linear combinations that are white noise will be associated to a small variance, and linear combinations close to nonstationary will be associated to a large variance. This association suggests looking for linear combinations of large or small variance, and it is well known that they will be given by the eigenvectors $\boldsymbol{m}_i$ in

$$\boldsymbol{\Gamma}_y(0)\boldsymbol{m}_i = \lambda_i \boldsymbol{m}_i$$

and the corresponding eigenvalues, λ_i, will be the variances of the linear combinations. In the particular case in which exact dimension reduction can be obtained, because one of the series is a linear combination of the others, this fact will be revealed by a zero eigenvalue in this covariance matrix $\boldsymbol{\Gamma}_y(0)$, and the linear combination will be given by the corresponding eigenvector. This approach can be extended to the nonstationary case. Suppose $\boldsymbol{y}_t$ is nonstationary $I(d)$. Then, following Peña and Poncela (2006), we define the generalized covariance matrices by

$$\boldsymbol{C}(k) = \frac{1}{T^{2d}} \sum (\boldsymbol{y}_{t-k} - \overline{\boldsymbol{y}})(\boldsymbol{y}_t - \overline{\boldsymbol{y}})',$$

where $\overline{\boldsymbol{y}} = T^{-1} \sum \boldsymbol{y}_t$. The solutions of

$$\boldsymbol{C}(0)\boldsymbol{m}_i = \lambda_i \boldsymbol{m}_i$$

will provide the interesting linear combinations: those that link to large eigenvalues may define the nonstationary components, and those that link to small eigenvalues may define the stationary components. However, note that principal components are not invariant under scale transformation of the variables and we may, by changing the scale, make the variance of a stationary component much larger than the one of a nonstationary one. For this reason, principal components in a time series can be useful when all the series have a common scale of measurement, but are less justified otherwise.

28.2.2 The Box and Tiao canonical analysis

Box and Tiao (1977) proposed to find linear combinations of a stationary time series with maximum predictability, and called the procedure canonical analysis. We will refer to this procedure as BT analysis. Let $x_{1t} = \widehat{x}_{1t-1}(1) + u_t$, where $\widehat{x}_{1t-1}(1)$ is the one step ahead prediction and u_t is the forecast error. Let σ_x^2 be the variance of x_{1t}, and σ_u^2 the variance of u. These authors define the predictability by

$$q = \frac{\sigma_x^2 - \sigma_u^2}{\sigma_x^2} = 1 - \sigma_x^{-2}\sigma_u^2. \tag{28.1}$$

Thus, a white noise series has a predictability equal to zero and a nonstationary process has a predictability close to one. For instance, an AR(1) has $\sigma_x^2 = \sigma_u^2/(1-\phi^2)$ and $q = \phi^2$. If $\phi \to 1$, then $q \to 1$. This measure can be interpreted as a generalized determination coefficient. A vector time series model implies a decomposition of the form

$$\boldsymbol{y}_t = \widehat{\boldsymbol{y}}_{t-1}(1) + \boldsymbol{\varepsilon}_t,$$

where $\widehat{\boldsymbol{y}}_{t-1}(1)$ is now the vector of one-step-ahead predictions and $\boldsymbol{\varepsilon}_t$ the forecast error. As these terms are uncorrelated, we can also split the covariance matrix, $\boldsymbol{\Gamma}_y(0)$, as

$$\boldsymbol{\Gamma}_y(0) = \boldsymbol{F}_y(0) + \boldsymbol{\Sigma},$$

where $E(\boldsymbol{\varepsilon}_t\boldsymbol{\varepsilon}_t') = \boldsymbol{\Sigma}$ and $E\left[(\widehat{\boldsymbol{y}}_{t-1}(1) - \boldsymbol{\mu})(\widehat{\boldsymbol{y}}_{t-1}(1) - \boldsymbol{\mu})'\right] = \boldsymbol{F}_y(0)$. It can be shown that the linear combinations of maximum predictability are defined by the largest eigenvectors of the predictability matrix

$$\boldsymbol{Q} = \boldsymbol{I} - \boldsymbol{\Gamma}_y(0)^{-1}\boldsymbol{\Sigma}, \tag{28.2}$$

and that the eigenvalues give the predictability of these linear combinations. Note that (28.2) reduces to (28.1) for scalar time series. If h linear combinations are white noise, this matrix will have h eigenvalues equal to zero and if r linear combinations approach the nonstationary case, $\boldsymbol{Q}$ will have r eigenvalues close to one. This analysis can be seen as (1) a generalized principal components approach for time series, and (2) a canonical correlation analysis between the vector of variables $\boldsymbol{y}_t$ and its lags. To illustrate the first interpretation we will use that, as the eigenvectors $\boldsymbol{m}_i$ of $\boldsymbol{Q}$ must satisfy $\boldsymbol{Q}\boldsymbol{m}_i = (\boldsymbol{I} - \boldsymbol{\Gamma}_y(0)^{-1}\boldsymbol{\Sigma})\boldsymbol{m}_i$ $= \lambda_i \boldsymbol{m},$ then $\boldsymbol{\Gamma}_y(0)^{-1}\boldsymbol{\Sigma}\boldsymbol{m}_i = (1 - \lambda_i)\boldsymbol{m}_i$, and also

$$\boldsymbol{\Sigma}^{-1}\boldsymbol{\Gamma}_y(0)\boldsymbol{m}_i = \alpha_i \boldsymbol{m}_i, \tag{28.3}$$

where $\alpha_i = (1-\lambda_i)^{-1}$. Note that in the matrix $\boldsymbol{\Sigma}^{-1}\boldsymbol{\Gamma}_y(0)$ the eigenvectors that link to eigenvalues equal to one define white noise components and those that link to a large eigenvalue define nonstationary components. In the particular case $\boldsymbol{\Sigma} = \sigma^2\boldsymbol{I},$ that is, the noises are uncorrelated with the same variance, the BT analysis is a principal component analysis of the vector time series. For instance, the linear combination of maximum predictability is the first principal component of the data. In the general case where $\boldsymbol{\Sigma}$ is a positive definite covariance matrix, calling $\boldsymbol{\Sigma} = \boldsymbol{A}\boldsymbol{D}\boldsymbol{A}'$ to the spectral decomposition of the noise covariance matrix, from (28.3) we have

$$(\boldsymbol{D}^{-1/2}\boldsymbol{A}'\boldsymbol{\Gamma}_y(0)\boldsymbol{A}\boldsymbol{D}^{-1/2})(\boldsymbol{D}^{1/2}\boldsymbol{A}'\boldsymbol{m}_i\ ) = \alpha_i(\boldsymbol{D}^{1/2}\boldsymbol{A}'\boldsymbol{m}_i),$$

and the BT analysis can be interpreted as: (a) transforming the vector of time series by $\boldsymbol{s}_t = \boldsymbol{D}^{-1/2}\boldsymbol{A}'\boldsymbol{y}_t$, so that the noise covariance of the transformed time series is the identity; (b) computing the principal components of $\boldsymbol{s}_t$, let us call them $\boldsymbol{v}_i$; and (c) transforming back the principal components by $\boldsymbol{m}_i = \boldsymbol{A}\boldsymbol{D}^{-1/2}$ $\boldsymbol{v}_i$. To obtain the canonical correlation analysis interpretation note that the canonical correlation coefficients between $\boldsymbol{y}_t$ and $\boldsymbol{y}_t^* = (\boldsymbol{y}'_{t-1}, \ldots, \boldsymbol{y}'_{t-k})'$ are given by the nonzero eigenvalues of the matrix

$$\boldsymbol{M} = \boldsymbol{\Gamma}_y^{-1}(0)\boldsymbol{\Gamma}_{yy^*}(k)\boldsymbol{\Gamma}_{y^*}^{-1}(0)\boldsymbol{\Gamma}'_{yy^*}(k) \tag{28.4}$$

where, assuming to simplify that $E(\boldsymbol{y}_t) = 0$, we have $\boldsymbol{\Gamma}_y(0) = E(\boldsymbol{y}_t\boldsymbol{y}'_t)$, $\boldsymbol{\Gamma}_{y^*}(0) =$ $E(\boldsymbol{y}_t^*\boldsymbol{y}_t^{*\prime})$ and $\boldsymbol{\Gamma}_{yy^*}(k) = E(\boldsymbol{y}_t\boldsymbol{y}_t^{*\prime})$. Let

$$\boldsymbol{\Gamma}_{y|y^*} = E\left[(\boldsymbol{y}_t - \widehat{\boldsymbol{\beta}}\boldsymbol{y}_t^*)(\boldsymbol{y}_t - \widehat{\boldsymbol{\beta}}\boldsymbol{y}_t^*)'\right]$$

be the residual covariance matrix of a multivariate regression equation between $\boldsymbol{y}_t$ and $\boldsymbol{y}_t^* = (\boldsymbol{y}'_{t-1}, \ldots, \boldsymbol{y}'_{t-k})'$. As $\widehat{\boldsymbol{\beta}} = \boldsymbol{\Gamma}_{yy^*}\ (k)\boldsymbol{\Gamma}_{y^*}^{-1}(0)$, we have

$$\boldsymbol{\Gamma}_{y|y^*} = \boldsymbol{\Gamma}_y(0) - \boldsymbol{\Gamma}_{yy^*}(k)\boldsymbol{\Gamma}_{y^*}^{-1}(0)\boldsymbol{\Gamma}'_{yy^*}(k) \tag{28.5}$$

and inserting $\boldsymbol{\Gamma}_{yy^*}(k)\boldsymbol{\Gamma}_{y^*}^{-1}(0)\boldsymbol{\Gamma}'_{yy^*}(k) = \boldsymbol{\Gamma}_y(0) - \boldsymbol{\Gamma}_{y|y^*}$ in (28.4), the $\boldsymbol{M}$ matrix can be written as

$$\boldsymbol{M} = \boldsymbol{I} - \boldsymbol{\Gamma}_y^{-1}\mathbf{(0)}\boldsymbol{\Gamma}_{y|y^*}$$

which is equivalent to the predictability matrix $\boldsymbol{Q}$ defined in (28.2). Thus, the linear combinations of maximum predictability are equivalent to the linear combinations of maximum correlation between the present and the past.

As an illustration, consider the VAR(1) model

$$\boldsymbol{y}_t = \boldsymbol{\Phi}\boldsymbol{y}_{t-1} + \boldsymbol{\varepsilon}_t. \tag{28.6}$$

Then $\boldsymbol{\Gamma}_y(0) = \boldsymbol{\Phi}\boldsymbol{\Gamma}_y(0)\boldsymbol{\Phi}' + \boldsymbol{\Sigma}$, $\boldsymbol{\Gamma}'_y(1) = \boldsymbol{\Phi}\boldsymbol{\Gamma}_y(0)$ and the matrix $\boldsymbol{Q}$ given by (28.2) can also be written as $\boldsymbol{Q} = \boldsymbol{I} - \boldsymbol{\Gamma}_y^{-1}(0)(\boldsymbol{\Gamma}_y(0) - \boldsymbol{\Phi}\boldsymbol{\Gamma}_y(0)\boldsymbol{\Phi}')$, or

$$\boldsymbol{Q} = \boldsymbol{\Gamma}_y^{-1}(0)\boldsymbol{\Phi}\boldsymbol{\Gamma}_y(0)\boldsymbol{\Phi}'$$

which implies

$$\boldsymbol{Q} = \boldsymbol{\Gamma}_y^{-1}(0)\boldsymbol{\Gamma}'_y(1)\boldsymbol{\Gamma}_y^{-1}(0)\boldsymbol{\Gamma}_y(1).$$

This matrix is the standard canonical correlation matrix whose eigenvalues are the canonical correlations between $\boldsymbol{y}_t$ and $\boldsymbol{y}_{t-1}$. A zero canonical correlation defines a linear combination that is white noise and a close to one canonical correlation defines a close to nonstationary component.

28.2.3 Reduced rank models

An alternative procedure for finding linear combinations with useful properties for model simplification are the reduced rank models; see Robinson (1973), Ahn and Reinsel (1990), Reinsel and Ahn (1992), and Reinsel and Velu (1998). Suppose for simplicity that a vector of time series is fitted by the VAR(1) model (28.6) and suppose that $\boldsymbol{\Phi} = \boldsymbol{A}_r\boldsymbol{B}_r$, where $\boldsymbol{A}_r$ is a full rank matrix of dimension $m \times r$, $(m > r)$, and $\boldsymbol{B}_r$ is also full rank with dimension $r \times m$. Denoting $\boldsymbol{z}_{t-1} = \boldsymbol{B}_r\boldsymbol{y}_{t-1}$, the model for the series can be written as

$$\boldsymbol{y}_t = \boldsymbol{A}_r\boldsymbol{z}_{t-1} + \boldsymbol{a}_t \tag{28.7}$$

and also, as $\boldsymbol{B}_r\boldsymbol{y}_t = \boldsymbol{B}_r\boldsymbol{A}_r\boldsymbol{z}_{t-1} + \boldsymbol{B}_r\boldsymbol{a}_t$, we have

$$\boldsymbol{z}_t = \boldsymbol{C}\boldsymbol{z}_{t-1} + \boldsymbol{u}_t \tag{28.8}$$

where $\boldsymbol{C} = \boldsymbol{B}_r\boldsymbol{A}_r$ is a $r \times r$ matrix and $\boldsymbol{u}_t = \boldsymbol{B}_r\boldsymbol{a}_t$. This is like a factor model with r factors $\boldsymbol{z}_{t-1}$ that follow an AR(1) model. An important implication from this model is that there exist $m - r$ linear combinations which are white noise. Denoting $\boldsymbol{A}_{m-r,\perp}$ for the orthogonal complement of $\boldsymbol{A}_r$, defined as the $m \times (m - r)$ matrix such that

$$\boldsymbol{A}'_{m-r,\perp}\boldsymbol{A}_r = \mathbf{0},$$

the $m-r$ linear combinations $\boldsymbol{A}'_{m-r,} \boldsymbol{y}_t$ are white noise, or, in other words, there must be $m-r$ zero canonical correlations between $\boldsymbol{y}_t$ and $\boldsymbol{y}_{t-1}$. These ideas can be generalized to general VAR(p) models. We can write

$$\boldsymbol{y}_t = \boldsymbol{F}\boldsymbol{y}_t^* + \boldsymbol{a}_t,$$

where $\boldsymbol{y}_t^* = (\boldsymbol{y}'_{t-1}, \ldots, \boldsymbol{y}'_{t-k})'$ and $\boldsymbol{F} = (\Phi_1, \ldots, \Phi_k)$. Then, as before, if $\boldsymbol{F}$ has reduced rank, $\boldsymbol{F} = \boldsymbol{A}_r\boldsymbol{B}_r$, we have

$$\boldsymbol{y}_t = \boldsymbol{A}_r\boldsymbol{z}_t + \boldsymbol{a}_t,$$

where $\boldsymbol{z}_t = \boldsymbol{B}_r\ \boldsymbol{y}_t$. The implication of this model is that the canonical correlations between $\boldsymbol{y}_t$ and $\boldsymbol{y}_t^*$ will have as many zero canonical correlations as white noise combinations. Also, it can be shown that the number of canonical correlations equal to one is the number of nonstationary linear combinations of the vector.

28.2.4 The scalar component models

Tiao and Tsay (1989) presented the concept of scalar component models as simplifying tools in VARMA models. A scalar component model is a linear combination of the vector time series that follows a simpler structure than the vector itself. These authors define SCM as follows. Assume that we can write $\boldsymbol{y}_t = \sum \boldsymbol{\Psi}_i\boldsymbol{a}_{t-i}$, where $\boldsymbol{a}_t$ is white noise. We will say that $x_t = \boldsymbol{v}'_0\boldsymbol{y}_t$ follows a SCM($p_{1,}q_1$) if there exist p_1 vectors $m \times 1$, $\boldsymbol{v}_1, \ldots, \boldsymbol{v}_{p_1}$ such that (i) $\boldsymbol{v}_{p_1}$ is nonzero when $p_1 > 0$, and (ii) the linear combination of $\boldsymbol{y}_t, \boldsymbol{y}_{t-1}, \ldots, \boldsymbol{y}_{t-p_1}$ given by $m_t = \boldsymbol{v}'_0\boldsymbol{y}_t + \sum_{l=1}^{p_1} \boldsymbol{v}'_l\boldsymbol{y}_{t-l}$ satisfies

$$E(\boldsymbol{a}_{t-j}m_t)\begin{cases} \neq 0 & \text{if } j = q_1 \\ = 0 & \text{if } j > q_1 \end{cases}.$$

The above definition implies the following restriction among the autocovariance matrices of $\boldsymbol{y}_t$:

$$\boldsymbol{\Gamma}_y(k)\boldsymbol{v}_0 + \boldsymbol{\Gamma}_y(k-1)\boldsymbol{v}_1 + \cdots + \boldsymbol{\Gamma}_y(k-p_1)\boldsymbol{v}_{p_1} = \boldsymbol{0}, \text{ for } l > q_1. \tag{28.9}$$

Of particular interest are SCM(0,0), which are white noise, and SCM(1,0), which can define a particular type of common trends. See Peña, Tiao and Tsay (2001) for a simple introduction to the use of SCM for model simplification. To find out the number of scalar component models, Tiao and Tsay (1989) proposed a chi-square test based on canonical correlation ideas for the rank of extended second moment matrices, which will be discussed in Section 28.3.

28.2.5 Dynamic factor models

A generalization of the idea of linear combinations with useful properties is the dynamic factor model. In this model, the m-dimensional vector of an observed time series is generated by a set of r nonobserved common factors and m specific components as follows:

$$\underset{m \times 1}{\boldsymbol{y}_t} = \underset{m \times r}{\boldsymbol{P}} \; \underset{r \times 1}{\boldsymbol{f}_t} + \underset{m \times 1}{\boldsymbol{n}_t}, \tag{28.10}$$

where $\boldsymbol{f}_t$ is the r-dimensional vector of common factors, $\boldsymbol{P}$ is the factor loading matrix, and $\boldsymbol{n}_t$ is the vector of specific components. Thus, all the common dynamic structure comes through the common factors, $\boldsymbol{f}_t$, whereas the vector $\boldsymbol{n}_t$ explains the specific dynamics for each component. If there is no specific dynamic structure, $\boldsymbol{n}_t$ is reduced to white noise. We assume linear time series models for the latent variable $\boldsymbol{f}_t$ and the noise $\boldsymbol{n}_t$. In particular, using the VARIMA(p, d, q) representation, the latent variable will be given by

$$\underset{r \times r}{\boldsymbol{\Phi}(B)} \; \underset{r \times 1}{\boldsymbol{f}_t} = \underset{r \times r}{\boldsymbol{\Theta}(B)} \; \underset{r \times 1}{\boldsymbol{a}_t}, \tag{28.11}$$

where B is the backshift operator, such that $B\boldsymbol{y}_t = \boldsymbol{y}_{t-1}$, and (i) the $r \times r$ matrix $\boldsymbol{\Phi}(B) = \boldsymbol{I} - \boldsymbol{\Phi}_1 B - \cdots - \boldsymbol{\Phi}_p B^p$ has the roots of the determinantal equation$|\boldsymbol{\Phi}(B)| = 0$ on or outside the unit circle; (ii) the $r \times r$ matrix $\boldsymbol{\Theta}(B) = \boldsymbol{I} - \boldsymbol{\Theta}_1 B - \cdots - \boldsymbol{\Theta}_q B^q$ has the roots of the determinantal equation$|\boldsymbol{\Theta}(B)| = 0$ outside the unit circle; and (iii) $\boldsymbol{a}_t \sim N_r(\boldsymbol{0}, \boldsymbol{\Sigma}_a)$ is serially uncorrelated, $\mathrm{E}(\boldsymbol{a}_t \boldsymbol{a}'_{t-h}) = \boldsymbol{0}$, $h \neq 0$. The noise, $\boldsymbol{n}_t$, also follows the VARMA model

$$\boldsymbol{\Phi}_n(B)\boldsymbol{n}_t = \boldsymbol{\Theta}_n(B)\boldsymbol{\varepsilon}_t, \tag{28.12}$$

where $\boldsymbol{\Phi}_n(B)$ and $\boldsymbol{\Theta}_n(B)$ are $m \times m$ diagonal matrices with $\boldsymbol{\Phi}_n(B) = \boldsymbol{I} - \boldsymbol{\Phi}_{n1} B - \cdots - \boldsymbol{\Phi}_{np} B^p$ and $\boldsymbol{\Theta}_n(B) = \boldsymbol{I} - \boldsymbol{\Theta}_{n1} B - \cdots - \boldsymbol{\Theta}_{nq} B^q$. The most interesting case is when the specific component is stationary so that the possible nonstationary dynamic structure in the vector of time series is due to the common factors. In this case the roots of the determinantal equations $|\boldsymbol{\Phi}_n(B)| = 0$ and $|\boldsymbol{\Theta}_n(B)| = 0$ are outside the unit circle. Therefore, each component follows a univariate ARMA(p_i, q_i), $i = 1, 2, \ldots, m$, being $p = \max(p_i)$ and $q = \max(q_i)$, $i = 1, 2, \ldots, m$. The sequence of vectors $\boldsymbol{\varepsilon}_t$ are normally distributed, with zero mean and diagonal covariance matrix $\boldsymbol{\Sigma}_\varepsilon$. We assume that the noises from the common factors and specific components are also uncorrelated for all lags, that is, $\forall h\ \mathrm{E}(\boldsymbol{a}_t \boldsymbol{\varepsilon}'_{t-h}) = \boldsymbol{0}.$ When $\boldsymbol{n}_t$ is white noise and the factors are stationary, models (28.10) and (28.11) are the factor model studied by Peña and Box (1987). The model as stated is not identified and we can choose either $\boldsymbol{\Sigma}_{\boldsymbol{a}} = \boldsymbol{I}$ or $\boldsymbol{P}'\boldsymbol{P} = \boldsymbol{I},$ although the model is not yet identified under rotations. Harvey (1989) imposes the additional condition that $p_{ij} = 0$ for $j > i$, where $\boldsymbol{P} = [p_{ij}]$.

Note that this factor model is very general and includes other formulations presented in the literature. For instance, Molenaar *et al.* (1992) have proposed a model of the form

$$\boldsymbol{y}_t = \sum_{i=0}^{s} \boldsymbol{P}_i \boldsymbol{f}_{t-i} + \boldsymbol{n}_t.$$

Letting $\boldsymbol{f}_t^* = (\boldsymbol{I} + \boldsymbol{P}_0^{-1}\boldsymbol{P}_1 B + \cdots + \boldsymbol{P}_0^{-1}\boldsymbol{P}_s B^s)\boldsymbol{f}_t = \varphi(B)\boldsymbol{f}_t$, we can write this model as $\boldsymbol{y}_t = \boldsymbol{P}_0\boldsymbol{f}_t^* + \boldsymbol{\varepsilon}_t$ where the new factors follow a different VARMA model. The factor model has an interesting implication in terms of canonical correlation. Suppose that there is no specific components so that the model is

$$\boldsymbol{y}_t = \boldsymbol{P}\boldsymbol{f}_t + \boldsymbol{\varepsilon}_t;$$

then, denoting $\boldsymbol{P}'_{\perp}$ for the $(m-r) \times m$ matrix which defines the null space of $\boldsymbol{P}$, such that $\boldsymbol{P}'_{\perp}\boldsymbol{P} = \boldsymbol{0}$, we have

$$\boldsymbol{P}'_{\perp}\boldsymbol{y}_t = \boldsymbol{P}'_{\perp}\boldsymbol{\varepsilon}_t$$

and there must be $m - r$ zero canonical correlations between $\boldsymbol{y}_t$ and $\boldsymbol{y}_t^*$.

28.2.6 State space models

State space models have been studied by Akaike (1974), Aoki (1987), Hannan and Deistler (1988), Harvey (1989), and Durbin and Koopman (2001), among others. They are defined by a measurement equation

$$\boldsymbol{y}_t = \boldsymbol{C}\boldsymbol{z}_t + \boldsymbol{\varepsilon}_t,$$

where $\boldsymbol{C}$ is $m\times\ s$, $\boldsymbol{z}_t$ is the $s \times 1$ state vector and $\boldsymbol{\varepsilon}_t$, $m \times 1$, is the innovation vector with $\mathrm{E}(\boldsymbol{\varepsilon}_t) = 0$, $\mathrm{E}(\boldsymbol{\varepsilon}_t\boldsymbol{\varepsilon}'_t) = \boldsymbol{\Sigma}_\epsilon$ $\mathrm{E}(\boldsymbol{\varepsilon}_t\boldsymbol{\varepsilon}'_\tau) = \boldsymbol{0}$ if $t \neq \tau$. The transition equation is

$$\boldsymbol{z}_t = \boldsymbol{G}\boldsymbol{z}_{t-1} + \boldsymbol{u}_t$$

with $\mathrm{E}(\boldsymbol{u}_t) = \boldsymbol{0}$, $\mathrm{E}(\boldsymbol{u}_t\boldsymbol{u}'_t) = \boldsymbol{\Sigma}_u$ and $\mathrm{E}(\boldsymbol{u}_t\boldsymbol{u}'_\tau) = \boldsymbol{0}$ if $t \neq \tau$. Although any VARMA model can be written in the state space form and we can always obtain the VARMA form of a state space representation, the state space formulation has the advantage of being defined in terms of the state vector which is the key component for dimension reduction. In fact, Akaike (1974) introduced canonical correlation in time series in order to find the dimension of the state space vector. For instance, we may have a dynamic factor model by

$$\boldsymbol{y}_t = \boldsymbol{C}\boldsymbol{z}_t + \boldsymbol{\varepsilon}_t,$$

where $\boldsymbol{C}$ is $m \times r$ and

$$\boldsymbol{z}_t = \boldsymbol{z}_{t-1} + \boldsymbol{\beta} + \boldsymbol{u}_t.$$

This is the common trends model because the vector of dimension m is generated by r factors that follow a random walk with a drift model. Note that the state vector coincides with the factor. The VARMA form of this model is

$$\nabla \boldsymbol{y}_t = \boldsymbol{C}(\boldsymbol{\beta} + \boldsymbol{u}_t) + \nabla \boldsymbol{\varepsilon}_t = \boldsymbol{c} + (\boldsymbol{I} - \boldsymbol{\Theta} B)\boldsymbol{a}_t$$

and the observed series will follow a VARIMA(1,1). However, in this formulation, the factor is completely lost and, as shown by Peña and Box (1987), fitting an ARIMA model to an observed time series generated from this model may be a difficult task because of the lack of identification of the parameter matrices. An additional advantage of the state space approach is that it allows for dimension reduction in some of the time series components and not in the others. See Casals *et al.* (2002) for useful structural decompositions in the state space approach. Suppose that the state vector is written as including the trend and the cycle of the time series as

$$\boldsymbol{y}_t = \boldsymbol{A}\boldsymbol{T}_t + \boldsymbol{B}\boldsymbol{s}_t + \boldsymbol{\varepsilon}_t,$$

where $\boldsymbol{A}$ is $m \times r$ and $\boldsymbol{B}$ is $m \times c$ where $r \leq m$ and $c \leq m$. Then, if $\boldsymbol{A}'_{m-r,\perp}\boldsymbol{A} = \mathbf{0}$ and $\boldsymbol{B}'_{m-c,\perp}\boldsymbol{B} = \mathbf{0},$ we have

$$\boldsymbol{A}'_{m-r,\perp}\boldsymbol{y}_t = \boldsymbol{A}'_{m-r,\perp}\boldsymbol{B}\boldsymbol{s}_t + \boldsymbol{A}'_{m-r,\perp}\boldsymbol{\varepsilon}_t$$

and

$$\boldsymbol{B}'_{m-c,\perp}\boldsymbol{y}_t = \boldsymbol{B}'_{m-c,\perp}\boldsymbol{A}\boldsymbol{T}_t + \boldsymbol{B}'_{m-c,\perp}\boldsymbol{\varepsilon}_t,$$

and we may have some linear combinations free from the trend and others free from the cycle. It could be that some of them are white noise if there are common vectors in the null space of the matrices $\boldsymbol{A}$ and $\boldsymbol{B}$.

28.2.7 Some conclusions

We have seen that canonical analysis plays a key role in all of the dimension reduction procedures for a time series. If $h \geq 1$ linear combinations are white noise, there is only dynamics in $m - h$ dimensions and this implies h zero canonical correlations between $\boldsymbol{y}_t$ and $\boldsymbol{y}_t^*$. Also, for integrated processes, an important simplification tool is finding linear combinations which are stationary. If there is cointegration and $h \geq 1$ linear combinations are stationary, then $m - h$ canonical correlations between $\boldsymbol{y}_t$ and $\boldsymbol{y}_t^*$ will be equal to one. It is interesting to understand the relationship between canonical analysis and principal components in time series. We have shown that when $\boldsymbol{\Sigma} = \sigma^2\boldsymbol{I},$ principal components and canonical analysis leads to similar conclusions. This is similar to the relationship between factor analysis and principal components in the static case. If the specific innovations of all the time series have the same variance and are uncorrelated, then a zero eigenvalue in the canonical correlation analysis of the series and its past values will be equivalent to an eigenvalue equal to one in the standardized principal components (SPC) of the matrix $\sigma^{-2}\mathbf{\Gamma}_y(0)$. Also a canonical correlation close to one will be equivalent to a large eigenvalue in the SPC. In practice, with nonstationary time series, the elements of $\mathbf{\Gamma}_y(0)$ will be much larger than those of $\mathbf{\Sigma}$, and this matrix often has diagonal elements of similar sizes and larger than the off-diagonal elements. In this case, the principal component matrix $\mathbf{\Gamma}_y(0)$ will be similar to the canonical correlation matrix $\mathbf{\Sigma}^{-1}\mathbf{\Gamma}_y(0)$, and both approaches will lead to similar results when applied for finding the cointegration or the factor space.

28.3 Dimension Reduction Tests

We present in this section tests for dimension reduction based on canonical correlation coefficients. Other related tests are the principal components tests by Stock and Watson (1988) and Harris (1997). An alternative way to decide about the dimension of the system is by using model selection criteria, such as AIC, BIC, and others. The relative advantages of these two approaches require more research before a clear recommendation can be made.

28.3.1 A test for zero canonical correlation coefficients

Let $\boldsymbol{y}_t^* = (\boldsymbol{y}_{t-1}', \ldots, \boldsymbol{y}_{t-k}')'$ be a $km \times 1$ vector of lag values of the series. We want to test if there exist linear combinations of $\boldsymbol{y}_t$ that are uncorrelated to linear combinations of $\boldsymbol{y}_t^*$ or, in other words, if there are zero canonical correlation coefficients between the two sets of variables. This test will allow us to find the least predictable components in the canonical analysis of Box-Tiao, the rank r in the reduced rank model, and can also be used to test for the number of factors in the dynamic factor model. Suppose that the null hypothesis is that there are h zero canonical coefficients. Note that if we accept the presence of h zero coefficients we must accept the presence of $h-1$. Thus, the test must be done sequentially starting with $h = 0$ and increasing h until $m-1$. The alternative hypothesis will be that there are less than h zero canonical correlation coefficients, and the test is: $H_0 : h$ $(h = 0, 1, \ldots, m-1)$ zero correlation coefficients versus H_1 : less than h zero correlation coefficients. The standard multivariate test for h zero canonical correlation coefficients is

$$L = -\{(T - mk) + g(m,k)\} \sum_{j=1}^{h} \log(1 - \widehat{\lambda}_j), \tag{28.13}$$

where $g(m,k) = (mk - m - 1)/2$ is a correction factor to improve the asymptotic distribution of the test statistic and $\widehat{\lambda}_1 \leq \widehat{\lambda}_2 \leq \cdots \leq \widehat{\lambda}_j \leq \cdots \leq \widehat{\lambda}_m$ are the ordered eigenvalues of

$$\widehat{\boldsymbol{M}}_k = \boldsymbol{C}_y^{-1}\boldsymbol{C}_{yy^*}\boldsymbol{C}_{y^*}^{-1}\boldsymbol{C}_{y^*y}, \tag{28.14}$$

where

$$\boldsymbol{C}_y = T^{-1}\sum_{t=2}^{T}(\boldsymbol{y}_t\boldsymbol{y}_t'),$$

$$\boldsymbol{C}_{yy^*} = T^{-1}\sum_{t=k}^{T}(\boldsymbol{y}_t\boldsymbol{y}_t^{*\prime}),$$

$$\boldsymbol{C}_{y^*} = T^{-1}\sum_{t=k}^{T}(\boldsymbol{y}_t^*\boldsymbol{y}_t^{*\prime}),$$

and L it is distributed asymptotically as $\chi^2_{h(mk-(m-h))}$. This test can be derived as a likelihood ratio test [see, e.g., Rechner (1995)]. It has a simple interpretation as a Box-Pierce test on the canonical correlation coefficients as under the null

$$L \simeq T\sum_{j=1}^{m-h}\widehat{\lambda}_j = T\sum_{j=1}^{m-h}\widehat{\rho}_j^2$$

where $\widehat{\rho}_j^2$ are the canonical correlations. This test has been used in reduced rank models [see Reinsel and Velu (1998)] to test for the dimension of the reduced rank matrix.

A modification of the previous test was proposed by Tiao and Tsay (1989) in order to test for SCM. Let $\boldsymbol{Y}^*_{h,t-j-1} = (\boldsymbol{y}'_{t-j-1},\ldots,\boldsymbol{y}'_{t-h-j-1})'$ and $\boldsymbol{Y}_{k,t} = (\boldsymbol{y}'_t,\ldots,\boldsymbol{y}'_{t-k})'$be $(h+1)m\times 1$ and $(k+1)m\times 1$, respectively, vectors of lag values of the series for $h \geq k \geq j \geq 0$. The purpose is to test the number of zero eigenvalues or zero canonical correlations between $\boldsymbol{Y}^*_{h,t-j-1}$ and $\boldsymbol{Y}_{k,t}$ that is determined by the rank of the lag second moment matrices $\boldsymbol{y}_t$ and the Yule-Walker equations of the overall process for $\boldsymbol{y}_t$. The test statistic is

$$TT = -(T-h-j)\sum_{j=i}^{s}\log\left(1-\frac{\widehat{\lambda}_j}{d_j}\right), \tag{28.15}$$

where s is the number of zero canonical correlations between $\boldsymbol{Y}^*_{h,t-j-1}$ and $\boldsymbol{Y}_{k,t}$ and $d_j/(T-h-j)$ is the sample variance of the two canonical variates whose sample canonical correlation is given by $\widehat{\lambda}_j$. Under the null hypothesis of s zero canonical correlations, the test statistic follows a chi-squared with $s((h-k)\times m+s)$ degrees of freedom.

28.3.2 A nonstandard test for canonical correlations

Suppose that in an $I(1)$ process we are interested in finding the number of nonstationary dimensions r or the number of independent linear combinations which are stationary, $m-r$, which is the cointegration dimension. We say that the components of a nonstationary $I(d)$ time series vector $\boldsymbol{y}_t$ are cointegrated if there exists a linear combination of them that is $I(d-b)$, where $b>0$, $d \geq b$ and d and b belong to the set of the natural numbers. The most interesting case is when the series are $I(1)$ but some linear combinations are $I(0)$ or stationary. A cointegration test in this case tries to determine how many independent linear combinations of the series can be considered as stationary. To simplify the exposition, suppose the VAR(1) given by (28.6). If all the roots of $|\boldsymbol{I} - \boldsymbol{\Phi}B| = 0$ are equal to one, all the eigenvalues of the matrix $\boldsymbol{\Phi}$ are equal to one and all the eigenvalues of the matrix

$$\boldsymbol{\Pi} = \boldsymbol{\Phi} - \boldsymbol{I}$$

are equal to zero. Note that this does not imply that $\boldsymbol{\Pi}$ is a zero matrix because it may not be symmetric. If the series are stationary, all the roots of $|\boldsymbol{I} - \boldsymbol{\Phi}B| = 0$ are inside the unit circle and the matrix $\boldsymbol{\Pi}$ is a full rank matrix. Cointegration represents the intermediate situation in which the series are nonstationary, but some linear combinations are stationary. Suppose that the matrix $\boldsymbol{\Phi}$ has r eigenvalues equal to one, or, equivalently, the matrix $\boldsymbol{\Pi}$ has r eigenvalues equal to zero. These properties can be applied to the error correction formulation of the VAR(1) obtained subtracting $\boldsymbol{y}_{t-1}$ from both sides of (28.6). Then

$$\nabla \boldsymbol{y}_t = \boldsymbol{\Pi} \boldsymbol{y}_{t-1} + \varepsilon_t. \tag{28.16}$$

If all the series are nonstationary, but there is no cointegration, $\boldsymbol{\Pi}$ is a null rank matrix; if all of them are stationary, $\boldsymbol{\Pi}$ is a full rank matrix and if there is cointegration the matrix $\boldsymbol{\Pi}$ must be rank deficient. Then, if $rank(\boldsymbol{\Pi}) = m-r$, we can write

$$\boldsymbol{\Pi} = \boldsymbol{A}_{m-r} \boldsymbol{B}_{m-r}$$

and the r linear combinations

$$\boldsymbol{A}'_{r,\perp} \nabla \boldsymbol{y}_t = \boldsymbol{A}'_{r,\perp} \varepsilon_t \tag{28.17}$$

must be white noise. Note that the cointegration relations are given by $\boldsymbol{z}_t = \boldsymbol{B}_{m-r} \boldsymbol{y}_t$. To see this, multiplying (28.16) by $\boldsymbol{B}_{m-r}$, we have

$$\nabla \boldsymbol{z}_t = \boldsymbol{B}_{m-r} \boldsymbol{A}_{m-r} \boldsymbol{z}_{t-1} + \boldsymbol{B}_{m-r} \varepsilon_t$$

and as $\boldsymbol{B}_{m-r} \boldsymbol{A}_{m-r}$ is a squared full rank matrix of dimension $m-r$, $\boldsymbol{z}_t$ must be stationary. We may build a test of cointegration by searching for zero canonical correlations between $\nabla \boldsymbol{y}_t$ and $\boldsymbol{y}_{t-1}$. Let $0 \leq \widehat{\lambda}_1 \leq \cdots \leq \widehat{\lambda}_m \leq 1$ be the eigenvalues of the matrix

$$\boldsymbol{M}_2 = \boldsymbol{S}_{11}^{-1}\boldsymbol{S}_{10}\boldsymbol{S}_{00}^{-1}\boldsymbol{S}_{01},$$

where

$$\boldsymbol{S}_{11} = T^{-1}\sum_{t=1}^{T}\nabla\boldsymbol{y}_t\nabla\boldsymbol{y}_t',$$

$$\boldsymbol{S}_{10} = T^{-1}\sum_{t=1}^{T}\nabla\boldsymbol{y}_t\boldsymbol{y}_{t-1}',$$

$$\boldsymbol{S}_{00} = T^{-1}\sum_{t=1}^{T}\boldsymbol{y}_{t-1}\boldsymbol{y}_{t-1}'.$$

Then the statistic for testing that there are r zero canonical correlations, or $m-r$ cointegration relations, is

$$L_{m-r} = -T\sum_{j=1}^{r}\log(1-\widehat{\lambda}_j). \tag{28.18}$$

This is the cointegration test for $I(1)$ variables developed by Johansen (1991, 1995) for VAR processes, which has become very popular in econometrics. The distribution of the test is nonstandard because although the linear combinations $\boldsymbol{A}_{r,\perp}'\nabla\boldsymbol{y}_t$ are white noise and uncorrelated to $\boldsymbol{z}_{t-1} = \boldsymbol{B}_{m-r}\boldsymbol{y}_{t-1}$, these linear combinations are not white noise. The percentiles of the distribution have been tabulated by simulation. Note that we could also test for zero canonical correlations between $\nabla\boldsymbol{y}_t$ and $\nabla\boldsymbol{y}_{t-1}$, $\nabla\boldsymbol{y}_{t-2}, \ldots$ since by (28.17) there are r linear combinations of $\nabla\boldsymbol{y}_t$ that are white noise. For instance, if we want to search for zero canonical correlations between $\nabla\boldsymbol{y}_t$ and its first lag $\nabla\boldsymbol{y}_{t-1}$, we will find zero canonical correlations between each sample and also within each sample of the variables due to (28.17). In this particular case, the asymptotic distribution of the test statistic is χ^2 since under the null hypothesis the smallest r canonical variates are white noise.

The generalization of the test for VAR(p) is straightforward. Suppose

$$\boldsymbol{y}_t = \boldsymbol{\Phi}_1\boldsymbol{y}_{t-1} + \cdots + \boldsymbol{\Phi}_p\boldsymbol{y}_{t-p} + \boldsymbol{\varepsilon}_t,$$

where $\boldsymbol{\varepsilon}_t \sim N(\boldsymbol{0},\boldsymbol{\Sigma})$. The process is nonstationary if some of the roots of the determinantal equation $|\boldsymbol{\Phi}(\boldsymbol{B})| = 0$ are on the unit circle, which implies that the matrix $\boldsymbol{I}-\sum_{i=1}^{p}\boldsymbol{\Phi}_i = -\boldsymbol{\Pi}$ is rank deficient. In order to use this property, we write the VAR model in the error correction form

$$\nabla\boldsymbol{y}_t = \boldsymbol{\Pi}\boldsymbol{y}_{t-1} + \sum_{i=1}^{p-1}\boldsymbol{\Gamma}_i\nabla\boldsymbol{y}_{t-i} + \boldsymbol{\varepsilon}_t, \tag{28.19}$$

where

$$\boldsymbol{\Pi}=\sum_{i=1}^{p}\boldsymbol{\Phi}_i-\boldsymbol{I},\text{ and }\boldsymbol{\Gamma}_i=\sum_{j=i+1}^{p}\boldsymbol{\Phi}_j. \tag{28.20}$$

Then, if rank($\boldsymbol{\Pi}$)$= m-r$, this matrix can be written as $\boldsymbol{\Pi}=\boldsymbol{A}_{m-r}\boldsymbol{B}_{m-r}$ and there will be $m-r$ cointegration relationships and r zero canonical correlations between $\nabla\boldsymbol{y}_t^*=\nabla\boldsymbol{y}_t-\sum_{i=1}^{p-1}\boldsymbol{\Gamma}_i\nabla\boldsymbol{y}_{t-i}$ and $\boldsymbol{y}_{t-1}^*=\boldsymbol{y}_{t-1}-\sum_{i=1}^{p-1}\boldsymbol{\Gamma}_i\nabla\boldsymbol{y}_{t-i}$. Note that, by (28.19), the r linear combinations

$$\nabla\boldsymbol{A}_{r,\perp}'(\boldsymbol{y}_t-\sum_{i=1}^{p-1}\boldsymbol{\Gamma}_i\nabla\boldsymbol{y}_{t-i})=\boldsymbol{A}_{r\perp}'\boldsymbol{\varepsilon}_t$$

are white noise, where $\boldsymbol{A}_{r,\perp}$ is the orthogonal complement of $\boldsymbol{A}_{m-r}$, that is $\boldsymbol{A}_{r,\perp}'\boldsymbol{A}_{m-r}=\boldsymbol{0}$. The $m-r$ linear combinations given by $\boldsymbol{B}_{m-r}\boldsymbol{y}_t$ are I(0). Thus, the test uses the residuals of a regression of $\nabla\boldsymbol{y}_t$ and $\boldsymbol{y}_{t-1}$ on the lags of the first differences and then looks at the canonical correlation between these two sets or residuals. As before, the test is done sequentially assuming 0 cointegration relations at the initial stage and going up to $m-1$ cointegration relations. The (nonstandard) critical values can be taken from Johansen (1995). Reinsel and Ahn (1992) have proposed a similar test for the number of unit roots in reduced rank autoregression models.

It is interesting to analyze this test when is applied to the dynamic factor model. Assuming that the factors are integrated with $d=1$, and follows the model

$$\underset{r\times r}{(1-B)\boldsymbol{\Phi}^*(B)}\quad\underset{r\times 1}{\boldsymbol{f}_t}\quad=\quad\underset{r\times r}{\boldsymbol{\Theta}(B)}\quad\underset{r\times 1}{\boldsymbol{a}_t}, \tag{28.21}$$

with $\boldsymbol{\Phi}^*(B)$ having all its roots outside the unit circle. Then

$$\boldsymbol{f}_t=\boldsymbol{f}_{t-1}+(\boldsymbol{\Phi}^*(B))^{-1}\boldsymbol{\Theta}(B)\quad\boldsymbol{a}_t. \tag{28.22}$$

From (28.10), we obtain

$$\boldsymbol{f}_t=\boldsymbol{P}^+(\boldsymbol{y}_t-\boldsymbol{n}_t), \tag{28.23}$$

where $\boldsymbol{P}^+=(\boldsymbol{P}'\boldsymbol{P})^{-1}\boldsymbol{P}$, $r\times m$, is the Moore-Penrose inverse matrix of $\boldsymbol{P}$, and from (28.10), (28.23) and (28.22) we can write

$$\boldsymbol{y}_t=\boldsymbol{P}\boldsymbol{P}^+(\boldsymbol{y}_{t-1}-\boldsymbol{n}_{t-1})+\boldsymbol{P}\left(\boldsymbol{\Phi}^*(B)\right)^{-1}\boldsymbol{\Theta}(B)\boldsymbol{a}_t+\boldsymbol{n}_t$$

and subtracting $\boldsymbol{y}_{t-1}$, we have

$$(1-B)\boldsymbol{y}_t=-(\boldsymbol{I}-\boldsymbol{P}\boldsymbol{P}^+)\boldsymbol{y}_{t-1}+\boldsymbol{P}\left(\boldsymbol{\Phi}^*(B)\right)^{-1}\boldsymbol{\Theta}(B)\boldsymbol{a}_t+\boldsymbol{n}_t-\boldsymbol{P}\boldsymbol{P}^+\boldsymbol{n}_{t-1}. \tag{28.24}$$

This is the error correction form implied by the factor model. Notice now that $\boldsymbol{P}\boldsymbol{P}^+=\boldsymbol{P}(\boldsymbol{P}'\boldsymbol{P})^{-1}\boldsymbol{P}'$ is a projection matrix, such that rank($\boldsymbol{P}\boldsymbol{P}^+$) $=r$ and it has all its eigenvalues equal one or zero since it is an idempotent matrix. Therefore, rank $(\boldsymbol{I}-\boldsymbol{P}\boldsymbol{P}^{+})=m-r$. The matrix $(\boldsymbol{I}-\boldsymbol{P}\boldsymbol{P}^{+})$ plays the role of the $\boldsymbol{\Pi}=\boldsymbol{A}\boldsymbol{B}$ matrix in the cointegration analysis and the test of r common factors is equivalent to the test of $m-r$ cointegration relations. However, in order to use Johansen's cointegration test, we have to assume that the process followed by $\boldsymbol{y}_t$ can be approximated by an unrestricted VAR. When the true model is the dynamic factor model usually we also have MA structure.

28.3.3 A canonical correlation test for factor models

Canonical correlation tests for factor models have been proposed by Hu and Chou (2004) and Peña and Poncela (2006). In this subsection, we review the latest one. Suppose the factor model without specific components is

$$\boldsymbol{y}_t=\boldsymbol{P}\boldsymbol{f}_t+\varepsilon_t. \tag{28.25}$$

Then, as shown by Peña and Box (1987), denoting $\boldsymbol{\Gamma}_f(k)$ for the covariance matrix of order k of the factors and assuming stationarity we have, for $k\neq 0$

$$\boldsymbol{\Gamma}_y(k)=\boldsymbol{P}\boldsymbol{\Gamma}_f(k)\boldsymbol{P}' \tag{28.26}$$

and $\text{rank}(\boldsymbol{\Gamma}_y(k))=\text{rank}(\boldsymbol{\Gamma}_f(k))$. Because (28.26) is true for all $k\neq 0$, there exists a $m\times(m-r)$ matrix $\boldsymbol{P}_\perp$, such that for all $k\neq 0$,

$$\boldsymbol{\Gamma}_y(k)\boldsymbol{P}_\perp=\boldsymbol{P}\boldsymbol{\Gamma}_f(k)\boldsymbol{P}'\boldsymbol{P}_\perp=\boldsymbol{0}. \tag{28.27}$$

The condition in (28.27) also implies that the $m-r$ independent linear combinations of the observed series given by $\boldsymbol{P}'_\perp\boldsymbol{y}_t$ are cross and serially uncorrelated for all lags $k\neq 0$. Therefore, the number of zero canonical correlations between $\boldsymbol{y}_{t-k}$ and $\boldsymbol{y}_t$ is given by the number of zero eigenvalues of the matrix $\boldsymbol{M}(k)$ defined as

$$\boldsymbol{M}(k)=\left[E(\boldsymbol{y}_t\boldsymbol{y}_t')\right]^{-1}E(\boldsymbol{y}_t\boldsymbol{y}_{t-k}')\left[E(\boldsymbol{y}_{t-k}\boldsymbol{y}_{t-k}')\right]^{-1}E(\boldsymbol{y}_{t-k}\boldsymbol{y}_t') \tag{28.28}$$

and since $\text{rank}(\boldsymbol{M}(k))=\text{rank}(\boldsymbol{\Gamma}_y(k))=r$, this number is $m-r$. Thus, the number of common factors, r, is equivalent to the number of nonzero canonical correlations between $\boldsymbol{y}_{t-k}$ and $\boldsymbol{y}_t$.

Consider now the finite sample case in which T observations are available. The squared sample canonical correlations between $\boldsymbol{y}_{t-k}$ and $\boldsymbol{y}_t$ are the eigenvalues of

$$\widehat{\boldsymbol{M}}_1(k)=\left[\sum_{t=k+1}^{T}(\boldsymbol{y}_t\boldsymbol{y}_t')\right]^{-1}\sum_{t=k+1}^{T}(\boldsymbol{y}_t\boldsymbol{y}_{t-k}')\left[\sum_{t=k+1}^{T}(\boldsymbol{y}_{t-k}\boldsymbol{y}_{t-k}')\right]^{-1}\sum_{t=k+1}^{T}(\boldsymbol{y}_{t-k}\boldsymbol{y}_t'). \tag{28.29}$$

In Peña and Poncela (2006), it has been shown that, given $\widehat{\lambda}_1 < \widehat{\lambda}_2 \leq \cdots \leq \widehat{\lambda}_m$, the ordered eigenvalues of the matrix $\widehat{\boldsymbol{M}}_1(k)$ given by (28.29), the statistic

$$S_{m-r} = -(T-k) \sum_{j=1}^{m-r} \log(1 - \widehat{\lambda}_j) \tag{28.30}$$

is asymptotically a $\chi^2_{(m-r)^2}$, both for stationary and nonstationary series. Note that we obtain standard distribution because: (1) $\boldsymbol{P}'_{\perp}\boldsymbol{y}_t$ and $\boldsymbol{P}'_{\perp}\boldsymbol{y}_{t-1}$ are uncorrelated, and (2) both $\boldsymbol{P}'_{\perp}\boldsymbol{y}_t$ and $\boldsymbol{P}'_{\perp}\boldsymbol{y}_{t-1}$ are white noise.

The result of this lemma is in the line of Robinson (1973) to test for zero canonical correlation of stationary time series. This result was modified by Tiao and Tsay (1989) to test for SCM, dividing each eigenvalue by the maximum possible variance that the sample cross correlation might have in the case of SCM. In our case, the variance of the cross correlation associated to white noise canonical variates is correctly specified as $1/(T-k)$. Hu and Chou (2004) proposed a similar test using several-second moment matrices simultaneously but instead of using canonical correlation between $\boldsymbol{y}_t$ and its past and future in order to check the rank of the second moment matrices, they use canonical correlation twice: once between $\boldsymbol{y}_t$ and its past and future in order to define past and future canonical variates, and a second time between $\boldsymbol{y}_t$ and the canonical variates define in the previous step. This means that while we are interested in the rank of the matrices defined in (28.26), they test for the rank of the matrices defined by $\boldsymbol{Q} = \boldsymbol{M}\boldsymbol{\Gamma}_y(k)\boldsymbol{M}'$ which have $m-r$ eigenvalues equal to zero if $\boldsymbol{M} = [\boldsymbol{P}'_{\perp}\ \boldsymbol{P}]$. Note that the test presented in this section leads to standard distribution in contrast to the ones presented in 28.3.2, as Johansen (1988) test for the cointegration rank of a VAR model and Reinsel and Ahn (1992) for the number of unit roots in reduced rank regression models.

28.4 Real Data Analysis

We study seven monthly stock indexes from November 1990 until April 2000. The indexes are (by alphabetical order as they are collected in the vector of time $\boldsymbol{y}_t$) DAX-30 from Germany, Dow Jones Composite (DJCOM) from the United States, FTSE from United Kingdom, NASDAQ, New York Stock Exchange (NYSE), Standard and Poor's 500 (SP500) from the United States, and the Canadian TSE. In order to correct for heteroskedasticity, we take the natural log of all the indexes. Plots of the logs of these indices are shown in Figure 28.1.

We apply the common factors canonical correlation test of Section 28.3.3 and obtain the results shown in Table 28.1. We have used up to 18 lags to show

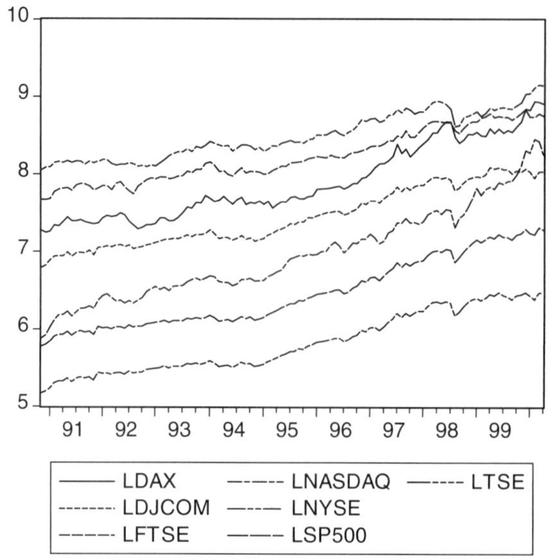

Figure 28.1: Logs of the monthly stock indexes

that the number of identified factors does not depend on the upper bound used for the number of lags. The statistics have already been divided by their critical values, so that a number greater than 1 means that we reject the null hypothesis of a maximum of r common factors at the usual 5% significance level, while a number smaller than 1 means that we cannot reject the null hypothesis of a maximum of r common factors. We present the results after lag two because some small correlations are found for lags 1 and 2. The outcome of the test indicates that a maximum of six common factors cannot be rejected.

To obtain the factors we build the generalized covariance matrices for lags one to five and extract the eigenvectors associated to the first common six eigenvalues of each matrix, see Tables 28.2, 28.3, and 28.4.

The first common factor is a weighted mean of all the indexes, and it can be interpreted as the general level of the world stock indexes. The second factor differentiates the behavior of the NASDAQ, the NYSE and the SP500 from the Canadian TSE and the British FTSE. The third factor separates the NASDAQ and British FTSE from the others. The fourth and sixth common factors are mainly assigned to a single index to characterize its differential performance (the fourth common factor to the German DAX and the sixth to the Chicago's SP500). Finally, the fifth common factor differentiates the British FTSE from the TSE.

In order to obtain the dynamics of the factors, we can perform univariate analysis over the linear combinations of the stock indexes given by the common

Table 28.1: Outcome of the test of Section 28.3.3 for the number of factors. The statistics have already been divided by their critical value, so that an outcome greater than 1 means that the null of a maximum of r common factors is rejected at the 5% significance level, while an outcome smaller than 1 means that the null of a maximum of r common factors cannot be rejected at the 5% significance level

	lag k															
r	3	4	5	6	7	8	9	10	11	12	13	14	15	16	17	18
0	32.6	30.5	29.5	29.0	28.0	27.0	26.3	25.7	25.8	25.4	25.1	24.6	24.2	24.9	25.0	24.7
1	18.8	16.9	16.1	15.9	15.0	13.8	13.0	12.4	12.8	12.6	12.2	11.6	11.2	11.9	12.2	12.0
2	12.8	10.7	10.0	9.9	9.0	7.8	6.9	6.4	6.6	6.7	6.5	6.4	6.7	7.4	7.3	6.9
3	10.0	7.8	6.9	7.2	7.0	6.3	5.4	4.5	4.1	4.1	4.4	4.9	5.3	5.5	5.6	5.4
4	6.2	3.9	3.4	4.5	4.2	3.3	2.4	1.8	1.7	1.7	1.8	2.1	2.8	3.1	3.3	3.7
5	2.5	1.5	1.6	3.2	3.4	1.8	1.0	1.2	1.1	1.0	1.0	1.1	1.2	1.0	1.2	1.3
6	0.7	0.05	0.5	0.5	0.05	0.09	0.10	0.2	0.7	0.5	0.04	0.02	0.04	0.08	0.2	0.06

Table 28.2: Eigenvectors associated to the first and second eigenvalues for the first five generalized covariance matrices of the stock indexes data

1st eigenvector				
lag k				
1	2	3	4	5
0.40	0.40	0.40	0.40	0.40
0.38	0.38	0.38	0.38	0.38
0.42	0.42	0.42	0.42	0.42
0.36	0.36	0.36	0.36	0.36
0.30	0.30	0.30	0.30	0.30
0.33	0.33	0.33	0.33	0.33
0.43	0.43	0.43	0.43	0.43

2nd eigenvector				
lag k				
1	2	3	4	5
0.10	0.10	0.10	0.10	0.107
−0.09	−0.08	−0.06	−0.05	−0.05
−0.36	−0.36	−0.35	−0.35	−0.35
0.64	0.62	0.60	0.58	0.57
0.21	0.22	0.24	0.25	0.26
0.28	0.29	0.31	0.31	0.32
−0.56	−0.57	−0.59	−0.60	−0.60

Table 28.3: Eigenvectors associated to the third and fourth eigenvalues for the first five generalized covariance matrices of the stock indexes data

3rd eigenvector				
lag k				
1	2	3	4	5
0.36	0.62	0.77	0.78	0.73
0.20	0.04	−0.10	−0.13	−0.12
−0.10	−0.21	−0.28	−0.33	−0.42
−0.66	−0.62	−0.52	−0.49	−0.48
0.39	0.29	0.17	0.15	0.20
0.30	0.20	0.08	0.05	0.06
−0.37	−0.25	−0.12	−0.04	−0.05
4th eigenvector				
lag k				
1	2	3	4	5
0.88	0.86	0.87	0.86	0.84
−0.28	−0.22	−0.30	−0.35	−0.36
−0.21	−0.24	−0.25	−0.24	−0.27
−0.29	−0.38	−0.25	−0.16	−0.12
0.09	0.00	−0.09	−0.16	0.20
−0.06	−0.01	−0.10	−0.15	−0.16
−0.02	−0.05	0.04	0.09	0.15

Table 28.4: Eigenvectors associated to the fifth and sixth eigenvalues for the first five generalized covariance matrices of the stock indexes data

5th eigenvector				
lag k				
1	2	3	4	5
0.09	0.07	0.02	0.01	0.09
−0.11	−0.12	−0.11	−0.13	−0.18
0.76	0.76	0.76	0.74	0.67
0.12	0.13	0.16	0.21	0.29
−0.31	−0.35	−0.39	−0.44	−0.54
−0.13	−0.09	−0.05	−0.06	−0.14
−0.52	−0.50	−0.48	−0.44	−0.34
6th eigenvector				
lag k				
1	2	3	4	5
0.17	0.17	0.18	0.18	0.19
0.41	0.27	0.23	0.21	0.20
0.06	−0.04	−0.16	−0.22	−0.20
0.11	0.11	0.11	0.12	0.13
0.33	0.42	0.38	0.35	0.35
−0.81	−0.82	−0.81	−0.81	−0.81
−0.15	−0.18	−0.26	−0.30	−0.27

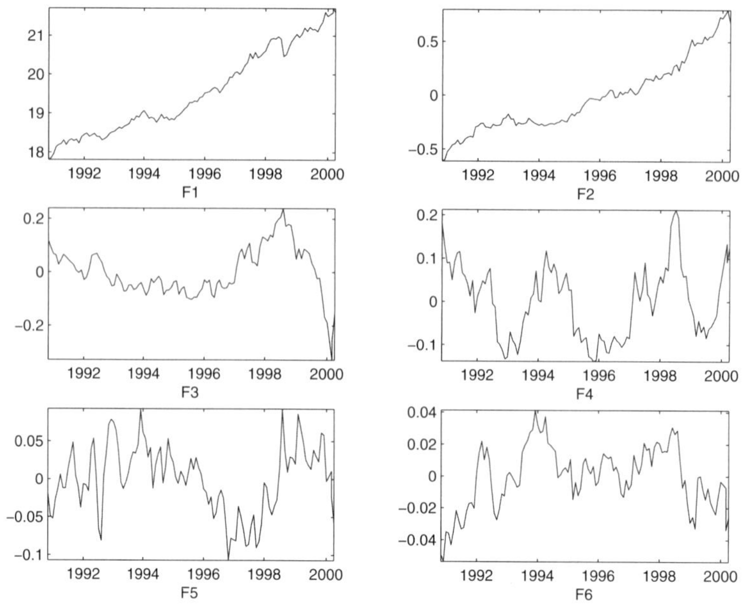

Figure 28.2: Plots of the six common factors of the stock indexes

eigenvectors (we have chosen the eigenvector associated to the generalized covariance matrix of lag one because it is built with more data than the remaining ones). Plots of the six common factors are shown in Figure 28.2.

From the plot, we see that the first two common factors, and possibly the third one, are nonstationary. In fact, if we apply the augmented Dickey-Fuller unit root test with automatic lag selection to minimize the Schwarz information criterion, we cannot reject a unit root for the first three factors, the p-value of the test for the fourth factor is 0.0688 and it is clearly rejected for the fifth and sixth common factors. The first three factors are random walks. The fourth factor could be considered as an AR(1) with autoregressive parameter very close to one (it is estimated as 0.9). The fifth and sixth common factors can be modeled as stationary autoregressive processes of order 2 and 1, respectively.

This analysis shows that the dimension of the nonstationary subspace for the seven stock indexes can be reduced to 3 or, at most, 4.

From this analysis, we expect to find three or four cointegration relations (number of series minus number of common trends) if we apply Johansen's cointegration test. All the selection criteria (Akaike, Schwartz, Hannan-Quin, maximum likelihood, and forecasting prediction error) indicate the order of the VAR process in levels should be 1. With this in mind, we perform Johansen's cointegration test and for a significance level of $\alpha = 0.05$. Assuming that there is no deterministic trends in the data and using the trace statistic of Section

28.3.2, we found four cointegration relations, which is in agreement with the factor analysis results. It is interesting to check that if we assume a deterministic trend, which we believe is a rather unusual fact with economic data [see, Peña (1995)], the number of cointegration relationships found is zero. In order to check the robustness of this conclusion, we perform the maximum eigenvalue test of Johansen, which tests the null hypothesis of s cointegrating relations against the alternative of $s+1$ cointegrating relations. This test statistic is computed as (being the eigenvalues as the same ones as in Section 28.3.2)

$$L(s|s+1) = -T\log(1-\widehat{\lambda}_{s+1}), \tag{28.31}$$

but now we found zero cointegration relations. This result is also obtained if we assume the rather unlikely assumption of deterministic trends in the data. When computing the roots of the companion matrix of the VAR process, one root very close to 1 (estimated as 0.998) and three (a real one and a pair of complex ones) of modulus 0.97 and 0.93 are found. The remaining roots are not close to 1. This might explain why the different versions of the tests detect from 0 to 4 cointegration relations, depending on the assumptions made in order to perform the test.

28.5 Concluding Remarks

We have shown in this chapter that canonical correlation analysis between the present and past values of the time series is a very powerful tool for dimension reduction. This approach allows a unified view of many of the procedures proposed for dimension reduction, including principal components, the canonical analysis of Box and Tiao, the reduced rank models of Reinsel *et al.*, the scalar component models of Tiao and Tsay, the Dynamic Factor model, and state space models. Canonical correlation offers also a unifying view for dimension reduction tests and will lead to similar results than principal components tests when the innovation covariance matrix of the time series is close to a scalar matrix $\sigma^2\boldsymbol{I}$.

Acknowledgments. This research has been supported by Cátedra BBVA de Calidad, DGES projects SEJ2004-03303 and BEC2002-00081, and CAM project 06/HSE/0016/2004.

References

1. Ahn, S. K., and Reinsel, G. C. (1988). Nested reduced-rank autoregressive models for multiple time series, *Journal of the American Statistical Association*, **83**, 849–856.

2. Ahn, S. K., and Reinsel, G. C. (1990). Estimation for partially nonstationary multivariate autoregressive models, *Journal of the American Statistical Association*, **85**, 813–823.

3. Ahn, S. K. (1997). Inference of vector autoregressive models with cointegration and scalar components, *Journal of the American Statistical Association*, **92**, 350–356.

4. Akaike, H. (1974). Markovian representation of stochastic processes and its application to the analysis of autoregressive moving average processes, *Annals of the Institute of Statistical Mathematics*, **26**, 363–387.

5. Anderson, T. W. (1963). The use of factor analysis in the statistical analysis of multiple time series, *Psychometrika*, **28**, 1–25.

6. Aoki, M. (1987). *State Space Modeling of Time Series*, Springer-Verlag, Heidelberg.

7. Banarjee, A. , Dolado, J., Galbraith, J. W., and Hendry, D. (1993). *Cointegration, Error Correction, and the Econometric Analysis of Nonstationary Data*, Oxford University Press, Oxford.

8. Box, G., and Tiao, G. (1977). A canonical analysis of multiple time series, *Biometrika*, **64**, 355–365.

9. Brillinger, D. R. (1981). *Time Series Data Analysis and Theory*, Expanded edition, Holden-Day, San Francisco.

10. Casals J., Jerez, M., and Sotoca, S. (2002). An exact multivariate model-based structural decomposition, *Journal of the American Statistical Association*, **97**, 533–564.

11. Durbin, J., and Koopman, S. K. (2001). *Time Series Analysis by State Space Models*, Oxford University Press, Oxford.

12. Engle, R. F., and Granger, C. W. J. (1987). Cointegration and error correction: Representation, estimation and testing, *Econometrica*, **55**, 251–276.

13. Escribano, A. and Peña, D. (1994). Cointegration and common factors, *Journal of Time Series Analysis*, **15**, 577–586.

14. Forni, M., Hallin, M., Lippi, M., and Reichlin, L. (2000). The generalized dynamic factor model: Identification and estimation, *The Review of Economic and Statistics*, **82**, 540–554.

15. Geweke, J. F., and Singleton K. J. (1981). Maximum likelihood confirmatory analysis of economic time series, *International Economic Review*, **22**, 37–54.

16. Hannan, E. J., and Deistler, M. (1988). *The Statistical Analysis of Linear Systems*, John Wiley & Sons, New York.

17. Harvey, A. (1989). *Forecasting Structural Time Series Models and the Kalman Filter*, Second edition, Cambridge University Press, Cambridge.

18. Harvey, A. (2001). Testing in unobserved components models, *Journal of Forecasting*, **20**, 1–19.

19. Harris, D. (1997). Principal components analysis of cointegrated time series, *Econometric Theory*, **13**, 529–557.

20. Hu, Y., and Chou, R. (2004). On the Peña-Box model, *Journal of Time Series Analysis*, **25**, 811–830.

21. Johansen, S. (1988). Statistical analysis of cointegration vectors, *Journal of Economic Dynamics and Control*, **12**, 231–254.

22. Johansen, S. (1991). Estimation and hypothesis testing of cointegration vectors in Gaussian vector autoregressive models, *Econometrica*, **59**, 1551–1580.

23. Johansen, S. (1995). *Likelihood-Based Inference in Cointegrated Vector Autoregressive Models*, Oxford University Press, Oxford.

24. Molenaar, P. C. M., De Gooijer, J. G., and Schmitz, B. (1992). Dynamic factor analysis of nonstationary multivariate time series, *Psychometrika*, **57**, 333–349.

25. Peña, D. (1995). Forecasting growth with time series models, *Journal of Forecasting*, **14**, 97–105.

26. Peña, D., and Box, G. (1987). Identifying a simplifying structure in time series, *Journal of the American Statistical Association*, **82**, 836–843.

27. Peña, D., and Poncela, P. (2004). Forecasting with nonstationary dynamic factor models, *Journal of Econometrics*, **119**, 291–321.

28. Peña, D., and Poncela, P. (2006). Nonstationary dynamic factor analysis, *Journal of Statistical Planning and Inference*, (to appear).

29. Peña, D., Tiao. G. C., and Tsay, R. S. (2001). *A Course in Time Series Analysis*, John Wiley & Sons, New York.

30. Priestley, M. B., Rao, T. S., and Tong, J. (1974). Applications of principal component analysis and factor analysis in the identification of multivariable systems, *IEEE Transactions on Automation and Control*, **19**, 703–704.

31. Quenouille, M. H. (1968). *The Analysis of Multiple Time Series*, Charles Griffin, London.

32. Rechner, A. C. (1995). *Methods of Multivariate Analysis*, John Wiley & Sons, New York.

33. Reinsel G. C., and Ahn, S. K. (1992). Vector autoregressive models with unit roots and reduced rank structure: estimation, likelihood ratio tests and forecasting, *Journal of Time Series Analysis*, **13**, 353–375.

34. Reinsel, G. C., and Velu, R. P. (1998). *Multivariate Reduced-Rank Regression*, Springer-Verlag, New York.

35. Robinson, P. M. (1973). Generalized canonical analysis for time series, *Journal of Multivariate Analysis*, **3**, 141–160.

36. Shumway, R. H., and Stoffer, D. S. (2000). *Time Series Analysis and Its Applications*, Springer-Verlag, New York.

37. Stock, J. H., and Watson, M. W. (1988). Testing for common trends, *Journal of the American Statistical Association*, **83**, 1097–1107.

38. Tiao, G. C., and Tsay, R. S. (1989). Model specification in multivariate time series, *Journal of the Royal Statistical Society, Series B*, **51**, 157–213.

39. Tsay, R. S., and Tiao, G. C. (1990). Asymptotic properties of multivariate nonstationary processes with applications to autoregressions, *The Annals of Statistics*, **18**, 220–250.

40. Velu, R. P., Reinsel, G. C., and Wichern, D. W. (1986). Reduced rank models for multiple time series, *Biometrika*, **73**, 105–118.

29

The Hat Problem and Some Variations

Wenge Guo,[1] **Subramanyam Kasala,**[2] **M. Bhaskara Rao,**[1] **and Brian Tucker**[3]

[1] *University of Cincinnati, Cincinnati, OH, USA*
[2] *University of North Carolina, Wilmington, NC, USA*
[3] *Pracs Institute, Fargo, ND, USA*

Abstract: The hat problem arose in the context of computational complexity. It started as a puzzle, but the problem has been found to have connections with coding theory and has reached the research frontier of mathematics, statistics and computer science. In this article, some variations of the hat problem are presented along with their solutions. An application is indicated.

Keywords and phrases: Computational complexity, optimization, strategy, winning probability

29.1 Introduction

The "hat problem" has been making rounds in mathematics, statistics, and computer science departments for quite some time. The problem straddles all these disciplines. For a technical description of the problem, see Buhler (2002). For a popular article on the problem, see Robinson (2001). The original hat problem appeared in Todd Ebert's thesis in computer science in connection with complexity theory. A version of the problem can be found in Ebert and Vollmer (2000). It is interesting to note that how this purely recreational problem has come to the research frontier with many problems yet unsolved. A simple version of the problem involves three participants and two colors. Three friends (Brenda, Glenda, Miranda say) are planning to participate in a game show in which a big prize can be won collectively. The host of the game show places a hat on each of the participants. The hat is either black or red. The choice of the colors is random and the placements are independent. What this means is that all the eight configurations of hats, listed in Table 29.1, on the heads of the participants are equally likely. Each participant can see the colors of the hats of her teammates but has no idea what the color of her hat is. The host asks each of the teammates separately what the color of her hat is. A teammate can guess the color of her hat, red (R) or black (B), or pass (P). The other

Table 29.1: List of all configurations (three people and two colors)

Configuration	Brenda	Glenda	Miranda	Probability
1	Red	Red	Red	1/8
2	Red	Red	Black	1/8
3	Red	Black	Red	1/8
4	Red	Black	Black	1/8
5	Black	Red	Red	1/8
6	Black	Red	Black	1/8
7	Black	Black	Red	1/8
8	Black	Black	Black	1/8

members of the team will not know what her response is. They can win the prize collectively if at least one of the teammates guesses the color and whoever guesses must be right. For example, if every one passes, they cannot win the prize. If only one guesses the color and the others pass, the one who guesses must be right in order to win the prize. If two guess the color and the other passes, both the guesses must be right in order to win the prize. If all three guess, all guesses must be right in order to win the prize. Before participating in the game show, the teammates can get into a huddle and formulate a strategy of responses. The basic question is: What is the best strategy of responses so as to maximize the chances of winning the prize.

Let us analyze a couple of strategies. One simple strategy is that every one guesses. If this is the case, the chances of winning the prize are 1/8. Another strategy is that one elects to guess and the others decide to pass. Winning the prize now solely depends on the one who elects to guess. The chances of winning the prize are then 50 percent. Is there a strategy that will improve the chances of winning the prize to more than 50 percent? It is not obvious. In order to improve the chances of winning, it seems that only one of the teammates should guess the color and the others should pass, but who guesses and who passes should be based on what actually they see on the stage. Consider the following strategy.

Instructions to Brenda

a. If the colors of hats of your teammates are both red, say that the color of your hat is black.

b. If the colors of hats of your teammates are both black, say that the color of your hat is red.

c. If the colors of hats of your teammates are different, then pass.

The same instructions are given to Glenda and Miranda.

Table 29.2: Actual configurations along with responses and outcomes (three people and two colors)

Actual Configuration			Responses			Outcome
Brenda	Glenda	Miranda	Brenda	Glenda	Miranda	
R	R	R	B	B	B	Loss
R	R	B	P	P	B	Win
R	B	R	P	B	P	Win
R	B	B	R	P	P	Win
B	R	R	B	P	P	Win
B	R	B	P	R	P	Win
B	B	R	P	P	R	Win
B	B	B	R	R	R	Loss

Under this strategy, let us evaluate the chances of winning the prize. The details are provided in Table 29.2.

It is now clear that the chances of winning the prize under this strategy are 75 percent. One can also show that there is no way one can improve the chances of winning to more than 75 percent. For future reference, let us call this strategy as *Strategy O*.

There are two main objectives we want to pursue in this chapter. One is to extend the hat problem to the case of three colors and three teammates. We will present an optimal strategy the teammates can pursue that will maximize the chances of winning the prize collectively. The other is to stay within the environment of two colors and three teammates, but the eight configurations that are possible are not equally likely. More precisely, we will be given a probability distribution on the set of all hat configurations and the task is to determine an optimal strategy that will maximize the probability of winning the prize. We will also present some other variations of the hat problem. Finally, we will end the paper with a number of open questions.

29.2 Hamming Codes

The hat problem has a close connection with "Covering Codes." In this section, the connection is explained in a rudimentary fashion.

Covering and packing are two of the most intriguing problems in mathematics useful in engineering. A *packing problem* in the traditional Euclidean space is to ask for the maximal number of identical nonintersecting spheres in a large volume. As an example, suppose we have a box with dimensions 1 meter $\times$ 1 meter $\times$ 1 meter. We want to pack the box with identical balls of radius 10 centimeters. In what way should we pack the box so as to accommodate the maximum number of balls? On the other hand, a *covering problem* in an Euclidean space asks for the minimal number of identical spheres to cover a specified volume.

A discrete analogue of the covering problem involves the so-called *Hamming space.* For a fixed positive integer n, it is the set of all n-tuples where each component in any n-tuple is either zero or one. The elements of the Hamming space are called *points.* Any nonempty subset of the Hamming space is called a *code* and its elements are called *codewords.* The *Hamming distance* between any two points is the number of components at which the points differ. The Hamming distance is a non-negative integer from zero to n. The *minimum distance* of a code is the smallest of the pairwise distances between its codewords. Let x be a point in the Hamming space and $r > 0$. A sphere of radius r with center at x in the Hamming space consists of all points within distance r from the center x.

Covering problem: Given n and r, what is the smallest number of spheres of radius r so that every point in the Hamming space belong to at least one of the spheres?

Example. Suppose $n = 3$ and $r = 1$.

Hamming Space: 000, 001, 010, 011, 100, 101, 110, 111

S(000, 1) = Sphere with center at 000 and radius one = 000, 001, 010, 100

S(111, 1) = Sphere with center at 111 and radius one = 111, 110, 101, 011

Every point in the Hamming space belongs to one of these two spheres. In other words, these two spheres cover the whole space. This covering is minimal.

Any such minimal covering gives rise to an optimal strategy in the hat problem. Identify $0 = R$ and $1 = B$. Let L be the set of centers of the spheres and W its complement. In this example, $L = 000, 111$ and $W = 001, 010, 100, 110, 101, 011$. A strategy S now can be developed such that for this strategy the set of losing configurations is L and the set of winning configurations is W. We begin with instructions to the teammates that make up the strategy S. The teammates Brenda, Glenda, and Miranda are ordered as they are mentioned and instructions to them proceed in that order. To begin with, they should be appraised with the notation $0 = R$ and $1 = B$, and also with the sets L and W.

Instructions to Brenda

Suppose you see 00. (This means that Brenda sees red hats on both Glenda and Miranda.) If there is a unique $u \in \{0, 1\}$ such that $u00 \in W$, say that the color of your hat is u. Otherwise, pass. Here, u is unique and in fact, $u = 1$.

Suppose you see 01. (This means that Brenda sees a red hat on Glenda and a black hat on Miranda.) If there is a unique $u \in \{0, 1\}$ such that $u01 \in W$, then say that the color of your hat is u. Otherwise, pass. Here, u is not unique. As a matter of fact, 001 and 101 both belong to W. In this case, you should pass.

Suppose you see 10. (This means that Brenda sees a black hat on Glenda and a red hat on Miranda.) If there is a unique $u \in \{0, 1\}$ such that $u10 \in W$, then say that the color of your hat is u. Otherwise, pass. Here, u is not unique. As a matter of fact, 010 and 110 both belong to W. In this case, you should pass.

Suppose you see 11. (This means that Brenda sees black hats on both Glenda and Miranda.) If there is a unique $u \in \{0, 1\}$ such that $u11 \in W$, then say that the color of your hat is u. Otherwise, pass. Here u is unique. As a matter of fact, $u = 0$.

Instructions to Glenda

Suppose you see 00. (This means that Glenda sees red hats on both Brenda and Miranda.) If there is a unique $u \in \{0, 1\}$ such that $0u0 \in W$, say that the color of your hat is u. Otherwise, pass. Here, u is unique and in fact, $u = 1$.

Suppose you see 01. (This means that Glenda sees a red hat on Brenda and a black hat on Miranda.) If there is a unique $u \in \{0, 1\}$ such that $0u1 \in W$, then say that the color of your hat is u. Otherwise, pass. Here, u is not unique. As a matter of fact, 001 and 011 both belong to W. In this case, you should pass.

Suppose you see 10. (This means that Glenda sees a black hat on Brenda and a red hat on Miranda.) If there is a unique $u \in \{0, 1\}$ such that $1u0 \in W$, then say that the color of your hat is u. Otherwise, pass. Here, u is not unique. As a matter of fact, 100 and 110 both belong to W. In this case, you should pass.

Suppose you see 11. (This means that Glenda sees black hats on both Brenda and Miranda.) If there is a unique $u \in \{0, 1\}$ such that $1u1 \in W$, then say that the color of your hat is u. Otherwise, pass. Here, u is unique. As a matter of fact, $u = 0$.

By now, the tone of instructions should be clear. Instructions to Miranda follow in the same tone.

In the general case of two colors and n participants, we look at the corresponding Hamming space and a minimal cover. An optimal strategy is built based on the minimal cover in the same way as outlined above. For a connection between the hat problem and minimal covers, see Lenstra and Seroussi (2004). For a comprehensive discussion of Hamming space and covers, see Cohen *et al.* (1997).

29.3 Three Teammates and Three Colors

We now consider the case of three teammates and three colors. Each of the teammates is fitted with a hat, which is red (R), black (B), or green (G) by the host. All the 27 configurations of hats are equally likely. Each participant can see the color of the hat each of her teammates has but cannot see the color of her own hat. Each participant is required to guess the color of her hat or pass. In order to win the prize collectively, at least one team mate should guess the color of her hat and whoever guesses must be right. What is the best strategy that will maximize the probability of winning the prize?

Let us formulate the problem mathematically. Brenda can see the colors of the hats of her teammates. What she sees is: RR, RB, BR, BB, RG, GR, GG, BG, or GB on Glenda and Miranda, respectively. She needs to respond: R, B, G, or P (Pass). Formally, we can introduce a map from the set of all possible hat configurations she sees on her teammates to the set of all possible responses. Thus, an *instruction* is a map f described by,

$$\{RR, RB, BR, BB, RG, GR, GG, BG, GB\} \xrightarrow{f} \{R, B, G, P\}.$$

Let $\mathbf{F}$ be the collection of all instructions. The cardinality of the set $\mathbf{F}$ is $4^9 = 262{,}144$. A *strategy* is a triplet $S = (f_1, f_2, f_3)$, where each f_i is a member of $\mathbf{F}$. Using the strategy S means that Brenda follows the instruction f_1, Glenda f_2, and Miranda f_3. Let S be the collection of all strategies. The cardinality of S is $4^{27} \approx 1.8 * 10^{16}$. For any given strategy, one can work out the probability of winning the prize. A complete enumeration of all strategies along with winning probability using a computer in order to find an optimal strategy is not feasible.

We restrict ourselves to symmetric strategies. A strategy $S = (f_1, f_2, f_3)$ is said to be symmetric if $f_1 = f_2 = f_3$. This means that all participants follow the same instructions. The total number of symmetric strategies is $262, 144$. This number is manageable by a computer. We have written a program that enumerates all symmetric strategies and computes the corresponding winning probabilities. We have identified optimal strategies from the list. There are several. A careful scrutiny of the optimal strategies led us to synthesize verbally what the instructions should be.

Designate one of the colors as "primary" and another color as "secondary." For example, we may take red as primary and black as secondary. The instructions to the participants are centered on these designations.

Table 29.3: List of all configurations along with responses under the symmetric strategy S (next page) and outcomes (three people and three colors)

Actual Configuration			Responses			Outcome
Brenda	Glenda	Miranda	Brenda	Glenda	Miranda	
R	R	R	B	B	B	Loss
R	R	B	P	P	B	Win
R	B	R	P	B	P	Win
R	B	B	R	P	P	Win
B	R	R	B	P	P	Win
B	R	B	P	R	P	Win
B	B	R	P	P	R	Win
B	B	B	R	R	R	Loss
R	R	G	P	P	B	Loss
R	G	R	P	B	P	Loss
R	G	G	R	P	P	Win
G	R	R	B	P	P	Loss
G	R	G	P	R	P	Win
G	G	R	P	P	R	Win
G	G	G	R	R	R	Loss
B	B	G	R	R	R	Loss
B	G	B	R	R	R	Loss
B	G	G	R	R	R	Loss
G	G	B	R	R	R	Loss
G	B	G	R	R	R	Loss
G	B	B	R	R	R	Loss
R	B	G	R	P	P	Win
R	G	B	R	P	P	Win
B	R	G	P	R	P	Win
B	G	R	P	P	R	Win
G	R	B	P	R	P	Win
G	B	R	P	P	R	Win

Instructions to Brenda

1. If both the colors you see are primary, say that the color of your hat is the secondary color.

2. If only one of the colors you see is primary, then pass.

3. If none of the colors you see is primary, say that the color of your hat is the primary color.

If the primary color is red and the secondary coloris black, mathematically, instructions to Brenda can be spelled out as follows.

$$\begin{gathered}\mathrm{f(RR) = B;}\\ \mathrm{f(RB) = P;\ f(BR) = P;\ f(RG) = P;\ f(GR) = P;\ f(BB) = R;}\\ \mathrm{f(BG) = R;\ f(GB) = R;\ f(GG) = R.}\end{gathered}$$

The same instructions are given to Glenda and Miranda. If they adopt this symmetric strategy $S = (f, f, f)$, the chances of winning the prize are 15/27. In Table 29.3 we outline all possible hat configurations and responses following the optimal symmetric strategy described above. In 15 cases out of 27, the teammates can win the prize. This is an optimal strategy among all symmetric strategies. In Section 29.5, we will show that this symmetric strategy is indeed optimal among all strategies.

29.4 Three Teammates and m Colors

The problem outlined in Section 29.3 can be generalized to the case of $m(\geq 3)$ colors. The number of participants remains the same. Each participant is fitted with a hat whose color is one of the m colors given. Let $C_1, C_2, \ldots, C_m$ be the colors that are used in the game. The total number of configurations of hats is m^3. As in Section 29.3, we confine our attention to symmetric strategies $S = (f, f, f)$, where f is any *instruction*, that is, f is a map from the set

$$\{(x, y)\,; x, y \in \{C_1, C_2, \ldots, C_m\}\}$$

into the set

$$\{C_1, C_2, \ldots, C_m, P\},$$

where the symbol P stands for "Pass." The vector (x, y) stands for the colors of the hats any participant will see on her teammates. When the host asks a participant about the color of her hat, she needs to respond $C_1, C_2, \ldots, C_m$, or P. An optimal strategy uses the following instruction f for each participant. To begin with, declare one of the colors as "primary" and one of the remaining colors as "secondary."

Table 29.4: List of all configurations of hats and winning ones (three people and m colors)

Configurations	Cardinality	No. Winning Configurations
(C_1, C_1, C_1)	1	0
$(C_1, C_1, C_j), j = 2, 3, \ldots, m$	$m - 1$	1
$(C_1, C_j, C_1), j = 2, 3, \ldots, m$	$m - 1$	1
$(C_j, C_1, C_1), j = 2, 3, \ldots, m$	$m - 1$	1
$(C_1, C_i, C_j), i, j = 2, 3, \ldots, m$	$(m - 1)^2$	$(m - 1)^2$
$(C_i, C_1, C_j), i, j = 2, 3, \ldots, m$	$(m - 1)^2$	$(m - 1)^2$
$(C_i, C_j, C_1), i, j = 2, 3, \ldots, m$	$(m - 1)^2$	$(m - 1)^2$
$(C_i, C_j, C_k), i, j, k = 2, 3, \ldots, m$	$(m - 1)^3$	0

Instructions (f) to any participant

1. If the colors of the hats of your teammates are both primary, you should say that the color of your hat is secondary color.

2. If only one of the colors of the hats of your teammates is primary, you should pass.

3. If none of the colors of the hats of your teammates is primary, you should say that the color your hat is the primary color.

Let us calculate the probability of winning the prize under the strategy $S = (f, f, f)$, where f is the instruction described above. For simplicity, let us declare that C_1 is the primary color and C_2 the secondary. We will make a complete list of all configurations of hats and then count how many of these configurations lead to winning the prize. To facilitate the calculations, form eight subsets of the set of all hat configurations based on the number of times the primary color C_1 is present in the configurations. The entire set of configurations is given in Table 29.4.

An explanation is in order on the above table. As an example, look at the hat configuration (C_1, C_i, C_j) for some $i, j = 2, 3, \ldots, m$. Under the instructions f outlined above, Brenda's response would be C_1, in which case she is right, and Glenda and Miranda would pass. Thus (C_1, C_i, C_j) would be a winning configuration under the strategy $S = (f, f, f)$. The total number of such hat configurations is $(m - 1)^2$, and as we have just observed, each one of them is a winning configuration. In totality, the team will win the prize in $3\,(m - 1)^2 + 3$ cases out of m^3 possible configurations. Hence the probability of winning the prize under the strategy S is given by

$$\frac{3(m-1)^2 + 3}{m^3}.$$

Let us contrast this strategy with the simple strategy, in which one of the participants chooses to guess the color of her hat while others choose to pass. Under this simple strategy, the probability of winning the prize is $\frac{1}{m}$. This probability is certainly less than $\frac{3(m-1)^2+3}{m^3}$.

We do not know that the strategy S, which is optimal in the set of all symmetric strategies, is optimal in the set of all strategies. However, the winning probability for S is very close to the upper bound, which will be discussed in the next section.

29.5 An Upper Bound for the Winning Probability

Let us consider the hat problem with q colors $C_1, C_2, \ldots, C_q$ and $n(\geq 3)$ participants. The participants are numbered serially from 1 to n. The *modus operandi* is similar to the basic hat problem. Each participant will be seeing the hats of the remaining $(n-1)$ participants. Her response is $C_1, C_2, \ldots, C_q$, or P(Pass). The set of all hat configurations is $\Sigma = \{C_1, C_2, \ldots, C_q\}^n$. Let $\mathbf{C} = \{C_1, C_2, \ldots, C_q, P\}$. An *instruction* to Participant No. i is a map f_i from the set

$$\{x_1 x_2 \cdots x_{i-1} x_{i+1} \cdots x_n : x_i \in \{C_1, C_2, \ldots, C_q\} \text{ for all } i\}$$

into the set $\mathbf{C}$. The entity $x_1 x_2 \cdots x_{i-1} x_{i+1} \cdots x_n$ stands as a generic symbol for the colors of the hats Participant No.i would see on her teammates and $f_i(x_1 x_2 \cdots x_{i-1} x_{i+1} \cdots x_n)$ is the response to the query what the color of her hat is. A *strategy* $S = (f_1, f_2, \ldots, f_n)$ is an n-tuple, where f_i is the instruction that Participant No. i follows, $i = 1, 2, \ldots, n$. For a given strategy S, we can check whether or not a configuration of hats is winning. Let W_S denote the set of all winning configurations under the strategy S and L_S losing configurations. Obviously, $\#W_S + \#L_S = q^n$. The objective is to find a strategy S for which $\#W_S$ is maximum, or equivalently, $\#L_S$ is minimum.

We will now work out an upper bound for $\#W_S$. For each $i = 1, 2, \ldots, n$, let

$$Q_i = \{x_1 x_2 \cdots x_{i-1} x_i x_{i+1} \cdots x_n \in \Sigma; f_i\left(x_1 x_2 \cdots x_{i-1} x_{i+1} \cdots x_n\right) \neq P\}.$$

Note that $Q_i = \phi$, the null set, if and only if $f_i \equiv P$, that is, as per the instruction f_i, Participant No. i passes all the time. It is now clear that Q_i is a multiple of q. Let $\#Q_i = q * t_i$, where t_i is a non-negative integer. Take any $x_1 x_2 \cdots x_{i-1} x_i x_{i+1} \cdots x_n$ in Q_i. Then, $x_1 x_2 \cdots x_{i-1} C_j x_{i+1} \cdots x_n \in Q_i$ for all $j = 1, 2, \ldots, q$. Of these q configurations, in only one configuration the guess by Participant No. i will be correct. Consequently, in t_i configurations from Q_i, the guesses by Participant No. i will be correct and in the remaining $(q-1)t_i$ configurations the guesses will be incorrect.

Let us interpret and understand all these entities in the context of the hat problem with two colors (R and B) and three participants. Suppose the instructions f_1, f_2 and f_3 to Brenda, Glenda, and Miranda, respectively, are:

Brenda	Glenda	Miranda
$f_1(RR) = $ P	$f_2(RR) = $ B	$f_3(RR) = $ R
$f_1(RB) = $ P	$f_2(RB) = $ P	$f_3(RB) = $ R
$f_1(BR) = $ P	$f_2(BR) = $ P	$f_3(BR) = $ R
$f_1(BB) = $ B	$f_2(BB) = $ R	$f_3(BB) = $ R

Then $Q_1 = \{$RBB, BBB$\}$, $Q_2 = \{$RRR, RBR, BRB, BBB$\}$, $Q_3 = \{$RRR, RRB, RBR, RBB, BRR, BRB, BBR, BBB$\}$. Further, $t_1 = 1$, $t_2 = 2$ and $t_3 = 4$. Of the two configurations in Q_1, if BBB is the configuration of hats, Brenda's guess will be correct. Of the four configurations in Q_2, Glenda's guess will be correct for each of the configurations RBR and BRB. Finally, of the eight configurations in Q_3, Miranda's guess will be correct for each of the configurations RRR, RBR, BRR, and BBR.

In the general case, the configurations in Q_i can be partitioned into two sets, one set Q_{i1} containing configurations in each of which Participant No. i's guess will be correct as per her instruction f_i and the other set Q_{i2} containing configurations in each of which Participant No. i's guess will be incorrect, with cardinalities t_i and $(q-1)t_i$, respectively. Now take any configuration from Σ. Let us determine whether or not it is a winning configuration as per the strategy $S = (f_1, f_2, \ldots, f_n)$. It is a winning configuration if at least one participant guessed correctly. Consequently,

$$W_S \subset Q_{i1} \cup Q_{i2} \cup \cdots \cup Q_{in}.$$

Therefore,

$$\#W_S \leq t_1 + t_2 + \cdots + t_n.$$

On the other hand, if at least one participant guesses wrongly under a given configuration, then it is a losing configuration. Therefore,

$$\#L_S \geq \frac{(q-1)(t_1 + t_2 + \cdots + t_n)}{n}.$$

Because $\#W_S + \#L_S = q^n$, it follows that

$$\#W_S \leq q^n - \frac{(q-1)(t_1 + t_2 + \cdots + t_n)}{n}.$$

Thus an upper bound for $\#W_S$ over all strategies S reduces to the following optimization problem:

$$Maximize \quad min\{t_1 + t_2 + \cdots + t_n, q^n - \frac{(q-1)(t_1 + t_2 + \cdots + t_n)}{n}\}$$

$$subject\ to\ the\ constraints \quad 0 \le t_i \le q^{n-1},\ i = 1, 2, \ldots, n.$$

Let us review the optimization problem vis-à-vis the hat problem with two colors, three participants, the strategy S spelled out above. Note that $W_S = \{\text{RBR, BRR, BBR}\}$; $\#W_S = 3$; $Q_{11} = \{\text{BBB}\}$; $Q_{21} = \{\text{RBR, BRB}\}$; $Q_{31} = \{\text{RRR, RBR, BRR, BBR}\}$; $\#W_S < t_1 + t_2 + t_3$; and $\#L_S = 5 > \frac{(t_1+t_2+t_3)}{3}$.

Let us now tackle the general optimization problem.

$$\begin{aligned}
&Maximize\ min\left\{t_1 + t_2 + \cdots + t_n, q^n - \frac{(q-1)(t_1 + t_2 + \cdots + t_n)}{n}\right\} \\
&\quad \le min\left\{max\{t_1 + t_2 + \cdots + t_n\}, q^n - max\left\{\frac{(q-1)(t_1 + t_2 + \cdots + t_n)}{n}\right\}\right\} \\
&\quad = min\left\{z, q^n - \frac{q-1}{n}z\right\} \\
&\quad = \frac{n}{n+q-1}q^n,
\end{aligned}$$

where $z = max\{t_1 + t_2 + \cdots + t_n\}$ and all the maximums are taken over all $t_1, t_2, \ldots, t_n$ subject to the constraints spelled out above. Consequently, an upper bound for the winning probability is given by $\frac{n}{n+q-1}$.

In the case of the hat problem with two colors and n participants, an upper bound for the winning probability is $\frac{n}{n+1}$. In particular, for the problem with two colors and three participants, an upper bound for the winning probability is 3/4. The strategy presented in Section 29.1 has the winning probability 3/4 and hence it is indeed optimal. In the case of the hat problem with two colors and 4 participants, an upper bound for the winning probability is 4/5. However, there is no strategy for which the winning probability is 4/5. This can be shown as follows. First of all, we show that there is a strategy S^* with winning probability 3/4. Suppose the four participants are: Brenda, Glenda, Miranda, and Yolanda. We instruct Brenda to pass. We instruct Glenda, Miranda, and Yolanda to ignore Brenda and play the game as though they are the only participants, and follow the three player optimal strategy. Under this strategy S^*, the probability of winning the prize is 3/4. Now, let S be any strategy. Its winning probability must be of the form m/16. Note that $3/4 = 12/16 < 4/5$ but $13/16 > 4/5$. Consequently, the winning probability under S has to be $\le 3/4$. Hence S^* is optimal for the game with four players and two colors.

For the hat problem with two colors and n participants, one can always find a strategy with winning probability 3/4. Instruct $(n-3)$ participants to pass all the time and the three remaining participants play the game as though they are the only participants.

For the hat problem with two colors and five participants, an upper bound with winning probability is 5/6. However, there is no strategy that achieves this winning probability. An optimal strategy in this case has a winning probability 3/4 only.

For the hat problem with two colors and n participants with n of the form $2^k - 1$, there is always an optimal strategy with winning probability $n/(n+1)$, the upper bound. For a description of an optimal strategy, see Buhler (2002).

If q (number of colors) $= 3$ and $n = 3$, the upper bound is 3/5. The strategy described in Section 29.3 has the winning probability 15/27. For any strategy S in this context, the winning probability must be of the form $m/27$. Note that $16/27 < 3/5$ but $17/27 > 3/5$. The question arises whether or not there is a strategy S with winning probability 16/27. Using a Coding Theory argument, which is not present here, we have shown that there is no strategy with winning probability 16/27. Consequently, the strategy presented in Section 29.3 is indeed optimal for the game with three colors and three participants.

In the case of the hat problem with m colors and three participants, the upper bound for winning probability is $3/(m+2)$. The symmetric strategy we have described in Section 29.4 has the winning probability $\frac{3(m-1)^2+3}{m^3}$. The difference between the upper bound and $\frac{3(m-1)^2+3}{m^3}$ is very small. As a matter of fact,

$$\frac{3}{(m+2)} - \frac{3(m-1)^2+3}{m^3} = \frac{6}{m^3},$$

which is close to zero even for moderate values of m. Consequently, we can say that the strategy presented in Section 29.4 is almost optimal.

29.6 General Distribution

We now work in the environment of two colors and three participants. The eight possible configurations of hats need not be equally likely. Let the distribution on the set of all configurations be given by

Configuration	Brenda	Glenda	Miranda	Probability
1	Red	Red	Red	p_1
2	Red	Red	Black	p_2
3	Red	Black	Red	p_3
4	Red	Black	Black	p_4
5	Black	Red	Red	p_5
6	Black	Red	Black	p_6
7	Black	Black	Red	p_7
8	Black	Black	Black	p_8

Given the distribution p_i's, the objective is to find an optimal strategy that maximizes the probability of winning the prize. For example, if $p_1 = 0 = p_8$, then there is a strategy that gives the probability of winning as unity no matter what the values of the other probabilities are. If $p_1 = 0.47 = p_8$ and $p_2 = p_3 = p_4 = p_5 = p_6 = p_7 = 0.01$, the strategy described in Section 29.1 is no longer optimal.

For a given distribution, one way to find an optimal strategy is to calculate the probability of winning the prize for each of the possible 531,441 strategies. From this collection of all strategies, we are able to identify 12 strategies and it is enough to calculate the probability of winning for each of these 12 strategies in order to determine an optimal strategy. The reasoning now follows.

Recall that an instruction to a participant is a map

$$f : \{RR, RB, BR, BB\} \to \{R, B, P\}.$$

A strategy is a triplet $S = (f_1, f_2, f_3)$, where f_1 is an instruction to Brenda, f_2 to Glenda, and f_3 to Miranda. Note that the total number of strategies is $81^3 = 531,441$. Given any strategy S, one can determine the set W_S of all winning configurations of hats. For example, if $f_1 \equiv R, f_2 \equiv R, and\, f_3 \equiv B$, then the only configuration that leads to the prize is RRB if the participants adopt the strategy $S = (f_1, f_2, f_3)$. Thus, $W_S = \{RRB\}$. We can now introduce a relation in the set of all strategies. Say that the strategy $S = (f_1, f_2, f_3)$ is at least as good as the strategy $T = (g_1, g_2, g_3)$ if $W_T \subseteq W_S$. Denote this relation by $T \leq S$. Given a choice between S and T, we would adopt the strategy S. The relation $\leq$ is transitive and reflexive. Consequently, it is a partial order.

We have written a computer program to make a complete list of all strategies along with their sets of winning configurations. A careful scrutiny of the list yields 12 maximal strategies. What this means in terms of the stipulated partial order is that given any strategy T one can find one of the maximal strategies S such that $W_T \subseteq W_S$. It is now transparent that for a given distribution on the set of all configurations, an optimal strategy is one of these 12 strategies. We will now give a list of all these 12 maximal strategies.

Maximal Strategy 1

Instruction to Brenda	Instruction to Glenda	Instruction to Miranda
$f_1(RR) = B$	$f_2 = f_1$	$f_3 = f_1$
$f_1(RB) = P$		
$f_1(BR) = P$		
$f_1(BB) = R$		

W_S = Winning set of configurations
= {RRB, RBR, RBB, BRR, BRB, BBR}

Note: This strategy is the same as the one described in Section 1. **Maximal**

Strategy 2

Instruction to Brenda	Instruction to Glenda	Instruction to Miranda
$f_1(RR) = R$	$f_2(RR) = P$	$f_3(RR) = P$
$f_1(RB) = P$	$f_2(RB) = R$	$f_3(RB) = R$
$f_1(BR) = P$	$f_2(BR) = B$	$f_3(BR) = B$
$f_1(BB) = B$	$f_2(BB) = P$	$f_3(BB) = P$

W_S = Winning set of configurations
= {RRR, RRB, RBR, BRB, BBR, BBB}

Maximal Strategy 3

Instruction to Brenda	Instruction to Glenda	Instruction to Miranda
$f_1(RR) = P$	$f_2(RR) = R$	$f_3(RR) = P$
$f_1(RB) = R$	$f_2(RB) = P$	$f_3(RB) = B$
$f_1(BR) = B$	$f_2(BR) = P$	$f_3(BR) = R$
$f_1(BB) = P$	$f_2(BB) = B$	$f_3(BB) = P$

W_S = Winning set of configurations
= {RRR, RRB, RBB, BRR, BBR, BBB}

Maximal Strategy 4

Instruction to Brenda	Instruction to Glenda	Instruction to Miranda
$f_1(RR) = P$	$f_2(RR) = P$	$f_3(RR) = R$
$f_1(RB) = B$	$f_2(RB) = B$	$f_3(RB) = P$
$f_1(BR) = R$	$f_2(BR) = R$	$f_3(BR) = P$
$f_1(BB) = P$	$f_2(BB) = B$	$f_3(BB) = B$

W_S = Winning set of configurations
= {RRR, RBR, RBB, BRR, BRB, BBB}

Maximal Strategy 5

Instruction to Brenda	Instruction to Glenda	Instruction to Miranda
$f_1(RR) = R$	$f_2(RR) = P$	$f_3(RR) = P$
$f_1(RB) = R$	$f_2(RB) = P$	$f_3(RB) = P$
$f_1(BR) = R$	$f_2(BR) = P$	$f_3(BR) = P$
$f_1(BB) = R$	$f_2(BB) = P$	$f_3(BB) = P$

W_S = Winning set of configurations
= {RRR, RRB, RBR, RBB}

Maximal Strategy 6

Instruction to Brenda	Instruction to Glenda	Instruction to Miranda
$f_1(RR) = B$	$f_2(RR) = P$	$f_3(RR) = P$
$f_1(RB) = B$	$f_2(RB) = P$	$f_3(RB) = P$
$f_1(BR) = B$	$f_2(BR) = P$	$f_3(BR) = P$
$f_1(BB) = B$	$f_2(BB) = P$	$f_3(BB) = P$

W_S = Winning set of configurations
= {BRR, BRB, BBR, BBB}

Maximal Strategy 7

Instruction to Brenda	Instruction to Glenda	Instruction to Miranda
$f_1(RR) = P$	$f_2(RR) = R$	$f_3(RR) = P$
$f_1(RB) = P$	$f_2(RB) = R$	$f_3(RB) = P$
$f_1(BR) = P$	$f_2(BR) = R$	$f_3(BR) = P$
$f_1(BB) = P$	$f_2(BB) = R$	$f_3(BB) = P$

W_S = Winning set of configurations
= {RRR, RRB, BRR, BRB}

Maximal Strategy 8

Instruction to Brenda	Instruction to Glenda	Instruction to Miranda
$f_1(RR) = P$	$f_2(RR) = B$	$f_3(RR) = P$
$f_1(RB) = P$	$f_2(RB) = B$	$f_3(RB) = P$
$f_1(BR) = P$	$f_2(BR) = B$	$f_3(BR) = P$
$f_1(BB) = P$	$f_2(BB) = B$	$f_3(BB) = P$

W_S = Winning set of configurations
= {RBR, RBB, BBR, BBB}

Maximal Strategy 9

Instruction to Brenda	Instruction to Glenda	Instruction to Miranda
$f_1(RR) = P$	$f_2(RR) = P$	$f_3(RR) = R$
$f_1(RB) = P$	$f_2(RB) = P$	$f_3(RB) = R$
$f_1(BR) = P$	$f_2(BR) = P$	$f_3(BR) = R$
$f_1(BB) = P$	$f_2(BB) = P$	$f_3(BB) = R$

W_S = Winning set of configurations
= {RRR, RBR, BRR, BBR}

Maximal Strategy 10

Instruction to Brenda	Instruction to Glenda	Instruction to Miranda
$f_1(RR) = P$	$f_2(RR) = P$	$f_3(RR) = B$
$f_1(RB) = P$	$f_2(RB) = P$	$f_3(RB) = B$
$f_1(BR) = P$	$f_2(BR) = P$	$f_3(BR) = B$
$f_1(BB) = P$	$f_2(BB) = P$	$f_3(BB) = B$

W_S = Winning set of configurations
= {RRB, RBB, BRB, BBB}

Maximal Strategy 11

Instruction to Brenda	Instruction to Glenda	Instruction to Miranda
$f_1(RR) = R$	$f_2(RR) = R$	$f_3(RR) = R$
$f_1(RB) = B$	$f_2(RB) = B$	$f_3(RB) = B$
$f_1(BR) = B$	$f_2(BR) = B$	$f_3(BR) = B$
$f_1(BB) = R$	$f_2(BB) = R$	$f_3(BB) = R$

W_S = Winning set of configurations
= {RRR, RBB, BRB, BBR}

Maximal Strategy 12

Instruction to Brenda	Instruction to Glenda	Instruction to Miranda
$f_1(RR) = B$	$f_2(RR) = B$	$f_3(RR) = B$
$f_1(RB) = R$	$f_2(RB) = R$	$f_3(RB) = R$
$f_1(BR) = R$	$f_2(BR) = R$	$f_3(BR) = R$
$f_1(BB) = B$	$f_2(BB) = B$	$f_3(BB) = B$

W_S = Winning set of configurations
= {RRB, RBR, BRR, BBB}

A summary of these strategies along with their winning and losing configurations is given in the following table.

Max. Str.	Config. RRR	RRB	RBR	RBB	BRR	BRB	BBR	BBB
1	L	W	W	W	W	W	W	L
2	W	W	W	L	L	W	W	W
3	W	W	L	W	W	L	W	W
4	W	L	W	W	W	W	L	W
5	W	W	W	W	L	L	L	L
6	L	L	L	L	W	W	W	W
7	W	W	L	L	W	W	L	L
8	L	L	W	W	L	L	W	W
9	W	L	W	L	W	L	W	L
10	L	W	L	W	L	W	L	W
11	W	L	L	W	L	W	W	L
12	L	W	W	L	W	L	L	W

Some comments are in order on these maximal strategies. No two of these strategies are comparable. This means that the winning set of configurations of any of these strategies is neither contained in nor contains the winning set of configurations of any one of the other strategies. In addition, an examination of the entire set of strategies yields other valuable information. There are no strategies with exactly three winning configurations or five winning configurations.

For any given distribution p_i's, the above table can be used to determine an optimal strategy that maximizes the probability of winning the prize. One simply calculates the probability of winning under each of these 12 strategies. Pick the one with maximum probability then. Let us look at the distribution $p_1 = p_8 = 0.47$ and $p_2 = p_3 = p_4 = p_5 = p_6 = p_7 = 0.01$. There are three optimal strategies available: Strategies 2, 3 and 4. The winning probability is 0.98.

These maximal strategies have certain symmetric or antisymmetric properties with respect to the configurations. For any strategy S, let WL_S (win-loss map) denote the map from the set {RRR, RRB, RBR, RBB, BRR, BRB, BBR, BBB} of all configurations into the set $\{W, L\}$ defined by

$$\begin{aligned} WL_S(\text{Configuration}) &= W \text{ if the configuration is a winning one,} \\ &= L \text{ if the configuration is a losing one.} \end{aligned}$$

A strategy S is symmetric if

$$\begin{aligned} WL_S(RRR) &= WL_S(BBB), \\ WL_S(RRB) &= WL_S(BBR), \\ WL_S(RBR) &= WL_S(BRB), \end{aligned}$$

and

$$WL_S(RBB) = WL_S(BRR).$$

If we flip R and B in the arguments of the map WL_S, the map remains invariant. We can now check that the maximal strategies 1, 2, 3, and 4 are symmetric. In addition, for each of these strategies, the number of winning configurations is six. The total number of symmetric strategies each with six winning configurations is four. We have exhausted all these strategies and they are indeed the first four strategies listed above.

A strategy S is antisymmetric if

$$\begin{aligned} WL_S(RRR) &\neq WL_S(BBB), \\ WL_S(RRB) &\neq WL_S(BBR), \\ WL_S(RBR) &\neq WL_S(BRB), \end{aligned}$$

and

$$WL_S(RBB) \neq WL_S(BRR).$$

The next eight strategies in the list are all antisymmetric. Each of these strategies has exactly four winning configurations. These strategies can be enumerated systematically by defining WL on the configurations RRR, RRB, RBR and RBB only.

Configurations	Win-Loss Maps							
	WL_1	WL_2	WL_3	WL_4	WL_5	WL_6	WL_7	WL_8
RRR	W	L	W	L	W	L	W	L
RRB	W	L	W	L	L	W	L	W
RBR	W	L	L	W	W	L	L	W
RBB	W	L	L	W	L	W	W	L

This is a complete enumeration of all antisymmetric strategies, each with four winning configurations.

The idea expounded so far can be extended to hat problems with n players and two colors. Consider, for example, four players and two colors. The total number of hat configurations is 16. Let $p_1, p_2, \ldots, p_{16}$ be any given probability distribution on the set of all configurations. The objective is to determine an optimal strategy that maximizes the winning probability. The total number of strategies is 3^{24}. A complete enumeration of all these strategies is outside the scope of any computer. However, one can write down maximal strategies for this problem. For example, the total number of maximal strategies each with 12 winning configurations is 28. (Note that no strategy will give more than 12 winning configurations.) All these strategies will have to be symmetric! The win-loss maps of all these strategies are obtained by selecting the configurations from {RRRR, RRRB, RRBR, RRBB, RBRR, RBRB, RBBR, RBBB} and assigning them the letter W. The win-loss maps can be completed by symmetry.

There are maximal strategies each with ten wining configurations and also each with eight winning configurations.

29.7 Other Variations

There are a number of variations of the hat problem considered in the literature. See Buhler (2002) for some of these. We would like to mention one new variation. Consider the hat problem with n colors and n participants. Each participant is fitted with a hat whose color is randomly picked from the given set of colors. Each participant can see the colors of hats of her teammates but does not know the colors of her hat. Each participant is asked separately to guess the color of her hat. No one is allowed to "pass." They can win collectively the prize if at least one of the guesses is correct. Is there a strategy of responses that will guarantee 100% chances of winning the prize? Yes, there is one. The reader may try to find one.

29.8 Some Open Problems

There are many open problems in the environment of traditional hat problem. Take the case of two colors and n participants. Optimal strategies are known for $n = 3, 4, 5, 6, 7$, and 8. Optimal strategy is known if $n = 2^k - 1$ for some positive integer $k \geq 2$. For all other cases, optimal strategies are not known. Take the case of three colors and n participants. Except for the case $n = 3$, which has been dealt in this paper, optimal strategies are not known. For the general cases of q colors and n participants, virtually nothing is known.

29.9 The Yeast Genome Problem

One of the most important problems in cell biology is to understand functionality of each and every gene of any living organism. A mammoth project, called the Deletion Project, is underway to study the DNA of the yeast organism. The genome of yeast organism has been completely mapped out. It has about 6,000 genes. Experiments on yeast cells, under the project, have the following basic ingredients:

1. Remove a gene from the cell.
2. Place the cell in a chamber at a set temperature.
3. Examine every one of the remaining cells whether or not it is active.

The data vector generated is of order 1×6000. Every entry in the vector, except one, is 0 (inactive) or 1 (active). The missing entry corresponds to the deleted gene. Repeat Steps 1, 2, and 3 with respect to every gene. At the set temperature, we will thus have 6,000 binary data vectors, each vector having exactly one blank space. The whole cell is also placed in the chamber without removing any of its genes. The data vector generated will not have any blanks. Using all these data vectors, one has to guess what would have been the role of the deleted gene had it been present in the cell. It is hoped that hat problem might provide some pointers.

References

1. Buhler, J. P. (2002). Hat tricks, *The Mathematical Intelligencer*, **24**, 44–49.

2. Cohen, G., Honkala, I., Litsyn, S., and Lobstein, A. (1997). *Covering Codes*, North-Holland, Amsterdam.

3. Ebert, T., and Vollmer, H. (2000). On the autoreducibility of random sequences, In *Proceedings of the 25th International Symposium on Mathematical Foundations of Computer Science*, Springer Lecture Notes in Computer Science, Vol. 1893, pp. 333–342.

4. Lenstra, H., and Seroussi, G. (2004). On hats and other covers, *Preprint*.

5. Robinson, S. (2001). Why mathematicians care about their hat color, *New York Times, Science Tuesday*, April 10, 2001.

Index